Math Fin Law 6

Mathematical Financial Law

Public Listed Firm Rule No.19905-23237

by

Steve Asikin

0

Steve Asikin ISBN 14: 978-1515158691, ISBN 10: 1515158691

MATH FIN LAW 6
Mathematical Financial Laws
Public Listed Firm Rule No. 19905-23237
By Steve Asikin

Creative Team:Steve Asikin

A Polimedia Paperback
First Published in Indonesia 2015
By Polimedia Publishing
This 1st edition published in 2015
Jalan Srengseng Sawah, Jagakarsa
Jakarta 12640, INDONESIA

ISBN-13: 978-1515158691
ISBN-10: 1515158691

1

PREFACE

Optimization models play an increasingly important role in financial decisions. Finance optimization process help companies with process optimization, including shrinking transaction processing costs, reducing error rates. Portfolio optimization is the process of choosing the proportions of various aspect as well as the return's sensitivity and among many measures of financial health. Many computational problems in finance can be solved by optimization techniques.

Against all beneficial of common financial ratio analyses, this mathematical law will do a synthesis, using all possible methods based on preset financial ratio as criteria. So then Accounting, prudential regulation and financial stability: elements of a synthesis set to make its the financial condition be effective and efficient. As a financial plan by constructing all decisions regarding each component affect the other components as Synthesis of best possible Financial Planning.

The nicest thing about it, is that only requires simple Junior Highschool Algebra. We just need to be consistent with its less number treatment of constellations. Its saddest part is that many professors and doctors are no longer able to understand Junior Highschool Algebra, and hiding in Accounting jargons. Please go back to its Introduction for glossary, at anywhere inside the book, so then we can still understand what are those mean with simple Algebraic Rule there. We try to avoid some sentences multiplied, added, eliminated divided, rooted or powered by another sentences, that make many traditional finance book successful (delivering few simple things, seen as difficult and scary), but less useful and seldom be accurate.

2

Steve Asikin ISBN 14: 978-1515158691, ISBN 10: 1515158691

TABLE OF CONTENTS

Steve Asikin ISBN 14: 978-1515158691, ISBN 10: 1515158691

INTRODUCTION:
The Math Fin Law Terminologies

Income Statement

S	V				
	M	F			
		O	I		
			B	T	
				A	D	Y
						H
					R

Balance Sheet

K	C	W	L	X	G
J					Z
P				Q	
N			E	U	
				R	

Please carefully look that all of 26 characters (**A** to **Z**) used once, except the **R**, that in Accountancy must be appeared in both diagrams at same identical value, with glossary (of well known abbreviations) in the next page. Only 2 (two) of them are historical as **U=E'** and **S'** means carried forward legal numbers from previous year.

The 13 (thirteen) common measures (**c, d, f, g, i, j, l, p, q, s, t, v,** and **y**), are put up front as desired criteria along the way, so then later analyst could be surprised.

In Balance Sheet, we found that **W= C+N= K+J+P+N= L+E= X+Q+U+R= G+Z+Q+U+R** as well, so then we put **W** in the middle of its T-Account balance.

In Income Statement, same applied as **S= V+F+I +T+Y+H+R= V+M= V+F+ O= V+F+I+B= V+F+I+T+A= V+F+I+B= V+F+I+T+D+R.**

Steve Asikin ISBN 14: 978-1515158691, ISBN 10: 1515158691

A= After Tax Income;

B= Before Tax Income,

C= Current Assets, **c**= C/X= Current Ratio,

D= Dividend Paid, **d**= D/A=Dividend Portion or Payout,

E= Equity or Capital, **E'**= Equity of Past Year,

F= Fixed Cost, **f**= F/S=Fixed Portion,

G= Goods or Trade Payables,

      g= 360G/V= Goods Payable Days,

H= Home Taken Dividend,

I= Interest Expense, **i**= I/S= Interest Portion,

J= Jobs or Account Receivable,

      j= 360J/S= Jobs or Trade Account ReceivableDays,

K= Kind of Cash,

L= Liabilities or Debt, **l**= L/E= Leverage or Gearing Ratio,

M= Margin of Contribution,

N= Non Current Assets,

O= Operational Surplus,

P= Procured Inventories,

      p= 360P/V= Procured Inventory Days,

Q= Quoted Longterm Liabilities,

      q= [C-P]/X= Quick or Acid Test Ratio,

R= Retained Earnings,

S= Sales or Revenues, **S'**= Sales of Past Year,

      s= [S/S']-1= Sales Growth,

T= Tax Paid, **t**= T/B= Tax Rate,

U= Utilized or Starting Capital,

V= Variable Cost, **v**= V/S= Variable Portion,

W= Wealth or Total Assets,

X= Xpress or Current Debt,

Y= Yielding Tax of Dividend,

      y= Y/D= Yielding Tax Rate of Dividend,

Z= Zero Trade Current Debt.

Steve Asikin ISBN 14: 978-1515158691, ISBN 10: 1515158691

CHAPTER-19:

Q= Quoted Longterm Debt Optimization
Continued from Book 1,2, 3, 4 and 5:

Rule-19905:

If both (**/**), (**U**), (**$**), (**f**), (**$'**), (**v**), (**s**), (**I**), (**T**), (**A**), (**d**)
and (**X**) are known, then its Quick or Acid Test Ratio
Planned is:

$$Q= /\{U+\$[1-f]-\$'v[1+s]-I-T-Ad\}-X$$

Rule-19906:

If both (**Q**), (**U**), (**$**), (**f**), (**$'**), (**v**), (**s**), (**I**), (**T**), (**A**), (**d**)
and (**X**) are known, then its Leverage or Gearing Ratio
Planned is:

$$/= [Q+X]/\{U+\$[1-f]-\$'v[1+s]-I-T-Ad\}$$

Rule-19907:

If both (**/**), (**Q**), (**$**), (**f**), (**$'**), (**v**), (**s**), (**I**), (**T**), (**A**), (**d**)
and (**X**) are known, then its Utilized or Starting
Capital must be:

$$U= \$'v[1+s]+I+T+Ad+[Q+X]//\$[1-f]$$

Rule-19908:

If both (**/**), (**U**), (**Q**), (**f**), (**$'**), (**v**), (**s**), (**I**), (**T**), (**A**), (**d**)
and (**X**) are known, then its Sales or Revenue Planned
is:

$$\$= \{\$'v[1+s]+I+T+Ad+[Q+X]//U\}/[1-f]$$

6

Steve Asikin ISBN 14: 978-1515158691, ISBN 10: 1515158691

Rule-19909:

If both (**/**), (**U**), (**$**), (**Q**), (**$'**), (**v**), (**s**), (**I**), (**T**), (**A**), (**d**) and (**X**) are known, then its Fixed Portion Planned is:

$$f= 1-\{\$'v[1+s]+I+T+Ad+[Q+X]/I\cdot U\}/\$$$

Rule-19910:

If both (**/**), (**U**), (**$**), (**f**), (**Q**), (**v**), (**s**), (**I**), (**T**), (**A**), (**d**) and (**X**) are known, then its Sales Past must be:

$$\$'= \{U+\$[1-f]-I-T-Ad-[Q+X]/I\}/\{v[1+s]\}$$

Rule-19911:

If both (**/**), (**U**), (**$**), (**f**), (**$'**), (**Q**), (**s**), (**I**), (**T**), (**A**), (**d**) and (**X**) are known, then its Variable Portion Planned is:

$$v= \{U+\$[1-f]-I-T-Ad-[Q+X]/I\}/\{\$'[1+s]\}$$

Rule-19912:

If both (**/**), (**U**), (**$**), (**f**), (**$'**), (**v**), (**Q**), (**I**), (**T**), (**A**), (**d**) and (**X**) are known, then its Sales Growth Planned is:

$$s= \$/\$'-1$$

or it can also be found as

$$s= \{U+\$[1-f]-I-T-Ad-[Q+X]/I\}/[\$'v]-1$$

Rule-19913:

If both (**/**), (**U**), (**$**), (**f**), (**$'**), (**v**), (**s**), (**Q**), (**T**), (**A**), (**d**) and (**X**) are known, then its Interest Expense Planned is:

$$I= U+\$[1-f]-\$'v[1+s]-T-Ad-[Q+X]/I$$

Steve Asikin ISBN 14: 978-1515158691, ISBN 10: 1515158691

Rule-19914:
    If both (**/**), (**U**), (**$**), (**f**), (**$'**), (**v**), (**s**), (**I**), (**Q**), (**A**), (**d**)
    and (**X**) are known, then its Tax Planned is:
    $$T= U+\$[1-f]-\$'v[1+s]-I-Ad-[Q+X]/I$$

Rule-19915:
    If both (**/**), (**U**), (**$**), (**f**), (**$'**), (**v**), (**s**), (**I**), (**T**), (**Q**), (**d**)
    and (**X**) are known, then its After Tax Income Planned
    is:
    $$A= \{U+\$[1-f]-\$'v[1+s]-I-T-[Q+X]/I\}/d$$

Rule-19916:
    If both (**/**), (**U**), (**$**), (**f**), (**$'**), (**v**), (**s**), (**I**), (**T**), (**A**), (**Q**)
    and (**X**) are known, then its Dividend Payout Planned
    is:
    $$d= \{U+\$[1-f]-\$'v[1+s]-I-T-[Q+X]/I\}/A$$

Rule-19917:
    If both (**/**), (**U**), (**$**), (**f**), (**$'**), (**v**), (**s**), (**I**), (**T**), (**A**), (**d**)
    and (**Q**) are known, then its Xpress or Current Debt
    Planned is:
    $$X= I\{U+\$[1-f]-\$'v[1+s]-I-T-Ad\}-Q$$

Rule-19918:
    If both (**/**), (**U**), (**$**), (**f**), (**$'**), (**v**), (**s**), (**I**), (**T**), (**A**), (**d**),
    (**P**), (**c**) and (**q**) are known, then its Quoted Longterm
    Debt Planned is:
    $$Q= I\{U+\$[1-f]-\$'v[1+s]-I-T-Ad\}-P/[c-q]$$

Steve Asikin ISBN 14: 978-1515158691, ISBN 10: 1515158691

Rule-19919:

If both **(Q)**, **(U)**, **($)**, **(f)**, **($')**, **(v)**, **(s)**, **(I)**, **(T)**, **(A)**, **(d)**, **(P)**, **(c)** and **(q)** are known, then its Leverage or Gearing Ratio Planned is:

$$\textit{l} = \{Q + P/[c-q]\}/\{U + \$[1-f] - \$'v[1+s] - I - T - Ad\}$$

Rule-19920:

If both **(l)**, **(Q)**, **($)**, **(f)**, **($')**, **(v)**, **(s)**, **(I)**, **(T)**, **(A)**, **(d)**, **(P)**, **(c)** and **(q)** are known, then its Utilized or Starting Capital must be:

$$U = \$'v[1+s] + I + T + Ad + \{Q + P/[c-q]\}/\textit{l} - \$[1-f]$$

Rule-19921:

If both **(l)**, **(U)**, **(Q)**, **(f)**, **($')**, **(v)**, **(s)**, **(I)**, **(T)**, **(A)**, **(d)**, **(P)**, **(c)** and **(q)** are known, then its Sales or Revenue Planned is:

$$\$ = (\$'v[1+s] + I + T + Ad + \{Q + P/[c-q]\}/\textit{l} - U)/[1-f]$$

Rule-19922:

If both **(l)**, **(U)**, **($)**, **(Q)**, **($')**, **(v)**, **(s)**, **(I)**, **(T)**, **(A)**, **(d)**, **(P)**, **(c)** and **(q)** are known, then its Fixed Portion Planned is:

$$f = 1 - (\$'v[1+s] + I + T + Ad + \{Q + P/[c-q]\}/\textit{l} - U)/\$$$

Rule-19923:

If both **(l)**, **(U)**, **($)**, **(f)**, **(Q)**, **(v)**, **(s)**, **(I)**, **(T)**, **(A)**, **(d)**, **(P)**, **(c)** and **(q)** are known, then its Sales Past must be:

$$\$' = (U + \$[1-f] - I - T - Ad - \{Q + P/[c-q]\}/\textit{l})/\{v[1+s]\}$$

Steve Asikin ISBN 14: 978-1515158691, ISBN 10: 1515158691

Rule-19924:

If both $(\textit{\textbf{I}})$, $(\textbf{U})$, $(\textbf{S})$, $(\textbf{f})$, $(\textbf{S'})$, $(\textbf{Q})$, $(\textbf{s})$, $(\textbf{I})$, $(\textbf{T})$, $(\textbf{A})$, $(\textbf{d})$, $(\textbf{P})$, $(\textbf{c})$ and $(\textbf{q})$ are known, then its Variable Portion Planned is:

$$v= (U+S[1-f]-I-T-Ad-\{Q+P/[c-q]\}/I)/\{S'[1+s]\}$$

Rule-19925:

If both $(\textit{\textbf{I}})$, $(\textbf{U})$, $(\textbf{S})$, $(\textbf{f})$, $(\textbf{S'})$, $(\textbf{v})$, $(\textbf{Q})$, $(\textbf{I})$, $(\textbf{T})$, $(\textbf{A})$, $(\textbf{d})$, $(\textbf{P})$, $(\textbf{c})$ and $(\textbf{q})$ are known, then its Sales Growth Planned is:

$$s= S/S'-1$$

or it can also be found as

$$s= (U+S[1-f]-I-T-Ad-\{Q+P/[c-q]\}/I)/[S'v]-1$$

Rule-19926:

If both $(\textit{\textbf{I}})$, $(\textbf{U})$, $(\textbf{S})$, $(\textbf{f})$, $(\textbf{S'})$, $(\textbf{v})$, $(\textbf{s})$, $(\textbf{Q})$, $(\textbf{T})$, $(\textbf{A})$, $(\textbf{d})$, $(\textbf{P})$, $(\textbf{c})$ and $(\textbf{q})$ are known, then its Interest Expense Planned is:

$$I= U+S[1-f]-S'v[1+s]-T-Ad-\{Q+P/[c-q]\}/I$$

Rule-19927:

If both $(\textit{\textbf{I}})$, $(\textbf{U})$, $(\textbf{S})$, $(\textbf{f})$, $(\textbf{S'})$, $(\textbf{v})$, $(\textbf{s})$, $(\textbf{I})$, $(\textbf{Q})$, $(\textbf{A})$, $(\textbf{d})$, $(\textbf{P})$, $(\textbf{c})$ and $(\textbf{q})$ are known, then its Tax Planned is:

$$T= U+S[1-f]-S'v[1+s]-I-Ad-\{Q+P/[c-q]\}/I$$

Rule-19928:

If both $(\textit{\textbf{I}})$, $(\textbf{U})$, $(\textbf{S})$, $(\textbf{f})$, $(\textbf{S'})$, $(\textbf{v})$, $(\textbf{s})$, $(\textbf{I})$, $(\textbf{T})$, $(\textbf{Q})$, $(\textbf{d})$, $(\textbf{P})$, $(\textbf{c})$ and $(\textbf{q})$ are known, then its After Tax Income Planned is:

$$A= (U+S[1-f]-S'v[1+s]-I-T-\{Q+P/[c-q]\}/I)/d$$

Steve Asikin ISBN 14: 978-1515158691, ISBN 10: 1515158691

Rule-19929:

If both (I), (U), (S), (f), (S'), (v), (s), (I), (T), (A), (Q), (P), (c) and (q) are known, then its Dividend Payout Planned is:

$$d = (U+S[1-f]-S'v[1+s]-I-T-\{Q+P/[c-q]\}/I)/A$$

Rule-19930:

If both (I), (U), (S), (f), (S'), (v), (s), (I), (T), (A), (d), (Q), (c) and (q) are known, then its Procured Inventory Planned is:

$$P = [c-q][I\{U+S[1-f]-S'v[1+s]-I-T-Ad\}-Q]$$

Rule-19931:

If both (I), (U), (S), (f), (S'), (v), (s), (I), (T), (A), (d), (P), (Q) and (q) are known, then its Current Ratio Planned is:

$$c = q+P/[I\{U+S[1-f]-S'v[1+s]-I-T-Ad\}-Q]$$

Rule-19932:

If both (I), (U), (S), (f), (S'), (v), (s), (I), (T), (A), (d), (P), (c) and (Q) are known, then its Quick or Acid Test Ratio Planned is:

$$q = c-P/[I\{U+S[1-f]-S'v[1+s]-I-T-Ad\}-Q]$$

Rule-19933:

If both (I), (U), (S), (f), (S'), (v), (s), (I), (T), (A), (d), (V), (p), (c) and (q) are known, then its Quoted Longterm Debt Planned is:

$$Q = I\{U+S[1-f]-S'v[1+s]-I-T-Ad\}-Vp/\{360[c-q]\}$$

Steve Asikin ISBN 14: 978-1515158691, ISBN 10: 1515158691

Rule-19934:

If both (**Q**), (**U**), (**$**), (**f**), (**$'**), (**v**), (**s**), (**I**), (**T**), (**A**), (**d**), (**V**), (**p**), (**c**) and (**q**) are known, then its Leverage or Gearing Ratio Planned is:

$$l = (Q + Vp/\{360[c-q]\})/\{U + \$[1-f] - \$'v[1+s] - I - T - Ad\}$$

Rule-19935:

If both (**l**), (**Q**), (**$**), (**f**), (**$'**), (**v**), (**s**), (**I**), (**T**), (**A**), (**d**), (**V**), (**p**), (**c**) and (**q**) are known, then its Utilized or Starting Capital must be:

$$U = \$'v[1+s] + I + T + Ad + (Q + Vp/\{360[c-q]\})/l - \$[1-f]$$

Rule-19936:

If both (**l**), (**U**), (**Q**), (**f**), (**$'**), (**v**), (**s**), (**I**), (**T**), (**A**), (**d**), (**V**), (**p**), (**c**) and (**q**) are known, then its Sales or Revenue Planned is:

$$\$ = [\$'v[1+s] + I + T + Ad + (Q + Vp/\{360[c-q]\})/l - U] / [1-f]$$

Rule-19937:

If both (**l**), (**U**), (**$**), (**Q**), (**$'**), (**v**), (**s**), (**I**), (**T**), (**A**), (**d**), (**V**), (**p**), (**c**) and (**q**) are known, then its Fixed Portion Planned is:

$$f = 1 - [\$'v[1+s] + I + T + Ad + (Q + Vp/\{360[c-q]\})/l - U]/\$$$

Rule-19938:

If both (**l**), (**U**), (**$**), (**f**), (**Q**), (**v**), (**s**), (**I**), (**T**), (**A**), (**d**), (**V**), (**p**), (**c**) and (**q**) are known, then its Sales Past must be:

$$\$' = [U + \$[1-f] - I - T - Ad - (Q + Vp/\{360[c-q]\})/l] / \{v[1+s]\}$$

12

Steve Asikin ISBN 14: 978-1515158691, ISBN 10: 1515158691

Rule-19939:

If both $(\textit{\textbf{I}})$, $(\textbf{U})$, $(\textbf{\$})$, $(\textbf{f})$, $(\textbf{\$'})$, $(\textbf{Q})$, $(\textbf{s})$, $(\textbf{I})$, $(\textbf{T})$, $(\textbf{A})$, $(\textbf{d})$, $(\textbf{V})$, $(\textbf{p})$, $(\textbf{c})$ and $(\textbf{q})$ are known, then its Variable Portion Planned is:

$$v = V/\$$$

or it can also be found as

$$v = [U+\$[1-f]-I-T-Ad-(Q+Vp/\{360[c-q]\})/\textit{I}] /\{\$'[1+s]\}$$

Rule-19940:

If both $(\textit{\textbf{I}})$, $(\textbf{U})$, $(\textbf{\$})$, $(\textbf{f})$, $(\textbf{\$'})$, $(\textbf{v})$, $(\textbf{Q})$, $(\textbf{I})$, $(\textbf{T})$, $(\textbf{A})$, $(\textbf{d})$, $(\textbf{V})$, $(\textbf{p})$, $(\textbf{c})$ and $(\textbf{q})$ are known, then its Sales Growth Planned is:

$$s = \$/\$'-1$$

or it can also be found as

$$s = [U+\$[1-f]-I-T-Ad-(Q+Vp/\{360[c-q]\})/\textit{I}]/[\$'v]-1$$

Rule-19941:

If both $(\textit{\textbf{I}})$, $(\textbf{U})$, $(\textbf{\$})$, $(\textbf{f})$, $(\textbf{\$'})$, $(\textbf{v})$, $(\textbf{s})$, $(\textbf{Q})$, $(\textbf{T})$, $(\textbf{A})$, $(\textbf{d})$, $(\textbf{V})$, $(\textbf{p})$, $(\textbf{c})$ and $(\textbf{q})$ are known, then its Interest Expense Planned is:

$$I = U+\$[1-f]-\$'v[1+s]-T-Ad-(Q+Vp/\{360[c-q]\})/\textit{I}$$

Rule-19942:

If both $(\textit{\textbf{I}})$, $(\textbf{U})$, $(\textbf{\$})$, $(\textbf{f})$, $(\textbf{\$'})$, $(\textbf{v})$, $(\textbf{s})$, $(\textbf{I})$, $(\textbf{Q})$, $(\textbf{A})$, $(\textbf{d})$, $(\textbf{V})$, $(\textbf{p})$, $(\textbf{c})$ and $(\textbf{q})$ are known, then its Tax Planned is:

$$T = U+\$[1-f]-\$'v[1+s]-I-Ad-(Q+Vp/\{360[c-q]\})/\textit{I}$$

Steve Asikin ISBN 14: 978-1515158691, ISBN 10: 1515158691

Rule-19943:

If both (I), (U), $(\$)$, (f), $(\$')$, (v), (s), (I), (T), (Q), (d), (V), (p), (c) and (q) are known, then its After Tax Income Planned is:

$$A= [U+\$[1-f]-\$'v[1+s]-I-T -(Q+Vp/\{360[c-q]\})/I]/d$$

Rule-19944:

If both (I), (U), $(\$)$, (f), $(\$')$, (v), (s), (I), (T), (A), (Q), (V), (p), (c) and (q) are known, then its Dividend Payout Planned is:

$$d= [U+\$[1-f]-\$'v[1+s]-I-T -(Q+Vp/\{360[c-q]\})/I]/A$$

Rule-19945:

If both (I), (U), $(\$)$, (f), $(\$')$, (v), (s), (I), (T), (A), (d), (Q), (p), (c) and (q) are known, then its Variable Cost Planned is:

$$V= 360[c-q][I\{U+\$[1-f]-\$'v[1+s]-I-T-Ad\}-Q]/p$$

Rule-19946:

If both (I), (U), $(\$)$, (f), $(\$')$, (v), (s), (I), (T), (A), (d), (V), (Q), (c) and (q) are known, then its Procured Inventory Days Planned:

$$p= 360[c-q][I\{U+\$[1-f]-\$'v[1+s]-I-T-Ad\}-Q]/V$$

Rule-19947:

If both (I), (U), $(\$)$, (f), $(\$')$, (v), (s), (I), (T), (A), (d), (V), (p), (Q) and (q) are known, then its Current Ratio Planned is:

$$c= q+Vp/\{360[I\{U+\$[1-f]-\$'v[1+s]-I-T-Ad\}-Q]\}$$

Steve Asikin ISBN 14: 978-1515158691, ISBN 10: 1515158691

Rule-19948:

If both (I), (U), (S), (f), (S'), (v), (s), (I), (T), (A), (d), (V), (p), (c) and (Q) are known, then its Quick or Acid Test Ratio Planned is:

$$q= c\text{-}Vp/\{360[I\{U+S[1\text{-}f]\text{-}S'v[1+s]\text{-}I\text{-}T\text{-}Ad\}\text{-}Q]\}$$

Rule-19949:

If both (I), (U), (S), (f), (S'), (v), (s), (I), (T), (A), (d), (p), (c) and (q) are known, then its Quoted Longterm Debt Planned is:

$$Q= I\{U+S[1\text{-}f]\text{-}S'v[1+s]\text{-}I\text{-}T\text{-}Ad\}\text{-}Svp/\{360[c\text{-}q]\}$$

Rule-19950:

If both (Q), (U), (S), (f), (S'), (v), (s), (I), (T), (A), (d), (p), (c) and (q) are known, then its Leverage or Gearing Ratio Planned is:

$$I= (Q+Svp/\{360[c\text{-}q]\})$$
$$/\{U+S[1\text{-}f]\text{-}S'v[1+s]\text{-}I\text{-}T\text{-}Ad\}$$

Rule-19951:

If both (I), (Q), (S), (f), (S'), (v), (s), (I), (T), (A), (d), (p), (c) and (q) are known, then its Utilized or Starting Capital must be:

$$U= S'v[1+s]+I+T+Ad+(Q+Svp/\{360[c\text{-}q]\})/I\text{-}S[1\text{-}f]$$

Rule-19952:

If both (I), (U), (Q), (f), (S'), (v), (s), (I), (T), (A), (d), (p), (c) and (q) are known, then its Sales or Revenue Planned is:

$$S= (I\{U\text{-}S'v[1+s]\text{-}I\text{-}T\text{-}Ad\}\text{-}S)/(vp/\{360[c\text{-}q]\}\text{-}I[1\text{-}f])$$

Steve Asikin ISBN 14: 978-1515158691, ISBN 10: 1515158691

Rule-19953:

If both (f), $(\mathbf{U})$, $(\$)$, $(\mathbf{Q})$, $(\$')$, $(\mathbf{v})$, $(\mathbf{s})$, $(\mathbf{I})$, $(\mathbf{T})$, $(\mathbf{A})$, $(\mathbf{d})$, $(\mathbf{p})$, $(\mathbf{c})$ and $(\mathbf{q})$ are known, then its Fixed Portion Planned is:

$$f= 1-[\$'v[1+s]+I+T+Ad$$
$$+(Q+\$vp/\{360[c-q]\})/fU]/\$$$

Rule-19954:

If both (f), $(\mathbf{U})$, $(\$)$, $(\mathbf{f})$, $(\mathbf{Q})$, $(\mathbf{v})$, $(\mathbf{s})$, $(\mathbf{I})$, $(\mathbf{T})$, $(\mathbf{A})$, $(\mathbf{d})$, $(\mathbf{p})$, $(\mathbf{c})$ and $(\mathbf{q})$ are known, then its Sales Past must be:

$$\$'= [U+\$[1-f]-I-T-Ad-(Q+\$vp/\{360[c-q]\})/f]$$
$$/\{v[1+s]\}$$

Rule-19955:

If both (f), $(\mathbf{U})$, $(\$)$, $(\mathbf{f})$, $(\$')$, $(\mathbf{Q})$, $(\mathbf{s})$, $(\mathbf{I})$, $(\mathbf{T})$, $(\mathbf{A})$, $(\mathbf{d})$, $(\mathbf{p})$, $(\mathbf{c})$ and $(\mathbf{q})$ are known, then its Variable Portion Planned is:

$$v= (f\{U+\$[1-f]-I-T-Ad\}-Q)$$
$$/(\$p/\{360[c-q]\}-\$'f[1+s])$$

Rule-19956:

If both (f), $(\mathbf{U})$, $(\$)$, $(\mathbf{f})$, $(\$')$, $(\mathbf{v})$, $(\mathbf{Q})$, $(\mathbf{I})$, $(\mathbf{T})$, $(\mathbf{A})$, $(\mathbf{d})$, $(\mathbf{p})$, $(\mathbf{c})$ and $(\mathbf{q})$ are known, then its Sales Growth Planned is:

$$s= \$/\$'-1$$

or it can also be found as

$$s= [U+\$[1-f]-I-T-Ad-(Q+\$vp/\{360[c-q]\})/f]$$
$$/[\$'v]-1$$

Steve Asikin ISBN 14: 978-1515158691, ISBN 10: 1515158691

Rule-19957:

If both **(I)**, **(U)**, **(S)**, **(f)**, **(S')**, **(v)**, **(s)**, **(Q)**, **(T)**, **(A)**, **(d)**, **(p)**, **(c)** and **(q)** are known, then its Interest Expense Planned is:

$$I= U+S[1-f]-S'v[1+s]-T-Ad-(Q+Svp/\{360[c-q]\})/I$$

Rule-19958:

If both **(I)**, **(U)**, **(S)**, **(f)**, **(S')**, **(v)**, **(s)**, **(I)**, **(Q)**, **(A)**, **(d)**, **(p)**, **(c)** and **(q)** are known, then its Tax Planned is:

$$T= U+S[1-f]-S'v[1+s]-I-Ad-(Q+Svp/\{360[c-q]\})/I$$

Rule-19959:

If both **(I)**, **(U)**, **(S)**, **(f)**, **(S')**, **(v)**, **(s)**, **(I)**, **(T)**, **(Q)**, **(d)**, **(p)**, **(c)** and **(q)** are known, then its After Tax Income Planned is:

$$A= [U+S[1-f]-S'v[1+s]-I-T$$
$$-(Q+Svp/\{360[c-q]\})/I]/d$$

Rule-19960:

If both **(I)**, **(U)**, **(S)**, **(f)**, **(S')**, **(v)**, **(s)**, **(I)**, **(T)**, **(A)**, **(Q)**, **(p)**, **(c)** and **(q)** are known, then its Dividend Payout Planned is:

$$d= [U+S[1-f]-S'v[1+s]-I-T$$
$$-(Q+Svp/\{360[c-q]\})/I]/A$$

Rule-19961:

If both **(I)**, **(U)**, **(S)**, **(f)**, **(S')**, **(v)**, **(s)**, **(I)**, **(T)**, **(A)**, **(d)**, **(Q)**, **(c)** and **(q)** are known, then its Procured Inventory Days Planned:

$$p= 360[c-q][I\{U+S[1-f]-S'v[1+s]-I-T-Ad\}-Q]/[Sv]$$

Steve Asikin ISBN 14: 978-1515158691, ISBN 10: 1515158691

<u>Rule-19962</u>:

If both (**/**), (**U**), (**S**), (**f**), (**S'**), (**v**), (**s**), (**I**), (**T**), (**A**), (**d**), (**p**), (**Q**) and (**q**) are known, then its Current Ratio Planned is:

$$c = q + Svp / \{360[/\{U+S[1-f]-S'v[1+s]-I-T-Ad\}-Q]\}$$

<u>Rule-19963</u>:

If both (**/**), (**U**), (**S**), (**f**), (**S'**), (**v**), (**s**), (**I**), (**T**), (**A**), (**d**), (**p**), (**c**) and (**Q**) are known, then its Quick or Acid Test Ratio Planned is:

$$q = c - Svp / \{360[/\{U+S[1-f]-S'v[1+s]-I-T-Ad\}-Q]\}$$

<u>Rule-19964</u>:

If both (**/**), (**U**), (**S**), (**f**), (**S'**), (**v**), (**s**), (**I**), (**T**), (**A**), (**d**), (**p**), (**c**) and (**q**) are known, then its Quoted Longterm Debt Planned is:

$$Q = /\{U+S[1-f]-S'v[1+s]-I-T-Ad\}$$
$$-S'vp[1+s]/\{360[c-q]\}$$

<u>Rule-19965</u>:

If both (**Q**), (**U**), (**S**), (**f**), (**S'**), (**v**), (**s**), (**I**), (**T**), (**A**), (**d**), (**p**), (**c**) and (**q**) are known, then its Leverage or Gearing Ratio Planned is:

$$/ = (Q+S'vp[1+s]/\{360[c-q]\})$$
$$/\{U+S[1-f]-S'v[1+s]-I-T-Ad\}$$

<u>Rule-19966</u>:

If both (**/**), (**Q**), (**S**), (**f**), (**S'**), (**v**), (**s**), (**I**), (**T**), (**A**), (**d**), (**p**), (**c**) and (**q**) are known, then its Utilized or Starting Capital must be:

$$U = S'v[1+s]+I+T+Ad$$
$$+(Q+S'vp[1+s]/\{360[c-q]\})//S[1-f]$$

Steve Asikin ISBN 14: 978-1515158691, ISBN 10: 1515158691

Rule-19967:

If both $(\textbf{\textit{l}})$, $(\textbf{U})$, $(\textbf{Q})$, $(\textbf{f})$, $(\textbf{S'})$, $(\textbf{v})$, $(\textbf{s})$, $(\textbf{I})$, $(\textbf{T})$, $(\textbf{A})$, $(\textbf{d})$, $(\textbf{p})$, $(\textbf{c})$ and $(\textbf{q})$ are known, then its Sales or Revenue Planned is:

$$S= (Q-\textit{l}\{U-S'v[1+s]-I-T-Ad\}$$
$$+S'vp[1+s]/\{360[c-q]\})/\{\textit{l}[1-f]\}$$

Rule-19968:

If both $(\textbf{\textit{l}})$, $(\textbf{U})$, $(\textbf{S})$, $(\textbf{Q})$, $(\textbf{S'})$, $(\textbf{v})$, $(\textbf{s})$, $(\textbf{I})$, $(\textbf{T})$, $(\textbf{A})$, $(\textbf{d})$, $(\textbf{p})$, $(\textbf{c})$ and $(\textbf{q})$ are known, then its Fixed Portion Planned is:

$$f= 1-[S'v[1+s]+I+T+Ad$$
$$+(Q+S'vp[1+s]/\{360[c-q]\})/\textit{l}U]/S$$

Rule-19969:

If both $(\textbf{\textit{l}})$, $(\textbf{U})$, $(\textbf{S})$, $(\textbf{f})$, $(\textbf{Q})$, $(\textbf{v})$, $(\textbf{s})$, $(\textbf{I})$, $(\textbf{T})$, $(\textbf{A})$, $(\textbf{d})$, $(\textbf{p})$, $(\textbf{c})$ and $(\textbf{q})$ are known, then its Sales Past must be:

$$S'= (\textit{l}\{U+S[1-f]-I-T-Ad\}-Q)$$
$$/[v[1+s](p/\{360[c-q]\}-\textit{l})]$$

Rule-19970:

If both $(\textbf{\textit{l}})$, $(\textbf{U})$, $(\textbf{S})$, $(\textbf{f})$, $(\textbf{S'})$, $(\textbf{Q})$, $(\textbf{s})$, $(\textbf{I})$, $(\textbf{T})$, $(\textbf{A})$, $(\textbf{d})$, $(\textbf{p})$, $(\textbf{c})$ and $(\textbf{q})$ are known, then its Variable Portion Planned is:

$$v= (\textit{l}\{U+S[1-f]-I-T-Ad\}-Q)$$
$$/[S'[1+s](p/\{360[c-q]\}-\textit{l})]$$

Steve Asikin ISBN 14: 978-1515158691, ISBN 10: 1515158691

Rule-19971:

If both (I), (U), (S), (f), (S'), (v), (Q), (I), (T), (A), (d), (p), (c) and (q) are known, then its Sales Growth Planned is:

$s = S/S' - 1$

or it can also be found as

$s = (I\{U + S[1-f] - I - T - Ad\} - Q)/[S'v(p/\{360[c-q]\} - I)] - 1$

Rule-19972:

If both (I), (U), (S), (f), (S'), (v), (s), (Q), (T), (A), (d), (p), (c) and (q) are known, then its Interest Expense Planned is:

$I = U + S[1-f] - S'v[1+s] - T - Ad$
$\quad -(Q + S'vp[1+s]/\{360[c-q]\})/I$

Rule-19973:

If both (I), (U), (S), (f), (S'), (v), (s), (I), (Q), (A), (d), (p), (c) and (q) are known, then its Tax Planned is:

$T = U + S[1-f] - S'v[1+s] - I - Ad$
$\quad -(Q + S'vp[1+s]/\{360[c-q]\})/I$

Rule-19974:

If both (I), (U), (S), (f), (S'), (v), (s), (I), (T), (Q), (d), (p), (c) and (q) are known, then its After Tax Income Planned is:

$A = [U + S[1-f] - S'v[1+s] - I - T$
$\quad -(Q + S'vp[1+s]/\{360[c-q]\})/I]/d$

Steve Asikin ISBN 14: 978-1515158691, ISBN 10: 1515158691

Rule-19975:

If both (I), (U), (S), (f), (S'), (v), (s), (I), (T), (A), (Q), (p), (c) and (q) are known, then its Dividend Payout Planned is:

$$d= [U+S[1-f]-S'v[1+s]-I-T$$
$$-(Q+S'vp[1+s]/\{360[c-q]\})/I]/A$$

Rule-19976:

If both (I), (U), (S), (f), (S'), (v), (s), (I), (T), (A), (d), (Q), (c) and (q) are known, then its Procured Inventory Days Planned:

$$p= 360[c-q][I\{U+S[1-f]-S'v[1+s]-I-T-Ad\}-Q]$$
$$/\{S'v[1+s]\}$$

Rule-19977:

If both (I), (U), (S), (f), (S'), (v), (s), (I), (T), (A), (d), (p), (Q) and (q) are known, then its Current Ratio Planned is:

$$c= q+S'vp[1+s]$$
$$/\{360[I\{U+S[1-f]-S'v[1+s]-I-T-Ad\}-Q]\}$$

Rule-19978:

If both (I), (U), (S), (f), (S'), (v), (s), (I), (T), (A), (d), (p), (c) and (Q) are known, then its Quick or Acid Test Ratio Planned is:

$$q= c-S'vp[1+s]$$
$$/\{360[I\{U+S[1-f]-S'v[1+s]-I-T-Ad\}-Q]\}$$

Steve Asikin ISBN 14: 978-1515158691, ISBN 10: 1515158691

Rule-19979:
> If both (**/**), (**U**), (**$**), (**f**), (**$'**), (**v**), (**s**), (**I**), (**t**), (**D**) and
> (**X**) are known, then its Quoted Longterm Debt
> Planned is:
> **Q**= **/U**+{**$**[1-**f**]-**$'v**[1+**s**]-**I**}[1-**t**]-**D**)-**X**

Rule-19980:
> If both (**Q**), (**U**), (**$**), (**f**), (**$'**), (**v**), (**s**), (**I**), (**t**), (**D**) and
> (**X**) are known, then its Leverage or Gearing Ratio
> Planned is:
> **/**= [**Q**+**X**]/(**U**+{**$**[1-**f**]-**$'v**[1+**s**]-**I**}[1-**t**]-**D**)

Rule-19981:
> If both (**/**), (**Q**), (**$**), (**f**), (**$'**), (**v**), (**s**), (**I**), (**t**), (**D**) and
> (**X**) are known, then its Utilized or Starting Capital
> must be:
> **U**= **D**+[**Q**+**X**]/**/**{**$**[1-**f**]-**$'v**[1+**s**]-**I**}[1-**t**]

Rule-19982:
> If both (**/**), (**U**), (**Q**), (**f**), (**$'**), (**v**), (**s**), (**I**), (**t**), (**D**) and
> (**X**) are known, then its Sales or Revenue Planned is:
> **$**= (**$'v**[1+**s**]+**I**+{**D**+[**Q**+**X**]/**/**−**U**}/[1-**t**])/[1-**f**]

Rule-19983:
> If both (**/**), (**U**), (**$**), (**Q**), (**$'**), (**v**), (**s**), (**I**), (**t**), (**D**) and
> (**X**) are known, then its Fixed Portion Planned is:
> **f**= 1-(**$'v**[1+**s**]+**I**+{**D**+[**Q**+**X**]/**/**−**U**}/[1-**t**])/**$**

Rule-19984:
> If both (**/**), (**U**), (**$**), (**f**), (**Q**), (**v**), (**s**), (**I**), (**t**), (**D**) and
> (**X**) are known, then its Sales Past must be:
> **$'**= (**$**[1-**f**]-**I**-{**D**+[**Q**+**X**]/**/**−**U**}/[1-**t**])/{**v**[1+**s**]}

22

Rule-19985:

If both (I), (U), (S), (f), (S'), (Q), (s), (I), (t), (D) and (X) are known, then its Variable Portion Planned is:

$$v= (S[1-f]-I-\{D+[Q+X]/I-U\}/[1-t])/\{S'[1+s]\}$$

Rule-19986:

If both (I), (U), (S), (f), (S'), (v), (Q), (I), (t), (D) and (X) are known, then its Sales Growth Planned is:

$$s= S/S'-1$$

or it can also be found as

$$s= (S[1-f]-I-\{D+[Q+X]/I-U\}/[1-t])/[S'v]-1$$

Rule-19987:

If both (I), (U), (S), (f), (S'), (v), (s), (Q), (t), (D) and (X) are known, then its Interest Expense Planned is:

$$I= S[1-f]-S'v[1+s]-\{D+[Q+X]/I-U\}/[1-t]$$

Rule-19988:

If both (I), (U), (S), (f), (S'), (v), (s), (I), (Q), (D) and (X) are known, then its Tax Rate Planned is:

$$t= 1-\{D+[Q+X]/I-U\}/\{S[1-f]-S'v[1+s]-I\}$$

Rule-19989:

If both (I), (U), (S), (f), (S'), (v), (s), (I), (t), (Q) and (X) are known, then its Dividend Planned is:

$$D= U+\{S[1-f]-S'v[1+s]-I\}[1-t]-[Q+X]/I$$

Rule-19990:

If both (I), (U), (S), (f), (S'), (v), (s), (I), (t), (D) and (Q) are known, then its Xpress or Current Debt Planned is:

$$X= I(U+\{S[1-f]-S'v[1+s]-I\}[1-t]-D)-Q$$

Steve Asikin ISBN 14: 978-1515158691, ISBN 10: 1515158691

Rule-19991:

If both (I), (U), (S), (f), (S'), (v), (s), (I), (t), (D), (P), (c) and (q) are known, then its Quoted Longterm Debt Planned is:

$$Q= I(U+\{S[1-f]-S'v[1+s]-I\}[1-t]-D)-P/[c-q]$$

Rule-19992:

If both (Q), (U), (S), (f), (S'), (v), (s), (I), (t), (D), (P), (c) and (q) are known, then its Leverage or Gearing Ratio Planned is:

$$I= \{Q+P/[c-q]\}/(U+\{S[1-f]-S'v[1+s]-I\}[1-t]-D)$$

Rule-19993:

If both (I), (Q), (S), (f), (S'), (v), (s), (I), (t), (D), (P), (c) and (q) are known, then its Utilized or Starting Capital must be:

$$U= D+\{Q+P/[c-q]\}/I\{S[1-f]-S'v[1+s]-I\}[1-t]$$

Rule-19994:

If both (I), (U), (Q), (f), (S'), (v), (s), (I), (t), (D), (P), (c) and (q) are known, then its Sales or Revenue Planned is:

$$S= [S'v[1+s]+I+(D+\{Q+P/[c-q]\}/I-U)/[1-t]]/[1-f]$$

Rule-19995:

If both (I), (U), (S), (Q), (S'), (v), (s), (I), (t), (D), (P), (c) and (q) are known, then its Fixed Portion Planned is:

$$f= 1-[S'v[1+s]+I+(D+\{Q+P/[c-q]\}/I-U)/[1-t]]/S$$

Steve Asikin ISBN 14: 978-1515158691, ISBN 10: 1515158691

Rule-19996:
   If both (**ʃ**), (**U**), (**$**), (**f**), (**Q**), (**v**), (**s**), (**I**), (**t**), (**D**), (**P**),
   (**c**) and (**q**) are known, then its Sales Past must be:
   $$\$'= [\$[1-f]-I-(D+\{Q+P/[c-q]\}/I-U)/[1-t]]/\{v[1+s]\}$$

Rule-19997:
   If both (**ʃ**), (**U**), (**$**), (**f**), (**$'**), (**Q**), (**s**), (**I**), (**t**), (**D**), (**P**),
   (**c**) and (**q**) are known, then its Variable Portion
   Planned is:
   $$v= [\$[1-f]-I-(D+\{Q+P/[c-q]\}/I-U)/[1-t]]/\{\$'[1+s]\}$$

Rule-19998:
   If both (**ʃ**), (**U**), (**$**), (**f**), (**$'**), (**v**), (**Q**), (**I**), (**t**), (**D**), (**P**),
   (**c**) and (**q**) are known, then its Sales Growth Planned
   is:
   $$s= \$/\$'-1$$
   or it can also be found as
   $$s= [\$[1-f]-I-(D+\{Q+P/[c-q]\}/I-U)/[1-t]]/[\$'v]-1$$

Rule-19999:
   If both (**ʃ**), (**U**), (**$**), (**f**), (**$'**), (**v**), (**s**), (**Q**), (**t**), (**D**), (**P**),
   (**c**) and (**q**) are known, then its Interest Expense
   Planned is:
   $$I= \$[1-f]-\$'v[1+s]-(D+\{Q+P/[c-q]\}/I-U)/[1-t]$$

Rule-20000:
   If both (**ʃ**), (**U**), (**$**), (**f**), (**$'**), (**v**), (**Q**), (**I**), (**t**), (**D**), (**P**),
   (**c**) and (**q**) are known, then its Tax Rate Planned is:
   $$t= 1-(D+\{Q+P/[c-q]\}/I-U)/\{\$[1-f]-\$'v[1+s]-I\}$$

Steve Asikin ISBN 14: 978-1515158691, ISBN 10: 1515158691

Rule-20001:

If both (I), (**U**), (**\$**), (**f**), (**\$'**), (**v**), (**s**), (**I**), (**t**), (**Q**), (**P**),
(**c**) and (**q**) are known, then its Dividend Planned is:

$$D= U+\{\$[1-f]-\$'v[1+s]-I\}[1-t]-\{Q+P/[c-q]\}//I$$

Rule-20002:

If both (I), (**U**), (**\$**), (**f**), (**\$'**), (**v**), (**s**), (**I**), (**t**), (**D**), (**Q**),
(**c**) and (**q**) are known, then its Procured Inventory
Planned is:

$$P= [c-q][\textit{I}U+\{\$[1-f]-\$'v[1+s]-I\}[1-t]-D)-Q]$$

Rule-20003:

If both (I), (**U**), (**\$**), (**f**), (**\$'**), (**v**), (**s**), (**I**), (**t**), (**D**), (**P**),
(**Q**) and (**q**) are known, then its Current Ratio Planned
is:

$$c= q+P/[\textit{I}U+\{\$[1-f]-\$'v[1+s]-I\}[1-t]-D)-Q]$$

Rule-20004:

If both (I), (**U**), (**\$**), (**f**), (**\$'**), (**v**), (**s**), (**I**), (**t**), (**D**), (**P**),
(**c**) and (**Q**) are known, then its Quick or Acid Test
Ratio Planned is:

$$q= c-P/[\textit{I}U+\{\$[1-f]-\$'v[1+s]-I\}[1-t]-D)-Q]$$

Rule-20005:

If both (I), (**U**), (**\$**), (**f**), (**\$'**), (**v**), (**s**), (**I**), (**t**), (**D**), (**V**),
(**p**), (**c**) and (**q**) are known, then its Quoted Longterm
Debt Planned is:

$$Q= \textit{I}U+\{\$[1-f]-\$'v[1+s]-I\}[1-t]-D)-Vp/\{360[c-q]\}$$

Steve Asikin ISBN 14: 978-1515158691, ISBN 10: 1515158691

Rule-20006:

If both **(Q)**, **(U)**, **($)**, **(f)**, **($')**, **(v)**, **(s)**, **(I)**, **(t)**, **(D)**, **(V)**, **(p)**, **(c)** and **(q)** are known, then its Leverage or Gearing Ratio Planned is:

$$l = (Q+Vp/\{360[c-q]\})$$
$$/(U+\{\$[1-f]-\$'v[1+s]-I\}[1-t]-D)$$

Rule-20007:

If both **(l)**, **(Q)**, **($)**, **(f)**, **($')**, **(v)**, **(s)**, **(I)**, **(t)**, **(D)**, **(V)**, **(p)**, **(c)** and **(q)** are known, then its Utilized or Starting Capital must be:

$$U = D+(Q+Vp/\{360[c-q]\})/l$$
$$-\{\$[1-f]-\$'v[1+s]-I\}[1-t]$$

Rule-20008:

If both **(l)**, **(U)**, **(Q)**, **(f)**, **($')**, **(v)**, **(s)**, **(I)**, **(t)**, **(D)**, **(V)**, **(p)**, **(c)** and **(q)** are known, then its Sales or Revenue Planned is:

$$\$ = \{\$'v[1+s]+I$$
$$+[D+(Q+Vp/\{360[c-q]\})/l-U]/[1-t]\}/[1-f]$$

Rule-20009:

If both **(l)**, **(U)**, **($)**, **(Q)**, **($')**, **(v)**, **(s)**, **(I)**, **(t)**, **(D)**, **(V)**, **(p)**, **(c)** and **(q)** are known, then its Fixed Portion Planned is:

$$f = 1-\{\$'v[1+s]+I$$
$$+[D+(Q+Vp/\{360[c-q]\})/l-U]/[1-t]\}/\$$$

Steve Asikin ISBN 14: 978-1515158691, ISBN 10: 1515158691

Rule-20010:

If both (**/**), (**U**), (**$**), (**f**), (**Q**), (**v**), (**s**), (**I**), (**t**), (**D**), (**V**), (**p**), (**c**) and (**q**) are known, then its Sales Past must be:

$\$' = \{\$[1\text{-}f]\text{-}I\text{-}[D+(Q+Vp/\{360[c\text{-}q]\})/I\text{-}U]/[1\text{-}t]\}$
$/\{v[1+s]\}$

Rule-20011:

If both (**/**), (**U**), (**$**), (**f**), (**$'**), (**Q**), (**s**), (**I**), (**t**), (**D**), (**V**), (**p**), (**c**) and (**q**) are known, then its Variable Portion Planned is:

$v = V/\$$

or it can also be found as

$v = \{\$[1\text{-}f]\text{-}I\text{-}[D+(Q+Vp/\{360[c\text{-}q]\})/I\text{-}U]/[1\text{-}t]\}$
$/\{\$'[1+s]\}$

Rule-20012:

If both (**/**), (**U**), (**$**), (**f**), (**$'**), (**v**), (**Q**), (**I**), (**t**), (**D**), (**V**), (**p**), (**c**) and (**q**) are known, then its Sales Growth Planned is:

$s = \$/\$'\text{-}1$

or it can also befound as

$s = \{\$[1\text{-}f]\text{-}I\text{-}[D+(Q+Vp/\{360[c\text{-}q]\})/I\text{-}U]/[1\text{-}t]\}$
$/[\$'v]\text{-}1$

Rule-20013:

If both (**/**), (**U**), (**$**), (**f**), (**$'**), (**v**), (**s**), (**Q**), (**t**), (**D**), (**V**), (**p**), (**c**) and (**q**) are known, then its Interest Expense Planned is:

$I = \$[1\text{-}f]\text{-}\$'v[1+s]$
$-[D+(Q+Vp/\{360[c\text{-}q]\})/I\text{-}U]/[1\text{-}t]$

Steve Asikin ISBN 14: 978-1515158691, ISBN 10: 1515158691

Rule-20014:

If both (I), (U), (S), (f), (S'), (v), (s), (I), (Q), (D), (V), (p), (c) and (q) are known, then its Tax Rate Planned is:

$$t= 1-[D+(Q+Vp/\{360[c-q]\})/I-U]$$
$$/\{S[1-f]-S'v[1+s]-I\}$$

Rule-20015:

If both (I), (U), (S), (f), (S'), (v), (s), (I), (t), (Q), (V), (p), (c) and (q) are known, then its Dividend Planned is:

$$D= U+\{S[1-f]-S'v[1+s]-I\}[1-t]$$
$$-(Q+Vp/\{360[c-q]\})/I$$

Rule-20016:

If both (I), (U), (S), (f), (S'), (v), (s), (I), (t), (D), (Q), (p), (c) and (q) are known, then its Variable Cost Planned is:

$$V= 360[c-q][I(U+\{S[1-f]-S'v[1+s]-I\}[1-t]-D)-Q]/p$$

Rule-20017:

If both (I), (U), (S), (f), (S'), (v), (s), (I), (t), (D), (V), (Q), (c) and (q) are known, then its Procured Inventory Days Planned:

$$p= 360[c-q][I(U+\{S[1-f]-S'v[1+s]-I\}[1-t]-D)-Q]/V$$

Rule-20018:

If both (I), (U), (S), (f), (S'), (v), (s), (I), (t), (D), (V), (p), (Q) and (q) are known, then its Current Ratio Planned is:

$$c= q+Vp/\{360[I(U+\{S[1-f]$$
$$-S'v[1+s]-I\}[1-t]-D)-Q]\}$$

Steve Asikin ISBN 14: 978-1515158691, ISBN 10: 1515158691

Rule-20019:

If both (I), (U), (S), (f), (S'), (v), (s), (I), (t), (D), (V), (p), (c) and (Q) are known, then its Quick or Acid Test Ratio Planned is:

$$q = c - Vp / \{360[I(U + \{S[1-f]-S'v[1+s]-I\}[1-t]-D)-Q]\}$$

Rule-20020:

If both (I), (U), (S), (f), (S'), (v), (s), (I), (t), (D), (p), (c) and (q) are known, then its Quoted Longterm Debt Planned is:

$$Q = I(U + \{S[1-f]-S'v[1+s]-I\}[1-t]-D) \\ -Svp / \{360[c-q]\}$$

Rule-20021:

If both (Q), (U), (S), (f), (S'), (v), (s), (I), (t), (D), (p), (c) and (q) are known, then its Leverage or Gearing Ratio Planned is:

$$I = (Q + Svp / \{360[c-q]\}) \\ /(U + \{S[1-f]-S'v[1+s]-I\}[1-t]-D)$$

Rule-20022:

If both (I), (Q), (S), (f), (S'), (v), (s), (I), (t), (D), (p), (c) and (q) are known, then its Utilized or Starting Capital must be:

$$U = D + (Q + Svp / \{360[c-q]\})/I \\ -\{S[1-f]-S'v[1+s]-I\}[1-t]$$

Steve Asikin ISBN 14: 978-1515158691, ISBN 10: 1515158691

Rule-20023:

If both (**/**), (**U**), (**Q**), (**f**), (**S'**), (**v**), (**s**), (**I**), (**t**), (**D**), (**p**), (**c**) and (**q**) are known, then its Sales or Revenue Planned is:

$$S= [/\!(U-\{S'v[1+s]+I\}[1-t]-D)-Q]$$
$$/(vp/\{360[c-q]\}+ S/\!(1-f][1-t]\}$$

Rule-20024:

If both (**/**), (**U**), (**S**), (**Q**), (**S'**), (**v**), (**s**), (**I**), (**t**), (**D**), (**p**), (**c**) and (**q**) are known, then its Fixed Portion Planned is:

$$f= 1-\{S'v[1+s]+I$$
$$+[D+(Q+Svp/\{360[c-q]\})/\!-U]/[1-t]\}/S$$

Rule-20025:

If both (**/**), (**U**), (**S**), (**f**), (**Q**), (**v**), (**s**), (**I**), (**t**), (**D**), (**p**), (**c**) and (**q**) are known, then its Sales Past must be:

$$S'= \{S[1-f]-I-[D+(Q+Svp/\{360[c-q]\})/\!-U]$$
$$/[1-t]\}/\{v[1+s]\}$$

Rule-20026:

If both (**/**), (**U**), (**S**), (**f**), (**S'**), (**Q**), (**s**), (**I**), (**t**), (**D**), (**p**), (**c**) and (**q**) are known, then its Variable Portion Planned is:

$$v= [/\!(U+\{S[1-f]-I\}[1-t]-D)-Q]$$
$$/(Sp/\{360[c-q]\}+S'/\!(1+s][1-t])$$

Steve Asikin ISBN 14: 978-1515158691, ISBN 10: 1515158691

Rule-20027:

If both (I), (**U**), (**S**), (**f**), (**S'**), (**v**), (**Q**), (**I**), (**t**), (**D**), (**p**), (**c**) and (**q**) are known, then its Sales Growth Planned is:

$s = S/S' - 1$

or it can also be found as

$s = \{S[1-f]-I-[D+(Q+Svp/\{360[c-q]\})/I-U]$
$/[1-t]\}/[S'v]-1$

Rule-20028:

If both (I), (**U**), (**S**), (**f**), (**S'**), (**v**), (**s**), (**Q**), (**t**), (**D**), (**p**), (**c**) and (**q**) are known, then its Interest Expense Planned is:

$I = S[1-f]-S'v[1+s]-[D+(Q+Svp/\{360[c-q]\})/I-U]$
$/[1-t]$

Rule-20029:

If both (I), (**U**), (**S**), (**f**), (**S'**), (**v**), (**s**), (**I**), (**Q**), (**D**), (**p**), (**c**) and (**q**) are known, then its Tax Rate Planned is:

$t = 1-[D+(Q+Svp/\{360[c-q]\})/I-U]$
$/\{S[1-f]-S'v[1+s]-I\}$

Rule-20030:

If both (I), (**U**), (**S**), (**f**), (**S'**), (**v**), (**s**), (**I**), (**t**), (**Q**), (**p**), (**c**) and (**q**) are known, then its Dividend Planned is:

$D = U+\{S[1-f]-S'v[1+s]-I\}[1-t]$
$-(Q+Svp/\{360[c-q]\})/I$

Steve Asikin ISBN 14: 978-1515158691, ISBN 10: 1515158691

Rule-20031:

If both $(\textbf{\textit{l}})$, $(\textbf{U})$, $(\textbf{S})$, $(\textbf{f})$, $(\textbf{S'})$, $(\textbf{v})$, $(\textbf{s})$, $(\textbf{I})$, $(\textbf{t})$, $(\textbf{D})$, $(\textbf{Q})$, $(\textbf{c})$ and $(\textbf{q})$ are known, then its Procured Inventory Days Planned:

$$\textbf{p} = 360[\textbf{c-q}][\textbf{\textit{l}U} + \{\textbf{S}[1\textbf{-f}]\textbf{-S'v}[1\textbf{+s}]\textbf{-I}\}[1\textbf{-t}]\textbf{-D})\textbf{-Q}]$$
$$/[\textbf{Sv}]$$

Rule-20032:

If both $(\textbf{\textit{l}})$, $(\textbf{U})$, $(\textbf{S})$, $(\textbf{f})$, $(\textbf{S'})$, $(\textbf{v})$, $(\textbf{s})$, $(\textbf{I})$, $(\textbf{t})$, $(\textbf{D})$, $(\textbf{p})$, $(\textbf{Q})$ and $(\textbf{q})$ are known, then its Current Ratio Planned is:

$$\textbf{c} = \textbf{q+Svp}$$
$$/\{360[\textbf{\textit{l}U} + \{\textbf{S}[1\textbf{-f}]\textbf{-S'v}[1\textbf{+s}]\textbf{-I}\}[1\textbf{-t}]\textbf{-D})\textbf{-Q}]\}$$

Rule-20033:

If both $(\textbf{\textit{l}})$, $(\textbf{U})$, $(\textbf{S})$, $(\textbf{f})$, $(\textbf{S'})$, $(\textbf{v})$, $(\textbf{s})$, $(\textbf{I})$, $(\textbf{t})$, $(\textbf{D})$, $(\textbf{p})$, $(\textbf{c})$ and $(\textbf{Q})$ are known, then its Quick or Acid Test Ratio Planned is:

$$\textbf{q} = \textbf{c-Svp}$$
$$/\{360[\textbf{\textit{l}U} + \{\textbf{S}[1\textbf{-f}]\textbf{-S'v}[1\textbf{+s}]\textbf{-I}\}[1\textbf{-t}]\textbf{-D})\textbf{-Q}]\}$$

Rule-20034:

If both $(\textbf{\textit{l}})$, $(\textbf{U})$, $(\textbf{S})$, $(\textbf{f})$, $(\textbf{S'})$, $(\textbf{v})$, $(\textbf{s})$, $(\textbf{I})$, $(\textbf{t})$, $(\textbf{D})$, $(\textbf{p})$, $(\textbf{c})$ and $(\textbf{q})$ are known, then its Quoted Longterm Debt Planned is:

$$\textbf{Q} = \textbf{\textit{l}U} + \{\textbf{S}[1\textbf{-f}]\textbf{-S'v}[1\textbf{+s}]\textbf{-I}\}[1\textbf{-t}]\textbf{-D})$$
$$\textbf{-S'vp}[1\textbf{+s}]/\{360[\textbf{c-q}]\}$$

Steve Asikin ISBN 14: 978-1515158691, ISBN 10: 1515158691

Rule-20035:

If both $(\mathbf{Q})$, $(\mathbf{U})$, $(\mathbf{S})$, $(\mathbf{f})$, $(\mathbf{S'})$, $(\mathbf{v})$, $(\mathbf{s})$, $(\mathbf{I})$, $(\mathbf{t})$, $(\mathbf{D})$, $(\mathbf{p})$, $(\mathbf{c})$ and $(\mathbf{q})$ are known, then its Leverage or Gearing Ratio Planned is:

$$\mathit{l} = (\mathbf{Q} + \mathbf{S'vp}[1+\mathbf{s}]/\{360[\mathbf{c}-\mathbf{q}]\})$$
$$/(\mathbf{U} + \{\mathbf{S}[1-\mathbf{f}] - \mathbf{S'v}[1+\mathbf{s}] - \mathbf{I}\}[1-\mathbf{t}] - \mathbf{D})$$

Rule-20036:

If both (l), $(\mathbf{Q})$, $(\mathbf{S})$, $(\mathbf{f})$, $(\mathbf{S'})$, $(\mathbf{v})$, $(\mathbf{s})$, $(\mathbf{I})$, $(\mathbf{t})$, $(\mathbf{D})$, $(\mathbf{p})$, $(\mathbf{c})$ and $(\mathbf{q})$ are known, then its Utilized or Starting Capital must be:

$$\mathbf{U} = \mathbf{D} + (\mathbf{Q} + \mathbf{S'vp}[1+\mathbf{s}]/\{360[\mathbf{c}-\mathbf{q}]\})/\mathit{l}$$
$$- \{\mathbf{S}[1-\mathbf{f}] - \mathbf{S'v}[1+\mathbf{s}] - \mathbf{I}\}[1-\mathbf{t}]$$

Rule-20037:

If both (l), $(\mathbf{U})$, $(\mathbf{Q})$, $(\mathbf{f})$, $(\mathbf{S'})$, $(\mathbf{v})$, $(\mathbf{s})$, $(\mathbf{I})$, $(\mathbf{t})$, $(\mathbf{D})$, $(\mathbf{p})$, $(\mathbf{c})$ and $(\mathbf{q})$ are known, then its Sales or Revenue Planned is:

$$\mathbf{S} = [\mathbf{Q} - \mathit{l}(\mathbf{U} - \{\mathbf{S'v}[1+\mathbf{s}] + \mathbf{I}\}[1-\mathbf{t}] - \mathbf{D})$$
$$+ \mathbf{S'vp}[1+\mathbf{s}]/\{360[\mathbf{c}-\mathbf{q}]\}]/\{\mathit{l}[1-\mathbf{f}]\,[1-\mathbf{t}]\}$$

Rule-20038:

If both (l), $(\mathbf{U})$, $(\mathbf{S})$, $(\mathbf{Q})$, $(\mathbf{S'})$, $(\mathbf{v})$, $(\mathbf{s})$, $(\mathbf{I})$, $(\mathbf{t})$, $(\mathbf{D})$, $(\mathbf{p})$, $(\mathbf{c})$ and $(\mathbf{q})$ are known, then its Fixed Portion Planned is:

$$\mathbf{f} = 1 - \{\mathbf{S'v}[1+\mathbf{s}] + \mathbf{I} + [\mathbf{D} + (\mathbf{Q} + \mathbf{S'vp}[1+\mathbf{s}]/\{360[\mathbf{c}-\mathbf{q}]\})/\mathit{l}$$
$$- \mathbf{U}]/[1-\mathbf{t}]\}/\mathbf{S}$$

Steve Asikin ISBN 14: 978-1515158691, ISBN 10: 1515158691

Rule-20039:

If both (**/**), (**U**), (**S**), (**f**), (**Q**), (**v**), (**s**), (**I**), (**t**), (**D**), (**p**), (**c**) and (**q**) are known, then its Sales Past must be:

$S' = [/U + \{S[1-f]-I\}[1-t]-D)-Q]$
$/[v[1+s](p/\{360[c-q]\}+/[1-t])]$

Rule-20040:

If both (**/**), (**U**), (**S**), (**f**), (**S'**), (**Q**), (**s**), (**I**), (**t**), (**D**), (**p**), (**c**) and (**q**) are known, then its Variable Portion Planned is:

$v = [/U + \{S[1-f]-I\}[1-t]-D)-Q]$
$/[S'[1+s](p/\{360[c-q]\}+/[1-t])]$

Rule-20041:

If both (**/**), (**U**), (**S**), (**f**), (**S'**), (**v**), (**Q**), (**I**), (**t**), (**D**), (**p**), (**c**) and (**q**) are known, then its Sales Growth Planned is:

$s = S/S' - 1$

or it can also be found as

$s = [/U + \{S[1-f]-I\}[1-t]-D)-Q]$
$/[S'v(p/\{360[c-q]\}+/[1-t])]-1$

Rule-20042:

If both (**/**), (**U**), (**S**), (**f**), (**S'**), (**v**), (**s**), (**Q**), (**t**), (**D**), (**p**), (**c**) and (**q**) are known, then its Interest Expense Planned is:

$I = S[1-f]-S'v[1+s]$
$-[D+(Q+S'vp[1+s]/\{360[c-q]\})//-U]/[1-t]$

35

Rule-20043:

    If both (**/**), (**U**), (**\$**), (**f**), (**\$'**), (**v**), (**s**), (**I**), (**Q**), (**D**), (**p**), (**c**) and (**q**) are known, then its Tax Rate Planned is:

$$t= 1-[\mathbf{D}+(\mathbf{Q}+\mathbf{\$'vp}[1+\mathbf{s}]/\{360[\mathbf{c}-\mathbf{q}]\})/\mathbf{/}-\mathbf{U}]$$
$$/\{\mathbf{\$}[1-\mathbf{f}]-\mathbf{\$'v}[1+\mathbf{s}]-\mathbf{I}\}$$

Rule-20044:

    If both (**/**), (**U**), (**\$**), (**f**), (**\$'**), (**v**), (**s**), (**I**), (**t**), (**Q**), (**p**), (**c**) and (**q**) are known, then its Dividend Payout Planned is:

$$D= \mathbf{U}+\{\mathbf{\$}[1-\mathbf{f}]-\mathbf{\$'v}[1+\mathbf{s}]-\mathbf{I}\}[1-\mathbf{t}]$$
$$-(\mathbf{Q}+\mathbf{\$'vp}[1+\mathbf{s}]/\{360[\mathbf{c}-\mathbf{q}]\})/\mathbf{/}$$

Rule-20045:

    If both (**/**), (**U**), (**\$**), (**f**), (**\$'**), (**v**), (**s**), (**I**), (**t**), (**D**), (**Q**), (**c**) and (**q**) are known, then its Procured Inventory Days Planned:

$$p= 360[\mathbf{c}-\mathbf{q}][\mathbf{/}(\mathbf{U}+\{\mathbf{\$}[1-\mathbf{f}]-\mathbf{\$'v}[1+\mathbf{s}]-\mathbf{I}\}[1-\mathbf{t}]-\mathbf{D})-\mathbf{Q}]$$
$$/\{\mathbf{\$'v}[1+\mathbf{s}]\}$$

Rule-20046:

    If both (**/**), (**U**), (**\$**), (**f**), (**\$'**), (**v**), (**s**), (**I**), (**t**), (**D**), (**p**), (**Q**) and (**q**) are known, then its Current Ratio Planned is:

$$c= \mathbf{q}+\mathbf{\$'vp}[1+\mathbf{s}]$$
$$/\{360[\mathbf{/}(\mathbf{U}+\{\mathbf{\$}[1-\mathbf{f}]-\mathbf{\$'v}[1+\mathbf{s}]-\mathbf{I}\}[1-\mathbf{t}]-\mathbf{D})-\mathbf{Q}]\}$$

Steve Asikin ISBN 14: 978-1515158691, ISBN 10: 1515158691

Rule-20047:

If both (f), (**U**), (**$**), (**f**), (**$'**), (**v**), (**s**), (**I**), (**t**), (**D**), (**p**), (**c**) and (**Q**) are known, then its Quick or Acid Test Ratio Planned is:

$q= c\text{-}\$'vp[1+s]$

$/\{360[\!\!\!f U+\{\$[1\text{-}f]\text{-}\$'v[1+s]\text{-}I\}[1\text{-}t]\text{-}D)\text{-}Q]\}$

Rule-20048:

If both (f), (**U**), (**$**), (**f**), (**$'**), (**v**), (**s**), (**I**), (**t**), (**A**), (**d**) and (**X**) are known, then its Quoted Longterm Debt Planned is:

$Q= f U+\{\$[1\text{-}f]\text{-}\$'v[1+s]\text{-}I\}[1\text{-}t]\text{-}Ad)\text{-}X$

Rule-20049:

If both (**Q**), (**U**), (**$**), (**f**), (**$'**), (**v**), (**s**), (**I**), (**t**), (**A**), (**d**) and (**X**) are known, then its Leverage or Gearing Ratio Planned is:

$f= [Q+X]/(U+\{\$[1\text{-}f]\text{-}\$'v[1+s]\text{-}I\}[1\text{-}t]\text{-}Ad)$

Rule-20050:

If both (f), (**Q**), (**$**), (**f**), (**$'**), (**v**), (**s**), (**I**), (**t**), (**A**), (**d**) and (**X**) are known, then its Utilized or Starting Capital must be:

$U= Ad+[Q+X]/f\{\$[1\text{-}f]\text{-}\$'v[1+s]\text{-}I\}[1\text{-}t]$

Rule-20051:

If both (f), (**U**), (**Q**), (**f**), (**$'**), (**v**), (**s**), (**I**), (**t**), (**A**), (**d**) and (**X**) are known, then its Sales or Revenue Planned is:

$\$= (\$'v[1+s]+I+\{Ad+[Q+X]/f\text{-}U\}/[1\text{-}t])/[1\text{-}f]$

Steve Asikin ISBN 14: 978-1515158691, ISBN 10: 1515158691

Rule-20052:

    If both (f), (U), $(\$)$, (Q), $(\$')$, (v), (s), (I), (t), (A), (d) and (X) are known, then its Fixed Portion Planned is:

$$f= 1-(\$'v[1+s]+I+\{Ad+[Q+X]/I-U\}/[1-t])/\$$$

Rule-20053:

    If both (f), (U), $(\$)$, (f), (Q), (v), (s), (I), (t), (A), (d) and (X) are known, then its Sales Past must be:

$$\$'= (\$[1-f]-I-\{Ad+[Q+X]/I-U\}/[1-t])/\{v[1+s]\}$$

Rule-20054:

If both (f), (U), $(\$)$, (f), $(\$')$, (Q), (s), (I), (t), (A), (d) and (X) are known, then its Variable Portion Planned is:

$$v= (\$[1-f]-I-\{Ad+[Q+X]/I-U\}/[1-t])/\{\$'[1+s]\}$$

Rule-20055:

    If both (f), (U), $(\$)$, (f), $(\$')$, (v), (Q), (I), (t), (A), (d) and (X) are known, then its Sales Growth Planned is:

$$s= \$/\$'-1$$

        or it can also be found as

$$s= (\$[1-f]-I-\{Ad+[Q+X]/I-U\}/[1-t])/[\$'v]-1$$

Rule-20056:

    If both (f), (U), $(\$)$, (f), $(\$')$, (v), (s), (Q), (t), (A), (d) and (X) are known, then its Interest Expense Planned is:

$$I= \$[1-f]-\$'v[1+s]-\{Ad+[Q+X]/I-U\}/[1-t]$$

Rule-20057:

    If both (f), (U), $(\$)$, (f), $(\$')$, (v), (s), (I), (Q), (A), (d) and (X) are known, then its Tax Rate Planned is:

$$t= 1-\{Ad+[Q+X]/I-U\}/\{\$[1-f]-\$'v[1+s]-I\}$$

Steve Asikin ISBN 14: 978-1515158691, ISBN 10: 1515158691

Rule-20058:

If both (I), (U), (S), (f), (S'), (v), (s), (I), (t), (Q), (d) and (X) are known, then its After Tax Income Planned is:

$$A= (U+\{S[1-f]-S'v[1+s]-I\}[1-t]-[Q+X]/I)/d$$

Rule-20059:

If both (I), (U), (S), (f), (S'), (v), (s), (I), (t), (A), (Q) and (X) are known, then its Dividend Portion Planned is:

$$d= (U+\{S[1-f]-S'v[1+s]-I\}[1-t]-[Q+X]/I)/A$$

Rule-20060:

If both (I), (U), (S), (f), (S'), (v), (s), (I), (t), (A), (d) and (Q) are known, then its Xpress or Current Debt Planned is:

$$X= I(U+\{S[1-f]-S'v[1+s]-I\}[1-t]-Ad)-Q$$

Rule-20061:

If both (I), (U), (S), (f), (S'), (v), (s), (I), (t), (A), (d), (P), (c) and (q) are known, then its Quoted Longterm Debt Planned is:

$$Q= I(U+\{S[1-f]-S'v[1+s]-I\}[1-t]-Ad)-P/[c-q]$$

Rule-20062:

If both (Q), (U), (S), (f), (S'), (v), (s), (I), (t), (A), (d), (P), (c) and (q) are known, then its Leverage or Gearing Ratio Planned is:

$$I= \{Q+P/[c-q]\}/(U+\{S[1-f]-S'v[1+s]-I\}[1-t]-Ad)$$

Steve Asikin ISBN 14: 978-1515158691, ISBN 10: 1515158691

Rule-20063:
> If both (**/**), (**Q**), (**$**), (**f**), (**$'**), (**v**), (**s**), (**I**), (**t**), (**A**), (**d**),
> (**P**), (**c**) and (**q**) are known, then its Utilized or
> Starting Capital must be:
> $$U = Ad + \{Q + P/[c-q]\}//\{\$[1-f]-\$'v[1+s]-I\}[1-t]$$

Rule-20064:
> If both (**/**), (**U**), (**Q**), (**f**), (**$'**), (**v**), (**s**), (**I**), (**t**), (**A**), (**d**),
> (**P**), (**c**) and (**q**) are known, then its Sales or Revenue
> Planned is:
> $$\$ = [\$'v[1+s]+I+(Ad+\{Q+P/[c-q]\}//-U)/[1-t]]/[1-f]$$

Rule-20065:
> If both (**/**), (**U**), (**$**), (**Q**), (**$'**), (**v**), (**s**), (**I**), (**t**), (**A**), (**d**),
> (**P**), (**c**) and (**q**) are known, then its Fixed Portion
> Planned is:
> $$f = 1-[\$'v[1+s]+I+(Ad+\{Q+P/[c-q]\}//-U)/[1-t]]/\$$$

Rule-20066:
> If both (**/**), (**U**), (**$**), (**f**), (**Q**), (**v**), (**s**), (**I**), (**t**), (**A**), (**d**),
> (**P**), (**c**) and (**q**) are known, then its Sales Past must
> be:
> $$\$' = [\$[1-f]-I-(Ad+\{Q+P/[c-q]\}//-U)/[1-t]]$$
> $$/\{v[1+s]\}$$

Rule-20067:
> If both (**/**), (**U**), (**$**), (**f**), (**$'**), (**Q**), (**s**), (**I**), (**t**), (**A**), (**d**),
> (**P**), (**c**) and (**q**) are known, then its Variable Portion
> Planned is:
> $$v = [\$[1-f]-I-(Ad+\{Q+P/[c-q]\}//-U)/[1-t]]$$
> $$/\{\$'[1+s]\}$$

Steve Asikin ISBN 14: 978-1515158691, ISBN 10: 1515158691

Rule-20068:

If both (I), (U), (S), (f), (S'), (v), (Q), (I), (t), (A), (d), (P), (c) and (q) are known, then its Sales Growth Planned is:

$s= S/S'-1$

or it can also be found as

$s= [S[1-f]-I-(Ad+\{Q+P/[c-q]\}/I-U)/[1-t]]/[S'v]-1$

Rule-20069:

If both (I), (U), (S), (f), (S'), (v), (s), (Q), (t), (A), (d), (P), (c) and (q) are known, then its Interest Expense Planned is:

$I= S[1-f]-S'v[1+s]-(Ad+\{Q+P/[c-q]\}/I-U)/[1-t]$

Rule-20070:

If both (I), (U), (S), (f), (S'), (v), (s), (I), (Q), (A), (d), (P), (c) and (q) are known, then its Tax Rate Planned is:

$t= 1-(Ad+\{Q+P/[c-q]\}/I-U)/\{S[1-f]-S'v[1+s]-I\}$

Rule-20071:

If both (I), (U), (S), (f), (S'), (v), (s), (I), (t), (Q), (d), (P), (c) and (q) are known, then its After Tax Income Planned is:

$A= (U+\{S[1-f]-S'v[1+s]-I\}[1-t]-\{Q+P/[c-q]\}/I)/d$

Rule-20072:

If both (I), (U), (S), (f), (S'), (v), (s), (I), (t), (A), (Q), (P), (c) and (q) are known, then its Dividend Payout Planned is:

$d= (U+\{S[1-f]-S'v[1+s]-I\}[1-t]-\{Q+P/[c-q]\}/I)/A$

Steve Asikin ISBN 14: 978-1515158691, ISBN 10: 1515158691

Rule-20073:

If both (**/**), (**U**), (**$**), (**f**), (**$'**), (**v**), (**s**), (**I**), (**t**), (**A**), (**d**), (**Q**), (**c**) and (**q**) are known, then its Procured Inventory Planned is:

$$\mathbf{P} = [\mathbf{c}\text{-}\mathbf{q}][\textit{/}\mathbf{U} + \{\mathbf{\$}[1\text{-}\mathbf{f}]\text{-}\mathbf{\$'v}[1\text{+}\mathbf{s}]\text{-}\mathbf{I}\}[1\text{-}\mathbf{t}]\text{-}\mathbf{Ad})\text{-}\mathbf{Q}]$$

Rule-20074:

If both (**/**), (**U**), (**$**), (**f**), (**$'**), (**v**), (**s**), (**I**), (**t**), (**A**), (**d**), (**P**), (**Q**) and (**q**) are known, then its Current Ratio Planned is:

$$\mathbf{c} = \mathbf{q} + \mathbf{P}/[\textit{/}\mathbf{U} + \{\mathbf{\$}[1\text{-}\mathbf{f}]\text{-}\mathbf{\$'v}[1\text{+}\mathbf{s}]\text{-}\mathbf{I}\}[1\text{-}\mathbf{t}]\text{-}\mathbf{Ad})\text{-}\mathbf{Q}]$$

Rule-20075:

If both (**/**), (**U**), (**$**), (**f**), (**$'**), (**v**), (**s**), (**I**), (**t**), (**A**), (**d**), (**P**), (**c**) and (**Q**) are known, then its Quick or Acid Test Ratio Planned is:

$$\mathbf{q} = \mathbf{c} - \mathbf{P}/[\textit{/}\mathbf{U} + \{\mathbf{\$}[1\text{-}\mathbf{f}]\text{-}\mathbf{\$'v}[1\text{+}\mathbf{s}]\text{-}\mathbf{I}\}[1\text{-}\mathbf{t}]\text{-}\mathbf{Ad})\text{-}\mathbf{Q}]$$

Rule-20076:

If both (**/**), (**U**), (**$**), (**f**), (**$'**), (**v**), (**s**), (**I**), (**t**), (**A**), (**d**), (**V**), (**p**), (**c**) and (**q**) are known, then its Quoted Longterm Debt Planned is:

$$\mathbf{Q} = \textit{/}\mathbf{U} + \{\mathbf{\$}[1\text{-}\mathbf{f}]\text{-}\mathbf{\$'v}[1\text{+}\mathbf{s}]\text{-}\mathbf{I}\}[1\text{-}\mathbf{t}]\text{-}\mathbf{Ad})$$
$$-\mathbf{Vp}/\{360[\mathbf{c}\text{-}\mathbf{q}]\}$$

Rule-20077:

If both (**Q**), (**U**), (**$**), (**f**), (**$'**), (**v**), (**s**), (**I**), (**t**), (**A**), (**d**), (**V**), (**p**), (**c**) and (**q**) are known, then its Leverage or Gearing Ratio Planned is:

$$\textit{/} = (\mathbf{Q} + \mathbf{Vp}/\{360[\mathbf{c}\text{-}\mathbf{q}]\})$$
$$/(\mathbf{U} + \{\mathbf{\$}[1\text{-}\mathbf{f}]\text{-}\mathbf{\$'v}[1\text{+}\mathbf{s}]\text{-}\mathbf{I}\}[1\text{-}\mathbf{t}]\text{-}\mathbf{Ad})$$

Steve Asikin ISBN 14: 978-1515158691, ISBN 10: 1515158691

Rule-20078:

If both (I), (**Q**), (**$**), (**f**), (**$'**), (**v**), (**s**), (**I**), (**t**), (**A**), (**d**), (**V**), (**p**), (**c**) and (**q**) are known, then its Utilized or Starting Capital must be:

$$U = Ad + (Q + Vp/\{360[c-q]\})/I$$
$$-\{\$[1-f]-\$'v[1+s]-I\}[1-t]$$

Rule-20079:

If both (I), (**U**), (**Q**), (**f**), (**$'**), (**v**), (**s**), (**I**), (**t**), (**A**), (**d**), (**V**), (**p**), (**c**) and (**q**) are known, then its Sales or Revenue Planned is:

$$\$ = \{\$'v[1+s]+I+[Ad+(Q+Vp/\{360[c-q]\})/I-U]$$
$$/[1-t]\}/[1-f]$$

Rule-20080:

If both (I), (**U**), (**$**), (**Q**), (**$'**), (**v**), (**s**), (**I**), (**t**), (**A**), (**d**), (**V**), (**p**), (**c**) and (**q**) are known, then its Fixed Portion Planned is:

$$f = 1-\{\$'v[1+s]+I+[Ad+(Q+Vp/\{360[c-q]\})/I-U]$$
$$/[1-t]\}/\$$$

Rule-20081:

If both (I), (**U**), (**$**), (**f**), (**Q**), (**v**), (**s**), (**I**), (**t**), (**A**), (**d**), (**V**), (**p**), (**c**) and (**q**) are known, then its Sales Past must be:

$$\$' = \{\$[1-f]-I-[Ad+(Q+Vp/\{360[c-q]\})/I-U]/[1-t]\}$$
$$/\{v[1+s]\}$$

Steve Asikin ISBN 14: 978-1515158691, ISBN 10: 1515158691

Rule-20082:

If both (I), (U), (S), (f), (S'), (Q), (s), (I), (t), (A), (d), (V), (p), (c) and (q) are known, then its Variable Portion Planned is:

$v = V/S$

or it can also be found as

$v = \{S[1-f]-I-[Ad+(Q+Vp/\{360[c-q]\})/I-U]/[1-t]\} / \{S'[1+s]\}$

Rule-20083:

If both (I), (U), (S), (f), (S'), (v), (Q), (I), (t), (A), (d), (V), (p), (c) and (q) are known, then its Sales Growth Planned is:

$s = S/S'-1$

or it can also be found as

$s = \{S[1-f]-I-[Ad+(Q+Vp/\{360[c-q]\})/I-U]/[1-t]\} / [S'v]-1$

Rule-20084:

If both (I), (U), (S), (f), (S'), (v), (s), (Q), (t), (A), (d), (V), (p), (c) and (q) are known, then its Interest Expense Planned is:

$I = S[1-f]-S'v[1+s]- [Ad+(Q+Vp/\{360[c-q]\})/I-U] / [1-t]$

Rule-20085:

If both (I), (U), (S), (f), (S'), (v), (s), (I), (Q), (A), (d), (V), (p), (c) and (q) are known, then its Tax Rate Planned is:

$t = 1-[Ad+(Q+Vp/\{360[c-q]\})/I-U] / \{S[1-f]-S'v[1+s]-I\}$

Steve Asikin ISBN 14: 978-1515158691, ISBN 10: 1515158691

Rule-20086:

If both (**/**), (**U**), (**$**), (**f**), (**$'**), (**v**), (**s**), (**I**), (**t**), (**Q**), (**d**), (**V**), (**p**), (**c**) and (**q**) are known, then its After Tax Income Planned is:

$$A= [U+\{\$[1-f]-\$'v[1+s]-I\}[1-t]$$
$$-(Q+Vp/\{360[c-q]\})//]/d$$

Rule-20087:

If both (**/**), (**U**), (**$**), (**f**), (**$'**), (**v**), (**s**), (**I**), (**t**), (**A**), (**Q**), (**V**), (**p**), (**c**) and (**q**) are known, then its Dividend Payout Planned is:

$$d= [U+\{\$[1-f]-\$'v[1+s]-I\}[1-t]$$
$$-(Q+Vp/\{360[c-q]\})//]/A$$

Rule-20088:

If both (**/**), (**U**), (**$**), (**f**), (**$'**), (**v**), (**s**), (**I**), (**t**), (**A**), (**d**), (**Q**), (**p**), (**c**) and (**q**) are known, then its Variable Cost Planned is:

$$V= 360[c-q]$$
$$[/(U+\{\$[1-f]-\$'v[1+s]-I\}[1-t]-Ad)-Q]/p$$

Rule-20089:

If both (**/**), (**U**), (**$**), (**f**), (**$'**), (**v**), (**s**), (**I**), (**t**), (**A**), (**d**), (**V**), (**Q**), (**c**) and (**q**) are known, then its Procured Inventory Days Planned:

$$p= 360[c-q]$$
$$[/(U+\{\$[1-f]-\$'v[1+s]-I\}[1-t]-Ad)-Q]/V$$

Steve Asikin ISBN 14: 978-1515158691, ISBN 10: 1515158691

Rule-20090:

If both (I), (U), (S), (f), (S'), (v), (s), (I), (t), (A), (d), (V), (p), (Q) and (q) are known, then its Current Ratio Planned is:

$$c = q + Vp$$
$$/\{360[I(U + \{S[1-f] - S'v[1+s] - I\}[1-t] - Ad) - Q]\}$$

Rule-20091:

If both (I), (U), (S), (f), (S'), (v), (s), (I), (t), (A), (d), (V), (p), (c) and (Q) are known, then its Quick oe Acid Test Ratio Planned is:

$$q = c - Vp$$
$$/\{360[I(U + \{S[1-f] - S'v[1+s] - I\}[1-t] - Ad) - Q]\}$$

Rule-20092:

If both (I), (U), (S), (f), (S'), (v), (s), (I), (t), (A), (d), (p), (c) and (q) are known, then its Quoted Longterm Debt Planned is:

$$Q = I(U + \{S[1-f] - S'v[1+s] - I\}[1-t] - Ad)$$
$$- Svp/\{360[c-q]\}$$

Rule-20093:

If both (Q), (U), (S), (f), (S'), (v), (s), (I), (t), (A), (d), (p), (c) and (q) are known, then its Leverage or Gearing Ratio Planned is:

$$I = (Q + Svp/\{360[c-q]\})/(U + \{S[1-f] - S'v[1+s] - I\}[1-t] - Ad)$$

Steve Asikin ISBN 14: 978-1515158691, ISBN 10: 1515158691

Rule-20094:

If both (*I*), (**Q**), (**S**), (**f**), (**S'**), (**v**), (**s**), (**I**), (**t**), (**A**), (**d**), (**p**), (**c**) and (**q**) are known, then its Utilized or Starting Capital must be:

$$U= Ad+(Q+Svp/\{360[c\text{-}q]\})/I$$
$$-\{S[1\text{-}f]\text{-}S'v[1+s]\text{-}I\}[1\text{-}t]$$

Rule-20095:

If both (*I*), (**U**), (**Q**), (**f**), (**S'**), (**v**), (**s**), (**I**), (**t**), (**A**), (**d**), (**p**), (**c**) and (**q**) are known, then its Sales or Revenue Planned is:

$$S= [I(U\text{-}S'v[1+s]\text{-}I\}[1\text{-}t]\text{-}Ad)\text{-}Q]$$
$$/(vp/\{360[c\text{-}q]\}\text{-}I[1\text{-}f]\,[1\text{-}t])$$

Rule-20096:

If both (*I*), (**U**), (**S**), (**Q**), (**S'**), (**v**), (**s**), (**I**), (**t**), (**A**), (**d**), (**p**), (**c**) and (**q**) are known, then its Fixed Portion Planned is:

$$f= 1\text{-}\{S'v[1+s]+I+[Ad+(Q+Svp/\{360[c\text{-}q]\})/I\text{-}U]$$
$$/[1\text{-}t]\}/S$$

Rule-20097:

If both (*I*), (**U**), (**S**), (**f**), (**Q**), (**v**), (**s**), (**I**), (**t**), (**A**), (**d**), (**p**), (**c**) and (**q**) are known, then its Sales Past must be:

$$S'= \{S[1\text{-}f]\text{-}I\text{-}[Ad+(Q+Svp/\{360[c\text{-}q]\})/I\text{-}U]$$
$$/[1\text{-}t]\}/\{v[1+s]\}$$

Steve Asikin ISBN 14: 978-1515158691, ISBN 10: 1515158691

Rule-20098:

If both $(\textit{f})$, $(\textbf{U})$, $(\textbf{S})$, $(\textbf{f})$, $(\textbf{S'})$, $(\textbf{Q})$, $(\textbf{s})$, $(\textbf{I})$, $(\textbf{t})$, $(\textbf{A})$, $(\textbf{d})$, $(\textbf{p})$, $(\textbf{c})$ and $(\textbf{q})$ are known, then its Variable Portion Planned is:

$$v= [\textit{f}\textbf{U}+\{\textbf{S}[1\text{-}\textbf{f}]\text{-}\textbf{I}\}[1\text{-}\textbf{t}]\text{-}\textbf{Ad})\text{-}\textbf{Q}]$$
$$/(\textbf{Sp}/\{360[\textbf{c}\text{-}\textbf{q}]\}+\textbf{S'}\textit{f}[1+\textbf{s}][1\text{-}\textbf{t}])$$

Rule-20099:

If both $(\textit{f})$, $(\textbf{U})$, $(\textbf{S})$, $(\textbf{f})$, $(\textbf{S'})$, $(\textbf{v})$, $(\textbf{Q})$, $(\textbf{I})$, $(\textbf{t})$, $(\textbf{A})$, $(\textbf{d})$, $(\textbf{p})$, $(\textbf{c})$ and $(\textbf{q})$ are known, then its Sales Growth Planned is:

$$s= \textbf{S}/\textbf{S'}\text{-}1$$

or it can also be found as

$$s= \{\textbf{S}[1\text{-}\textbf{f}]\text{-}\textbf{I}\text{-}[\textbf{Ad}+(\textbf{Q}+\textbf{Svp}/\{360[\textbf{c}\text{-}\textbf{q}]\})/\textit{f}\text{-}\textbf{U}]/[1\text{-}\textbf{t}]\}$$
$$/[\textbf{S'v}]\text{-}1$$

Rule-20100:

If both $(\textit{f})$, $(\textbf{U})$, $(\textbf{S})$, $(\textbf{f})$, $(\textbf{S'})$, $(\textbf{v})$, $(\textbf{s})$, $(\textbf{Q})$, $(\textbf{t})$, $(\textbf{A})$, $(\textbf{d})$, $(\textbf{p})$, $(\textbf{c})$ and $(\textbf{q})$ are known, then its Interest Expense Planned is:

$$I= \textbf{S}[1\text{-}\textbf{f}]\text{-}\textbf{S'v}[1+\textbf{s}]\text{-}[\textbf{Ad}+(\textbf{Q}+\textbf{Svp}/\{360[\textbf{c}\text{-}\textbf{q}]\})/\textit{f}\text{-}\textbf{U}]$$
$$/[1\text{-}\textbf{t}]$$

Rule-20101:

If both $(\textit{f})$, $(\textbf{U})$, $(\textbf{S})$, $(\textbf{f})$, $(\textbf{S'})$, $(\textbf{v})$, $(\textbf{s})$, $(\textbf{I})$, $(\textbf{Q})$, $(\textbf{A})$, $(\textbf{d})$, $(\textbf{p})$, $(\textbf{c})$ and $(\textbf{q})$ are known, then its Tax Rate Planned is:

$$t= 1\text{-}[\textbf{Ad}+(\textbf{Q}+\textbf{Svp}/\{360[\textbf{c}\text{-}\textbf{q}]\})/\textit{f}\text{-}\textbf{U}]$$
$$/\{\textbf{S}[1\text{-}\textbf{f}]\text{-}\textbf{S'v}[1+\textbf{s}]\text{-}\textbf{I}\}$$

Steve Asikin ISBN 14: 978-1515158691, ISBN 10: 1515158691

Rule-20102:

If both (**∫**), (**U**), (**$**), (**f**), (**$'**), (**v**), (**s**), (**I**), (**t**), (**Q**), (**d**), (**p**), (**c**) and (**q**) are known, then its After Tax Income Planned is:

$$A = [U + \{\$[1-f] - \$'v[1+s] - I\}[1-t] - (Q + \$vp / \{360[c-q]\}) / \mathbf{∫} / d$$

Rule-20103:

If both (**∫**), (**U**), (**$**), (**f**), (**$'**), (**v**), (**s**), (**I**), (**t**), (**A**), (**Q**), (**p**), (**c**) and (**q**) are known, then its Dividend Payout Planned is:

$$d = [U + \{\$[1-f] - \$'v[1+s] - I\}[1-t] - (Q + \$vp / \{360[c-q]\}) / \mathbf{∫} / A$$

Rule-20104:

If both (**∫**), (**U**), (**$**), (**f**), (**$'**), (**v**), (**s**), (**I**), (**t**), (**A**), (**d**), (**Q**), (**c**) and (**q**) are known, then its Procured Inventory Days Planned:

$$p = 360[c-q]$$
$$[\mathbf{∫}U + \{\$[1-f] - \$'v[1+s] - I\}[1-t] - Ad) - Q] / [\$v]$$

Rule-20105:

If both (**∫**), (**U**), (**$**), (**f**), (**$'**), (**v**), (**s**), (**I**), (**t**), (**A**), (**d**), (**p**), (**Q**) and (**q**) are known, then its Current Ratio Planned is:

$$c = q + \$vp$$
$$/ \{360[\mathbf{∫}U + \{\$[1-f] - \$'v[1+s] - I\}[1-t] - Ad) - Q]\}$$

Steve Asikin ISBN 14: 978-1515158691, ISBN 10: 1515158691

Rule-20106:

If both (**/**), (**U**), (**$**), (**f**), (**$'**), (**v**), (**s**), (**I**), (**t**), (**A**), (**d**), (**p**), (**c**) and (**Q**) are known, then its Current Ratio Planned is:

$$q= c\text{-}$up$$
$$/\{360[\textbf{/}(\textbf{U}+\{\$[1\text{-}f]\text{-}\$'v[1+s]\text{-}I\}[1\text{-}t]\text{-}Ad)\text{-}Q]\}$$

Rule-20107:

If both (**/**), (**U**), (**$**), (**f**), (**$'**), (**v**), (**s**), (**I**), (**t**), (**A**), (**d**), (**p**), (**c**) and (**q**) are known, then its Quoted Longterm Debt Planned is:

$$Q= \textbf{/}(\textbf{U}+\{\$[1\text{-}f]\text{-}\$'v[1+s]\text{-}I\}[1\text{-}t]\text{-}Ad)$$
$$-\$'vp[1+s]/\{360[c\text{-}q]\}$$

Rule-20108:

If both (**Q**), (**U**), (**$**), (**f**), (**$'**), (**v**), (**s**), (**I**), (**t**), (**A**), (**d**), (**p**), (**c**) and (**q**) are known, then its Leverage or Gearing Ratio Planned is:

$$\textbf{/} = (\textbf{Q}+\$'vp[1+s]/\{360[c\text{-}q]\})$$
$$/(\textbf{U}+\{\$[1\text{-}f]\text{-}\$'v[1+s]\text{-}I\}[1\text{-}t]\text{-}Ad)$$

Rule-20109:

If both (**/**), (**Q**), (**$**), (**f**), (**$'**), (**v**), (**s**), (**I**), (**t**), (**A**), (**d**), (**p**), (**c**) and (**q**) are known, then its Utilized or Starting Capital must be:

$$U= Ad+(\textbf{Q}+\$'vp[1+s]/\{360[c\text{-}q]\})/\textbf{/}$$
$$-\{\$[1\text{-}f]\text{-}\$'v[1+s]\text{-}I\}[1\text{-}t]$$

Steve Asikin ISBN 14: 978-1515158691, ISBN 10: 1515158691

<u>Rule-20110</u>:

If both (**∫**), (**U**), (**Q**), (**f**), (**S'**), (**v**), (**s**), (**I**), (**t**), (**A**), (**d**), (**p**), (**c**) and (**q**) are known, then its Sales or Revenue Planned is:

$$S= [Q-∫(U+\{ -S'v[1+s]-I\}[1-t]-Ad)$$
$$+S'vp[1+s]/\{360[c-q]\}]/\{∫[1-f][1-t]\}$$

<u>Rule-20111</u>:

If both (**∫**), (**U**), (**S**), (**Q**), (**S'**), (**v**), (**s**), (**I**), (**t**), (**A**), (**d**), (**p**), (**c**) and (**q**) are known, then its Fixed Portion Planned is:

$$f= 1-\{S'v[1+s]+I+[Ad$$
$$+(Q+S'vp[1+s]/\{360[c-q]\})/∫-U]/[1-t]\}/S$$

<u>Rule-20112</u>:

If both (**∫**), (**U**), (**S**), (**f**), (**Q**), (**v**), (**s**), (**I**), (**t**), (**A**), (**d**), (**p**), (**c**) and (**q**) are known, then its Sales Past must be:

$$S'= [∫(U+\{S[1-f]-I\}[1-t]-Ad)-Q]$$
$$/[v[1+s](p/\{360[c-q]\}+∫[1-t])]$$

<u>Rule-20113</u>:

If both (**∫**), (**U**), (**S**), (**f**), (**S'**), (**Q**), (**s**), (**I**), (**t**), (**A**), (**d**), (**p**), (**c**) and (**q**) are known, then its Variable Portion Planned is:

$$v= [∫(U+\{S[1-f]-I\}[1-t]-Ad)-Q]$$
$$/[S'[1+s](p/\{360[c-q]\}+∫[1-t])]$$

Steve Asikin ISBN 14: 978-1515158691, ISBN 10: 1515158691

Rule-20114:

If both (**/**), (**U**), (**\$**), (**f**), (**\$'**), (**v**), (**Q**), (**I**), (**t**), (**A**), (**d**), (**p**), (**c**) and (**q**) are known, then its Sales Growth Planned is:

$s= \$/\$'-1$

or it can also be found as

$s= [/(U+\{\$[1-f]-I\}[1-t]-Ad)-Q]$
$/[\$'v(p/\{360[c-q]\}+/[1-t])]-1$

Rule-20115:

If both (**/**), (**U**), (**\$**), (**f**), (**\$'**), (**v**), (**s**), (**Q**), (**t**), (**A**), (**d**), (**p**), (**c**) and (**q**) are known, then its Interest Expense Planned is:

$I= \$[1-f]-\$'v[1+s]-[Ad$
$+(Q+\$'vp[1+s]/\{360[c-q]\})/-U]/[1-t]$

Rule-20116:

If both (**/**), (**U**), (**\$**), (**f**), (**\$'**), (**v**), (**s**), (**I**), (**Q**), (**A**), (**d**), (**p**), (**c**) and (**q**) are known, then its Tax Rate Planned is:

$t= 1-[Ad+(Q+\$'vp[1+s]/\{360[c-q]\})/-U]$
$/\{\$[1-f]-\$'v[1+s]-I\}$

Rule-20117:

If both (**/**), (**U**), (**\$**), (**f**), (**\$'**), (**v**), (**s**), (**I**), (**t**), (**Q**), (**d**), (**p**), (**c**) and (**q**) are known, then its After Tax Income Planned is:

$A= [U+\{\$[1-f]-\$'v[1+s]-I\}[1-t]$
$-(Q+\$'vp[1+s]/\{360[c-q]\})//]/d$

Steve Asikin ISBN 14: 978-1515158691, ISBN 10: 1515158691

Rule-20118:

If both (**/**), (**U**), (**S**), (**f**), (**S'**), (**v**), (**s**), (**I**), (**t**), (**A**), (**Q**), (**p**), (**c**) and (**q**) are known, then its Dividend Payout Planned is:

$$d= [U+\{S[1-f]-S'v[1+s]-I\}[1-t]$$
$$-(Q+S'vp[1+s]/\{360[c-q]\})/I]/A$$

Rule-20119:

If both (**/**), (**U**), (**S**), (**f**), (**S'**), (**v**), (**s**), (**I**), (**t**), (**A**), (**d**), (**Q**), (**c**) and (**q**) are known, then its Procured Inventory Days Planned:

$$p= 360[c-q][I(U+\{S[1-f]-S'v[1+s]-I\}[1-t]-Ad)-Q]$$
$$/\{S'v[1+s]\}$$

Rule-20120:

If both (**/**), (**U**), (**S**), (**f**), (**S'**), (**v**), (**s**), (**I**), (**t**), (**A**), (**d**), (**p**), (**Q**) and (**q**) are known, then its Current Ratio Planned is:

$$c= q+S'vp[1+s]$$
$$/\{360[I(U+\{S[1-f]-S'v[1+s]-I\}[1-t]-Ad)-Q]\}$$

Rule-20121:

If both (**/**), (**U**), (**S**), (**f**), (**S'**), (**v**), (**s**), (**I**), (**t**), (**A**), (**d**), (**p**), (**c**) and (**q**) are known, then its Quoted Longterm Debt Planned is:

$$q= c-S'vp[1+s]/\{360[I(U+\{S[1-f]-S'v[1+s]-I\}$$
$$[1-t]-Ad)-Q]\}$$

Steve Asikin ISBN 14: 978-1515158691, ISBN 10: 1515158691

Rule-20122:

If both (I), (**U**), (**\$**), (**f**), (**\$'**), (**v**), (**s**), (**I**), (**t**), (**d**) and (**X**) are known, then its Quoted Longterm Debt Planned is:

$$Q = I(U + \{\$[1-f] - \$'v[1+s] - I\}[1-t][1-d]) - X$$

Rule-20123:

If both (**Q**), (**U**), (**\$**), (**f**), (**\$'**), (**v**), (**s**), (**I**), (**t**), (**d**) and (**X**) are known, then its Leverage or Gearing Planned is:

$$I = [Q+X]/(U + \{\$[1-f] - \$'v[1+s] - I\}[1-t][1-d])$$

Rule-20124:

If both (I), (**Q**), (**\$**), (**f**), (**\$'**), (**v**), (**s**), (**I**), (**t**), (**d**) and (**X**) are known, then its Utilized ofr Starting Capital must be:

$$U = [Q+X]/I - \{\$[1-f] - \$'v[1+s] - I\}[1-t][1-d]$$

Rule-20125:

If both (I), (**U**), (**Q**), (**f**), (**\$'**), (**v**), (**s**), (**I**), (**t**), (**d**) and (**X**) are known, then its Sales or Revenue Planned is:

$$\$ = (\$'v[1+s] + I + \{[Q+X]/I - U\}/\{[1-t][1-d]\})/[1-f]$$

Rule-20126:

If both (I), (**U**), (**\$**), (**Q**), (**\$'**), (**v**), (**s**), (**I**), (**t**), (**d**) and (**X**) are known, then its Fixed Portion Planned is:

$$f = 1 - (\$'v[1+s] + I + \{[Q+X]/I - U\}/\{[1-t][1-d]\})/\$$$

Rule-20127:

If both (I), (**U**), (**\$**), (**f**), (**Q**), (**v**), (**s**), (**I**), (**t**), (**d**) and (**X**) are known, then its Sales Past must be:

$$\$' = (\$[1-f] - I - \{[Q+X]/I - U\}/\{[1-t][1-d]\})/\{v[1+s]\}$$

Steve Asikin ISBN 14: 978-1515158691, ISBN 10: 1515158691

Rule-20128:

If both (f), (U), $(\$)$, (f), $(\$')$, (Q), (s), (I), (t), (d) and (X) are known, then its Variable Portion Planned is:

$$v= (\$[1\text{-}f]\text{-}I\text{-}\{[Q+X]/I\text{-}U\}/\{[1\text{-}t][1\text{-}d]\})/\{\$'[1+s]\}$$

Rule-20129:

If both (f), (U), $(\$)$, (f), $(\$')$, (v), (Q), (I), (t), (d) and (X) are known, then its Sales Growth Planned is:

$$s= \$/\$'\text{-}1$$

or it can also be found as

$$s= (\$[1\text{-}f]\text{-}I\text{-}\{[Q+X]/I\text{-}U\}/\{[1\text{-}t][1\text{-}d]\})/[\$'v]\text{-}1$$

Rule-20130:

If both (f), (U), $(\$)$, (f), $(\$')$, (v), (s), (Q), (t), (d) and (X) are known, then its Interest Expense Planned is:

$$I= \$[1\text{-}f]\text{-}\$'v[1+s]\text{-}\{[Q+X]/I\text{-}U\}/\{[1\text{-}t][1\text{-}d]\}$$

Rule-20131:

If both (f), (U), $(\$)$, (f), $(\$')$, (v), (s), (I), (Q), (d) and (X) are known, then its Tax Rate Planned is:

$$t= 1\text{-}\{[Q+X]/I\text{-}U\}/(\{\$[1\text{-}f]\text{-}\$'v[1+s]\text{-}I\}[1\text{-}d])$$

Rule-20132:

If both (f), (U), $(\$)$, (f), $(\$')$, (v), (s), (I), (t), (Q) and (X) are known, then its Dividend Payout Planned is:

$$d= 1\text{-}\{[Q+X]/I\text{-}U\}/(\{\$[1\text{-}f]\text{-}\$'v[1+s]\text{-}I\}[1\text{-}t])$$

Steve Asikin ISBN 14: 978-1515158691, ISBN 10: 1515158691

Rule-20133:
If both (f), (U), (S), (f), (S'), (v), (s), (I), (t), (d) and (Q) are known, then its Xpress or Current Debt Planned is:
$$X= f(U+\{S[1-f]-S'v[1+s]-I\}[1-t][1-d])-Q$$

Rule-20134:
If both (f), (U), (S), (f), (S'), (v), (s), (I), (t), (d), (P), (c) and (q) are known, then its Quoted Longterm Debt Planned is:
$$Q= f(U+\{S[1-f]-S'v[1+s]-I\}[1-t][1-d])-P/[c-q]$$

Rule-20135:
If both (Q), (U), (S), (f), (S'), (v), (s), (I), (t), (d), (P), (c) and (q) are known, then its Leverage or Gearing Ratio Planned is:
$$f= \{Q+P/[c-q]\}/(U+\{S[1-f]-S'v[1+s]-I\}[1-t][1-d])$$

Rule-20136:
If both (f), (Q), (S), (f), (S'), (v), (s), (I), (t), (d), (P), (c) and (q) are known, then its Utilized or Starting Capital must be:
$$U= \{Q+P/[c-q]\}/f\{S[1-f]-S'v[1+s]-I\}[1-t][1-d]$$

Rule-20137:
If both (f), (U), (Q), (f), (S'), (v), (s), (I), (t), (d), (P), (c) and (q) are known, then its Sales or Revenue Planned is:
$$S= [S'v[1+s]+I+(\{Q+P/[c-q]\}/f-U)$$
$$/\{[1-t][1-d]\}]/[1-f]$$

Steve Asikin ISBN 14: 978-1515158691, ISBN 10: 1515158691

Rule-20138:

If both (**/**), (**U**), (**\$**), (**Q**), (**\$'**), (**v**), (**s**), (**I**), (**t**), (**d**), (**P**), (**c**) and (**q**) are known, then its Fixed Cost Planned is:

$$f= 1-[\$'v[1+s]+I+(\{Q+P/[c-q]\}/I-U)$$
$$/\{[1-t][1-d]\}]/\$$$

Rule-20139:

If both (**/**), (**U**), (**\$**), (**f**), (**Q**), (**v**), (**s**), (**I**), (**t**), (**d**), (**P**), (**c**) and (**q**) are known, then its Sales Past must be:

$$\$'= [\$[1-f]-I-(\{Q+P/[c-q]\}/I-U)$$
$$/\{[1-t][1-d]\}]/\{v[1+s]\}$$

Rule-20140:

If both (**/**), (**U**), (**\$**), (**f**), (**\$'**), (**Q**), (**s**), (**I**), (**t**), (**d**), (**P**), (**c**) and (**q**) are known, then its Variable Portion Planned is:

$$v= [\$[1-f]-I-(\{Q+P/[c-q]\}/I-U)/\{[1-t][1-d]\}]$$
$$/\{\$'[1+s]\}$$

Rule-20141:

If both (**/**), (**U**), (**\$**), (**f**), (**\$'**), (**v**), (**s**), (**I**), (**t**), (**d**), (**P**), (**c**) and (**q**) are known, then its Quoted Longterm Debt Planned is:

$$s= \$/\$'-1$$

or it can also be found

$$s= [\$[1-f]-I-(\{Q+P/[c-q]\}/I-U)/\{[1-t][1-d]\}]$$
$$/ [\$'v]-1$$

Steve Asikin ISBN 14: 978-1515158691, ISBN 10: 1515158691

Rule-20142:
> If both (**/**), (**U**), (**$**), (**f**), (**$'**), (**v**), (**s**), (**Q**), (**t**), (**d**), (**P**),
> (**c**) and (**q**) are known, then its Interest Expense
> Planned is:
> $I = \$[1-f]-\$'v[1+s]-(\{Q+P/[c-q]\}//-U)/\{[1-t][1-d]\}$

Rule-20143:
> If both (**/**), (**U**), (**$**), (**f**), (**$'**), (**v**), (**s**), (**I**), (**Q**), (**d**), (**P**),
> (**c**) and (**q**) are known, then its Tax Rate Planned is:
> $t = 1-(\{Q+P/[c-q]\}//-U)/(\{\$[1-f]-\$'v[1+s]-I\}[1-d])$

Rule-20144:
> If both (**/**), (**U**), (**$**), (**f**), (**$'**), (**v**), (**s**), (**I**), (**t**), (**d**), (**P**),
> (**c**) and (**q**) are known, then its Quoted Longterm Debt
> Planned is:
> $d = 1-(\{Q+P/[c-q]\}//-U)/(\{\$[1-f]-\$'v[1+s]-I\}[1-t])$

Rule-20145:
> If both (**/**), (**U**), (**$**), (**f**), (**$'**), (**v**), (**s**), (**I**), (**t**), (**d**), (**Q**),
> (**c**) and (**q**) are known, then its Procured Inventory
> Planned is:
> $P = [c-q][/U+\{\$[1-f]-\$'v[1+s]-I\}[1-t][1-d])-Q]$

Rule-20146:
> If both (**/**), (**U**), (**$**), (**f**), (**$'**), (**v**), (**s**), (**I**), (**t**), (**d**), (**P**),
> (**Q**) and (**q**) are known, then its Current Ratio Planned
> is:
> $c = q+P/[/U+\{\$[1-f]-\$'v[1+s]-I\}[1-t][1-d])-Q]$

Steve Asikin ISBN 14: 978-1515158691, ISBN 10: 1515158691

Rule-20147:

If both (I), (U), (S), (f), (S'), (v), (s), (I), (t), (d), (P), (c) and (Q) are known, then its Quick or Acid Test Ratio Planned is:

$$q = c-P/[I(U+\{S[1-f]-S'v[1+s]-I\}[1-t][1-d])-Q]$$

Rule-20148:

If both (I), (U), (S), (f), (S'), (v), (s), (I), (t), (d), (V), (p), (c) and (q) are known, then its Quoted Longterm Debt Planned is:

$$Q = I(U+\{S[1-f]-S'v[1+s]-I\}[1-t][1-d])$$
$$-Vp/\{360[c-q]\}$$

Rule-20149:

If both (Q), (U), (S), (f), (S'), (v), (s), (I), (t), (d), (V), (p), (c) and (q) are known, then its Leverage or Gearing Ratio Planned is:

$$I = (Q+Vp/\{360[c-q]\})$$
$$/ (U+\{S[1-f]-S'v[1+s]-I\}[1-t][1-d])$$

Rule-20150:

If both (I), (Q), (S), (f), (S'), (v), (s), (I), (t), (d), (V), (p), (c) and (q) are known, then its Utilized or Starting Capital must be:

$$U = (Q+Vp/\{360[c-q]\})/I$$
$$-\{S[1-f]-S'v[1+s]-I\}[1-t][1-d]$$

Steve Asikin ISBN 14: 978-1515158691, ISBN 10: 1515158691

Rule-20151:

If both (**/**), (**U**), (**Q**), (**f**), (**S'**), (**v**), (**s**), (**I**), (**t**), (**d**), (**V**), (**p**), (**c**) and (**q**) are known, then its Sales or Revenue Planned is:

$$S= \{S'v[1+s]+I+[(Q+Vp/\{360[c-q]\})/I-U]$$
$$/\{[1-t][1-d]\}\}/[1-f]$$

Rule-20152:

If both (**/**), (**U**), (**S**), (**Q**), (**S'**), (**v**), (**s**), (**I**), (**t**), (**d**), (**V**), (**p**), (**c**) and (**q**) are known, then its Fixed Portion Planned is:

$$f= 1-\{S'v[1+s]+I$$
$$+[(Q+Vp/\{360[c-q]\})/I-U]/\{[1-t][1-d]\}\}/S$$

Rule-20153:

If both (**/**), (**U**), (**S**), (**f**), (**Q**), (**v**), (**s**), (**I**), (**t**), (**d**), (**V**), (**p**), (**c**) and (**q**) are known, then its Sales Past must be:

$$S'= \{S[1-f]-I-[(Q+Vp/\{360[c-q]\})/I-U]$$
$$/\{[1-t][1-d]\}\}/\{v[1+s]\}$$

Rule-20154:

If both (**/**), (**U**), (**S**), (**f**), (**S'**), (**Q**), (**s**), (**I**), (**t**), (**d**), (**V**), (**p**), (**c**) and (**q**) are known, then its Quoted Longterm Debt Planned is:

$$v= V/S$$

or it can also be found as

$$v= \{S[1-f]-I-[(Q+Vp/\{360[c-q]\})/I-U]$$
$$/\{[1-t][1-d]\}\}/\{S'[1+s]\}$$

Steve Asikin ISBN 14: 978-1515158691, ISBN 10: 1515158691

Rule-20155:

If both (**/**), (**U**), (**$**), (**f**), (**$'**), (**v**), (**Q**), (**I**), (**t**), (**d**), (**V**), (**p**), (**c**) and (**q**) are known, then its Sales Growth Planned is:

$s = \$/\$'-1$

      or it can also be found as

$s = \{\$[1\text{-}f]\text{-}I\text{-}[(Q+Vp/\{360[c\text{-}q]\})/I\text{-}U]$
$/\{[1\text{-}t][1\text{-}d]\}\}/[\$'v]\text{-}1$

Rule-20156:

If both (**/**), (**U**), (**$**), (**f**), (**$'**), (**v**), (**s**), (**Q**), (**t**), (**d**), (**V**), (**p**), (**c**) and (**q**) are known, then its Leverage or Gearing Ratio Planned is:

$I = \$[1\text{-}f]\text{-}\$'v[1+s]$
$-[(Q+Vp/\{360[c\text{-}q]\})/I\text{-}U]/\{[1\text{-}t][1\text{-}d]\}$

Rule-20157:

If both (**/**), (**U**), (**$**), (**f**), (**$'**), (**v**), (**s**), (**I**), (**Q**), (**d**), (**V**), (**p**), (**c**) and (**q**) are known, then its Tax Rate Planned is:

$t = 1\text{-}[(Q+Vp/\{360[c\text{-}q]\})/I\text{-}U]$
$/(\{\$[1\text{-}f]\text{-}\$'v[1+s]\text{-}I\}[1\text{-}d])$

Rule-20158:

If both (**/**), (**U**), (**$**), (**f**), (**$'**), (**v**), (**s**), (**I**), (**t**), (**Q**), (**V**), (**p**), (**c**) and (**q**) are known, then its Dividend Payout Planned is:

$d = 1\text{-}[(Q+Vp/\{360[c\text{-}q]\})/I\text{-}U]$
$/(\{\$[1\text{-}f]\text{-}\$'v[1+s]\text{-}I\}[1\text{-}t])$

Steve Asikin ISBN 14: 978-1515158691, ISBN 10: 1515158691

Rule-20159:

If both $(\textit{l})$, $(\textbf{U})$, $(\textbf{\$})$, $(\textbf{f})$, $(\textbf{\$'})$, $(\textbf{v})$, $(\textbf{s})$, $(\textbf{I})$, $(\textbf{t})$, $(\textbf{d})$, $(\textbf{Q})$, $(\textbf{p})$, $(\textbf{c})$ and $(\textbf{q})$ are known, then its Variable Cost Planned is:

$$V = 360[c\text{-}q]$$
$$[\textit{l}U + \{\$[1\text{-}f]\text{-}\$'v[1+s]\text{-}I\}[1\text{-}t][1\text{-}d])\text{-}Q]/p$$

Rule-20160:

If both $(\textit{l})$, $(\textbf{U})$, $(\textbf{\$})$, $(\textbf{f})$, $(\textbf{\$'})$, $(\textbf{v})$, $(\textbf{s})$, $(\textbf{I})$, $(\textbf{t})$, $(\textbf{d})$, $(\textbf{V})$, $(\textbf{Q})$, $(\textbf{c})$ and $(\textbf{q})$ are known, then its Procured Inventory Days Planned:

$$p = 360[c\text{-}q]$$
$$[\textit{l}U + \{\$[1\text{-}f]\text{-}\$'v[1+s]\text{-}I\}[1\text{-}t][1\text{-}d])\text{-}Q]/V$$

Rule-20161:

If both $(\textit{l})$, $(\textbf{U})$, $(\textbf{\$})$, $(\textbf{f})$, $(\textbf{\$'})$, $(\textbf{v})$, $(\textbf{s})$, $(\textbf{I})$, $(\textbf{t})$, $(\textbf{d})$, $(\textbf{V})$, $(\textbf{p})$, $(\textbf{Q})$ and $(\textbf{q})$ are known, then its Current Ratio Planned is:

$$c = q + Vp/\{360[\textit{l}U + \{\$[1\text{-}f]\text{-}\$'v[1+s]\text{-}I\}$$
$$[1\text{-}t][1\text{-}d])\text{-}Q]\}$$

Rule-20162:

If both $(\textit{l})$, $(\textbf{U})$, $(\textbf{\$})$, $(\textbf{f})$, $(\textbf{\$'})$, $(\textbf{v})$, $(\textbf{s})$, $(\textbf{I})$, $(\textbf{t})$, $(\textbf{d})$, $(\textbf{V})$, $(\textbf{p})$, $(\textbf{c})$ and $(\textbf{Q})$ are known, then its Quick or Acid Test Ratio Planned is:

$$q = c - Vp/\{360[\textit{l}U + \{\$[1\text{-}f]\text{-}\$'v[1+s]\text{-}I\}$$
$$[1\text{-}t][1\text{-}d])\text{-}Q]\}$$

Steve Asikin ISBN 14: 978-1515158691, ISBN 10: 1515158691

Rule-20163:

If both (*I*), (**U**), (**S**), (**f**), (**S'**), (**v**), (**s**), (**I**), (**t**), (**d**), (**p**), (**c**) and (**q**) are known, then its Quoted Longterm Debt Planned is:

$$Q = I(U + \{S[1-f]-S'v[1+s]-I\}[1-t][1-d])$$
$$-Svp/\{360[c-q]\}$$

Rule-20164:

If both (**Q**), (**U**), (**S**), (**f**), (**S'**), (**v**), (**s**), (**I**), (**t**), (**d**), (**p**), (**c**) and (**q**) are known, then its Leverage or Gearing Ratio Planned is:

$$I = (Q + Svp/\{360[c-q]\})$$
$$/(U + \{S[1-f]-S'v[1+s]-I\}[1-t][1-d])$$

Rule-20165:

If both (*I*), (**Q**), (**S**), (**f**), (**S'**), (**v**), (**s**), (**I**), (**t**), (**d**), (**p**), (**c**) and (**q**) are known, then its Utilized or Starting Capital must be:

$$U = (Q + Svp/\{360[c-q]\})/I$$
$$-\{S[1-f]-S'v[1+s]-I\}[1-t][1-d]$$

Rule-20166:

If both (*I*), (**U**), (**Q**), (**f**), (**S'**), (**v**), (**s**), (**I**), (**t**), (**d**), (**p**), (**c**) and (**q**) are known, then its Sales or Revenue Planned is:

$$S = [I(U-\{S'v[1+s]+I\}[1-t][1-d])-Q]$$
$$/(vp/\{360[c-q]\}-I[1-f][1-t][1-d])$$

Steve Asikin ISBN 14: 978-1515158691, ISBN 10: 1515158691

Rule-20167:

If both (l), (U), (S), (Q), (S'), (v), (s), (l), (t), (d), (p), (c) and (q) are known, then its Fixed Portion Planned is:

$$f = 1 - \{S'v[1+s] + l + [(Q + Svp/\{360[c-q]\})/l - U]$$
$$/\{[1-t][1-d]\}\}/S$$

Rule-20168:

If both (l), (U), (S), (f), (Q), (v), (s), (l), (t), (d), (p), (c) and (q) are known, then its Sales Past must be:

$$S' = \{S[1-f] - l - [(Q + Svp/\{360[c-q]\})/l - U]$$
$$/\{[1-t][1-d]\}\}/\{v[1+s]\}$$

Rule-20169:

If both (l), (U), (S), (f), (S'), (Q), (s), (l), (t), (d), (p), (c) and (q) are known, then its Variable Portion Planned is:

$$v = [l U + \{S[1-f] - l\}[1-t][1-d]) - Q]$$
$$/(Sp/\{360[c-q]\} + S'/[1+s][1-t][1-d])$$

Rule-20170:

If both (l), (U), (S), (f), (S'), (v), (Q), (l), (t), (d), (p), (c) and (q) are known, then its Sales Growth Planned is:

$$s = S/S' - 1$$

or it can also be found as

$$s = \{S[1-f] - l - [(Q + Svp/\{360[c-q]\})/l - U]$$
$$/\{[1-t][1-d]\}\}/[S'v] - 1$$

Steve Asikin ISBN 14: 978-1515158691, ISBN 10: 1515158691

Rule-20171:
  If both (I), (U), (S), (f), (S'), (v), (s), (Q), (t), (d), (p), (c) and (q) are known, then its Interest Expense Planned is:
  $$I = S[1-f] - S'v[1+s] - [(Q+Svp/\{360[c-q]\})/I-U]$$
  $$/\{[1-t][1-d]\}$$

Rule-20172:
  If both (I), (U), (S), (f), (S'), (v), (s), (I), (Q), (d), (p), (c) and (q) are known, then its Tax Rate Planned is:
  $$t = 1 - [(Q+Svp/\{360[c-q]\})/I-U]$$
  $$/(\{S[1-f]-S'v[1+s]-I\}[1-d])$$

Rule-20173:
  If both (I), (U), (S), (f), (S'), (v), (s), (I), (t), (Q), (p), (c) and (q) are known, then its Dividend Payout Planned is:
  $$d = 1 - [(Q+Svp/\{360[c-q]\})/I-U]$$
  $$/(\{S[1-f]-S'v[1+s]-I\}[1-t])$$

Rule-20174:
  If both (I), (U), (S), (f), (S'), (v), (s), (I), (t), (d), (Q), (c) and (q) are known, then its Procured Inventory Days Planned:
  $$p = 360[c-q]$$
  $$[I U + \{S[1-f]-S'v[1+s]-I\}[1-t][1-d]) - Q]/[Sv]$$

Steve Asikin ISBN 14: 978-1515158691, ISBN 10: 1515158691

Rule-20175:

If both (I), (U), (S), (f), (S'), (v), (s), (I), (t), (d), (p), (Q) and (q) are known, then its Current Ratio Planned is:

$$c = q + Svp / \{360[I U + \{S[1-f] - S'v[1+s] - I\} [1-t][1-d]) - Q]\}$$

Rule-20176:

If both (I), (U), (S), (f), (S'), (v), (s), (I), (t), (d), (p), (c) and (q) are known, then its Quoted Longterm Debt Planned is:

$$q = c - Svp / \{360[I U + \{S[1-f] - S'v[1+s] - I\} [1-t][1-d]) - Q]\}$$

Rule-20177:

If both (I), (U), (S), (f), (S'), (v), (s), (I), (t), (d), (p), (c) and (q) are known, then its Quoted Longterm Debt Planned is:

$$Q = I U + \{S[1-f] - S'v[1+s] - I\} [1-t][1-d]) \\ - S'vp[1+s] / \{360[c-q]\}$$

Rule-20178:

If both (Q), (U), (S), (f), (S'), (v), (s), (I), (t), (d), (p), (c) and (q) are known, then its Leverage or Gearing Ratio Planned is:

$$I = (Q + S'vp[1+s] / \{360[c-q]\}) \\ / (U + \{S[1-f] - S'v[1+s] - I\} [1-t][1-d])$$

Steve Asikin ISBN 14: 978-1515158691, ISBN 10: 1515158691

Rule-20179:

If both (**/**), (**Q**), (**$**), (**f**), (**$'**), (**v**), (**s**), (**I**), (**t**), (**d**), (**p**), (**c**) and (**q**) are known, then its Utilized or Starting Capital must be:

$$\mathbf{U}= (\mathbf{Q}+\mathbf{\$'vp}[1+s]/\{360[\mathbf{c\text{-}q}]\})/\mathbf{/}$$
$$-\{\mathbf{\$}[1\text{-}\mathbf{f}]\text{-}\mathbf{\$'v}[1+s]\text{-}\mathbf{I}\}[1\text{-}\mathbf{t}][1\text{-}\mathbf{d}]$$

Rule-20180:

If both (**/**), (**U**), (**Q**), (**f**), (**$'**), (**v**), (**s**), (**I**), (**t**), (**d**), (**p**), (**c**) and (**q**) are known, then its Sales or Revenue Planned is:

$$\mathbf{\$}= [\mathbf{Q}\text{-}\mathbf{/}(\mathbf{U}\text{-}\{\mathbf{\$'v}[1+s]+\mathbf{I}\}[1\text{-}\mathbf{t}][1\text{-}\mathbf{d}])$$
$$+\mathbf{\$'vp}[1+s]/\{360[\mathbf{c\text{-}q}]\}]/\{\mathbf{/}[1\text{-}\mathbf{f}][1\text{-}\mathbf{t}][1\text{-}\mathbf{d}]\}$$

Rule-20181:

If both (**/**), (**U**), (**$**), (**Q**), (**$'**), (**v**), (**s**), (**I**), (**t**), (**d**), (**p**), (**c**) and (**q**) are known, then its Fixed Portion Planned is:

$$\mathbf{f}= 1\text{-}\{\mathbf{\$'v}[1+s]+\mathbf{I}+[(\mathbf{Q}+\mathbf{\$'vp}[1+s]/\{360[\mathbf{c\text{-}q}]\})/\mathbf{/}\text{-}\mathbf{U}]$$
$$/\{[1\text{-}\mathbf{t}][1\text{-}\mathbf{d}]\}\}/\mathbf{\$}$$

Rule-20182:

If both (**/**), (**U**), (**$**), (**f**), (**Q**), (**v**), (**s**), (**I**), (**t**), (**d**), (**p**), (**c**) and (**q**) are known, then its Sales Past must be:

$$\mathbf{\$'}= [\mathbf{/}(\mathbf{U}+\{\mathbf{\$}[1\text{-}\mathbf{f}]\text{-}\mathbf{I}\}[1\text{-}\mathbf{t}][1\text{-}\mathbf{d}])\text{-}\mathbf{Q}]$$
$$/[\mathbf{v}[1+s](\mathbf{p}/\{360[\mathbf{c\text{-}q}]\}+\mathbf{\$'}/[1+s]\;[1\text{-}\mathbf{t}][1\text{-}\mathbf{d}])]$$

Steve Asikin ISBN 14: 978-1515158691, ISBN 10: 1515158691

Rule-20183:

If both (I), (U), (S), (f), (S'), (Q), (s), (I), (t), (d), (p), (c) and (q) are known, then its Variable Portion Planned is:

$$v= [I(U+\{S[1-f]-I\}[1-t][1-d])-Q]$$
$$/[S'[1+s](p/\{360[c-q]\}+S'I[1+s] [1-t][1-d])]$$

Rule-20184:

If both (I), (U), (S), (f), (S'), (v), (Q), (I), (t), (d), (p), (c) and (q) are known, then its Sales Growth Planned is:

$$s= S/S'-1$$

or it can also be found as

$$s=[I(U+\{S[1-f]-I\}[1-t][1-d])-Q]$$
$$/[S'v(p/\{360[c-q]\}+S'I[1+s] [1-t][1-d])]-1$$

Rule-20185:

If both (I), (U), (S), (f), (S'), (v), (s), (Q), (t), (d), (p), (c) and (q) are known, then its Interest Expense Planned is:

$$I= S[1-f]-S'v[1+s]-[(Q+S'vp[1+s]/\{360[c-q]\})/I-U]$$
$$/\{[1-t][1-d]\}$$

Rule-20186:

If both (I), (U), (S), (f), (S'), (v), (s), (I), (Q), (d), (p), (c) and (q) are known, then its Tax Rate Planned is:

$$t= 1-[(Q+S'vp[1+s]/\{360[c-q]\})/I-U]$$
$$/(\{S[1-f]-S'v[1+s]-I\}[1-d])$$

Steve Asikin ISBN 14: 978-1515158691, ISBN 10: 1515158691

Rule-20187:

If both (f), (U), (S), (f), (S'), (v), (s), (I), (t), (Q), (p), (c) and (q) are known, then its Dividend Payout Planned is:

$$d= 1-[(Q+S'vp[1+s]/\{360[c-q]\})/I-U]$$
$$/(\{S[1-f]-S'v[1+s]-I\}[1-t])$$

Rule-20188:

If both (f), (U), (S), (f), (S'), (v), (s), (I), (t), (d), (Q), (c) and (q) are known, then its Procured Inventory Days Planned:

$$p= 360[c-q][I(U+\{S[1-f]-S'v[1+s]-I\}[1-t][1-d])-Q]$$
$$/\{S'v[1+s]\}$$

Rule-20189:

If both (f), (U), (S), (f), (S'), (v), (s), (I), (t), (d), (p), (Q) and (q) are known, then its Current Ratio Planned is:

$$c= q+S'vp[1+s]/\{360[I(U+\{S[1-f]-S'v[1+s]-I\}$$
$$[1-t][1-d])-Q]\}$$

Rule-20190:

If both (f), (U), (S), (f), (S'), (v), (s), (I), (t), (d), (p), (c) and (Q) are known, then its Quick or Acid Test Ratio Planned is:

$$q= c-S'vp[1+s]/\{360[I(U+\{S[1-f]-S'v[1+s]-I\}$$
$$[1-t][1-d])-Q]\}$$

Steve Asikin ISBN 14: 978-1515158691, ISBN 10: 1515158691

Rule-20191:

If both $(\textbf{\textit{I}})$, $(\textbf{U})$, $(\textbf{\$})$, $(\textbf{f})$, $(\textbf{i})$, $(\textbf{\$'})$, $(\textbf{v})$, $(\textbf{s})$, $(\textbf{T})$, $(\textbf{D})$ and $(\textbf{X})$ are known, then its Quoted Longterm Debt Planned is:

$$\textbf{Q} = \textit{I}\{\textbf{U} + \$[1-\textbf{f}-\textbf{i}] - \$'\textbf{v}[1+\textbf{s}] - \textbf{T} - \textbf{D}\} - \textbf{X}$$

Rule-20192:

If both $(\textbf{Q})$, $(\textbf{U})$, $(\textbf{\$})$, $(\textbf{f})$, $(\textbf{i})$, $(\textbf{\$'})$, $(\textbf{v})$, $(\textbf{s})$, $(\textbf{T})$, $(\textbf{D})$ and $(\textbf{X})$ are known, then its Leverage or Gearing Ratio Planned is:

$$\textit{I} = [\textbf{Q}+\textbf{X}]/\{\textbf{U} + \$[1-\textbf{f}-\textbf{i}] - \$'\textbf{v}[1+\textbf{s}] - \textbf{T} - \textbf{D}\}$$

Rule-20193:

If both $(\textbf{\textit{I}})$, $(\textbf{Q})$, $(\textbf{\$})$, $(\textbf{f})$, $(\textbf{i})$, $(\textbf{\$'})$, $(\textbf{v})$, $(\textbf{s})$, $(\textbf{T})$, $(\textbf{D})$ and $(\textbf{X})$ are known, then its Utilized or Starting Capital must be:

$$\textbf{U} = \$'\textbf{v}[1+\textbf{s}] + \textbf{T} + \textbf{D} + [\textbf{Q}+\textbf{X}]/\textit{I} - \$[1-\textbf{f}-\textbf{i}]$$

Rule-20194:

If both $(\textbf{\textit{I}})$, $(\textbf{U})$, $(\textbf{Q})$, $(\textbf{f})$, $(\textbf{i})$, $(\textbf{\$'})$, $(\textbf{v})$, $(\textbf{s})$, $(\textbf{T})$, $(\textbf{D})$ and $(\textbf{X})$ are known, then its Sales or Revenue Planned is:

$$\$ = \{\$'\textbf{v}[1+\textbf{s}] + \textbf{T} + \textbf{D} + [\textbf{Q}+\textbf{X}]/\textit{I} - \textbf{U}\}/[1-\textbf{f}-\textbf{i}]$$

Rule-20195:

If both $(\textbf{\textit{I}})$, $(\textbf{U})$, $(\textbf{\$})$, $(\textbf{Q})$, $(\textbf{i})$, $(\textbf{\$'})$, $(\textbf{v})$, $(\textbf{s})$, $(\textbf{T})$, $(\textbf{D})$ and $(\textbf{X})$ are known, then its Fixed Portion Planned is:

$$\textbf{f} = 1 - \textbf{i} - \{\$'\textbf{v}[1+\textbf{s}] + \textbf{T} + \textbf{D} + [\textbf{Q}+\textbf{X}]/\textit{I} - \textbf{U}\}/\$$$

Steve Asikin ISBN 14: 978-1515158691, ISBN 10: 1515158691

Rule-20196:
   If both (**I**), (**U**), (**S**), (**f**), (**Q**), (**S'**), (**v**), (**s**), (**T**), (**D**) and
   (**X**) are known, then its Interest Portion Planned is:
   $i= 1-f-\{S'v[1+s]+T+D+[Q+X]/IU\}/S$

Rule-20197:
   If both (**I**), (**U**), (**S**), (**f**), (**i**), (**Q**), (**v**), (**s**), (**T**), (**D**) and
   (**X**) are known, then its Sales Past must be:
   $S'= \{U+S[1-f-i]+T+D+[Q+X]/I\}/\{v[1+s]\}$

Rule-20198:
   If both (**I**), (**U**), (**S**), (**f**), (**i**), (**S'**), (**Q**), (**s**), (**T**), (**D**) and
   (**X**) are known, then its Variable Portion Planned is:
   $v= \{U+S[1-f-i]+T+D+[Q+X]/I\}/\{S'[1+s]\}$

Rule-20199:
   If both (**I**), (**U**), (**S**), (**f**), (**i**), (**S'**), (**v**), (**Q**), (**T**), (**D**) and
   (**X**) are known, then its Sales Growth Planned is:
   $s= S/S'-1$
                  or it can also be found as
   $s= \{U+S[1-f-i]+T+D+[Q+X]/I\}/[S'v]-1$

Rule-20200:
   If both (**I**), (**U**), (**S**), (**f**), (**i**), (**S'**), (**v**), (**s**), (**Q**), (**D**) and
   (**X**) are known, then its Tax Planned is:
   $T= U+S[1-f-i]-S'v[1+s]-D-[Q+X]/I$

Rule-20201:
   If both (**I**), (**U**), (**S**), (**f**), (**i**), (**S'**), (**v**), (**s**), (**T**), (**Q**) and
   (**X**) are known, then its Dividend Planned is:
   $D= U+S[1-f-i]-S'v[1+s]-T-[Q+X]/I$

Steve Asikin ISBN 14: 978-1515158691, ISBN 10: 1515158691

Rule-20202:

If both (l), (**U**), (**$**), (**f**), (**i**), (**$'**), (**v**), (**s**), (**T**), (**D**) and (**Q**) are known, then its Xpress or Current Debt Planned is:

$$\mathbf{X} = \mathit{l}\{\mathbf{U} + \mathbf{\$}[1\text{-}\mathbf{f}\text{-}\mathbf{i}]\text{-}\mathbf{\$'v}[1+\mathbf{s}]\text{-}\mathbf{T}\text{-}\mathbf{D}\}\text{-}\mathbf{Q}$$

Rule-20203:

If both (l), (**U**), (**$**), (**f**), (**i**), (**$'**), (**v**), (**s**), (**T**), (**D**), (**P**), (**c**) and (**q**) are known, then its Quoted Longterm Debt Planned is:

$$\mathbf{Q} = \mathit{l}\{\mathbf{U} + \mathbf{\$}[1\text{-}\mathbf{f}\text{-}\mathbf{i}]\text{-}\mathbf{\$'v}[1+\mathbf{s}]\text{-}\mathbf{T}\text{-}\mathbf{D}\}\text{-}\mathbf{P}/[\mathbf{c}\text{-}\mathbf{q}]$$

Rule-20204:

If both (**Q**), (**U**), (**$**), (**f**), (**i**), (**$'**), (**v**), (**s**), (**T**), (**D**), (**P**), (**c**) and (**q**) are known, then its Leverage or Gearing Ratio Planned is:

$$\mathit{l} = \{\mathbf{Q} + \mathbf{P}/[\mathbf{c}\text{-}\mathbf{q}]\}/\{\mathbf{U} + \mathbf{\$}[1\text{-}\mathbf{f}\text{-}\mathbf{i}]\text{-}\mathbf{\$'v}[1+\mathbf{s}]\text{-}\mathbf{T}\text{-}\mathbf{D}\}$$

Rule-20205:

If both (l), (**Q**), (**$**), (**f**), (**i**), (**$'**), (**v**), (**s**), (**T**), (**D**), (**P**), (**c**) and (**q**) are known, then its Utilized or Starting Capital must be:

$$\mathbf{U} = \mathbf{\$'v}[1+\mathbf{s}] + \mathbf{T} + \mathbf{D} + \{\mathbf{Q} + \mathbf{P}/[\mathbf{c}\text{-}\mathbf{q}]\}/\mathit{l}\text{-}\mathbf{\$}[1\text{-}\mathbf{f}\text{-}\mathbf{i}]$$

Rule-20206:

If both (l), (**U**), (**Q**), (**f**), (**i**), (**$'**), (**v**), (**s**), (**T**), (**D**), (**P**), (**c**) and (**q**) are known, then its Sales or Revenue Planned is:

$$\mathbf{\$} = (\mathbf{\$'v}[1+\mathbf{s}] + \mathbf{T} + \mathbf{D} + \{\mathbf{Q} + \mathbf{P}/[\mathbf{c}\text{-}\mathbf{q}]\}/\mathit{l}\text{-}\mathbf{U})/[1\text{-}\mathbf{f}\text{-}\mathbf{i}]$$

Steve Asikin ISBN 14: 978-1515158691, ISBN 10: 1515158691

Rule-20207:

 If both (f), (U), (S), (Q), (i), (S'), (v), (s), (T), (D), (P), (c) and (q) are known, then its Fixed Portion Planned is:

$$f= 1-i-(S'v[1+s]+T+D+\{Q+P/[c-q]\}/fU)/S$$

Rule-20208:

 If both (f), (U), (S), (f), (Q), (S'), (v), (s), (T), (D), (P), (c) and (q) are known, then its Interest Portion Planned is:

$$i= 1-f-(S'v[1+s]+T+D+\{Q+P/[c-q]\}/fU)/S$$

Rule-20209:

 If both (f), (U), (S), (f), (i), (Q), (v), (s), (T), (D), (P), (c) and (q) are known, then its Sales Past must be:

$$S'= (U+S[1-f-i]+T+D+\{Q+P/[c-q]\}/f)/\{v[1+s]\}$$

Rule-20210:

 If both (f), (U), (S), (f), (i), (S'), (Q), (s), (T), (D), (P), (c) and (q) are known, then its Variable Portion Planned is:

$$v= (U+S[1-f-i]+T+D+\{Q+P/[c-q]\}/f)/\{S'[1+s]\}$$

Rule-20211:

 If both (f), (U), (S), (f), (i), (S'), (v), (Q), (T), (D), (P), (c) and (q) are known, then its Sales Growth Planned is:

$$s= S/S'-1$$

  or it can also be found as

$$s= (U+S[1-f-i]+T+D+\{Q+P/[c-q]\}/f)/[S'v]-1$$

Steve Asikin ISBN 14: 978-1515158691, ISBN 10: 1515158691

Rule-20212:

If both (I), (U), (S), (f), (i), (S'), (v), (s), (Q), (D), (P), (c) and (q) are known, then its Tax Planned is:

$T= U+S[1-f-i]-S'v[1+s]-D-\{Q+P/[c-q]\}/I$

Rule-20213:

If both (I), (U), (S), (f), (i), (S'), (v), (s), (T), (Q), (P), (c) and (q) are known, then its Dividend Planned is:

$D= U+S[1-f-i]-S'v[1+s]-T-\{Q+P/[c-q]\}/I$

Rule-20214:

If both (I), (U), (S), (f), (i), (S'), (v), (s), (T), (D), (Q), (c) and (q) are known, then its Procured Inventory Planned is:

$P= [c-q](I\{U+S[1-f-i]-S'v[1+s]-T-D\}-Q)$

Rule-20215:

If both (I), (U), (S), (f), (i), (S'), (v), (s), (T), (D), (P), (Q) and (q) are known, then its Current Ratio Planned is:

$c= q+P/(I\{U+S[1-f-i]-S'v[1+s]-T-D\}-Q)$

Rule-20216:

If both (I), (U), (S), (f), (i), (S'), (v), (s), (T), (D), (P), (c) and (Q) are known, then its Quick or Acid Test Ratio Planned is:

$q= c-P/(I\{U+S[1-f-i]-S'v[1+s]-T-D\}-Q)$

Steve Asikin ISBN 14: 978-1515158691, ISBN 10: 1515158691

Rule-20217:

If both (f), (U), (S), (f), (i), (S'), (v), (s), (T), (D), (V), (p), (c) and (q) are known, then its Quoted Longterm Debt Planned is:

$Q = f\{U+S[1-f-i]-S'v[1+s]-T-D\}-Vp/\{360[c-q]\}$

Rule-20218:

If both (Q), (U), (S), (f), (i), (S'), (v), (s), (T), (D), (V), (p), (c) and (q) are known, then its Leverage or Gearing Ratio Planned is:

$f = (Q+Vp/\{360[c-q]\})/\{U+S[1-f-i]-S'v[1+s]-T-D\}$

Rule-20219:

If both (f), (Q), (S), (f), (i), (S'), (v), (s), (T), (D), (V), (p), (c) and (q) are known, then its Utilized or Starting Capital must be:

$U = S'v[1+s]+T+D+(Q+Vp/\{360[c-q]\})/f-S[1-f-i]$

Rule-20220:

If both (f), (U), (Q), (f), (i), (S'), (v), (s), (T), (D), (V), (p), (c) and (q) are known, then its Sales Revenue Planned is:

$S = [S'v[1+s]+T+D+(Q+Vp/\{360[c-q]\})/f-U]/[1-f-i]$

Rule-20221:

If both (f), (U), (S), (Q), (i), (S'), (v), (s), (T), (D), (V), (p), (c) and (q) are known, then its Fixed Portion Planned is:

$f = 1-i-[S'v[1+s]+T+D+(Q+Vp/\{360[c-q]\})/f-U]/S$

Steve Asikin ISBN 14: 978-1515158691, ISBN 10: 1515158691

Rule-20222:

If both (I), (**U**), (**$**), (**f**), (**Q**), (**$'**), (**v**), (**s**), (**T**), (**D**), (**V**), (**p**), (**c**) and (**q**) are known, then its Interest Portion Planned is:

$$i= 1-f-[\$'v[1+s]+T+D+(Q+Vp/\{360[c-q]\})/IU]/\$$$

Rule-20223:

If both (I), (**U**), (**$**), (**f**), (**i**), (**Q**), (**v**), (**s**), (**T**), (**D**), (**V**), (**p**), (**c**) and (**q**) are known, then its Sales Past must be:

$$\$'= [U+\$[1-f-i]+T+D+(Q+Vp/\{360[c-q]\})/I]$$
$$/\{v[1+s]\}$$

Rule-20224:

If both (I), (**U**), (**$**), (**f**), (**i**), (**$'**), (**v**), (**s**), (**T**), (**D**), (**V**), (**p**), (**c**) and (**q**) are known, then its Quoted Longterm Debt Planned is:

$$v= V/\$$$

or it can also be found as

$$v= [U+\$[1-f-i]+T+D+(Q+Vp/\{360[c-q]\})/I]$$
$$/\{\$'[1+s]\}$$

Rule-20225:

If both (I), (**U**), (**$**), (**f**), (**i**), (**$'**), (**v**), (**Q**), (**T**), (**D**), (**V**), (**p**), (**c**) and (**q**) are known, then its Sales Growth Planned is:

$$s= \$/\$'-1$$

or it can also be found as

$$s= [U+\$[1-f-i]+T+D+(Q+Vp/\{360[c-q]\})/I]$$
$$/[\$'v]-1$$

Steve Asikin ISBN 14: 978-1515158691, ISBN 10: 1515158691

Rule-20226:

If both (I), (U), (S), (f), (i), (S'), (v), (s), (Q), (D), (V), (p), (c) and (q) are known, then its Tax Planned is:

$$T= U+S[1-f-i]-S'v[1+s]-D-(Q+Vp/\{360[c-q]\})/I$$

Rule-20227:

If both (I), (U), (S), (f), (i), (S'), (v), (s), (T), (Q), (V), (p), (c) and (q) are known, then its Dividend Planned is:

$$D= U+S[1-f-i]-S'v[1+s]-T-(Q+Vp/\{360[c-q]\})/I$$

Rule-20228:

If both (I), (U), (S), (f), (i), (S'), (v), (s), (T), (D), (Q), (p), (c) and (q) are known, then its Variable Cost Planned is:

$$V= 360[c-q](I\{U+S[1-f-i]-S'v[1+s]-T-D\}-Q)/p$$

Rule-20229:

If both (I), (U), (S), (f), (i), (S'), (v), (s), (T), (D), (V), (Q), (c) and (q) are known, then its Procured Inventory Days Planned:

$$p= 360[c-q](I\{U+S[1-f-i]-S'v[1+s]-T-D\}-Q)/V$$

Rule-20230:

If both (I), (U), (S), (f), (i), (S'), (v), (s), (T), (D), (V), (p), (Q) and (q) are known, then its Current Ratio Planned is:

$$c= q+Vp/[360(I\{U+S[1-f-i]-S'v[1+s]-T-D\}-Q)]$$

Steve Asikin ISBN 14: 978-1515158691, ISBN 10: 1515158691

Rule-20231:

If both (I), (U), (S), (f), (i), (S'), (v), (s), (T), (D), (V), (p), (c) and (Q) are known, then its Quick or Acid Test Ratio Planned is:

$q= c-Vp/[360(I\{U+S[1-f-i]-S'v[1+s]-T-D\}-Q)]$

Rule-20232:

If both (I), (U), (S), (f), (i), (S'), (v), (s), (T), (D), (p), (c) and (q) are known, then its Quoted Longterm Debt Planned is:

$Q= I\{U+S[1-f-i]-S'v[1+s]-T-D\}-Svp/\{360[c-q]\}$

Rule-20233:

If both (I), (U), (S), (f), (i), (S'), (v), (s), (T), (D), (p), (c) and (q) are known, then its Quoted Longterm Debt Planned is:

$I= (Q+Svp/\{360[c-q]\})/\{U+S[1-f-i]-S'v[1+s]-T-D\}$

Rule-20234:

If both (I), (Q), (S), (f), (i), (S'), (v), (s), (T), (D), (p), (c) and (q) are known, then its Utilized or Starting Capital must be:

$U= S'v[1+s]+T+D+(Q+Svp/\{360[c-q]\})/I-S[1-f-i]$

Rule-20235:

If both (I), (U), (Q), (f), (i), (S'), (v), (s), (T), (D), (p), (c) and (q) are known, then its Sales or Revenue Planned is:

$S= (I\{U-S'v[1+s]-T-D\}-Q)/(vp/\{360[c-q]\}-I[1-f-i])$

Steve Asikin ISBN 14: 978-1515158691, ISBN 10: 1515158691

Rule-20236:

If both (f), (U), (S), (Q), (i), (S'), (v), (s), (T), (D), (p), (c) and (q) are known, then its Fixed Portion Planned is:

$$f = 1-i-[S'v[1+s]+T+D+(Q+Svp/\{360[c-q]\})/fU]/S$$

Rule-20237:

If both (f), (U), (S), (f), (Q), (S'), (v), (s), (T), (D), (p), (c) and (q) are known, then its Interest Portion Planned is:

$$i = 1-f-[S'v[1+s]+T+D+(Q+Svp/\{360[c-q]\})/fU]/S$$

Rule-20238:

If both (f), (U), (S), (f), (i), (Q), (v), (s), (T), (D), (p), (c) and (q) are known, then its Sales Past must be:

$$S' = [U+S[1-f-i]+T+D+(Q+Svp/\{360[c-q]\})/f]$$
$$/\{v[1+s]\}$$

Rule-20239:

If both (f), (U), (S), (f), (i), (S'), (Q), (s), (T), (D), (p), (c) and (q) are known, then its Variable Portion Planned is:

$$v = (f\{U+S[1-f-i]-T-D\}-Q)$$
$$/(Sp/\{360[c-q]\}+S'f[1+s])$$

Steve Asikin ISBN 14: 978-1515158691, ISBN 10: 1515158691

Rule-20240:

If both (I), (U), (S), (f), (i), (S'), (v), (Q), (T), (D), (p), (c) and (q) are known, then its Sales Growth Planned is:

$s= S/S'-1$

or it can also be found as

$s= [U+S[1-f-i]+T+D+(Q+Svp/\{360[c-q]\})/I]$
$/[S'v]-1$

Rule-20241:

If both (I), (U), (S), (f), (i), (S'), (v), (s), (Q), (D), (p), (c) and (q) are known, then its Tax Planned is:

$T= U+S[1-f-i]-S'v[1+s]-D-(Q+Svp/\{360[c-q]\})/I$

Rule-20242:

If both (I), (U), (S), (f), (i), (S'), (v), (s), (T), (Q), (p), (c) and (q) are known, then its Dividend Planned is:

$D= U+S[1-f-i]-S'v[1+s]-T-(Q+Svp/\{360[c-q]\})/I$

Rule-20243:

If both (I), (U), (S), (f), (i), (S'), (v), (s), (T), (D), (Q), (c) and (q) are known, then its Procured Inventory Days Planned:

$p= 360[c-q](I\{U+S[1-f-i]-S'v[1+s]-T-D\}-Q)/[Sv]$

Rule-20244:

If both (I), (U), (S), (f), (i), (S'), (v), (s), (T), (D), (p), (Q) and (q) are known, then its Current Ratio Planned is:

$c= q+Svp/[360(I\{U+S[1-f-i]-S'v[1+s]-T-D\}-Q)]$

Steve Asikin ISBN 14: 978-1515158691, ISBN 10: 1515158691

Rule-20245:

If both (f), (U), $(\$)$, (f), (i), $(\$')$, (v), (s), (T), (D), (p), (c) and (Q) are known, then its Quick or Acid Test Ratio Planned is:

$$q= c-\$vp/[360(f\{U+\$[1-f-i]-\$'v[1+s]-T-D\}-Q)]$$

Rule-20246:

If both (f), (U), $(\$)$, (f), (i), $(\$')$, (v), (s), (T), (D), (p), (c) and (q) are known, then its Quoted Longterm Debt Planned is:

$$Q= f\{U+\$[1-f-i]-\$'v[1+s]-T-D\}$$
$$-\$'vp[1+s]/\{360[c-q]\}$$

Rule-20247:

If both (Q), (U), $(\$)$, (f), (i), $(\$')$, (v), (s), (T), (D), (p), (c) and (q) are known, then its Leverage or Gearing Ratio Planned is:

$$f= (Q+\$'vp[1+s]/\{360[c-q]\})$$
$$/\{U+\$[1-f-i]-\$'v[1+s]-T-D\}$$

Rule-20248:

If both (f), (Q), $(\$)$, (f), (i), $(\$')$, (v), (s), (T), (D), (p), (c) and (q) are known, then its Utilized or Starting Capital must be:

$$U= \$'v[1+s]+T+D$$
$$+(Q+\$'vp[1+s]/\{360[c-q]\})/f\$[1-f-i]$$

Steve Asikin ISBN 14: 978-1515158691, ISBN 10: 1515158691

Rule-20249:

If both $(\textbf{/})$, $(\textbf{U})$, $(\textbf{Q})$, $(\textbf{f})$, $(\textbf{i})$, $(\textbf{S'})$, $(\textbf{v})$, $(\textbf{s})$, $(\textbf{T})$, $(\textbf{D})$, $(\textbf{p})$, $(\textbf{c})$ and $(\textbf{q})$ are known, then its Sales or Revenue Planned is:

$$S = (Q - /\{U - S'v[1+s] - T - D\} + S'vp[1+s]/\{360[c-q]\}) / \{/[1-f-i]\}$$

Rule-20250:

If both $(\textbf{/})$, $(\textbf{U})$, $(\textbf{S})$, $(\textbf{Q})$, $(\textbf{i})$, $(\textbf{S'})$, $(\textbf{v})$, $(\textbf{s})$, $(\textbf{T})$, $(\textbf{D})$, $(\textbf{p})$, $(\textbf{c})$ and $(\textbf{q})$ are known, then its Fixed Portion Planned is:

$$f = 1 - i - [S'v[1+s] + T + D + (Q + S'vp[1+s]/\{360[c-q]\})//U]/S$$

Rule-20251:

If both $(\textbf{/})$, $(\textbf{U})$, $(\textbf{S})$, $(\textbf{f})$, $(\textbf{Q})$, $(\textbf{S'})$, $(\textbf{v})$, $(\textbf{s})$, $(\textbf{T})$, $(\textbf{D})$, $(\textbf{p})$, $(\textbf{c})$ and $(\textbf{q})$ are known, then its Interest Portion Planned is:

$$i = 1 - f - [S'v[1+s] + T + D + (Q + S'vp[1+s]/\{360[c-q]\})//U]/S$$

Rule-20252:

If both $(\textbf{/})$, $(\textbf{U})$, $(\textbf{S})$, $(\textbf{f})$, $(\textbf{i})$, $(\textbf{Q})$, $(\textbf{v})$, $(\textbf{s})$, $(\textbf{T})$, $(\textbf{D})$, $(\textbf{p})$, $(\textbf{c})$ and $(\textbf{q})$ are known, then its Sales Past must be:

$$S' = (/\{U + S[1-f-i] - T - D\} - Q) / [v[1+s](p/\{360[c-q]\} + /)]$$

Steve Asikin ISBN 14: 978-1515158691, ISBN 10: 1515158691

Rule-20253:

If both (I), (U), (S), (f), (i), (S'), (Q), (s), (T), (D), (p), (c) and (q) are known, then its Variable Portion Planned is:

$$v = (I\{U+S[1-f-i]-T-D\}-Q) / [S'[1+s](p/\{360[c-q]\}+I)]$$

Rule-20254:

If both (I), (U), (S), (f), (i), (S'), (v), (Q), (T), (D), (p), (c) and (q) are known, then its Sales Growth Planned is:

$$s = S/S'-1$$

or it can also be found as

$$s = (I\{U+S[1-f-i]-T-D\}-Q)/[S'v(p/\{360[c-q]\}+I)]-1$$

Rule-20255:

If both (I), (U), (S), (f), (i), (S'), (v), (s), (Q), (D), (p), (c) and (q) are known, then its Tax Planned is:

$$T = U+S[1-f-i]-S'v[1+s]-D -(Q+S'vp[1+s]/\{360[c-q]\})/I$$

Rule-20256:

If both (I), (U), (S), (f), (i), (S'), (v), (s), (T), (Q), (p), (c) and (q) are known, then its Dividend Planned is:

$$D = U+S[1-f-i]-S'v[1+s]-T -(Q+S'vp[1+s]/\{360[c-q]\})/I$$

Steve Asikin ISBN 14: 978-1515158691, ISBN 10: 1515158691

Rule-20257:

If both (I), (**U**), (**$**), (**f**), (**i**), (**$'**), (**v**), (**s**), (**T**), (**D**), (**Q**), (**c**) and (**q**) are known, then its Procured Inventory Days Planned:

$$\mathbf{p} = 360[\mathbf{c}\text{-}\mathbf{q}](I\{\mathbf{U}\text{+}\$[1\text{-}\mathbf{f}\text{-}\mathbf{i}]\text{-}\$'\mathbf{v}[1\text{+}\mathbf{s}]\text{-}\mathbf{T}\text{-}\mathbf{D}\}\text{-}\mathbf{Q}) / \{\$'\mathbf{v}[1\text{+}\mathbf{s}]\}$$

Rule-20258:

If both (I), (**U**), (**$**), (**f**), (**i**), (**$'**), (**v**), (**s**), (**T**), (**D**), (**p**), (**Q**) and (**q**) are known, then its Current Ratio Planned is:

$$\mathbf{c} = \mathbf{q}\text{+}\$'\mathbf{vp}[1\text{+}\mathbf{s}] / [360(I\{\mathbf{U}\text{+}\$[1\text{-}\mathbf{f}\text{-}\mathbf{i}]\text{-}\$'\mathbf{v}[1\text{+}\mathbf{s}]\text{-}\mathbf{T}\text{-}\mathbf{D}\}\text{-}\mathbf{Q})]$$

Rule-20259:

If both (I), (**U**), (**$**), (**f**), (**i**), (**$'**), (**v**), (**s**), (**T**), (**D**), (**p**), (**c**) and (**q**) are known, then its Quoted Longterm Debt Planned is:

$$\mathbf{q} = \mathbf{c}\text{-}\$'\mathbf{vp}[1\text{+}\mathbf{s}] / [360(I\{\mathbf{U}\text{+}\$[1\text{-}\mathbf{f}\text{-}\mathbf{i}]\text{-}\$'\mathbf{v}[1\text{+}\mathbf{s}]\text{-}\mathbf{T}\text{-}\mathbf{D}\}\text{-}\mathbf{Q})]$$

Rule-20260:

If both (I), (**U**), (**$**), (**f**), (**i**), (**$'**), (**v**), (**s**), (**T**), (**d**) and (**X**) are known, then its Quoted Longterm Debt Planned is:

$$\mathbf{Q} = I\mathbf{U}\text{+}\{\$[1\text{-}\mathbf{f}\text{-}\mathbf{i}]\text{-}\$'\mathbf{v}[1\text{+}\mathbf{s}]\text{-}\mathbf{T}\}[1\text{-}\mathbf{d}])\text{-}\mathbf{X}$$

Steve Asikin ISBN 14: 978-1515158691, ISBN 10: 1515158691

Rule-20261:
> If both (**Q**), (**U**), (**$**), (**f**), (**i**), (**$'**), (**v**), (**s**), (**T**), (**d**) and (**X**) are known, then its Leverage or Gearing Ratio Planned is:
>
> I= [**Q**+**X**]/(**U**+{**$**[1-**f**-**i**]-**$'v**[1+**s**]-**T**}[1-**d**])

Rule-20262:
> If both (**I**), (**Q**), (**$**), (**f**), (**i**), (**$'**), (**v**), (**s**), (**T**), (**d**) and (**X**) are known, then its Utilized or Starting Capital must be:
>
> **U**= [**Q**+**X**]/I{**$**[1-**f**-**i**]-**$'v**[1+**s**]-**T**}[1-**d**]

Rule-20263:
> If both (**I**), (**U**), (**Q**), (**f**), (**i**), (**$'**), (**v**), (**s**), (**T**), (**d**) and (**X**) are known, then its Sales or Revenue Planned is:
>
> **$**= (**$'v**[1+**s**]+**T**+{[**Q**+**X**]/I–**U**}/[1-**d**])/[1-**f**-**i**]

Rule-20264:
> If both (**I**), (**U**), (**$**), (**Q**), (**i**), (**$'**), (**v**), (**s**), (**T**), (**d**) and (**X**) are known, then its Fixed Portion Planned is:
>
> **f**= 1-**i**-(**$'v**[1+**s**]+**T**+{[**Q**+**X**]/I–**U**}/[1-**d**])/**$**

Rule-20265:
> If both (**I**), (**U**), (**$**), (**f**), (**Q**), (**$'**), (**v**), (**s**), (**T**), (**d**) and (**X**) are known, then its Interest Portion Planned is:
>
> **i**= 1-**f**-(**$'v**[1+**s**]+**T**+{[**Q**+**X**]/I–**U**}/[1-**d**])/**$**

Rule-20266:
> If both (**I**), (**U**), (**$**), (**f**), (**i**), (**Q**), (**v**), (**s**), (**T**), (**d**) and (**X**) are known, then its Sales Past must be:
>
> **$'**= (**$**[1-**f**-**i**]-**T**-{[**Q**+**X**]/I–**U**}/[1-**d**])/{**v**[1+**s**]}

Steve Asikin ISBN 14: 978-1515158691, ISBN 10: 1515158691

Rule-20267:

If both (**/**), (**U**), (**$**), (**f**), (**i**), (**$'**), (**Q**), (**s**), (**T**), (**d**) and (**X**) are known, then its Variable Portion Planned is:

$$v = (\$[1\text{-}f\text{-}i]\text{-}T\text{-}\{[Q+X]/I\text{-}U\}/[1\text{-}d])/\{\$'[1+s]\}$$

Rule-20268:

If both (**/**), (**U**), (**$**), (**f**), (**i**), (**$'**), (**v**), (**Q**), (**T**), (**d**) and (**X**) are known, then its Sales Growth Planned is:

$$s = \$/\$'\text{-}1$$

or it can also be found as

$$s = (\$[1\text{-}f\text{-}i]\text{-}T\text{-}\{[Q+X]/I\text{-}U\}/[1\text{-}d])/[\$'v]\text{-}1$$

Rule-20269:

If both (**/**), (**U**), (**$**), (**f**), (**i**), (**$'**), (**v**), (**s**), (**Q**), (**d**) and (**X**) are known, then its Tax Planned is:

$$T = \$[1\text{-}f\text{-}i]\text{-}\$'v[1+s]\text{-}\{[Q+X]/I\text{-}U\}/[1\text{-}d]$$

Rule-20270:

If both (**/**), (**U**), (**$**), (**f**), (**i**), (**$'**), (**v**), (**s**), (**T**), (**Q**) and (**X**) are known, then its Dividend Payout Planned is:

$$d = 1\text{-}\{[Q+X]/I\text{-}U\}/\{\$[1\text{-}f\text{-}i]\text{-}\$'v[1+s]\text{-}T\}$$

Rule-20271:

If both (**/**), (**U**), (**$**), (**f**), (**i**), (**$'**), (**v**), (**s**), (**T**), (**d**) and (**Q**) are known, then its Quick or Acid Test Ratio Planned is:

$$X = I(U+\{\$[1\text{-}f\text{-}i]\text{-}\$'v[1+s]\text{-}T\}[1\text{-}d])\text{-}Q$$

Steve Asikin ISBN 14: 978-1515158691, ISBN 10: 1515158691

Rule-20272:

> If both (I), (U), (S), (f), (i), (S'), (v), (s), (T), (d), (P), (c) and (q) are known, then its Quoted Longterm Debt Planned is:
>
> $$Q= I(U+\{S[1\text{-}f\text{-}i]\text{-}S'v[1+s]\text{-}T\}[1\text{-}d])\text{-}P/[c\text{-}q]$$

Rule-20273:

> If both (Q), (U), (S), (f), (i), (S'), (v), (s), (T), (d), (P), (c) and (q) are known, then its Leverage or Gearing Ratio Planned is:
>
> $$I= \{Q+P/[c\text{-}q]\}/(U+\{S[1\text{-}f\text{-}i]\text{-}S'v[1+s]\text{-}T\}[1\text{-}d])$$

Rule-20274:

> If both (I), (Q), (S), (f), (i), (S'), (v), (s), (T), (d), (P), (c) and (q) are known, then its Utilized or Starting Capital must be:
>
> $$U= \{Q+P/[c\text{-}q]\}/I\{S[1\text{-}f\text{-}i]\text{-}S'v[1+s]\text{-}T\}[1\text{-}d]$$

Rule-20275:

> If both (I), (U), (Q), (f), (i), (S'), (v), (s), (T), (d), (P), (c) and (q) are known, then its Sales or Revenue Planned is:
>
> $$S= [S'v[1+s]+T+(\{Q+P/[c\text{-}q]\}/I\text{--}U)/[1\text{-}d]]/[1\text{-}f\text{-}i]$$

Rule-20276:

> If both (I), (U), (S), (Q), (i), (S'), (v), (s), (T), (d), (P), (c) and (q) are known, then its Fixed Portion Planned is:
>
> $$f= 1\text{-}i\text{-}[S'v[1+s]+T+(\{Q+P/[c\text{-}q]\}/I\text{--}U)/[1\text{-}d]]/S$$

Steve Asikin ISBN 14: 978-1515158691, ISBN 10: 1515158691

Rule-20277:
  If both (I), (U), $(\$)$, (f), (Q), $(\$')$, (v), (s), (T), (d), (P), (c) and (q) are known, then its Interest Portion Planned is:
  $$i= 1-f-[\$'v[1+s]+T+(\{Q+P/[c-q]\}//-U)/[1-d]]/\$$$

Rule-20278:
  If both (I), (U), $(\$)$, (f), (i), (Q), (v), (s), (T), (d), (P), (c) and (q) are known, then its Sales Past must be:
  $$\$'= [\$[1-f-i]-T-(\{Q+P/[c-q]\}//-U)/[1-d]]/\{v[1+s]\}$$

Rule-20279:
  If both (I), (U), $(\$)$, (f), (i), $(\$')$, (Q), (s), (T), (d), (P), (c) and (q) are known, then its Variable Portion Planned is:
  $$v= [\$[1-f-i]-T-(\{Q+P/[c-q]\}//-U)/[1-d]]/\{\$'[1+s]\}$$

Rule-20280:
  If both (I), (U), $(\$)$, (f), (i), $(\$')$, (v), (Q), (T), (d), (P), (c) and (q) are known, then its Sales Growth Planned is:
  $$s= \$/\$'-1$$
  or it can also be found as
  $$s= [\$[1-f-i]-T-(\{Q+P/[c-q]\}//-U)/[1-d]]/[\$'v]-1$$

Rule-20281:
  If both (I), (U), $(\$)$, (f), (i), $(\$')$, (v), (s), (Q), (d), (P), (c) and (q) are known, then its Tax Planned is:
  $$T= \$[1-f-i]-\$'v[1+s]-(\{Q+P/[c-q]\}//-U)/[1-d]$$

Steve Asikin ISBN 14: 978-1515158691, ISBN 10: 1515158691

Rule-20282:

If both (**∫**), (**U**), (**$**), (**f**), (**i**), (**$'**), (**v**), (**s**), (**T**), (**Q**), (**P**), (**c**) and (**q**) are known, then its Dividend Payout Planned is:

$$d= 1-(\{Q+P/[c-q]\}/\!\!\int U)/\{\$[1-f-i]-\$'v[1+s]-T\}$$

Rule-20283:

If both (**∫**), (**U**), (**$**), (**f**), (**i**), (**$'**), (**v**), (**s**), (**T**), (**d**), (**Q**), (**c**) and (**q**) are known, then its Procured Inventory Planned is:

$$P= [c-q][\int U+\{\$[1-f-i]-\$'v[1+s]-T\}[1-d])-Q]$$

Rule-20284:

If both (**∫**), (**U**), (**$**), (**f**), (**i**), (**$'**), (**v**), (**s**), (**T**), (**d**), (**P**), (**Q**) and (**q**) are known, then its Current Ratio Planned is:

$$c= q+P/[\int U+\{\$[1-f-i]-\$'v[1+s]-T\}[1-d])-Q]$$

Rule-20285:

If both (**∫**), (**U**), (**$**), (**f**), (**i**), (**$'**), (**v**), (**s**), (**T**), (**d**), (**P**), (**c**) and (**Q**) are known, then its Quick or Acid Test Ratio Planned is:

$$q= c-P/[\int U+\{\$[1-f-i]-\$'v[1+s]-T\}[1-d])-Q]$$

Rule-20286:

If both (**∫**), (**U**), (**$**), (**f**), (**i**), (**$'**), (**v**), (**s**), (**T**), (**d**), (**V**), (**p**), (**c**) and (**q**) are known, then its Quoted Longterm Debt Planned is:

$$Q= \int U+\{\$[1-f-i]-\$'v[1+s]-T\}[1-d])-Vp/\{360[c-q]\}$$

Rule-20287:

    If both **(Q)**, **(U)**, **($)**, **(f)**, **(i)**, **($')**, **(v)**, **(s)**, **(T)**, **(d)**, **(V)**, **(p)**, **(c)** and **(q)** are known, then its Leverage or Gearing Ratio Planned is:

$$I= (Q+Vp/\{360[c-q]\})$$
$$/(U+\{\$[1-f-i]-\$'v[1+s]-T\}[1-d])$$

Rule-20288:

    If both **(I)**, **(Q)**, **($)**, **(f)**, **(i)**, **($')**, **(v)**, **(s)**, **(T)**, **(d)**, **(V)**, **(p)**, **(c)** and **(q)** are known, then its Utilized or Starting Capital must be:

$$U= (Q+Vp/\{360[c-q]\})/I$$
$$-\{\$[1-f-i]-\$'v[1+s]-T\}[1-d]$$

Rule-20289:

    If both **(I)**, **(U)**, **(Q)**, **(f)**, **(i)**, **($')**, **(v)**, **(s)**, **(T)**, **(d)**, **(V)**, **(p)**, **(c)** and **(q)** are known, then its Sales or Revenue Planned is:

$$\$= \{\$'v[1+s]+T+[(Q+Vp/\{360[c-q]\})/I-U]/[1-d]\}$$
$$/[1-f-i]$$

Rule-20290:

    If both **(I)**, **(U)**, **($)**, **(Q)**, **(i)**, **($')**, **(v)**, **(s)**, **(T)**, **(d)**, **(V)**, **(p)**, **(c)** and **(q)** are known, then its Fixed Portion Planned is:

$$f= 1-i-\{\$'v[1+s]+T+[(Q+Vp/\{360[c-q]\})/I-U]$$
$$/[1-d]\}/\$$$

Steve Asikin ISBN 14: 978-1515158691, ISBN 10: 1515158691

Rule-20291:

If both $(\textit{I})$, $(\textbf{U})$, $(\textbf{S})$, $(\textbf{f})$, $(\textbf{Q})$, $(\textbf{S'})$, $(\textbf{v})$, $(\textbf{s})$, $(\textbf{T})$, $(\textbf{d})$, $(\textbf{V})$, $(\textbf{p})$, $(\textbf{c})$ and $(\textbf{q})$ are known, then its Interest Portion Planned is:

$$\textbf{i} = 1 - \textbf{f} - \{\textbf{S'v}[1+\textbf{s}] + \textbf{T} + [(\textbf{Q}+\textbf{Vp}/\{360[\textbf{c-q}]\})/\textit{I-}\textbf{U}] /[1-\textbf{d}]\}/\textbf{S}$$

Rule-20292:

If both $(\textit{I})$, $(\textbf{U})$, $(\textbf{S})$, $(\textbf{f})$, $(\textbf{i})$, $(\textbf{Q})$, $(\textbf{v})$, $(\textbf{s})$, $(\textbf{T})$, $(\textbf{d})$, $(\textbf{V})$, $(\textbf{p})$, $(\textbf{c})$ and $(\textbf{q})$ are known, then its Sales Past must be:

$$\textbf{S'} = \{\textbf{S}[1-\textbf{f}-\textbf{i}] - \textbf{T} - [(\textbf{Q}+\textbf{Vp}/\{360[\textbf{c-q}]\})/\textit{I-}\textbf{U}]/[1-\textbf{d}]\} /\{\textbf{v}[1+\textbf{s}]\}$$

Rule-20293:

If both $(\textit{I})$, $(\textbf{U})$, $(\textbf{S})$, $(\textbf{f})$, $(\textbf{i})$, $(\textbf{S'})$, $(\textbf{Q})$, $(\textbf{s})$, $(\textbf{T})$, $(\textbf{d})$, $(\textbf{V})$, $(\textbf{p})$, $(\textbf{c})$ and $(\textbf{q})$ are known, then its Variable Portion Planned is:

$$\textbf{v} = \textbf{V}/\textbf{S}$$

or it can always be found as

$$\textbf{v} = \{\textbf{S}[1-\textbf{f}-\textbf{i}] - \textbf{T} - [(\textbf{Q}+\textbf{Vp}/\{360[\textbf{c-q}]\})/\textit{I-}\textbf{U}]/[1-\textbf{d}]\} /\{\textbf{S'}[1+\textbf{s}]\}$$

Rule-20294:

If both $(\textit{I})$, $(\textbf{U})$, $(\textbf{S})$, $(\textbf{f})$, $(\textbf{i})$, $(\textbf{S'})$, $(\textbf{v})$, $(\textbf{Q})$, $(\textbf{T})$, $(\textbf{d})$, $(\textbf{V})$, $(\textbf{p})$, $(\textbf{c})$ and $(\textbf{q})$ are known, then its Sales Growth Planned is:

$$\textbf{s} = \textbf{S}/\textbf{S'} - 1$$

or it can also be found as

$$\textbf{s} = \{\textbf{S}[1-\textbf{f}-\textbf{i}] - \textbf{T} - [(\textbf{Q}+\textbf{Vp}/\{360[\textbf{c-q}]\})/\textit{I-}\textbf{U}]/[1-\textbf{d}]\} /[\textbf{S'v}] - 1$$

Steve Asikin ISBN 14: 978-1515158691, ISBN 10: 1515158691

Rule-20295:

If both (f), (U), (S), (f), (i), (S'), (v), (s), (Q), (d), (V), (p), (c) and (q) are known, then its Tax Planned is:

$$T = S[1\text{-}f\text{-}i]\text{-}S'v[1+s]$$
$$-[(Q+Vp/\{360[c\text{-}q]\})/\textit{f-}U]/[1\text{-}d]$$

Rule-20296:

If both (f), (U), (S), (f), (i), (S'), (v), (s), (T), (Q), (V), (p), (c) and (q) are known, then its Dividend Payout Planned is:

$$d = 1\text{-}[(Q+Vp/\{360[c\text{-}q]\})/\textit{f-}U]$$
$$/\{S[1\text{-}f\text{-}i]\text{-}S'v[1+s]\text{-}T\}$$

Rule-20297:

If both (f), (U), (S), (f), (i), (S'), (v), (s), (T), (d), (Q), (p), (c) and (q) are known, then its Variable Cost Planned is:

$$V = 360[c\text{-}q][\textit{f}U+\{S[1\text{-}f\text{-}i]\text{-}S'v[1+s]\text{-}T\}[1\text{-}d])\text{-}Q]/p$$

Rule-20298:

If both (f), (U), (S), (f), (i), (S'), (v), (s), (T), (d), (V), (Q), (c) and (q) are known, then its Procured Inventory Days Planned:

$$p = 360[c\text{-}q][\textit{f}U+\{S[1\text{-}f\text{-}i]\text{-}S'v[1+s]\text{-}T\}[1\text{-}d])\text{-}Q]/V$$

Rule-20299:

If both (f), (U), (S), (f), (i), (S'), (v), (s), (T), (d), (V), (p), (Q) and (q) are known, then its Current Ratio Planned is:

$$c = q+Vp$$
$$/\{360[\textit{f}U+\{S[1\text{-}f\text{-}i]\text{-}S'v[1+s]\text{-}T\}[1\text{-}d])\text{-}Q]\}$$

Steve Asikin ISBN 14: 978-1515158691, ISBN 10: 1515158691

Rule-20300:

If both $(\textit{I})$, $(\textbf{U})$, $(\textbf{S})$, $(\textbf{f})$, $(\textbf{i})$, $(\textbf{S'})$, $(\textbf{v})$, $(\textbf{s})$, $(\textbf{T})$, $(\textbf{d})$, $(\textbf{V})$, $(\textbf{p})$, $(\textbf{c})$ and $(\textbf{Q})$ are known, then its Quick or Acid Test Ratio Planned is:

$$q = c - Vp/\{360[\textit{I}U + \{S[1-f-i]-S'v[1+s]-T\}[1-d])-Q]\}$$

Rule-20301:

If both $(\textit{I})$, $(\textbf{U})$, $(\textbf{S})$, $(\textbf{f})$, $(\textbf{i})$, $(\textbf{S'})$, $(\textbf{v})$, $(\textbf{s})$, $(\textbf{T})$, $(\textbf{d})$, $(\textbf{p})$, $(\textbf{c})$ and $(\textbf{q})$ are known, then its Quoted Longterm Debt Planned is:

$$Q = \textit{I}U + \{S[1-f-i]-S'v[1+s]-T\}[1-d])$$
$$-Svp/\{360[c-q]\}$$

Rule-20302:

If both $(\textbf{Q})$, $(\textbf{U})$, $(\textbf{S})$, $(\textbf{f})$, $(\textbf{i})$, $(\textbf{S'})$, $(\textbf{v})$, $(\textbf{s})$, $(\textbf{T})$, $(\textbf{d})$, $(\textbf{p})$, $(\textbf{c})$ and $(\textbf{q})$ are known, then its Leverage or Gearing Ratio Planned is:

$$\textit{I} = (Q + Svp/\{360[c-q]\})$$
$$/(U + \{S[1-f-i]-S'v[1+s]-T\}[1-d])$$

Rule-20303:

If both $(\textit{I})$, $(\textbf{Q})$, $(\textbf{S})$, $(\textbf{f})$, $(\textbf{i})$, $(\textbf{S'})$, $(\textbf{v})$, $(\textbf{s})$, $(\textbf{T})$, $(\textbf{d})$, $(\textbf{p})$, $(\textbf{c})$ and $(\textbf{q})$ are known, then its Utilized or Starting Capital must be:

$$U = (Q + Svp/\{360[c-q]\})/\textit{I}$$
$$-\{S[1-f-i]-S'v[1+s]-T\}[1-d]$$

Steve Asikin ISBN 14: 978-1515158691, ISBN 10: 1515158691

Rule-20304:

If both (I), (U), (S), (f), (i), (S'), (v), (s), (T), (d), (p), (c) and (q) are known, then its Quoted Longterm Debt Planned is:

$$S= [I(U-S'v[1+s]-T\}[1-d])-Q]$$
$$/(vp/\{360[c-q]\}-I[1-f-i]\ [1-d])$$

Rule-20305:

If both (I), (U), (S), (Q), (i), (S'), (v), (s), (T), (d), (p), (c) and (q) are known, then its Fixed Portion Planned is:

$$f= 1-i-\{S'v[1+s]+T+[(Q+Svp/\{360[c-q]\})/I-U]$$
$$/[1-d]\}/S$$

Rule-20306:

If both (I), (U), (S), (f), (Q), (S'), (v), (s), (T), (d), (p), (c) and (q) are known, then its Interest Portion Planned is:

$$i= 1-f-\{S'v[1+s]+T+[(Q+Svp/\{360[c-q]\})/I-U]$$
$$/[1-d]\}/S$$

Rule-20307:

If both (I), (U), (S), (f), (i), (Q), (v), (s), (T), (d), (p), (c) and (q) are known, then its Sales Past must be:

$$S'= \{S[1-f-i]-T-[(Q+Svp/\{360[c-q]\})/I-U]/[1-d]\}$$
$$/\{v[1+s]\}$$

Steve Asikin ISBN 14: 978-1515158691, ISBN 10: 1515158691

Rule-20308:

If both (**/**), (**U**), (**$**), (**f**), (**i**), (**$'**), (**Q**), (**s**), (**T**), (**d**), (**p**), (**c**) and (**q**) are known, then its Variable Portion Planned is:

$$v= [/U+\{\$[1\text{-}f\text{-}i]\text{-}T\}[1\text{-}d])\text{-}Q]$$
$$/(\$p/\{360[c\text{-}q]\}+\$'/[1+s]\ [1\text{-}d])$$

Rule-20309:

If both (**/**), (**U**), (**$**), (**f**), (**i**), (**$'**), (**v**), (**Q**), (**T**), (**d**), (**p**), (**c**) and (**q**) are known, then its Sales Growth Planned is:

$$s= \$/\$'\text{-}1$$

or it can also be found as

$$s= \{\$[1\text{-}f\text{-}i]\text{-}T\text{-}[(Q+\$vp/\{360[c\text{-}q]\})/\text{/-}U]/[1\text{-}d]\}$$
$$/[\$'v]\text{-}1$$

Rule-20310:

If both (**/**), (**U**), (**$**), (**f**), (**i**), (**$'**), (**v**), (**s**), (**Q**), (**d**), (**p**), (**c**) and (**q**) are known, then its Tax Planned is:

$$T= \$[1\text{-}f\text{-}i]\text{-}\$'v[1+s]$$
$$\text{-}[(Q+\$vp/\{360[c\text{-}q]\})/\text{/-}U]/[1\text{-}d]$$

Rule-20311:

If both (**/**), (**U**), (**$**), (**f**), (**i**), (**$'**), (**v**), (**s**), (**T**), (**Q**), (**p**), (**c**) and (**q**) are known, then its Dividend Payout Planned is:

$$d= 1\text{-}[(Q+\$vp/\{360[c\text{-}q]\})/\text{/-}U]$$
$$/\{\$[1\text{-}f\text{-}i]\text{-}\$'v[1+s]\text{-}T\}$$

95

Steve Asikin ISBN 14: 978-1515158691, ISBN 10: 1515158691

Rule-20312:

If both (I), (**U**), (**$**), (**f**), (**i**), (**$'**), (**v**), (**s**), (**T**), (**d**), (**Q**), (**c**) and (**q**) are known, then its Procured Inventory Days Planned:

$$\mathbf{p} = 360[\mathbf{c\text{-}q}][\mathbf{(U} + \{\mathbf{\$}[1\text{-}\mathbf{f\text{-}i}]\text{-}\mathbf{\$'v}[1+\mathbf{s}]\text{-}\mathbf{T}\}[1\text{-}\mathbf{d}]) \text{-}\mathbf{Q}] / [\mathbf{\$v}]$$

Rule-20313:

If both (I), (**U**), (**$**), (**f**), (**i**), (**$'**), (**v**), (**s**), (**T**), (**d**), (**p**), (**Q**) and (**q**) are known, then its Current Ratio Planned is:

$$\mathbf{c} = \mathbf{q} + \mathbf{\$vp} / \{360[\mathbf{(U} + \{\mathbf{\$}[1\text{-}\mathbf{f\text{-}i}]\text{-}\mathbf{\$'v}[1+\mathbf{s}]\text{-}\mathbf{T}\}[1\text{-}\mathbf{d}]) \text{-}\mathbf{Q}]\}$$

Rule-20314:

If both (I), (**U**), (**$**), (**f**), (**i**), (**$'**), (**v**), (**s**), (**T**), (**d**), (**p**), (**c**) and (**Q**) are known, then its Quick or Acid Test Ratio Planned is:

$$\mathbf{q} = \mathbf{c} \text{-} \mathbf{\$vp} / \{360[\mathbf{(U} + \{\mathbf{\$}[1\text{-}\mathbf{f\text{-}i}]\text{-}\mathbf{\$'v}[1+\mathbf{s}]\text{-}\mathbf{T}\}[1\text{-}\mathbf{d}]) \text{-}\mathbf{Q}]\}$$

Rule-20315:

If both (I), (**U**), (**$**), (**f**), (**i**), (**$'**), (**v**), (**s**), (**T**), (**d**), (**p**), (**c**) and (**q**) are known, then its Quoted Longterm Debt Planned is:

$$\mathbf{Q} = \mathbf{(U} + \{\mathbf{\$}[1\text{-}\mathbf{f\text{-}i}]\text{-}\mathbf{\$'v}[1+\mathbf{s}]\text{-}\mathbf{T}\}[1\text{-}\mathbf{d}]) \text{-}\mathbf{\$'vp}[1+\mathbf{s}] / \{360[\mathbf{c\text{-}q}]\}$$

Steve Asikin ISBN 14: 978-1515158691, ISBN 10: 1515158691

Rule-20316:

If both (**Q**), (**U**), (**$**), (**f**), (**i**), (**$'**), (**v**), (**s**), (**T**), (**d**), (**p**), (**c**) and (**q**) are known, then its Leverage or Gearing Ratio Planned is:

$$f= (Q+\$'vp[1+s]/\{360[c-q]\})$$
$$/(U+\{\$[1-f-i]-\$'v[1+s]-T\}[1-d])$$

Rule-20317:

If both (**f**), (**Q**), (**$**), (**f**), (**i**), (**$'**), (**v**), (**s**), (**T**), (**d**), (**p**), (**c**) and (**q**) are known, then its Utilized or Starting Capital must be:

$$U= (Q+\$'vp[1+s]/\{360[c-q]\})/f$$
$$-\{\$[1-f-i]-\$'v[1+s]-T\}[1-d]$$

Rule-20318:

If both (**f**), (**U**), (**Q**), (**f**), (**i**), (**$'**), (**v**), (**s**), (**T**), (**d**), (**p**), (**c**) and (**q**) are known, then its Sales or Revenue Planned is:

$$\$= [f\{U-\$'v[1+s]-T\}[1-d])-Q]$$
$$/(vp/\{360[c-q]\}-f[1-f-i] [1-d])$$

Rule-20319:

If both (**f**), (**U**), (**$**), (**Q**), (**i**), (**$'**), (**v**), (**s**), (**T**), (**d**), (**p**), (**c**) and (**q**) are known, then its Fixed Portion Planned is:

$$f= 1-i-\{\$'v[1+s]+T$$
$$+[(Q+\$'vp[1+s]/\{360[c-q]\})/f-U]/[1-d]\}/\$$$

Steve Asikin ISBN 14: 978-1515158691, ISBN 10: 1515158691

Rule-20320:

If both (**/**), (**U**), (**$**), (**f**), (**Q**), (**$'**), (**v**), (**s**), (**T**), (**d**), (**p**), (**c**) and (**q**) are known, then its Interest Portion Planned is:

$$i = 1-f-\{ \$'v[1+s]+T$$
$$+[(Q+\$'vp[1+s]/\{360[c-q]\})/\textit{/-}U]/[1-d]\}/\$$$

Rule-20321:

If both (**/**), (**U**), (**$**), (**f**), (**i**), (**Q**), (**v**), (**s**), (**T**), (**d**), (**p**), (**c**) and (**q**) are known, then its Sales Past must be:

$$\$' = [\textit{/}U+\{\$[1-f-i]-T\}[1-d])-Q]$$
$$/[v[1+s](p/\{360[c-q]\}+\textit{/}[1-d])]$$

Rule-20322:

If both (**/**), (**U**), (**$**), (**f**), (**i**), (**$'**), (**Q**), (**s**), (**T**), (**d**), (**p**), (**c**) and (**q**) are known, then its Variable Portion Planned is:

$$v = [\textit{/}U+\{\$[1-f-i]-T\}[1-d])-Q]$$
$$/[\$'[1+s](p/\{360[c-q]\}+\textit{/}[1-d])]$$

Rule-20323:

If both (**/**), (**U**), (**$**), (**f**), (**i**), (**$'**), (**v**), (**Q**), (**T**), (**d**), (**p**), (**c**) and (**q**) are known, then its Sales Growth Planned is:

$$s = \$/\$'-1$$

or it can also be found as

$$s = [\textit{/}U+\{\$[1-f-i]-T\}[1-d])-Q]$$
$$/[\$'v(p/\{360[c-q]\}+\textit{/}[1-d])]-1$$

Steve Asikin ISBN 14: 978-1515158691, ISBN 10: 1515158691

Rule-20324:

If both (I), (U), (S), (f), (i), (S'), (v), (s), (Q), (d), (p), (c) and (q) are known, then its Tax Planned is:

$$T = S[1-f-i]-S'v[1+s]$$
$$-[(Q+S'vp[1+s]/\{360[c-q]\})/I-U]/[1-d]$$

Rule-20325:

If both (I), (U), (S), (f), (i), (S'), (v), (s), (T), (Q), (p), (c) and (q) are known, then its Dividend Payout Planned is:

$$d = 1-[(Q+S'vp[1+s]/\{360[c-q]\})/I-U]$$
$$/\{S[1-f-i]-S'v[1+s]-T\}$$

Rule-20326:

If both (I), (U), (S), (f), (i), (S'), (v), (s), (T), (d), (Q), (c) and (q) are known, then its Procured Inventory Days Planned:

$$p = 360[c-q][I(U+\{S[1-f-i]-S'v[1+s]-T\}[1-d])-Q]$$
$$/\{S'v[1+s]\}$$

Rule-20327:

If both (I), (U), (S), (f), (i), (S'), (v), (s), (T), (d), (p), (Q) and (q) are known, then its Current Ratio Planned is:

$$c = q+S'vp[1+s]$$
$$/\{360[I(U+\{S[1-f-i]-S'v[1+s]-T\}[1-d])-Q]\}$$

Steve Asikin ISBN 14: 978-1515158691, ISBN 10: 1515158691

Rule-20328:

If both (f), (U), $(\$)$, (f), (i), $(\$')$, (v), (s), (T), (d), (p), (c) and (Q) are known, then its Quick or Acid Test Ratio Planned is:

$$q = c - \$'vp[1+s]$$
$$/\{360[f(U+\{\$[1-f-i]-\$'v[1+s]-T\}[1-d])-Q]\}$$

Rule-20329:

If both (f), (U), $(\$)$, (f), (i), $(\$')$, (v), (s), (T), (A), (d) and (X) are known, then its Quoted Longterm Debt Planned is:

$$Q = f\{U+\$[1-f-i]-\$'v[1+s]-T-Ad\}-X$$

Rule-20330:

If both (Q), (U), $(\$)$, (f), (i), $(\$')$, (v), (s), (T), (A), (d) and (X) are known, then its Leverage or Gearing Ratio Planned is:

$$f = [Q+X]/\{U+\$[1-f-i]-\$'v[1+s]-T-Ad\}$$

Rule-20331:

If both (f), (Q), $(\$)$, (f), (i), $(\$')$, (v), (s), (T), (A), (d) and (X) are known, then its Utilized or Starting Capital must be:

$$U = \$'v[1+s]+T+Ad+[Q+X]/f\$[1-f-i]$$

Rule-20332:

If both (f), (U), (Q), (f), (i), $(\$')$, (v), (s), (T), (A), (d) and (X) are known, then its Sales or Revenue Planned is:

$$\$ = \{\$'v[1+s]+T+Ad+[Q+X]/fU\}/[1-f-i]$$

Steve Asikin ISBN 14: 978-1515158691, ISBN 10: 1515158691

Rule-20333:

If both (**/**), (**U**), (**$**), (**Q**), (**i**), (**$'**), (**v**), (**s**), (**T**), (**A**), (**d**) and (**X**) are known, then its Fixed Portion Planned is:

$$f = 1-i-\{\$'v[1+s]+T+Ad+[Q+X]/I-U\}/\$$$

Rule-20334:

If both (**/**), (**U**), (**$**), (**f**), (**Q**), (**$'**), (**v**), (**s**), (**T**), (**A**), (**d**) and (**X**) are known, then its Interest Portion Planned is:

$$i = 1-f-\{\$'v[1+s]+T+Ad+[Q+X]/I-U\}/\$$$

Rule-20335:

If both (**/**), (**U**), (**$**), (**f**), (**i**), (**Q**), (**v**), (**s**), (**T**), (**A**), (**d**) and (**X**) are known, then its Sales Past must be:

$$\$' = \{U+\$[1-f-i]-T-Ad-[Q+X]/I\}/\{v[1+s]\}$$

Rule-20336:

If both (**/**), (**U**), (**$**), (**f**), (**i**), (**$'**), (**Q**), (**s**), (**T**), (**A**), (**d**) and (**X**) are known, then its Variable Portion Planned is:

$$v = \{U+\$[1-f-i]-T-Ad-[Q+X]/I\}/\{\$'[1+s]\}$$

Rule-20337:

If both (**/**), (**U**), (**$**), (**f**), (**i**), (**$'**), (**v**), (**Q**), (**T**), (**A**), (**d**) and (**X**) are known, then its Sales Growth Planned is:

$$s = \$/\$'-1$$

or it can also be found as

$$s = \{U+\$[1-f-i]-T-Ad-[Q+X]/I\}/[\$'v]-1$$

Steve Asikin ISBN 14: 978-1515158691, ISBN 10: 1515158691

Rule-20338:

If both (**/**), (**U**), (**$**), (**f**), (**i**), (**$'**), (**v**), (**s**), (**Q**), (**A**), (**d**) and (**X**) are known, then its Tax Planned is:

$$T= U+\$[1-f-i]-\$'v[1+s]-Ad-[Q+X]//$$

Rule-20339:

If both (**/**), (**U**), (**$**), (**f**), (**i**), (**$'**), (**v**), (**s**), (**T**), (**Q**), (**d**) and (**X**) are known, then its After Tax Income Planned is:

$$A= \{U+\$[1-f-i]-\$'v[1+s]-T- [Q+X]//\}/d$$

Rule-20340:

If both (**/**), (**U**), (**$**), (**f**), (**i**), (**$'**), (**v**), (**s**), (**T**), (**A**), (**Q**) and (**X**) are known, then its Dividend Payout Planned is:

$$d= \{U+\$[1-f-i]-\$'v[1+s]-T- [Q+X]//\}/A$$

Rule-20341:

If both (**/**), (**U**), (**$**), (**f**), (**i**), (**$'**), (**v**), (**s**), (**T**), (**A**), (**d**) and (**Q**) are known, then its Xpress or Current Debt Planned is:

$$X= /\{U+\$[1-f-i]-\$'v[1+s]-T-Ad\}-Q$$

Rule-20342:

If both (**/**), (**U**), (**$**), (**f**), (**i**), (**$'**), (**v**), (**s**), (**T**), (**A**), (**d**), (**P**), (**c**) and (**q**) are known, then its Quoted Longterm Debt Planned is:

$$Q= /\{U+\$[1-f-i]-\$'v[1+s]-T-Ad\}-P/[c-q]$$

Steve Asikin ISBN 14: 978-1515158691, ISBN 10: 1515158691

<u>Rule-20343</u>:

If both (**Q**), (**U**), (**S**), (**f**), (**i**), (**S'**), (**v**), (**s**), (**T**), (**A**), (**d**), (**P**), (**c**) and (**q**) are known, then its Leverage or Gearing Ratio Planned is:

$$I = \{Q+P/[c-q]\}/\{U+S[1-f-i]-S'v[1+s]-T-Ad\}$$

<u>Rule-20344</u>:

If both (**I**), (**Q**), (**S**), (**f**), (**i**), (**S'**), (**v**), (**s**), (**T**), (**A**), (**d**), (**P**), (**c**) and (**q**) are known, then its Utilized or Starting Capital must be:

$$U = S'v[1+s]+T+Ad+\{Q+P/[c-q]\}/I-S[1-f-i]$$

<u>Rule-20345</u>:

If both (**I**), (**U**), (**Q**), (**f**), (**i**), (**S'**), (**v**), (**s**), (**T**), (**A**), (**d**), (**P**), (**c**) and (**q**) are known, then its Sales or Revenue Planned is:

$$S = (S'v[1+s]+T+Ad+\{Q+P/[c-q]\}/I-U)/[1-f-i]$$

<u>Rule-20346</u>:

If both (**I**), (**U**), (**S**), (**f**), (**i**), (**S'**), (**v**), (**s**), (**T**), (**A**), (**d**), (**P**), (**c**) and (**q**) are known, then its Fixed Portion Planned is:

$$f = 1-i-(S'v[1+s]+T+Ad+\{Q+P/[c-q]\}/I-U)/S$$

<u>Rule-20347</u>:

If both (**I**), (**U**), (**S**), (**f**), (**Q**), (**S'**), (**v**), (**s**), (**T**), (**A**), (**d**), (**P**), (**c**) and (**q**) are known, then its Interest Portion Planned is:

$$i = 1-f-(S'v[1+s]+T+Ad+\{Q+P/[c-q]\}/I-U)/S$$

Steve Asikin ISBN 14: 978-1515158691, ISBN 10: 1515158691

Rule-20348:
    If both (I), (U), (S), (f), (i), (Q), (v), (s), (T), (A), (d), (P), (c) and (q) are known, then its Sales Past must be:
$$S' = (U+S[1-f-i]-T-Ad-\{Q+P/[c-q]\}/I)/\{v[1+s]\}$$

Rule-20349:
    If both (I), (U), (S), (f), (i), (S'), (Q), (s), (T), (A), (d), (P), (c) and (q) are known, then its Variable Portion Planned is:
$$v = (U+S[1-f-i]-T-Ad-\{Q+P/[c-q]\}/I)/\{S'[1+s]\}$$

Rule-20350:
    If both (I), (U), (S), (f), (i), (S'), (v), (Q), (T), (A), (d), (P), (c) and (q) are known, then its Sales Growth Planned is:
$$s = S/S' - 1$$
        or it can also be found as
$$s = (U+S[1-f-i]-T-Ad-\{Q+P/[c-q]\}/I)/[S'v] - 1$$

Rule-20351:
    If both (I), (U), (S), (f), (i), (S'), (v), (s), (Q), (A), (d), (P), (c) and (q) are known, then its Tax Rate Planned is:
$$T = U+S[1-f-i]-S'v[1+s]-Ad-[Q+X]/I$$

Rule-20352:
    If both (I), (U), (S), (f), (i), (S'), (v), (s), (T), (Q), (d), (P), (c) and (q) are known, then its After Tax Income Planned is:
$$A = (U+S[1-f-i]-S'v[1+s]-T-\{Q+P/[c-q]\}/I)/d$$

`

Steve Asikin ISBN 14: 978-1515158691, ISBN 10: 1515158691

Rule-20353:

If both (f), (U), $(\$)$, (f), (i), $(\$')$, (v), (s), (T), (A), (Q), (P), (c) and (q) are known, then its Dividend Payout Planned is:

$$d= (U+\$[1\text{-}f\text{-}i]\text{-}\$'v[1+s]\text{-}T\text{-}\{Q+P/[c\text{-}q]\}/f)/A$$

Rule-20354:

If both (f), (U), $(\$)$, (f), (i), $(\$')$, (v), (s), (T), (A), (d), (Q), (c) and (q) are known, then its Procured Inventory Planned is:

$$P= [c\text{-}q][f\{U+\$[1\text{-}f\text{-}i]\text{-}\$'v[1+s]\text{-}T\text{-}Ad\}\text{-}Q]$$

Rule-20355:

If both (f), (U), $(\$)$, (f), (i), $(\$')$, (v), (s), (T), (A), (d), (P), (Q) and (q) are known, then its Current Ratio Planned is:

$$c= q+P/[f\{U+\$[1\text{-}f\text{-}i]\text{-}\$'v[1+s]\text{-}T\text{-}Ad\}\text{-}Q]$$

Rule-20356:

If both (f), (U), $(\$)$, (f), (i), $(\$')$, (v), (s), (T), (A), (d), (P), (c) and (Q) are known, then its Quick or Acid Test Ratio Planned is:

$$q= c\text{-}P/[f\{U+\$[1\text{-}f\text{-}i]\text{-}\$'v[1+s]\text{-}T\text{-}Ad\}\text{-}Q]$$

Rule-20357:

If both (f), (U), $(\$)$, (f), (i), $(\$')$, (v), (s), (T), (A), (d), (V), (p), (c) and (q) are known, then its Quoted Longterm Debt Planned is:

$$Q= f\{U+\$[1\text{-}f\text{-}i]\text{-}\$'v[1+s]\text{-}T\text{-}Ad\}\text{-}Vp/\{360[c\text{-}q]\}$$

Steve Asikin ISBN 14: 978-1515158691, ISBN 10: 1515158691

Rule-20358:

If both **(Q)**, **(U)**, **($)**, **(f)**, **(i)**, **($')**, **(v)**, **(s)**, **(T)**, **(A)**, **(d)**, **(V)**, **(p)**, **(c)** and **(q)** are known, then its Leverage or Gearing Ratio Planned is:

$$\ell = (Q+Vp/\{360[c-q]\})/\{U+\$[1-f-i]-\$'v[1+s]-T-Ad\}$$

Rule-20359:

If both **(ℓ)**, **(Q)**, **($)**, **(f)**, **(i)**, **($')**, **(v)**, **(s)**, **(T)**, **(A)**, **(d)**, **(V)**, **(p)**, **(c)** and **(q)** are known, then its Utilized or Starting Capital must be:

$$U = \$'v[1+s]+T+Ad+(Q+Vp/\{360[c-q]\})/\ell-\$[1-f-i]$$

Rule-20360:

If both **(ℓ)**, **(U)**, **(Q)**, **(f)**, **(i)**, **($')**, **(v)**, **(s)**, **(T)**, **(A)**, **(d)**, **(V)**, **(p)**, **(c)** and **(q)** are known, then its Sales or Revenue Planned is:

$$\$ = [\$'v[1+s]+T+Ad+(Q+Vp/\{360[c-q]\})/\ell U]$$
$$/[1-f-i]$$

Rule-20361:

If both **(ℓ)**, **(U)**, **($)**, **(Q)**, **(i)**, **($')**, **(v)**, **(s)**, **(T)**, **(A)**, **(d)**, **(V)**, **(p)**, **(c)** and **(q)** are known, then its Fixed Portion Planned is:

$$f = 1-i-[\$'v[1+s]+T+Ad+(Q+Vp/\{360[c-q]\})/\ell U]/\$$$

Rule-20362:

If both **(ℓ)**, **(U)**, **($)**, **(f)**, **(Q)**, **($')**, **(v)**, **(s)**, **(T)**, **(A)**, **(d)**, **(V)**, **(p)**, **(c)** and **(q)** are known, then its Interest Portion Planned is:

$$i = 1-f-[\$'v[1+s]+T+Ad+(Q+Vp/\{360[c-q]\})/\ell U]/\$$$

Steve Asikin ISBN 14: 978-1515158691, ISBN 10: 1515158691

Rule-20363:

    If both (I), (U), (S), (f), (i), (Q), (v), (s), (T), (A), (d), (V), (p), (c) and (q) are known, then its Sales Past must be:

$$S' = [U+S[1-f-i]-T-Ad-(Q+Vp/\{360[c-q]\})/I] / \{v[1+s]\}$$

Rule-20364:

    If both (I), (U), (S), (f), (i), (S'), (Q), (s), (T), (A), (d), (V), (p), (c) and (q) are known, then its Quoted Longterm Debt Planned is:

$$v = V/S$$

        or it can also be found as

$$v = [U+S[1-f-i]-T-Ad-(Q+Vp/\{360[c-q]\})/I] / \{S'[1+s]\}$$

Rule-20365:

    If both (I), (U), (S), (f), (i), (S'), (v), (Q), (T), (A), (d), (V), (p), (c) and (q) are known, then its Sales Growth Planned is:

$$s = S/S' - 1$$

        or it can also be found as

$$s = [U+S[1-f-i]-T-Ad-(Q+Vp/\{360[c-q]\})/I]/[S'v] - 1$$

Rule-20366:

    If both (I), (U), (S), (f), (i), (S'), (v), (s), (Q), (A), (d), (V), (p), (c) and (q) are known, then its Tax Planned is:

$$T = U+S[1-f-i]-S'v[1+s]-Ad-(Q+Vp/\{360[c-q]\})/I$$

Steve Asikin ISBN 14: 978-1515158691, ISBN 10: 1515158691

Rule-20367:

If both (**/**), (**U**), (**$**), (**f**), (**i**), (**$'**), (**v**), (**s**), (**T**), (**Q**), (**d**), (**V**), (**p**), (**c**) and (**q**) are known, then its After Tax Income Planned is:

$$A= (U+\$[1-f-i]-\$'v[1+s]-T-(Q+Vp/\{360[c-q]\}))//\!/d$$

Rule-20368:

If both (**/**), (**U**), (**$**), (**f**), (**i**), (**$'**), (**v**), (**s**), (**T**), (**A**), (**Q**), (**V**), (**p**), (**c**) and (**q**) are known, then its Dividend Payout Planned is:

$$d= (U+\$[1-f-i]-\$'v[1+s]-T$$
$$-(Q+Vp/\{360[c-q]\})//\!/A$$

Rule-20369:

If both (**/**), (**U**), (**$**), (**f**), (**i**), (**$'**), (**v**), (**s**), (**T**), (**A**), (**d**), (**Q**), (**p**), (**c**) and (**q**) are known, then its Variable Cost Planned is:

$$V= 360[c-q][/\!\{U+\$[1-f-i]-\$'v[1+s]-T-Ad\}-Q]/p$$

Rule-20370:

If both (**/**), (**U**), (**$**), (**f**), (**i**), (**$'**), (**v**), (**s**), (**T**), (**A**), (**d**), (**V**), (**Q**), (**c**) and (**q**) are known, then its Procured Inventory Days Planned:

$$p= 360[c-q][/\!\{U+\$[1-f-i]-\$'v[1+s]-T-Ad\}-Q]/V$$

Rule-20371:

If both (**/**), (**U**), (**$**), (**f**), (**i**), (**$'**), (**v**), (**s**), (**T**), (**A**), (**d**), (**V**), (**p**), (**Q**) and (**q**) are known, then its Current Ratio Planned is:

$$c= q+Vp/\{360[/\!\{U+\$[1-f-i]-\$'v[1+s]-T-Ad\}-Q]\}$$

Steve Asikin ISBN 14: 978-1515158691, ISBN 10: 1515158691

Rule-20372:
> If both (*I*), (**U**), (**S**), (**f**), (**i**), (**S'**), (**v**), (**s**), (**T**), (**A**), (**d**),
> (**V**), (**p**), (**c**) and (**Q**) are known, then its Quick or Acid
> Test Ratio Planned is:
>
> $$q= c\text{-}Vp/\{360[\textit{I}\,U+S[1\text{-}f\text{-}i]\text{-}S'v[1+s]\text{-}T\text{-}Ad\}\text{-}Q]\}$$

Rule-20373:
> If both (*I*), (**U**), (**S**), (**f**), (**i**), (**S'**), (**v**), (**s**), (**T**), (**A**), (**d**),
> (**p**), (**c**) and (**q**) are known, then its Quoted Longterm
> Debt Planned is:
>
> $$Q= \textit{I}\,U+S[1\text{-}f\text{-}i]\text{-}S'v[1+s]\text{-}T\text{-}Ad\}\text{-}Svp/\{360[c\text{-}q]\}$$

Rule-20374:
> If both (**Q**), (**U**), (**S**), (**f**), (**i**), (**S'**), (**v**), (**s**), (**T**), (**A**), (**d**),
> (**p**), (**c**) and (**q**) are known, then its Interest Expense
> Planned is:
>
> $$\textit{I}= (Q+Svp/\{360[c\text{-}q]\})$$
> $$/\{U+S[1\text{-}f\text{-}i]\text{-}S'v[1+s]\text{-}T\text{-}Ad\}$$

Rule-20375:
> If both (*I*), (**Q**), (**S**), (**f**), (**i**), (**S'**), (**v**), (**s**), (**T**), (**A**), (**d**),
> (**p**), (**c**) and (**q**) are known, then its Utilized or Starting
> Capital must be:
>
> $$U= S'v[1+s]+T+Ad+(Q+Svp/\{360[c\text{-}q]\})/\textit{I}\,S[1\text{-}f\text{-}i]$$

Rule-20376:
> If both (*I*), (**U**), (**Q**), (**f**), (**i**), (**S'**), (**v**), (**s**), (**T**), (**A**), (**d**),
> (**p**), (**c**) and (**q**) are known, then its Sales or Revenue
> Planned is:
>
> $$S= (\textit{I}\,U\text{-}S'v[1+s]\text{-}T\text{-}Ad\}\text{-}Q)$$
> $$/(vp/\{360[c\text{-}q]\}\text{-}\textit{I}[1\text{-}f\text{-}i])$$

Steve Asikin ISBN 14: 978-1515158691, ISBN 10: 1515158691

Rule-20377:

If both (**/**), (**U**), (**$**), (**Q**), (**i**), (**$'**), (**v**), (**s**), (**T**), (**A**), (**d**), (**p**), (**c**) and (**q**) are known, then its Fixed Portion Planned is:

$$f= 1-i-[\$'v[1+s]+T+Ad$$
$$+(Q+\$vp/\{360[c-q]\})/\textit{/}U]/\$$$

Rule-20378:

If both (**/**), (**U**), (**$**), (**f**), (**Q**), (**$'**), (**v**), (**s**), (**T**), (**A**), (**d**), (**p**), (**c**) and (**q**) are known, then its Interest Portion Planned is:

$$i= 1-f-[\$'v[1+s]+T+Ad$$
$$+(Q+\$vp/\{360[c-q]\})/\textit{/}U]/\$$$

Rule-20379:

If both (**/**), (**U**), (**$**), (**f**), (**i**), (**Q**), (**v**), (**s**), (**T**), (**A**), (**d**), (**p**), (**c**) and (**q**) are known, then its Sales Past must be:

$$\$'= [U+\$[1-f-i]-T-Ad-(Q+\$vp/\{360[c-q]\})/\textit{/}]$$
$$/\{v[1+s]\}$$

Rule-20380:

If both (**/**), (**U**), (**$**), (**f**), (**i**), (**$'**), (**Q**), (**s**), (**T**), (**A**), (**d**), (**p**), (**c**) and (**q**) are known, then its Variable Portion Planned is:

$$v= (\textit{/}\{U+\$[1-f-i]-T-Ad\}-Q)$$
$$/(\$p/\{360[c-q]\}+\$'\textit{/}[1+s])$$

Steve Asikin ISBN 14: 978-1515158691, ISBN 10: 1515158691

Rule-20381:

> If both (I), (U), (S), (f), (i), (S'), (v), (Q), (T), (A), (d), (p), (c) and (q) are known, then its Sales Growth Planned is:
>
> $s = S/S' - 1$
>
> > or it can also be found as
>
> $s = [U+S[1-f-i]-T-Ad-(Q+Svp/\{360[c-q]\})/I]$
> $/[S'v]-1$

Rule-20382:

> If both (I), (U), (S), (f), (i), (S'), (v), (s), (Q), (A), (d), (p), (c) and (q) are known, then its Tax Planned is:
> $T = U+S[1-f-i]-S'v[1+s]-Ad-(Q+Svp/\{360[c-q]\})/I$

Rule-20383:

> If both (I), (U), (S), (f), (i), (S'), (v), (s), (T), (Q), (d), (p), (c) and (q) are known, then its After Tax Income Planned is:
> $A = (U+S[1-f-i]-S'v[1+s]-T$
> $-(Q+Svp/\{360[c-q]\})/I)/d$

Rule-20384:

> If both (I), (U), (S), (f), (i), (S'), (v), (s), (T), (A), (Q), (p), (c) and (q) are known, then its Dividend Payout Planned is:
> $d = (U+S[1-f-i]-S'v[1+s]-T$
> $-(Q+Svp/\{360[c-q]\})/I)/A$

Steve Asikin ISBN 14: 978-1515158691, ISBN 10: 1515158691

Rule-20385:

If both (**/**), (**U**), (**$**), (**f**), (**i**), (**$'**), (**v**), (**s**), (**T**), (**A**), (**d**), (**Q**), (**c**) and (**q**) are known, then its Procured Inventory Days Planned:

$$p= 360[c\text{-}q][/\{U\text{+}\$[1\text{-}f\text{-}i]\text{-}\$'v[1\text{+}s]\text{-}T\text{-}Ad\}\text{-}Q]/[\$v]$$

Rule-20386:

If both (**/**), (**U**), (**$**), (**f**), (**i**), (**$'**), (**v**), (**s**), (**T**), (**A**), (**d**), (**p**), (**Q**) and (**q**) are known, then its Current Ratio Planned is:

$$c= q\text{+}\$vp/\{360[/\{U\text{+}\$[1\text{-}f\text{-}i]\text{-}\$'v[1\text{+}s]\text{-}T\text{-}Ad\}\text{-}Q]\}$$

Rule-20387:

If both (**/**), (**U**), (**$**), (**f**), (**i**), (**$'**), (**v**), (**s**), (**T**), (**A**), (**d**), (**p**), (**c**) and (**Q**) are known, then its Quick or Acid Test Ratio Planned is:

$$q= c\text{-}\$vp/\{360[/\{U\text{+}\$[1\text{-}f\text{-}i]\text{-}\$'v[1\text{+}s]\text{-}T\text{-}Ad\}\text{-}Q]\}$$

Rule-20388:

If both (**/**), (**U**), (**$**), (**f**), (**i**), (**$'**), (**v**), (**s**), (**T**), (**A**), (**d**), (**p**), (**c**) and (**q**) are known, then its Quoted Longterm Debt Planned is:

$$Q= /\{U\text{+}\$[1\text{-}f\text{-}i]\text{-}\$'v[1\text{+}s]\text{-}T\text{-}Ad\}$$
$$-\$'vp[1\text{+}s]/\{360[c\text{-}q]\}$$

Steve Asikin ISBN 14: 978-1515158691, ISBN 10: 1515158691

Rule-20389:

If both $(\mathbf{Q})$, $(\mathbf{U})$, $(\mathbf{S})$, $(\mathbf{f})$, $(\mathbf{i})$, $(\mathbf{S'})$, $(\mathbf{v})$, $(\mathbf{s})$, $(\mathbf{T})$, $(\mathbf{A})$, $(\mathbf{d})$, $(\mathbf{p})$, $(\mathbf{c})$ and $(\mathbf{q})$ are known, then its Leverage or Gearing Ratio Planned is:

$$l = (\mathbf{Q} + \mathbf{S'vp}[1+\mathbf{s}]/\{360[\mathbf{c}-\mathbf{q}]\})$$
$$/\{\mathbf{U} + \mathbf{S}[1-\mathbf{f}-\mathbf{i}] - \mathbf{S'v}[1+\mathbf{s}] - \mathbf{T} - \mathbf{Ad}\}$$

Rule-20390:

If both (l), $(\mathbf{Q})$, $(\mathbf{S})$, $(\mathbf{f})$, $(\mathbf{i})$, $(\mathbf{S'})$, $(\mathbf{v})$, $(\mathbf{s})$, $(\mathbf{T})$, $(\mathbf{A})$, $(\mathbf{d})$, $(\mathbf{p})$, $(\mathbf{c})$ and $(\mathbf{q})$ are known, then its Utilized or Starting Capital must be:

$$\mathbf{U} = \mathbf{S'v}[1+\mathbf{s}] + \mathbf{T} + \mathbf{Ad}$$
$$+(\mathbf{Q} + \mathbf{S'vp}[1+\mathbf{s}]/\{360[\mathbf{c}-\mathbf{q}]\})/l - \mathbf{S}[1-\mathbf{f}-\mathbf{i}]$$

Rule-20391:

If both (l), $(\mathbf{U})$, $(\mathbf{Q})$, $(\mathbf{f})$, $(\mathbf{i})$, $(\mathbf{S'})$, $(\mathbf{v})$, $(\mathbf{s})$, $(\mathbf{T})$, $(\mathbf{A})$, $(\mathbf{d})$, $(\mathbf{p})$, $(\mathbf{c})$ and $(\mathbf{q})$ are known, then its Sales or Revenue Planned is:

$$\mathbf{S} = (\mathbf{Q} - l\{\mathbf{U} - \mathbf{S'v}[1+\mathbf{s}] - \mathbf{T} - \mathbf{Ad}\} + \mathbf{S'vp}[1+\mathbf{s}]/\{360[\mathbf{c}-\mathbf{q}]\})$$
$$/\{l[1-\mathbf{f}-\mathbf{i}]\}$$

Rule-20392:

If both (l), $(\mathbf{U})$, $(\mathbf{S})$, $(\mathbf{Q})$, $(\mathbf{i})$, $(\mathbf{S'})$, $(\mathbf{v})$, $(\mathbf{s})$, $(\mathbf{T})$, $(\mathbf{A})$, $(\mathbf{d})$, $(\mathbf{p})$, $(\mathbf{c})$ and $(\mathbf{q})$ are known, then its Fixed Portion Planned is:

$$\mathbf{f} = 1 - \mathbf{i} - [\mathbf{S'v}[1+\mathbf{s}] + \mathbf{T} + \mathbf{Ad}$$
$$+(\mathbf{Q} + \mathbf{S'vp}[1+\mathbf{s}]/\{360[\mathbf{c}-\mathbf{q}]\})/l - \mathbf{U}]/\mathbf{S}$$

Steve Asikin ISBN 14: 978-1515158691, ISBN 10: 1515158691

Rule-20393:

 If both (I), $(\mathbf{U})$, $(\mathbf{S})$, $(\mathbf{f})$, $(\mathbf{Q})$, $(\mathbf{S'})$, $(\mathbf{v})$, $(\mathbf{s})$, $(\mathbf{T})$, $(\mathbf{A})$, $(\mathbf{d})$, $(\mathbf{p})$, $(\mathbf{c})$ and $(\mathbf{q})$ are known, then its Interest Portion Planned is:

$$i = 1 - f - [S'v[1+s] + T + Ad + (Q + S'vp[1+s]/\{360[c-q]\})/IU]/S$$

Rule-20394:

 If both (I), $(\mathbf{U})$, $(\mathbf{S})$, $(\mathbf{f})$, $(\mathbf{i})$, $(\mathbf{Q})$, $(\mathbf{v})$, $(\mathbf{s})$, $(\mathbf{T})$, $(\mathbf{A})$, $(\mathbf{d})$, $(\mathbf{p})$, $(\mathbf{c})$ and $(\mathbf{q})$ are known, then its Sales Past must be:

$$S' = (I\{U + S[1-f-i] - T - Ad\} - Q) / [v[1+s](p/\{360[c-q]\} + I)]$$

Rule-20395:

 If both (I), $(\mathbf{U})$, $(\mathbf{S})$, $(\mathbf{f})$, $(\mathbf{i})$, $(\mathbf{S'})$, $(\mathbf{Q})$, $(\mathbf{s})$, $(\mathbf{T})$, $(\mathbf{A})$, $(\mathbf{d})$, $(\mathbf{p})$, $(\mathbf{c})$ and $(\mathbf{q})$ are known, then its Variable Portion Planned is:

$$v = (I\{U + S[1-f-i] - T - Ad\} - Q) / [S'[1+s](p/\{360[c-q]\} + I)]$$

Rule-20396:

 If both (I), $(\mathbf{U})$, $(\mathbf{S})$, $(\mathbf{f})$, $(\mathbf{i})$, $(\mathbf{S'})$, $(\mathbf{v})$, $(\mathbf{Q})$, $(\mathbf{T})$, $(\mathbf{A})$, $(\mathbf{d})$, $(\mathbf{p})$, $(\mathbf{c})$ and $(\mathbf{q})$ are known, then its Sales Growth Planned is:

$$s = S/S' - 1$$

  or it can also be found as

$$s = (I\{U + S[1-f-i] - T - Ad\} - Q) / [S'v(p/\{360[c-q]\} + I)] - 1$$

114

Steve Asikin ISBN 14: 978-1515158691, ISBN 10: 1515158691

Rule-20397:

If both (**/**), (**U**), (**$**), (**f**), (**i**), (**$'**), (**v**), (**s**), (**Q**), (**A**), (**d**), (**p**), (**c**) and (**q**) are known, then its Tax Planned is:

$$T = U + \$[1-f-i] - \$'v[1+s] - Ad$$
$$-(Q + \$'vp[1+s]/\{360[c-q]\})/\text{/}$$

Rule-20398:

If both (**/**), (**U**), (**$**), (**f**), (**i**), (**$'**), (**v**), (**s**), (**T**), (**Q**), (**d**), (**p**), (**c**) and (**q**) are known, then its After Tax Income Planned is:

$$A = (U + \$[1-f-i] - \$'v[1+s] - T$$
$$-(Q + \$'vp[1+s]/\{360[c-q]\})/\text{/}/d$$

Rule-20399:

If both (**/**), (**U**), (**$**), (**f**), (**i**), (**$'**), (**v**), (**s**), (**T**), (**A**), (**Q**), (**p**), (**c**) and (**q**) are known, then its Dividend Payout Planned is:

$$d = (U + \$[1-f-i] - \$'v[1+s] - T$$
$$-(Q + \$'vp[1+s]/\{360[c-q]\})/\text{/}/A$$

Rule-20400:

If both (**/**), (**U**), (**$**), (**f**), (**i**), (**$'**), (**v**), (**s**), (**T**), (**A**), (**d**), (**Q**), (**c**) and (**q**) are known, then its Procured Inventory Days Planned:

$$p = 360[c-q][\text{/}\{U + \$[1-f-i] - \$'v[1+s] - T - Ad\} - Q]$$
$$/\{\$'v[1+s]\}$$

Steve Asikin ISBN 14: 978-1515158691, ISBN 10: 1515158691

Rule-20401:

If both (**/**), (**U**), (**$**), (**f**), (**i**), (**$'**), (**v**), (**s**), (**T**), (**A**), (**d**), (**p**), (**Q**) and (**q**) are known, then its Current Ratio Planned is:

$$c = q + \$'vp[1+s]$$
$$/\{360[/\{U+\$[1-f-i]-\$'v[1+s]-T-Ad\}-Q]\}$$

Rule-20402:

If both (**/**), (**U**), (**$**), (**f**), (**i**), (**$'**), (**v**), (**s**), (**T**), (**A**), (**d**), (**p**), (**c**) and (**Q**) are known, then its Quick or Acid Test Ratio Planned is:

$$q = c - \$'vp[1+s]$$
$$/\{360[/\{U+\$[1-f-i]-\$'v[1+s]-T-Ad\}-Q]\}$$

Rule-20403:

If both (**/**), (**U**), (**$**), (**f**), (**i**), (**$'**), (**v**), (**s**), (**t**), (**D**) and (**X**) are known, then its Quoted Longterm Debt Planned is:

$$Q = /\{U+\{\$[1-f-i]-\$'v[1+s]\}[1-t]-D)-X$$

Rule-20404:

If both (**Q**), (**U**), (**$**), (**f**), (**i**), (**$'**), (**v**), (**s**), (**t**), (**D**) and (**X**) are known, then its Leverage or Gearing Ratio Planned is:

$$/= [Q+X]/(U+\{\$[1-f-i]-\$'v[1+s]\}[1-t]-D)$$

Rule-20405:

If both (**/**), (**Q**), (**$**), (**f**), (**i**), (**$'**), (**v**), (**s**), (**t**), (**D**) and (**X**) are known, then its Utilized or Starting Capital Planned is:

$$U = D+[Q+X]//\{\$[1-f-i]-\$'v[1+s]\}[1-t]$$

Steve Asikin ISBN 14: 978-1515158691, ISBN 10: 1515158691

Rule-20406:

If both (f), (U), (Q), (f), (i), $(\$')$, (v), (s), (t), (D) and (X) are known, then its Sales or RevenuePlanned is:
$$\$= (\$'v[1+s]+\{D+[Q+X]//-U\}/[1-t])/[1-f-i]$$

Rule-20407:

If both (f), (U), $(\$)$, (Q), (i), $(\$')$, (v), (s), (t), (D) and (X) are known, then its Fixed Portion Planned is:
$$f= 1-i-(\$'v[1+s]+\{D+[Q+X]//-U\}/[1-t])/\$$$

Rule-20408:

If both (f), (U), $(\$)$, (f), (Q), $(\$')$, (v), (s), (t), (D) and (X) are known, then its Interest Portion Planned is:
$$i= 1-f-(\$'v[1+s]+\{D+[Q+X]//-U\}/[1-t])/\$$$

Rule-20409:

If both (f), (U), $(\$)$, (f), (i), (Q), (v), (s), (t), (D) and (X) are known, then its Sales Past must be:
$$\$'= (\$[1-f-i]-\{D+[Q+X]//-U\}/[1-t])/\{v[1+s]\}$$

Rule-20410:

If both (f), (U), $(\$)$, (f), (i), $(\$')$, (Q), (s), (t), (D) and (X) are known, then its Variable Portion Planned is:
$$v= (\$[1-f-i]-\{D+[Q+X]//-U\}/[1-t])/\{\$'[1+s]\}$$

Rule-20411:

If both (f), (U), $(\$)$, (f), (i), $(\$')$, (v), (Q), (t), (D) and (X) are known, then its Sales Growth Planned is:
$$s= \$/\$'-1$$

or it can also be found as
$$s= (\$[1-f-i]-\{D+[Q+X]//-U\}/[1-t])/[\$'v]-1$$

Steve Asikin ISBN 14: 978-1515158691, ISBN 10: 1515158691

Rule-20412:

If both $(\textbf{\textit{l}})$, $(\textbf{U})$, $(\textbf{\$})$, $(\textbf{f})$, $(\textbf{i})$, $(\textbf{\$'})$, $(\textbf{v})$, $(\textbf{s})$, $(\textbf{Q})$, $(\textbf{D})$ and $(\textbf{X})$ are known, then its Tax Rate Planned is:

$$t = 1 - \{D + [Q+X]/l - U\}/\{\$[1-f-i] - \$'v[1+s]\}$$

Rule-20413:

If both $(\textbf{\textit{l}})$, $(\textbf{U})$, $(\textbf{\$})$, $(\textbf{f})$, $(\textbf{i})$, $(\textbf{\$'})$, $(\textbf{v})$, $(\textbf{s})$, $(\textbf{t})$, $(\textbf{Q})$ and $(\textbf{X})$ are known, then its Dividend Planned is:

$$D = U + \{\$[1-f-i] - \$'v[1+s]\}[1-t] - [Q+X]/l$$

Rule-20414:

If both $(\textbf{\textit{l}})$, $(\textbf{U})$, $(\textbf{\$})$, $(\textbf{f})$, $(\textbf{i})$, $(\textbf{\$'})$, $(\textbf{v})$, $(\textbf{s})$, $(\textbf{t})$, $(\textbf{D})$ and $(\textbf{Q})$ are known, then its Xpress or Current Debt Planned is:

$$X = l(U + \{\$[1-f-i] - \$'v[1+s]\}[1-t] - D) - Q$$

Rule-20415:

If both $(\textbf{\textit{l}})$, $(\textbf{U})$, $(\textbf{\$})$, $(\textbf{f})$, $(\textbf{i})$, $(\textbf{\$'})$, $(\textbf{v})$, $(\textbf{s})$, $(\textbf{t})$, $(\textbf{D})$, $(\textbf{P})$, $(\textbf{c})$ and $(\textbf{q})$ are known, then its Quoted Longterm Debt Planned is:

$$Q = l(U + \{\$[1-f-i] - \$'v[1+s]\}[1-t] - D) - P/[c-q]$$

Rule-20416:

If both $(\textbf{Q})$, $(\textbf{U})$, $(\textbf{\$})$, $(\textbf{f})$, $(\textbf{i})$, $(\textbf{\$'})$, $(\textbf{v})$, $(\textbf{s})$, $(\textbf{t})$, $(\textbf{D})$, $(\textbf{P})$, $(\textbf{c})$ and $(\textbf{q})$ are known, then its Leverage or Gearing Ratio Planned is:

$$l = \{Q + P/[c-q]\}/(U + \{\$[1-f-i] - \$'v[1+s]\}[1-t] - D)$$

Steve Asikin ISBN 14: 978-1515158691, ISBN 10: 1515158691

Rule-20417:

If both (*I*), (**Q**), (**S**), (**f**), (**i**), (**S'**), (**v**), (**s**), (**t**), (**D**), (**P**), (**c**) and (**q**) are known, then its Utilized or Starting Capital must be:

$$U = D + \{Q + P/[c-q]\}/I\{S[1-f-i] - S'v[1+s]\}[1-t]$$

Rule-20418:

If both (*I*), (**U**), (**Q**), (**f**), (**i**), (**S'**), (**v**), (**s**), (**t**), (**D**), (**P**), (**c**) and (**q**) are known, then its Sales or Revenue Planned is:

$$S = [S'v[1+s] + (D + \{Q + P/[c-q]\}/I - U)/[1-t]]/[1-f-i]$$

Rule-20419:

If both (*I*), (**U**), (**S**), (**Q**), (**i**), (**S'**), (**v**), (**s**), (**t**), (**D**), (**P**), (**c**) and (**q**) are known, then its Fixed Portion Planned is:

$$f = 1 - i - [S'v[1+s] + (D + \{Q + P/[c-q]\}/I - U)/[1-t]]/S$$

Rule-20420:

If both (*I*), (**U**), (**S**), (**f**), (**Q**), (**S'**), (**v**), (**s**), (**t**), (**D**), (**P**), (**c**) and (**q**) are known, then its Interest Portion Planned is:

$$i = 1 - f - [S'v[1+s] + (D + \{Q + P/[c-q]\}/I - U)/[1-t]]/S$$

Rule-20421:

If both (*I*), (**U**), (**S**), (**f**), (**i**), (**Q**), (**v**), (**s**), (**t**), (**D**), (**P**), (**c**) and (**q**) are known, then its Sales Past must be:

$$S' = [S[1-f-i] - (D + \{Q + P/[c-q]\}/I - U)/[1-t]]/\{v[1+s]\}$$

Steve Asikin ISBN 14: 978-1515158691, ISBN 10: 1515158691

Rule-20422:

If both (I), (**U**), (**$**), (**f**), (**i**), (**$'**), (**Q**), (**s**), (**t**), (**D**), (**P**), (**c**) and (**q**) are known, then its Variable Portion Planned is:

$v = [\$[1\text{-}f\text{-}i]\text{-}(D+\{Q+P/[c\text{-}q]\}/I\text{-}U)/[1\text{-}t]]/\{\$'[1+s]\}$

Rule-20423:

If both (I), (**U**), (**$**), (**f**), (**i**), (**$'**), (**v**), (**Q**), (**t**), (**D**), (**P**), (**c**) and (**q**) are known, then its Sales Growth Planned is:

$s = \$/\$'\text{-}1$

or it can also be found as

$s = [\$[1\text{-}f\text{-}i]\text{-}(D+\{Q+P/[c\text{-}q]\}/I\text{-}U)/[1\text{-}t]]/[\$'v]\text{-}1$

Rule-20424:

If both (I), (**U**), (**$**), (**f**), (**i**), (**$'**), (**v**), (**s**), (**Q**), (**D**), (**P**), (**c**) and (**q**) are known, then its Tax Rate Planned is:

$t = 1\text{-}(D+\{Q+P/[c\text{-}q]\}/I\text{-}U)/\{\$[1\text{-}f\text{-}i]\text{-}\$'v[1+s]\}$

Rule-20425:

If both (I), (**U**), (**$**), (**f**), (**i**), (**$'**), (**v**), (**s**), (**t**), (**Q**), (**P**), (**c**) and (**q**) are known, then its Dividend Planned is:

$D = U+\{\$[1\text{-}f\text{-}i]\text{-}\$'v[1+s]\}[1\text{-}t]\text{-}\{Q+P/[c\text{-}q]\}/I$

Rule-20426:

If both (I), (**U**), (**$**), (**f**), (**i**), (**$'**), (**v**), (**s**), (**t**), (**D**), (**Q**), (**c**) and (**q**) are known, then its Procured Inventory Planned is:

$P = [c\text{-}q](I U+\{\$[1\text{-}f\text{-}i]\text{-}\$'v[1+s]\}[1\text{-}t]\text{-}D)\text{-}Q)$

Steve Asikin ISBN 14: 978-1515158691, ISBN 10: 1515158691

Rule-20427:

If both (I), (**U**), (**$**), (**f**), (**i**), (**$'**), (**v**), (**s**), (**t**), (**D**), (**P**), (**Q**) and (**q**) are known, then its Current Ratio Planned is:

$$c= q+P/(\textit{I}U+\{\$[1\text{-}f\text{-}i]\text{-}\$'v[1+s]\}[1\text{-}t]\text{-}D)\text{-}Q)$$

Rule-20428:

If both (I), (**U**), (**$**), (**f**), (**i**), (**$'**), (**v**), (**s**), (**t**), (**D**), (**P**), (**c**) and (**Q**) are known, then its Quick or Acid Test Ratio Planned is:

$$q= c\text{-}P/(\textit{I}U+\{\$[1\text{-}f\text{-}i]\text{-}\$'v[1+s]\}[1\text{-}t]\text{-}D)\text{-}Q)$$

Rule-20429:

If both (I), (**U**), (**$**), (**f**), (**i**), (**$'**), (**v**), (**s**), (**t**), (**D**), (**V**), (**p**), (**c**) and (**q**) are known, then its Quoted Longterm Debt Planned is:

$$Q= \textit{I}U+\{\$[1\text{-}f\text{-}i]\text{-}\$'v[1+s]\}[1\text{-}t]\text{-}D)\text{-}Vp/\{360[c\text{-}q]\}$$

Rule-20430:

If both (**Q**), (**U**), (**$**), (**f**), (**i**), (**$'**), (**v**), (**s**), (**t**), (**D**), (**V**), (**p**), (**c**) and (**q**) are known, then its Leverage or Gearing Ratio Planned is:

$$\textit{I}= (Q+Vp/\{360[c\text{-}q]\})$$
$$/(U+\{\$[1\text{-}f\text{-}i]\text{-}\$'v[1+s]\}[1\text{-}t]\text{-}D)$$

Rule-20431:

If both (I), (**Q**), (**$**), (**f**), (**i**), (**$'**), (**v**), (**s**), (**t**), (**D**), (**V**), (**p**), (**c**) and (**q**) are known, then its Utilized or Starting Capital must be:

$$U= D+(Q+Vp/\{360[c\text{-}q]\})/\textit{I}$$
$$-\{\$[1\text{-}f\text{-}i]\text{-}\$'v[1+s]\}[1\text{-}t]$$

Steve Asikin ISBN 14: 978-1515158691, ISBN 10: 1515158691

Rule-20432:

If both (**/**), (**U**), (**Q**), (**f**), (**i**), (**S'**), (**v**), (**s**), (**t**), (**D**), (**V**), (**p**), (**c**) and (**q**) are known, then its Sales or Revenue Planned is:

$$S= \{S'v[1+s]+[D+(Q+Vp/\{360[c-q]\})/\text{/-}U]$$
$$/[1\text{-}t]\}/[1\text{-}f\text{-}i]$$

Rule-20433:

If both (**/**), (**U**), (**S**), (**Q**), (**i**), (**S'**), (**v**), (**s**), (**t**), (**D**), (**V**), (**p**), (**c**) and (**q**) are known, then its Fixed Cost Planned is:

$$f= 1\text{-}i\text{-}\{S'v[1+s]$$
$$+[D+(Q+Vp/\{360[c-q]\})/\text{/-}U]/[1\text{-}t]\}/S$$

Rule-20434:

If both (**/**), (**U**), (**S**), (**f**), (**Q**), (**S'**), (**v**), (**s**), (**t**), (**D**), (**V**), (**p**), (**c**) and (**q**) are known, then its Interest Portion Planned is:

$$i= 1\text{-}f\text{-}\{S'v[1+s]$$
$$+[D+(Q+Vp/\{360[c-q]\})/\text{/-}U]/[1\text{-}t]\}/S$$

Rule-20435:

If both (**/**), (**U**), (**S**), (**f**), (**i**), (**Q**), (**v**), (**s**), (**t**), (**D**), (**V**), (**p**), (**c**) and (**q**) are known, then its Sales Past must be:

$$S'= \{S[1\text{-}f\text{-}i]\text{-}[D+(Q+Vp/\{360[c-q]\})/\text{/-}U]$$
$$/[1\text{-}t]\}/\{v[1+s]\}$$

Steve Asikin ISBN 14: 978-1515158691, ISBN 10: 1515158691

Rule-20436:

> If both (I), (U), (S), (f), (i), (S'), (Q), (s), (t), (D), (V), (p), (c) and (q) are known, then its Variable Portion Planned is:
>
> $v= V/S$
>
> > or it can also be found as
>
> $v= \{S[1\text{-}f\text{-}i]\text{-}[D+(Q+Vp/\{360[c\text{-}q]\})/I\text{-}U]$
> $/[1\text{-}t]\}/\{S'[1+s]\}$

Rule-20437:

> If both (I), (U), (S), (f), (i), (S'), (v), (Q), (t), (D), (V), (p), (c) and (q) are known, then its Sales Growth Planned is:
>
> $s= S/S'\text{-}1$
>
> > or it can also be found as
>
> $s= \{S[1\text{-}f\text{-}i]\text{-}[D+(Q+Vp/\{360[c\text{-}q]\})/I\text{-}U]$
> $/[1\text{-}t]\}/[S'v]\text{-}1$

Rule-20438:

> If both (I), (U), (S), (f), (i), (S'), (v), (s), (Q), (D), (V), (p), (c) and (q) are known, then its Tax Rate Planned is:
>
> $t= 1\text{-}[D+(Q+Vp/\{360[c\text{-}q]\})/I\text{-}U]$
> $/\{S[1\text{-}f\text{-}i]\text{-}S'v[1+s]\}$

Rule-20439:

> If both (I), (U), (S), (f), (i), (S'), (v), (s), (t), (Q), (V), (p), (c) and (q) are known, then its Dividend Planned is:
>
> $D= U+\{S[1\text{-}f\text{-}i]\text{-}S'v[1+s]\}[1\text{-}t]$
> $-(Q+Vp/\{360[c\text{-}q]\})/I$

Steve Asikin ISBN 14: 978-1515158691, ISBN 10: 1515158691

Rule-20440:

If both (**/**), (**U**), (**S**), (**f**), (**i**), (**S'**), (**v**), (**s**), (**t**), (**D**), (**Q**), (**p**), (**c**) and (**q**) are known, then its Variable Cost Planned is:

$$V= 360[c-q](\text{/}U+\{S[1-f-i]-S'v[1+s]\}[1-t]-D)-Q)/p$$

Rule-20441:

If both (**/**), (**U**), (**S**), (**f**), (**i**), (**S'**), (**v**), (**s**), (**t**), (**D**), (**V**), (**Q**), (**c**) and (**q**) are known, then its Procured Inventory Days Planned:

$$p= 360[c-q](\text{/}U+\{S[1-f-i]-S'v[1+s]\}[1-t]-D)-Q)/V$$

Rule-20442:

If both (**/**), (**U**), (**S**), (**f**), (**i**), (**S'**), (**v**), (**s**), (**t**), (**D**), (**V**), (**p**), (**Q**) and (**q**) are known, then its Current Ratio Planned is:

$$c= q+Vp/[360(\text{/}U+\{S[1-f-i]-S'v[1+s]\}[1-t]-D)-Q)]$$

Rule-20443:

If both (**/**), (**U**), (**S**), (**f**), (**i**), (**S'**), (**v**), (**s**), (**t**), (**D**), (**V**), (**p**), (**c**) and (**Q**) are known, then its Quick or Acid Test Ratio Planned is:

$$q= c-Vp/[360(\text{/}U+\{S[1-f-i]-S'v[1+s]\}[1-t]-D)-Q)]$$

Rule-20444:

If both (**/**), (**U**), (**S**), (**f**), (**i**), (**S'**), (**v**), (**s**), (**t**), (**D**), (**p**), (**c**) and (**q**) are known, then its Quoted Longterm Debt Planned is:

$$Q= \text{/}U+\{S[1-f-i]-S'v[1+s]\}[1-t]-D)$$
$$-Svp/\{360[c-q]\}$$

Steve Asikin ISBN 14: 978-1515158691, ISBN 10: 1515158691

Rule-20445:
 If both (**Q**), (**U**), (**$**), (**f**), (**i**), (**$'**), (**v**), (**s**), (**t**), (**D**), (**p**), (**c**) and (**q**) are known, then its Leverage or Gearing Ratio Planned is:
$$\textit{l} = (\mathbf{Q} + \mathbf{\$vp}/\{360[\mathbf{c}\text{-}\mathbf{q}]\})$$
$$/(\mathbf{U} + \{\mathbf{\$}[1\text{-}\mathbf{f}\text{-}\mathbf{i}]\text{-}\mathbf{\$'v}[1\text{+}\mathbf{s}]\}[1\text{-}\mathbf{t}]\text{-}\mathbf{D})$$

Rule-20446:
 If both (**l**), (**Q**), (**$**), (**f**), (**i**), (**$'**), (**v**), (**s**), (**t**), (**D**), (**p**), (**c**) and (**q**) are known, then its Utilized or Starting Capital must be:
$$\mathbf{U} = \mathbf{D} + (\mathbf{Q} + \mathbf{\$vp}/\{360[\mathbf{c}\text{-}\mathbf{q}]\})/\textit{l}$$
$$-\{\mathbf{\$}[1\text{-}\mathbf{f}\text{-}\mathbf{i}]\text{-}\mathbf{\$'v}[1\text{+}\mathbf{s}]\}[1\text{-}\mathbf{t}]$$

Rule-20447:
 If both (**l**), (**U**), (**Q**), (**f**), (**i**), (**$'**), (**v**), (**s**), (**t**), (**D**), (**p**), (**c**) and (**q**) are known, then its Sales or Revenue Planned is:
$$\mathbf{\$} = (\textit{l}\{\mathbf{U}\text{-}\mathbf{\$'v}[1\text{+}\mathbf{s}][1\text{-}\mathbf{t}]\text{-}\mathbf{D}\}\text{-}\mathbf{Q})$$
$$/(\mathbf{vp}/\{360[\mathbf{c}\text{-}\mathbf{q}]\}\text{-}\textit{l}[1\text{-}\mathbf{f}\text{-}\mathbf{i}][1\text{-}\mathbf{t}])$$

Rule-20448:
 If both (**l**), (**U**), (**$**), (**Q**), (**i**), (**$'**), (**v**), (**s**), (**t**), (**D**), (**p**), (**c**) and (**q**) are known, then its Fixed Portion Planned is:
$$\mathbf{f} = 1\text{-}\mathbf{i}\text{-}\{\mathbf{\$'v}[1\text{+}\mathbf{s}]\text{+}[\mathbf{D}+(\mathbf{Q}+\mathbf{\$vp}/\{360[\mathbf{c}\text{-}\mathbf{q}]\})/\textit{l}\text{-}\mathbf{U}]$$
$$/[1\text{-}\mathbf{t}]\}/\mathbf{\$}$$

Steve Asikin ISBN 14: 978-1515158691, ISBN 10: 1515158691

Rule-20449:

If both (I), (U), (S), (f), (Q), (S'), (v), (s), (t), (D), (p), (c) and (q) are known, then its Interest Portion Planned is:

$$i= 1-f-\{S'v[1+s]+[D+(Q+Svp/\{360[c-q]\})/I-U] /[1-t]\}/S$$

Rule-20450:

If both (I), (U), (S), (f), (i), (Q), (v), (s), (t), (D), (p), (c) and (q) are known, then its Sales Past must be:

$$S'= \{S[1-f-i]-[D+(Q+Svp/\{360[c-q]\})/I-U] /[1-t]\}/\{v[1+s]\}$$

Rule-20451:

If both (I), (U), (S), (f), (i), (S'), (Q), (s), (t), (D), (p), (c) and (q) are known, then its Variable Portion Planned is:

$$v= (I\{U+S[1-f-i][1-t]-D\}-Q) /(Sp/\{360[c-q]\}+S'I[1+s][1-t])$$

Rule-20452:

If both (I), (U), (S), (f), (i), (S'), (v), (Q), (t), (D), (p), (c) and (q) are known, then its Sales Growth Planned is:

$$s= S/S'-1$$

or it can also be found as

$$s= \{S[1-f-i]-[D+(Q+Svp/\{360[c-q]\})/I-U]/[1-t]\} /[S'v]-1$$

Steve Asikin ISBN 14: 978-1515158691, ISBN 10: 1515158691

Rule-20453:

If both (**ʃ**), (**U**), (**$**), (**f**), (**i**), (**$'**), (**v**), (**s**), (**T**), (**D**), (**p**), (**c**) and (**q**) are known, then its Tax Rate Planned is:

$$t= 1-[D+(Q+Svp/\{360[c-q]\})/ʃ-U]$$
$$/\{$[1-f-i]-$'v[1+s]\}$$

Rule-20454:

If both (**ʃ**), (**U**), (**$**), (**f**), (**i**), (**$'**), (**v**), (**s**), (**t**), (**Q**), (**p**), (**c**) and (**q**) are known, then its Dividend Planned is:

$$D= U+\{$[1-f-i]-$'v[1+s]\}[1-t]$$
$$-(Q+Svp/\{360[c-q]\})/ʃ$$

Rule-20455:

If both (**ʃ**), (**U**), (**$**), (**f**), (**i**), (**$'**), (**v**), (**s**), (**t**), (**D**), (**Q**), (**c**) and (**q**) are known, then its Procured Inventory Days Planned:

$$p= 360[c-q](ʃU+\{$[1-f-i]-$'v[1+s]\}[1-t]-D)-Q)$$
$$/[Sv]$$

Rule-20456:

If both (**ʃ**), (**U**), (**$**), (**f**), (**i**), (**$'**), (**v**), (**s**), (**t**), (**D**), (**p**), (**Q**) and (**q**) are known, then its Current Ratio Planned is:

$$c= q+Svp$$
$$/[360(ʃU+\{$[1-f-i]-$'v[1+s]\}[1-t]-D)-Q)]$$

Steve Asikin ISBN 14: 978-1515158691, ISBN 10: 1515158691

Rule-20457:

If both (I), (U), $(\$)$, (f), (i), $(\$')$, (v), (s), (t), (D), (p), (c) and (Q) are known, then its Quick or Acid Test Ratio Planned is:

$$q = c\text{-}\$vp$$
$$/[360(I(U+\{\$[1\text{-}f\text{-}i]\text{-}\$'v[1+s]\}[1\text{-}t]\text{-}D)\text{-}Q)]$$

Rule-20458:

If both (I), (U), $(\$)$, (f), (i), $(\$')$, (v), (s), (t), (D), (p), (c) and (q) are known, then its Quoted Longterm Debt Planned is:

$$Q = I(U+\{\$[1\text{-}f\text{-}i]\text{-}\$'v[1+s]\}[1\text{-}t]\text{-}D)$$
$$-\$'vp[1+s]/\{360[c\text{-}q]\}$$

Rule-20459:

If both (Q), (U), $(\$)$, (f), (i), $(\$')$, (v), (s), (t), (D), (p), (c) and (q) are known, then its Leverage or Gearing Ratio Planned is:

$$I = (Q+\$'vp[1+s]/\{360[c\text{-}q]\})$$
$$/(U+\{\$[1\text{-}f\text{-}i]\text{-}\$'v[1+s]\}[1\text{-}t]\text{-}D)$$

Rule-20460:

If both (I), (Q), $(\$)$, (f), (i), $(\$')$, (v), (s), (t), (D), (p), (c) and (q) are known, then its Utilized or Starting Capital must be:

$$U = D+(Q+\$'vp[1+s]/\{360[c\text{-}q]\})/I$$
$$-\{\$[1\text{-}f\text{-}i]\text{-}\$'v[1+s]\}[1\text{-}t]$$

Steve Asikin ISBN 14: 978-1515158691, ISBN 10: 1515158691

Rule-20461:

If both (I), (U), (Q), (f), (i), (S'), (v), (s), (t), (D), (p), (c) and (q) are known, then its Sales or Revenue Planned is:

$$S = (Q - I\{U - S'v[1+s][1-t] - D$$
$$+ S'vp[1+s]/\{360[c-q]\})/\{I[1-f-i]$$

Rule-20462:

If both (I), (U), (S), (Q), (i), (S'), (v), (s), (t), (D), (p), (c) and (q) are known, then its Fixed Portion Planned is:

$$f = 1 - i - \{S'v[1+s] + [D + (Q + S'vp[1+s]/\{360[c-q]\})/I$$
$$- U]/[1-t]\}/S$$

Rule-20463:

If both (I), (U), (S), (f), (Q), (S'), (v), (s), (t), (D), (p), (c) and (q) are known, then its Interest Portion Planned is:

$$i = 1 - f - \{S'v[1+s] + [D + (Q + S'vp[1+s]/\{360[c-q]\})/I$$
$$- U]/[1-t]\}/S$$

Rule-20464:

If both (I), (U), (S), (f), (i), (Q), (v), (s), (t), (D), (p), (c) and (q) are known, then its Sales Past must be:

$$S' = (I\{U + S[1-f-i][1-t] - D\} - Q)$$
$$/[v[1+s](p/\{360[c-q]\} + I[1-t])]$$

Steve Asikin ISBN 14: 978-1515158691, ISBN 10: 1515158691

Rule-20465:

If both (*I*), (**U**), (**\$**), (**f**), (**i**), (**\$'**), (**Q**), (**s**), (**t**), (**D**), (**p**), (**c**) and (**q**) are known, then its Variable Portion Planned is:

$$v= (I\{U+\$[1\text{-}f\text{-}i][1\text{-}t]\text{-}D\}\text{-}Q)$$
$$/[\$'[1+s](p/\{360[c\text{-}q]\}+I[1\text{-}t])]$$

Rule-20466:

If both (*I*), (**U**), (**\$**), (**f**), (**i**), (**\$'**), (**v**), (**Q**), (**t**), (**D**), (**p**), (**c**) and (**q**) are known, then its Sales Growth Planned is:

$$s= \$/\$'\text{-}1$$

or it can also be found as

$$s= (I\{U+\$[1\text{-}f\text{-}i][1\text{-}t]\text{-}D\}\text{-}Q)$$
$$/[\$'v(p/\{360[c\text{-}q]\}+I[1\text{-}t])]\text{-}1$$

Rule-20467:

If both (*I*), (**U**), (**\$**), (**f**), (**i**), (**\$'**), (**v**), (**s**), (**Q**), (**D**), (**p**), (**c**) and (**q**) are known, then its Tax Rate Planned is:

$$t= 1\text{-}[D+(Q+\$'vp[1+s]/\{360[c\text{-}q]\})/I\text{-}U]$$
$$/\{\$[1\text{-}f\text{-}i]\text{-}\$'v[1+s]\}$$

Rule-20468:

If both (*I*), (**U**), (**\$**), (**f**), (**i**), (**\$'**), (**v**), (**s**), (**t**), (**Q**), (**p**), (**c**) and (**q**) are known, then its Dividend Planned is:

$$D= U+\{\$[1\text{-}f\text{-}i]\text{-}\$'v[1+s]\}[1\text{-}t]$$
$$\text{-}(Q+\$'vp[1+s]/\{360[c\text{-}q]\})/I$$

Steve Asikin ISBN 14: 978-1515158691, ISBN 10: 1515158691

Rule-20469:

If both (**ʃ**), (**U**), (**$**), (**f**), (**i**), (**$'**), (**v**), (**s**), (**t**), (**D**), (**Q**), (**c**) and (**q**) are known, then its Procured Inventory Days Planned:

$$p= 360[c-q](\text{ʃU}+\{\$[1-f-i]-\$'v[1+s]\}[1-t]-D)-Q)$$
$$/\{\$'v[1+s]\}$$

Rule-20470:

If both (**ʃ**), (**U**), (**$**), (**f**), (**i**), (**$'**), (**v**), (**s**), (**t**), (**D**), (**p**), (**Q**) and (**q**) are known, then its Current Ratio Planned is:

$$c= q+\$'vp[1+s]$$
$$/[360(\text{ʃU}+\{\$[1-f-i]-\$'v[1+s]\}[1-t]-D)-Q)]$$

Rule-20471:

If both (**ʃ**), (**U**), (**$**), (**f**), (**i**), (**$'**), (**v**), (**s**), (**t**), (**D**), (**p**), (**c**) and (**Q**) are known, then its Quick or Acid Test Ratio Planned is:

$$q= c-\$'vp[1+s]$$
$$/[360(\text{ʃU}+\{\$[1-f-i]-\$'v[1+s]\}[1-t]-D)-Q)]$$

Rule-20472:

If both (**ʃ**), (**U**), (**$**), (**f**), (**i**), (**$'**), (**v**), (**s**), (**t**), (**A**), (**d**) and (**X**) are known, then its Quoted Longterm Debt Planned is:

$$Q= \text{ʃU}+\{\$[1-f-i]-\$'v[1+s]\}[1-t]-Ad)-X$$

Steve Asikin ISBN 14: 978-1515158691, ISBN 10: 1515158691

Rule-20473:

If both (**Q**), (**U**), (**S**), (**f**), (**i**), (**S'**), (**v**), (**s**), (**t**), (**A**), (**d**) and (**X**) are known, then its Leverage or Gearing Ratio Planned is:

$$\text{/}= [Q+X]/(U+\{S[1\text{-}f\text{-}i]\text{-}S'v[1+s]\}[1\text{-}t]\text{-}Ad)$$

Rule-20474:

If both (**/**), (**Q**), (**S**), (**f**), (**i**), (**S'**), (**v**), (**s**), (**t**), (**A**), (**d**) and (**X**) are known, then its Utilized or Starting Capital must be:

$$U= Ad+[Q+X]/\text{/}\{S[1\text{-}f\text{-}i]\text{-}S'v[1+s]\}[1\text{-}t]$$

Rule-20475:

If both (**/**), (**U**), (**Q**), (**f**), (**i**), (**S'**), (**v**), (**s**), (**t**), (**A**), (**d**) and (**X**) are known, then its Sales or Revenue Planned is:

$$S= (\{Ad+[Q+X]/\text{/-}U\}/[1\text{-}t]+S'v[1+s])/[1\text{-}f\text{-}i]$$

Rule-20476:

If both (**/**), (**U**), (**S**), (**Q**), (**i**), (**S'**), (**v**), (**s**), (**t**), (**A**), (**d**) and (**X**) are known, then its Fixed Portion Planned is:

$$f= 1\text{-}i\text{-}(\{Ad+[Q+X]/\text{/-}U\}/[1\text{-}t]+S'v[1+s])/S$$

Rule-20477:

If both (**/**), (**U**), (**S**), (**f**), (**Q**), (**S'**), (**v**), (**s**), (**t**), (**A**), (**d**) and (**X**) are known, then its Interest Portion Planned is:

$$i= 1\text{-}f\text{-}(\{Ad+[Q+X]/\text{/-}U\}/[1\text{-}t]+S'v[1+s])/S$$

Steve Asikin ISBN 14: 978-1515158691, ISBN 10: 1515158691

Rule-20478:

If both $(\textit{f})$, $(\textbf{U})$, $(\textbf{S})$, $(\textbf{f})$, $(\textbf{i})$, $(\textbf{Q})$, $(\textbf{v})$, $(\textbf{s})$, $(\textbf{t})$, $(\textbf{A})$, $(\textbf{d})$ and $(\textbf{X})$ are known, then its Sales Past must be:

$$S' = (S[1\text{-}f\text{-}i]\text{-}\{Ad+[Q+X]/fU\}/[1\text{-}t]\})/\{v[1+s]\}$$

Rule-20479:

If both $(\textit{f})$, $(\textbf{U})$, $(\textbf{S})$, $(\textbf{f})$, $(\textbf{i})$, $(\textbf{S'})$, $(\textbf{Q})$, $(\textbf{s})$, $(\textbf{t})$, $(\textbf{A})$, $(\textbf{d})$ and $(\textbf{X})$ are known, then its Variable Portion Planned is:

$$v = (S[1\text{-}f\text{-}i]\text{-}\{Ad+[Q+X]/fU\}/[1\text{-}t]\})/\{S'[1+s]\}$$

Rule-20480:

If both $(\textit{f})$, $(\textbf{U})$, $(\textbf{S})$, $(\textbf{f})$, $(\textbf{i})$, $(\textbf{S'})$, $(\textbf{v})$, $(\textbf{Q})$, $(\textbf{t})$, $(\textbf{A})$, $(\textbf{d})$ and $(\textbf{X})$ are known, then its Sales Growth Planned is:

$$s = S/S'\text{-}1$$

or it can also be found as

$$s = (S[1\text{-}f\text{-}i]\text{-}\{Ad+[Q+X]/fU\}/[1\text{-}t]\})/[S'v]\text{-}1$$

Rule-20481:

If both $(\textit{f})$, $(\textbf{U})$, $(\textbf{S})$, $(\textbf{f})$, $(\textbf{i})$, $(\textbf{S'})$, $(\textbf{v})$, $(\textbf{s})$, $(\textbf{Q})$, $(\textbf{A})$, $(\textbf{d})$ and $(\textbf{X})$ are known, then its Tax Rate Planned is:

$$t = 1\text{-}\{Ad+[Q+X]/fU\}/\{S[1\text{-}f\text{-}i]\text{-}S'v[1+s]\}$$

Rule-20482:

If both $(\textit{f})$, $(\textbf{U})$, $(\textbf{S})$, $(\textbf{f})$, $(\textbf{i})$, $(\textbf{S'})$, $(\textbf{v})$, $(\textbf{s})$, $(\textbf{t})$, $(\textbf{Q})$, $(\textbf{d})$ and $(\textbf{X})$ are known, then its After Tax Income Planned is:

$$A = (U+\{S[1\text{-}f\text{-}i]\text{-}S'v[1+s]\}[1\text{-}t]\text{-}[Q+X]/f)/d$$

Steve Asikin ISBN 14: 978-1515158691, ISBN 10: 1515158691

Rule-20483:

If both (*l*), (**U**), (**$**), (**f**), (**i**), (**$'**), (**v**), (**s**), (**t**), (**A**), (**Q**) and (**X**) are known, then its Dividend Payout Planned is:

$$d= (U+\{\$[1-f-i]-\$'v[1+s]\}[1-t]-[Q+X]/l)/A$$

Rule-20484:

If both (*l*), (**U**), (**$**), (**f**), (**i**), (**$'**), (**v**), (**s**), (**t**), (**A**), (**d**) and (**Q**) are known, then its Xpress or Current Debt Planned is:

$$X= l(U+\{\$[1-f-i]-\$'v[1+s]\}[1-t]-Ad)-Q$$

Rule-20485:

If both (*l*), (**U**), (**$**), (**f**), (**i**), (**$'**), (**v**), (**s**), (**t**), (**A**), (**d**), (**P**), (**c**) and (**q**) are known, then its Quoted Longterm Debt Planned is:

$$Q= l(U+\{\$[1-f-i]-\$'v[1+s]\}[1-t]-Ad)-P/[c-q]$$

Rule-20486:

If both (**Q**), (**U**), (**$**), (**f**), (**i**), (**$'**), (**v**), (**s**), (**t**), (**A**), (**d**), (**P**), (**c**) and (**q**) are known, then its Leverage or Gearing Ratio Planned is:

$$l= \{Q+P/[c-q]\}/(U+\{\$[1-f-i]-\$'v[1+s]\}[1-t]-Ad)$$

Rule-20487:

If both (*l*), (**Q**), (**$**), (**f**), (**i**), (**$'**), (**v**), (**s**), (**t**), (**A**), (**d**), (**P**), (**c**) and (**q**) are known, then its Utilized or Starting Capital must be:

$$U= Ad+\{Q+P/[c-q]\}/l-\{\$[1-f-i]-\$'v[1+s]\}[1-t]$$

Steve Asikin ISBN 14: 978-1515158691, ISBN 10: 1515158691

Rule-20488:

    If both (f), (U), (Q), (f), (i), (S'), (v), (s), (t), (A), (d), (P), (c) and (q) are known, then its Sales or Revenue Planned is:

$$S= [(Ad+\{Q+P/[c-q]\}/fU)/[1-t]+S'v[1+s]]/[1-f-i]$$

Rule-20489:

    If both (f), (U), (S), (Q), (i), (S'), (v), (s), (t), (A), (d), (P), (c) and (q) are known, then its Fixed Portion Planned is:

$$f= 1-i-[(Ad+\{Q+P/[c-q]\}/fU)/[1-t]+S'v[1+s]]/S$$

Rule-20490:

    If both (f), (U), (S), (f), (Q), (S'), (v), (s), (t), (A), (d), (P), (c) and (q) are known, then its Interest Portion Planned is:

$$i= 1-f-[(Ad+\{Q+P/[c-q]\}/fU)/[1-t]+S'v[1+s]]/S$$

Rule-20491:

    If both (f), (U), (S), (f), (i), (Q), (v), (s), (t), (A), (d), (P), (c) and (q) are known, then its Sales Past must be:

$$S' = [S[1-f-i]-(Ad+\{Q+P/[c-q]\}/fU)/[1-t]\}] /\{v[1+s]\}$$

Rule-20492:

    If both (f), (U), (S), (f), (i), (S'), (Q), (s), (t), (A), (d), (P), (c) and (q) are known, then its Variable Portion Planned is:

$$v = [S[1-f-i]-(Ad+\{Q+P/[c-q]\}/fU)/[1-t]\}] /\{S'[1+s]\}$$

Steve Asikin ISBN 14: 978-1515158691, ISBN 10: 1515158691

Rule-20493:

If both (I), (U), (S), (f), (i), (S'), (v), (Q), (t), (A), (d), (P), (c) and (q) are known, then its Sales Growth Planned is:

$s = S/S' - 1$

or it can also be found as

$s = [S[1-f-i]-(Ad+\{Q+P/[c-q]\}/IU)/[1-t]\}]/[S'v]-1$

Rule-20494:

If both (I), (U), (S), (f), (i), (S'), (v), (s), (Q), (A), (d), (P), (c) and (q) are known, then its Tax Rate Planned is:

$t = 1-(Ad+\{Q+P/[c-q]\}/IU)/\{S[1-f-i]-S'v[1+s]\}$

Rule-20495:

If both (I), (U), (S), (f), (i), (S'), (v), (s), (t), (Q), (d), (P), (c) and (q) are known, then its After Tax Income Planned is:

$A = (U+\{S[1-f-i]-S'v[1+s]\}[1-t]-\{Q+P/[c-q]\}/I)/d$

Rule-20496:

If both (I), (U), (S), (f), (i), (S'), (v), (s), (t), (A), (Q), (P), (c) and (q) are known, then its Dividend Payout Planned is:

$d = (U+\{S[1-f-i]-S'v[1+s]\}[1-t]-\{Q+P/[c-q]\}/I)/A$

Rule-20497:

If both (I), (U), (S), (f), (i), (S'), (v), (s), (t), (A), (d), (Q), (c) and (q) are known, then its Procured Inventory Planned is:

$P = [c-q][IU+\{S[1-f-i]-S'v[1+s]\}[1-t]-Ad)-Q]$

Steve Asikin ISBN 14: 978-1515158691, ISBN 10: 1515158691

Rule-20498:

If both (**/**), (**U**), (**$**), (**f**), (**i**), (**$'**), (**v**), (**s**), (**t**), (**A**), (**d**), (**P**), (**Q**) and (**q**) are known, then its Current Ratio Planned is:

$$c= q+P/[/(U+\{\$[1\text{-}f\text{-}i]\text{-}\$'v[1+s]\}[1\text{-}t]\text{-}Ad)\text{-}Q]$$

Rule-20499:

If both (**/**), (**U**), (**$**), (**f**), (**i**), (**$'**), (**v**), (**s**), (**t**), (**A**), (**d**), (**P**), (**c**) and (**Q**) are known, then its Quick or Acid Test Ratio Planned is:

$$q= c\text{-}P/[/(U+\{\$[1\text{-}f\text{-}i]\text{-}\$'v[1+s]\}[1\text{-}t]\text{-}Ad)\text{-}Q]$$

Rule-20500:

If both (**/**), (**U**), (**$**), (**f**), (**i**), (**$'**), (**v**), (**s**), (**t**), (**A**), (**d**), (**V**), (**p**), (**c**) and (**q**) are known, then its Quoted Longterm Debt Planned is:

$$Q= /(U+\{\$[1\text{-}f\text{-}i]\text{-}\$'v[1+s]\}[1\text{-}t]\text{-}Ad)$$
$$-Vp/\{360[c\text{-}q]\}$$

Rule-20501:

If both (**Q**), (**U**), (**$**), (**f**), (**i**), (**$'**), (**v**), (**s**), (**t**), (**A**), (**d**), (**V**), (**p**), (**c**) and (**q**) are known, then its Leverage or Gearing Ratio Planned is:

$$/= (Q+Vp/\{360[c\text{-}q]\})$$
$$/(U+\{\$[1\text{-}f\text{-}i]\text{-}\$'v[1+s]\}[1\text{-}t]\text{-}Ad)$$

Steve Asikin ISBN 14: 978-1515158691, ISBN 10: 1515158691

Rule-20502:

If both (I), (Q), $(\$)$, (f), (i), $(\$')$, (v), (s), (t), (A), (d), (V), (p), (c) and (q) are known, then its Utilized or Starting Capital must be:

$$U = Ad + (Q + Vp / \{360[c-q]\}) / I$$
$$- \{\$[1-f-i] - \$'v[1+s]\}[1-t]$$

Rule-20503:

If both (I), (U), (Q), (f), (i), $(\$')$, (v), (s), (t), (A), (d), (V), (p), (c) and (q) are known, then its Sales or Revenue Planned is:

$$\$ = \{[Ad + (Q + Vp / \{360[c-q]\}) / I - U]$$
$$/ [1-t] + \$'v[1+s]\} / [1-f-i]$$

Rule-20504:

If both (I), (U), $(\$)$, (Q), (i), $(\$')$, (v), (s), (t), (A), (d), (V), (p), (c) and (q) are known, then its Fixed Portion Planned is:

$$f = 1 - i - \{[Ad + (Q + Vp / \{360[c-q]\}) / I - U]$$
$$/ [1-t] + \$'v[1+s]\} / \$$$

Rule-20505:

If both (I), (U), $(\$)$, (f), (Q), $(\$')$, (v), (s), (t), (A), (d), (V), (p), (c) and (q) are known, then its Interest Portion Planned is:

$$i = 1 - f - \{[Ad + (Q + Vp / \{360[c-q]\}) / I - U]$$
$$/ [1-t] + \$'v[1+s]\} / \$$$

Steve Asikin ISBN 14: 978-1515158691, ISBN 10: 1515158691

Rule-20506:

If both (I), (**U**), (**$**), (**f**), (**i**), (**Q**), (**v**), (**s**), (**t**), (**A**), (**d**), (**V**), (**p**), (**c**) and (**q**) are known, then its Sales Past must be:

$$\$' = \{\$[1\text{-}f\text{-}i]\text{-}[Ad+(Q+Vp/\{360[c\text{-}q]\})/IU]$$
$$/[1\text{-}t]\}\}/\{v[1+s]\}$$

Rule-20507:

If both (I), (**U**), (**$**), (**f**), (**i**), (**$'**), (**Q**), (**s**), (**t**), (**A**), (**d**), (**V**), (**p**), (**c**) and (**q**) are known, then its Variable Portion Planned is:

$$v = V/\$$$

or it can also be found as

$$v = \{\$[1\text{-}f\text{-}i]\text{-}[Ad+(Q+Vp/\{360[c\text{-}q]\})/IU]$$
$$/[1\text{-}t]\}\}/\{\$'[1+s]\}$$

Rule-20508:

If both (I), (**U**), (**$**), (**f**), (**i**), (**$'**), (**v**), (**Q**), (**t**), (**A**), (**d**), (**V**), (**p**), (**c**) and (**q**) are known, then its Sales Growth Planned is:

$$s = \$/\$' - 1$$

or it can also be found as

$$s = \{\$[1\text{-}f\text{-}i]\text{-}[Ad+(Q+Vp/\{360[c\text{-}q]\})/IU]$$
$$/[1\text{-}t]\}\}/[\$'v]\text{-}1$$

Rule-20509:

If both (I), (**U**), (**$**), (**f**), (**i**), (**$'**), (**v**), (**s**), (**Q**), (**A**), (**d**), (**V**), (**p**), (**c**) and (**q**) are known, then its Tax Rate Planned is:

$$t = 1\text{-}[Ad+(Q+Vp/\{360[c\text{-}q]\})/IU]$$
$$/\{\$[1\text{-}f\text{-}i]\text{-}\$'v[1+s]\}$$

Steve Asikin ISBN 14: 978-1515158691, ISBN 10: 1515158691

Rule-20510:

If both (**/**), (**U**), (**$**), (**f**), (**i**), (**$'**), (**v**), (**s**), (**t**), (**Q**), (**d**), (**V**), (**p**), (**c**) and (**q**) are known, then its After Tax Income Planned is:

$$A = [U + \{\$[1\text{-}f\text{-}i] \text{-} \$'v[1+s]\}[1\text{-}t]$$
$$-(Q+Vp/\{360[c\text{-}q]\})/\text{/}]/d$$

Rule-20511:

If both (**/**), (**U**), (**$**), (**f**), (**i**), (**$'**), (**v**), (**s**), (**t**), (**A**), (**Q**), (**V**), (**p**), (**c**) and (**q**) are known, then its Dividend Payout Planned is:

$$d = [U + \{\$[1\text{-}f\text{-}i] \text{-} \$'v[1+s]\}[1\text{-}t]$$
$$-(Q+Vp/\{360[c\text{-}q]\})/\text{/}]/A$$

Rule-20512:

If both (**/**), (**U**), (**$**), (**f**), (**i**), (**$'**), (**v**), (**s**), (**t**), (**A**), (**d**), (**Q**), (**p**), (**c**) and (**q**) are known, then its Variable Cost Planned is:

$$V = 360[c\text{-}q]$$
$$[\text{/}(U + \{\$[1\text{-}f\text{-}i] \text{-} \$'v[1+s]\}[1\text{-}t] \text{-} Ad) \text{-} Q]/p$$

Rule-20513:

If both (**/**), (**U**), (**$**), (**f**), (**i**), (**$'**), (**v**), (**s**), (**t**), (**A**), (**d**), (**V**), (**Q**), (**c**) and (**q**) are known, then its Procured Inventory Days Planned:

$$p = 360[c\text{-}q]$$
$$[\text{/}(U + \{\$[1\text{-}f\text{-}i] \text{-} \$'v[1+s]\}[1\text{-}t] \text{-} Ad) \text{-} Q]/V$$

Steve Asikin ISBN 14: 978-1515158691, ISBN 10: 1515158691

Rule-20514:

If both $(\textit{I})$, $(\textbf{U})$, $(\textbf{S})$, $(\textbf{f})$, $(\textbf{i})$, $(\textbf{S'})$, $(\textbf{v})$, $(\textbf{s})$, $(\textbf{t})$, $(\textbf{A})$, $(\textbf{d})$, $(\textbf{V})$, $(\textbf{p})$, $(\textbf{Q})$ and $(\textbf{q})$ are known, then its Current Ratio Planned is:

$$c = q + Vp / \{360[I(U + \{S[1-f-i]-S'v[1+s]\}[1-t]-Ad)-Q]\}$$

Rule-20515:

If both $(\textit{I})$, $(\textbf{U})$, $(\textbf{S})$, $(\textbf{f})$, $(\textbf{i})$, $(\textbf{S'})$, $(\textbf{v})$, $(\textbf{s})$, $(\textbf{t})$, $(\textbf{A})$, $(\textbf{d})$, $(\textbf{V})$, $(\textbf{p})$, $(\textbf{c})$ and $(\textbf{Q})$ are known, then its Quick or Acid Test Ratio Planned is:

$$q = c - Vp / \{360[I(U + \{S[1-f-i]-S'v[1+s]\}[1-t]-Ad)-Q]\}$$

Rule-20516:

If both $(\textit{I})$, $(\textbf{U})$, $(\textbf{S})$, $(\textbf{f})$, $(\textbf{i})$, $(\textbf{S'})$, $(\textbf{v})$, $(\textbf{s})$, $(\textbf{t})$, $(\textbf{A})$, $(\textbf{d})$, $(\textbf{p})$, $(\textbf{c})$ and $(\textbf{q})$ are known, then its Quoted Longterm Debt Planned is:

$$Q = I(U + \{S[1-f-i]-S'v[1+s]\}[1-t]-Ad) -Svp/\{360[c-q]\}$$

Rule-20517:

If both $(\textbf{Q})$, $(\textbf{U})$, $(\textbf{S})$, $(\textbf{f})$, $(\textbf{i})$, $(\textbf{S'})$, $(\textbf{v})$, $(\textbf{s})$, $(\textbf{t})$, $(\textbf{A})$, $(\textbf{d})$, $(\textbf{p})$, $(\textbf{c})$ and $(\textbf{q})$ are known, then its Leverage or Gearing Ratio Planned is:

$$I = (Q + Svp/\{360[c-q]\}) /(U + \{S[1-f-i]-S'v[1+s]\}[1-t]-Ad)$$

Steve Asikin ISBN 14: 978-1515158691, ISBN 10: 1515158691

Rule-20518:

If both (**/**), (**Q**), (**S**), (**f**), (**i**), (**S'**), (**v**), (**s**), (**t**), (**A**), (**d**), (**p**), (**c**) and (**q**) are known, then its Utilized or Starting Capital must be:

$$U = Ad + (Q + Svp/\{360[c-q]\})//$$
$$- \{S[1-f-i] - S'v[1+s]\}[1-t]$$

Rule-20519:

If both (**/**), (**U**), (**Q**), (**f**), (**i**), (**S'**), (**v**), (**s**), (**t**), (**A**), (**d**), (**p**), (**c**) and (**q**) are known, then its Sales or Revenue Planned is:

$$S = (/\{U - S'v[1+s][1-t] - Ad\} - Q)$$
$$/(vp/\{360[c-q]\} - /[1-f-i][1-t])$$

Rule-20520:

If both (**/**), (**U**), (**S**), (**Q**), (**i**), (**S'**), (**v**), (**s**), (**t**), (**A**), (**d**), (**p**), (**c**) and (**q**) are known, then its Fixed Portion Planned is:

$$f = 1 - i - \{[Ad + (Q + Svp/\{360[c-q]\})//U]$$
$$/[1-t] + S'v[1+s]\}/S$$

Rule-20521:

If both (**/**), (**U**), (**S**), (**f**), (**Q**), (**S'**), (**v**), (**s**), (**t**), (**A**), (**d**), (**p**), (**c**) and (**q**) are known, then its Interest Portion Planned is:

$$i = 1 - f - \{[Ad + (Q + Svp/\{360[c-q]\})//U]$$
$$/[1-t] + S'v[1+s]\}/S$$

Steve Asikin ISBN 14: 978-1515158691, ISBN 10: 1515158691

Rule-20522:

If both (I), (**U**), (**$**), (**f**), (**i**), (**Q**), (**v**), (**s**), (**t**), (**A**), (**d**), (**p**), (**c**) and (**q**) are known, then its Sales Past must be:

$$S' = \{S[1\text{-}f\text{-}i]\text{-}[Ad+(Q+Svp/\{360[c\text{-}q]\})/IU]$$
$$/[1\text{-}t]\}\}/\{v[1+s]\}$$

Rule-20523:

If both (I), (**U**), (**$**), (**f**), (**i**), (**$'**), (**Q**), (**s**), (**t**), (**A**), (**d**), (**p**), (**c**) and (**q**) are known, then its Variable Portion Planned is:

$$v = (I\{U+S[1\text{-}f\text{-}i][1\text{-}t]\text{-}Ad\}\text{-}Q)$$
$$/(Sp/\{360[c\text{-}q]\}+S'v/[1+s][1\text{-}t])$$

Rule-20524:

If both (I), (**U**), (**$**), (**f**), (**i**), (**$'**), (**v**), (**Q**), (**t**), (**A**), (**d**), (**p**), (**c**) and (**q**) are known, then its Sales Growth Planned is:

$$s = S/S'\text{-}1$$

or it can also be found

$$s = \{S[1\text{-}f\text{-}i]\text{-}[Ad+(Q+Svp/\{360[c\text{-}q]\})/IU]$$
$$/[1\text{-}t]\}\}/[S'v]\text{-}1$$

Rule-20525:

If both (I), (**U**), (**$**), (**f**), (**i**), (**$'**), (**v**), (**s**), (**Q**), (**A**), (**d**), (**p**), (**c**) and (**q**) are known, then its Tax Rate Planned is:

$$t = 1\text{-}[Ad+(Q+Svp/\{360[c\text{-}q]\})/IU]$$
$$/\{S[1\text{-}f\text{-}i]\text{-}S'v[1+s]\}$$

Steve Asikin ISBN 14: 978-1515158691, ISBN 10: 1515158691

Rule-20526:

If both (**/**), (**U**), (**$**), (**f**), (**i**), (**$'**), (**v**), (**s**), (**t**), (**Q**), (**d**), (**p**), (**c**) and (**q**) are known, then its After Tax Income Planned is:

$$A = [U + \{\$[1\text{-}f\text{-}i]\text{-}\$'v[1+s]\}[1\text{-}t]$$
$$-(Q + \$vp/\{360[c\text{-}q]\})/\text{/}]/d$$

Rule-20527:

If both (**/**), (**U**), (**$**), (**f**), (**i**), (**$'**), (**v**), (**s**), (**t**), (**A**), (**Q**), (**p**), (**c**) and (**q**) are known, then its Dividend Payout Planned is:

$$d = [U + \{\$[1\text{-}f\text{-}i]\text{-}\$'v[1+s]\}[1\text{-}t]$$
$$-(Q + \$vp/\{360[c\text{-}q]\})/\text{/}]/A$$

Rule-20528:

If both (**/**), (**U**), (**$**), (**f**), (**i**), (**$'**), (**v**), (**s**), (**t**), (**A**), (**d**), (**Q**), (**c**) and (**q**) are known, then its Procured Inventory Days Planned:

$$p = 360[c\text{-}q][\text{/}U + \{\$[1\text{-}f\text{-}i]\text{-}\$'v[1+s]\}[1\text{-}t]\text{-}Ad)\text{-}Q]$$
$$/[\$vp]$$

Rule-20529:

If both (**/**), (**U**), (**$**), (**f**), (**i**), (**$'**), (**v**), (**s**), (**t**), (**A**), (**d**), (**p**), (**Q**) and (**q**) are known, then its Current Ratio Planned is:

$$c = q + \$vp$$
$$/\{360[\text{/}U + \{\$[1\text{-}f\text{-}i]\text{-}\$'v[1+s]\}[1\text{-}t]\text{-}Ad)\text{-}Q]\}$$

Steve Asikin ISBN 14: 978-1515158691, ISBN 10: 1515158691

Rule-20530:

If both (I), (**U**), (**$**), (**f**), (**i**), (**$'**), (**v**), (**s**), (**t**), (**A**), (**d**),
(**p**), (**c**) and (**Q**) are known, then its Quick or Acid
Test Ratio Planned is:

q= c-$vp

$$/\{360[I(U+\{\$[1-f-i]-\$'v[1+s]\}[1-t]-Ad)-Q]\}$$

Rule-20531:

If both (I), (**U**), (**$**), (**f**), (**i**), (**$'**), (**v**), (**s**), (**t**), (**A**), (**d**),
(**p**), (**c**) and (**q**) are known, then its Quoted Longterm
Debt Planned is:

Q= I(**U**+{$[1-f-i]-$'v[1+s]}[1-t]-Ad)
-$'vp[1+s]/{360[c-q]}

Rule-20532:

If both (**Q**), (**U**), (**$**), (**f**), (**i**), (**$'**), (**v**), (**s**), (**t**), (**A**), (**d**),
(**p**), (**c**) and (**q**) are known, then its Leverage or
Gearing Ratio Planned is:

I= (Q+$'vp[1+s]/{360[c-q]})
/(U+{$[1-f-i]-$'v[1+s]}[1-t]-Ad)

Rule-20533:

If both (I), (**Q**), (**$**), (**f**), (**i**), (**$'**), (**v**), (**s**), (**t**), (**A**), (**d**),
(**p**), (**c**) and (**q**) are known, then its Utilized or Starting
Capital must be:

U= Ad+(Q+$'vp[1+s]/{360[c-q]})/$I$
-{$[1-f-i]-$'v[1+s]}[1-t]

Steve Asikin ISBN 14: 978-1515158691, ISBN 10: 1515158691

Rule-20534:
   If both (**/**), (**U**), (**Q**), (**f**), (**i**), (**S'**), (**v**), (**s**), (**t**), (**A**), (**d**),
   (**p**), (**c**) and (**q**) are known, then its Sales or Revenue
   Planned is:
   $$S= (Q-/\{U-S'v[1+s][1-t]-Ad\}$$
   $$-S'vp[1+s]/\{360[c-q]\})/\{/[1-f-i][1-t]\}$$

Rule-20535:
   If both (**/**), (**U**), (**S**), (**Q**), (**i**), (**S'**), (**v**), (**s**), (**t**), (**A**), (**d**),
   (**p**), (**c**) and (**q**) are known, then its Fixed Portion
   Planned is:
   $$f= 1-i-\{[Ad+(Q+S'vp[1+s]/\{360[c-q]\})//U]$$
   $$/[1-t]+S'v[1+s]\}/S$$

Rule-20536:
   If both (**/**), (**U**), (**S**), (**f**), (**Q**), (**S'**), (**v**), (**s**), (**t**), (**A**), (**d**),
   (**p**), (**c**) and (**q**) are known, then its Interest Portion
   Planned is:
   $$i= 1-f-\{[Ad+(Q+S'vp[1+s]/\{360[c-q]\})//U]$$
   $$/[1-t]+S'v[1+s]\}/S$$

Rule-20537:
   If both (**/**), (**U**), (**S**), (**f**), (**i**), (**Q**), (**v**), (**s**), (**t**), (**A**), (**d**),
   (**p**), (**c**) and (**q**) are known, then its Sales Past must be:
   $$S' = (/\{U+S[1-f-i][1-t]-Ad\}-Q)$$
   $$/[v[1+s](p/\{360[c-q]\}+/[1-t])]$$

Steve Asikin ISBN 14: 978-1515158691, ISBN 10: 1515158691

Rule-20538:

If both (**/**), (**U**), (**S**), (**f**), (**i**), (**S'**), (**Q**), (**s**), (**t**), (**A**), (**d**), (**p**), (**c**) and (**q**) are known, then its Variable Portion Planned is:

$$v = (\textit{/}\{U+S[1\text{-}f\text{-}i][1\text{-}t]\text{-}Ad\}\text{-}Q)$$
$$/[S'[1+s](p/\{360[c\text{-}q]\}+\textit{/}[1\text{-}t])]$$

Rule-20539:

If both (**/**), (**U**), (**S**), (**f**), (**i**), (**S'**), (**v**), (**Q**), (**t**), (**A**), (**d**), (**p**), (**c**) and (**q**) are known, then its Sales Growth Planned is:

$$s = S/S'\text{-}1$$

or it can also be found as

$$s = (\textit{/}\{U+S[1\text{-}f\text{-}i][1\text{-}t]\text{-}Ad\}\text{-}Q)$$
$$/[S'v(p/\{360[c\text{-}q]\}+\textit{/}[1\text{-}t])]\text{-}1$$

Rule-20540:

If both (**/**), (**U**), (**S**), (**f**), (**i**), (**S'**), (**v**), (**s**), (**Q**), (**A**), (**d**), (**p**), (**c**) and (**q**) are known, then its Tax Rate Planned is:

$$t = 1\text{-}[Ad+(Q+S'vp[1+s]/\{360[c\text{-}q]\})/\textit{/U}]$$
$$/\{S[1\text{-}f\text{-}i]\text{-}S'v[1+s]\}$$

Rule-20541:

If both (**/**), (**U**), (**S**), (**f**), (**i**), (**S'**), (**v**), (**s**), (**t**), (**Q**), (**d**), (**p**), (**c**) and (**q**) are known, then its After Tax Income Planned is:

$$A = [U+\{S[1\text{-}f\text{-}i]\text{-}S'v[1+s]\}[1\text{-}t]$$
$$\text{-}(Q+S'vp[1+s]/\{360[c\text{-}q]\})/\textit{/}]/d$$

Steve Asikin ISBN 14: 978-1515158691, ISBN 10: 1515158691

Rule-20542:

If both $(/)$, (U), (S), (f), (i), (S'), (v), (s), (t), (A), (Q), (p), (c) and (q) are known, then its Dividend Payout Planned is:

$$d= [U+\{S[1\text{-}f\text{-}i]\text{-}S'v[1+s]\}[1\text{-}t]$$
$$-(Q+S'vp[1+s]/\{360[c\text{-}q]\})//]/A$$

Rule-20543:

If both $(/)$, (U), (S), (f), (i), (S'), (v), (s), (t), (A), (d), (Q), (c) and (q) are known, then its Procured Inventory Days Planned:

$$p= 360[c\text{-}q][/U+\{S[1\text{-}f\text{-}i]\text{-}S'v[1+s]\}[1\text{-}t]\text{-}Ad)\text{-}Q]$$
$$/\{S'v[1+s]\}$$

Rule-20544:

If both $(/)$, (U), (S), (f), (i), (S'), (v), (s), (t), (A), (d), (p), (Q) and (q) are known, then its Current Ratio Planned is:

$$c= q+S'vp[1+s]$$
$$/\{360[/U+\{S[1\text{-}f\text{-}i]\text{-}S'v[1+s]\}[1\text{-}t]\text{-}Ad)\text{-}Q]\}$$

Rule-20545:

If both $(/)$, (U), (S), (f), (i), (S'), (v), (s), (t), (A), (d), (p), (c) and (Q) are known, then its Quick or Acid Test Ratio Planned is:

$$q= c\text{-}S'vp[1+s]$$
$$/\{360[/U+\{S[1\text{-}f\text{-}i]\text{-}S'v[1+s]\}[1\text{-}t]\text{-}Ad)\text{-}Q]\}$$

Steve Asikin ISBN 14: 978-1515158691, ISBN 10: 1515158691

Rule-20546:

If both (f), (U), (S), (f), (i), (S'), (v), (s), (t), (d) and (X) are known, then its Quoted Longterm Debt Planned is:
$$Q = f(U + \{S[1-f-i]-S'v[1+s]\}[1-t][1-d]) - X$$

Rule-20547:

If both (Q), (U), (S), (f), (i), (S'), (v), (s), (t), (d) and (X) are known, then its Leverage or Gearing Ratio Planned is:
$$f = [Q+X]/(U+\{S[1-f-i]-S'v[1+s]\}[1-t][1-d])$$

Rule-20548:

If both (f), (Q), (S), (f), (i), (S'), (v), (s), (t), (d) and (X) are known, then its Utilized or Starting Capital must be:
$$U = [Q+X]/f\{S[1-f-i]-S'v[1+s]\}[1-t][1-d]$$

Rule-20549:

If both (f), (U), (Q), (f), (i), (S'), (v), (s), (t), (d) and (X) are known, then its Sales or Revenue Planned is:
$$S = (S'v[1+s]+\{[Q+X]/f-U\}/\{[1-t][1-d]\})/[1-f-i]$$

Rule-20550:

If both (f), (U), (S), (Q), (i), (S'), (v), (s), (t), (d) and (X) are known, then its Fixed Portion Planned is:
$$f = 1-i-(S'v[1+s]+\{[Q+X]/f-U\}/\{[1-t][1-d]\})/S$$

Rule-20551:

If both (f), (U), (S), (f), (Q), (S'), (v), (s), (t), (d) and (X) are known, then its Interest Portion Planned is:
$$i = 1-f-(S'v[1+s]+\{[Q+X]/f-U\}/\{[1-t][1-d]\})/S$$

Steve Asikin ISBN 14: 978-1515158691, ISBN 10: 1515158691

Rule-20552:

    If both (f), $(\mathbf{U})$, $(\$)$, $(\mathbf{f})$, $(\mathbf{i})$, $(\mathbf{Q})$, $(\mathbf{v})$, $(\mathbf{s})$, $(\mathbf{t})$, $(\mathbf{d})$ and $(\mathbf{X})$ are known, then its Sales Past must be:

    $\$'= (\$[1\text{-}\mathbf{f}\text{-}\mathbf{i}]\text{-}\{[\mathbf{Q}\text{+}\mathbf{X}]/\mathit{f}\mathbf{U}\}/\{[1\text{-}\mathbf{t}][1\text{-}\mathbf{d}]\})/\{\mathbf{v}[1\text{+}\mathbf{s}]\}$

Rule-20553:

    If both (f), $(\mathbf{U})$, $(\$)$, $(\mathbf{f})$, $(\mathbf{i})$, $(\$')$, $(\mathbf{Q})$, $(\mathbf{s})$, $(\mathbf{t})$, $(\mathbf{d})$ and $(\mathbf{X})$ are known, then its Variable Portion Planned is:

    $\mathbf{v}= (\$[1\text{-}\mathbf{f}\text{-}\mathbf{i}]\text{-}\{[\mathbf{Q}\text{+}\mathbf{X}]/\mathit{f}\mathbf{U}\}/\{[1\text{-}\mathbf{t}][1\text{-}\mathbf{d}]\})/\{\$'[1\text{+}\mathbf{s}]\}$

Rule-20554:

    If both (f), $(\mathbf{U})$, $(\$)$, $(\mathbf{f})$, $(\mathbf{i})$, $(\$')$, $(\mathbf{v})$, $(\mathbf{Q})$, $(\mathbf{t})$, $(\mathbf{d})$ and $(\mathbf{X})$ are known, then its Sales Growth Planned is:

    $\mathbf{s}= \$/\$'\text{-}1$

        or it can also be found as

    $\mathbf{s}= (\$[1\text{-}\mathbf{f}\text{-}\mathbf{i}]\text{-}\{[\mathbf{Q}\text{+}\mathbf{X}]/\mathit{f}\mathbf{U}\}/\{[1\text{-}\mathbf{t}][1\text{-}\mathbf{d}]\})/[\$'\mathbf{v}]\text{-}1$

Rule-20555:

    If both (f), $(\mathbf{U})$, $(\$)$, $(\mathbf{f})$, $(\mathbf{i})$, $(\$')$, $(\mathbf{v})$, $(\mathbf{s})$, $(\mathbf{Q})$, $(\mathbf{d})$ and $(\mathbf{X})$ are known, then its Tax Rate Planned is:

    $\mathbf{t}= 1\text{-}\{[\mathbf{Q}\text{+}\mathbf{X}]/\mathit{f}\mathbf{U}\}/((\{\$[1\text{-}\mathbf{f}\text{-}\mathbf{i}]\text{-}\$'\mathbf{v}[1\text{+}\mathbf{s}]\}[1\text{-}\mathbf{d}])$

Rule-20556:

    If both (f), $(\mathbf{U})$, $(\$)$, $(\mathbf{f})$, $(\mathbf{i})$, $(\$')$, $(\mathbf{v})$, $(\mathbf{s})$, $(\mathbf{t})$, $(\mathbf{Q})$ and $(\mathbf{X})$ are known, then its Dividend Payout Planned is:

    $\mathbf{d}= 1\text{-}\{[\mathbf{Q}\text{+}\mathbf{X}]/\mathit{f}\mathbf{U}\}/((\{\$[1\text{-}\mathbf{f}\text{-}\mathbf{i}]\text{-}\$'\mathbf{v}[1\text{+}\mathbf{s}]\}[1\text{-}\mathbf{t}])$

Steve Asikin ISBN 14: 978-1515158691, ISBN 10: 1515158691

Rule-20557:

If both (**/**), (**U**), (**$**), (**f**), (**i**), (**$'**), (**v**), (**s**), (**t**), (**d**) and (**Q**) are known, then its Xpress or Current Debt Planned is:

$$X= \textit{/}(U+\{\$[1\text{-}f\text{-}i]\text{-}\$'v[1+s]\}[1\text{-}t][1\text{-}d])\text{-}Q$$

Rule-20558:

If both (**/**), (**U**), (**$**), (**f**), (**i**), (**$'**), (**v**), (**s**), (**t**), (**d**), (**P**), (**c**) and (**q**) are known, then its Quoted Longterm Debt Planned is:

$$Q= \textit{/}(U+\{\$[1\text{-}f\text{-}i]\text{-}\$'v[1+s]\}[1\text{-}t][1\text{-}d])\text{-}P/[c\text{-}q]$$

Rule-20559:

If both (**Q**), (**U**), (**$**), (**f**), (**i**), (**$'**), (**v**), (**s**), (**t**), (**d**), (**P**), (**c**) and (**q**) are known, then its Leverage or Gearing Ratio Planned is:

$$\textit{/}= \{Q+P/[c\text{-}q]\}/(U+\{\$[1\text{-}f\text{-}i]\text{-}\$'v[1+s]\}[1\text{-}t][1\text{-}d])$$

Rule-20560:

If both (**/**), (**Q**), (**$**), (**f**), (**i**), (**$'**), (**v**), (**s**), (**t**), (**d**), (**P**), (**c**) and (**q**) are known, then its Utilized or Starting Capital must be:

$$U=\{Q+P/[c\text{-}q]\}/\textit{/}\{\$[1\text{-}f\text{-}i]\text{-}\$'v[1+s]\}[1\text{-}t][1\text{-}d]$$

Rule-20561:

If both (**/**), (**U**), (**Q**), (**f**), (**i**), (**$'**), (**v**), (**s**), (**t**), (**d**), (**P**), (**c**) and (**q**) are known, then its Sales or Revenue Planned is:

$$\$= [\$'v[1+s]+(\{Q+P/[c\text{-}q]\}/\textit{/}U)/\{[1\text{-}t][1\text{-}d]\}]$$
$$/[1\text{-}f\text{-}i]$$

Steve Asikin ISBN 14: 978-1515158691, ISBN 10: 1515158691

Rule-20562:

  If both (**/**), (**U**), (**$**), (**f**), (**i**), (**$'**), (**v**), (**s**), (**t**), (**d**), (**P**),
  (**c**) and (**q**) are known, then its Quoted Longterm Debt
  Planned is:

  $$f = 1 - i - [\$'v[1+s] + (\{Q + P/[c-q]\}/\text{/-}U) / \{[1-t][1-d]\}]/\$$$

Rule-20563:

  If both (**/**), (**U**), (**$**), (**f**), (**Q**), (**$'**), (**v**), (**s**), (**t**), (**d**), (**P**),
  (**c**) and (**q**) are known, then its Interest Portion
  Planned is:

  $$i = 1 - f - [\$'v[1+s] + (\{Q + P/[c-q]\}/\text{/-}U) / \{[1-t][1-d]\}]/\$$$

Rule-20564:

  If both (**/**), (**U**), (**$**), (**f**), (**i**), (**Q**), (**v**), (**s**), (**t**), (**d**), (**P**),
  (**c**) and (**q**) are known, then its Sales Past must be:

  $$\$' = [\$[1-f-i] - (\{Q + P/[c-q]\}/\text{/-}U)/\{[1-t][1-d]\}] / \{v[1+s]\}$$

Rule-20565:

  If both (**/**), (**U**), (**$**), (**f**), (**i**), (**$'**), (**Q**), (**s**), (**t**), (**d**), (**P**),
  (**c**) and (**q**) are known, then its Variable Portion
  Planned is:

  $$v = [\$[1-f-i] - (\{Q + P/[c-q]\}/\text{/-}U)/\{[1-t][1-d]\}] / \{\$'[1+s]\}$$

Steve Asikin ISBN 14: 978-1515158691, ISBN 10: 1515158691

Rule-20566:

If both (I), (U), (S), (f), (i), (S'), (v), (Q), (t), (d), (P), (c) and (q) are known, then its Sales Growth Planned is:

$$s = S/S' - 1$$

> or it can also be found as

$$s = [S[1-f-i]-(\{Q+P/[c-q]\}/IU)/\{[1-t][1-d]\}]/[S'v]-1$$

Rule-20567:

If both (I), (U), (S), (f), (i), (S'), (v), (s), (Q), (d), (P), (c) and (q) are known, then its Tax Rate Planned is:

$$t = 1-(\{Q+P/[c-q]\}/IU)/(\{S[1-f-i]-S'v[1+s]\}[1-d])$$

Rule-20568:

If both (I), (U), (S), (f), (i), (S'), (v), (s), (t), (Q), (P), (c) and (q) are known, then its Dividend Portion Planned is:

$$d = 1-(\{Q+P/[c-q]\}/IU)/(\{S[1-f-i]-S'v[1+s]\}[1-t])$$

Rule-20569:

If both (I), (U), (S), (f), (i), (S'), (v), (s), (t), (d), (Q), (c) and (q) are known, then its Procured Inventory Planned is:

$$P = [c-q][IU+\{S[1-f-i]-S'v[1+s]\}[1-t][1-d])-Q]$$

Rule-20570:

If both (I), (U), (S), (f), (i), (S'), (v), (s), (t), (d), (P), (Q) and (q) are known, then its Current Ratio Planned is:

$$c = q-P/[IU+\{S[1-f-i]-S'v[1+s]\}[1-t][1-d])-Q]$$

Steve Asikin ISBN 14: 978-1515158691, ISBN 10: 1515158691

Rule-20571:

If both (I), $(\mathbf{U})$, $(\mathbf{S})$, $(\mathbf{f})$, $(\mathbf{i})$, $(\mathbf{S'})$, $(\mathbf{v})$, $(\mathbf{s})$, $(\mathbf{t})$, $(\mathbf{d})$, $(\mathbf{P})$, $(\mathbf{c})$ and $(\mathbf{Q})$ are known, then its Current Ratio Planned is:

$$q = c - P / [I(U + \{S[1-f-i] - S'v[1+s]\}[1-t][1-d]) - Q]$$

Rule-20572:

If both (I), $(\mathbf{U})$, $(\mathbf{S})$, $(\mathbf{f})$, $(\mathbf{i})$, $(\mathbf{S'})$, $(\mathbf{v})$, $(\mathbf{s})$, $(\mathbf{t})$, $(\mathbf{d})$, $(\mathbf{V})$, $(\mathbf{p})$, $(\mathbf{c})$ and $(\mathbf{q})$ are known, then its Quoted Longterm Debt Planned is:

$$Q = I(U + \{S[1-f-i] - S'v[1+s]\}[1-t][1-d]) - Vp / \{360[c-q]\}$$

Rule-20573:

If both $(\mathbf{Q})$, $(\mathbf{U})$, $(\mathbf{S})$, $(\mathbf{f})$, $(\mathbf{i})$, $(\mathbf{S'})$, $(\mathbf{v})$, $(\mathbf{s})$, $(\mathbf{t})$, $(\mathbf{d})$, $(\mathbf{V})$, $(\mathbf{p})$, $(\mathbf{c})$ and $(\mathbf{q})$ are known, then its Leverage or Gearing Ratio Planned is:

$$I = (Q + Vp / \{360[c-q]\}) / (U + \{S[1-f-i] - S'v[1+s]\}[1-t][1-d])$$

Rule-20574:

If both (I), $(\mathbf{U})$, $(\mathbf{S})$, $(\mathbf{f})$, $(\mathbf{i})$, $(\mathbf{S'})$, $(\mathbf{v})$, $(\mathbf{s})$, $(\mathbf{t})$, $(\mathbf{d})$, $(\mathbf{V})$, $(\mathbf{p})$, $(\mathbf{c})$ and $(\mathbf{q})$ are known, then its Quoted Longterm Debt Planned is:

$$U = (Q + Vp / \{360[c-q]\}) / I - \{S[1-f-i] - S'v[1+s]\}[1-t][1-d]$$

Steve Asikin ISBN 14: 978-1515158691, ISBN 10: 1515158691

Rule-20575:

If both (*I*), (**U**), (**Q**), (**f**), (**i**), (**S'**), (**v**), (**s**), (**t**), (**d**), (**V**), (**p**), (**c**) and (**q**) are known, then its Sales or Revenue Planned is:

$$S= \{S'v[1+s]+[(Q+Vp/\{360[c-q]\})/IU]$$
$$/\{[1-t][1-d]\}\}/[1-f-i]$$

Rule-20576:

If both (*I*), (**U**), (**S**), (**Q**), (**i**), (**S'**), (**v**), (**s**), (**t**), (**d**), (**V**), (**p**), (**c**) and (**q**) are known, then its Fixed Portion Planned is:

$$f= 1-i-\{S'v[1+s]+[(Q+Vp/\{360[c-q]\})/IU]$$
$$/\{[1-t][1-d]\}\}/S$$

Rule-20577:

If both (*I*), (**U**), (**S**), (**f**), (**Q**), (**S'**), (**v**), (**s**), (**t**), (**d**), (**V**), (**p**), (**c**) and (**q**) are known, then its Interest Portion Planned is:

$$i= 1-f-\{S'v[1+s]+[(Q+Vp/\{360[c-q]\})/IU]$$
$$/\{[1-t][1-d]\}\}/S$$

Rule-20578:

If both (*I*), (**U**), (**S**), (**f**), (**i**), (**Q**), (**v**), (**s**), (**t**), (**d**), (**V**), (**p**), (**c**) and (**q**) are known, then its Sales Past must be:

$$S'= \{S[1-f-i]-[(Q+Vp/\{360[c-q]\})/IU]$$
$$/\{[1-t][1-d]\}\}/\{v[1+s]\}$$

Steve Asikin ISBN 14: 978-1515158691, ISBN 10: 1515158691

Rule-20579:

If both (f), (U), $(\$)$, (f), (i), $(\$')$, (v), (s), (t), (d), (V), (p), (c) and (q) are known, then its Quoted Longterm Debt Planned is:

$v= V/\$$

or it can also be found as

$v= \{\$[1\text{-}f\text{-}i]\text{-}[(Q+Vp/\{360[c\text{-}q]\})/fU]$
$\quad /\{[1\text{-}t][1\text{-}d]\}\}/\{\$'[1+s]\}$

Rule-20580:

If both (f), (U), $(\$)$, (f), (i), $(\$')$, (v), (Q), (t), (d), (V), (p), (c) and (q) are known, then its Sales Growth Planned is:

$s= \$/\$'\text{-}1$

or it can also be found as

$s= \{\$[1\text{-}f\text{-}i]\text{-}[(Q+Vp/\{360[c\text{-}q]\})/fU]$
$\quad /\{[1\text{-}t][1\text{-}d]\}\}/[\$'v]\text{-}1$

Rule-20581:

If both (f), (U), $(\$)$, (f), (i), $(\$')$, (v), (s), (Q), (d), (V), (p), (c) and (q) are known, then its Tax Rate Planned is:

$t= 1\text{-}\{[(Q+Vp/\{360[c\text{-}q]\})/fU]$
$\quad / (\{\$[1\text{-}f\text{-}i]\text{-}\$'v[1+s]\}[1\text{-}t]\}$

Rule-20582:

If both (f), (U), $(\$)$, (f), (i), $(\$')$, (v), (s), (t), (Q), (V), (p), (c) and (q) are known, then its Dividend Payout Planned is:

$d= 1\text{-}\{[(Q+Vp/\{360[c\text{-}q]\})/fU]$
$\quad / (\{\$[1\text{-}f\text{-}i]\text{-}\$'v[1+s]\}[1\text{-}t]\}$

Steve Asikin ISBN 14: 978-1515158691, ISBN 10: 1515158691

Rule-20583:

If both (**/**), (**U**), (**$**), (**f**), (**i**), (**$'**), (**v**), (**s**), (**t**), (**d**), (**Q**), (**p**), (**c**) and (**q**) are known, then its Variable Cost Planned is:

$V = 360[c-q]$

$[(U+\{\$[1-f-i]-\$'v[1+s]\}[1-t][1-d])-Q]/p$

Rule-20584:

If both (**/**), (**U**), (**$**), (**f**), (**i**), (**$'**), (**v**), (**s**), (**t**), (**d**), (**V**), (**Q**), (**c**) and (**q**) are known, then its Procured Inventory Days Planned:

$p = 360[c-q]$

$[(U+\{\$[1-f-i]-\$'v[1+s]\}[1-t][1-d])-Q]/V$

Rule-20585:

If both (**/**), (**U**), (**$**), (**f**), (**i**), (**$'**), (**v**), (**s**), (**t**), (**d**), (**V**), (**p**), (**Q**) and (**q**) are known, then its Current Ratio Planned is:

$c = q - Vp/\{360[(U+\{\$[1-f-i]-\$'v[1+s]\}$
$[1-t][1-d])-Q]\}$

Rule-20586:

If both (**/**), (**U**), (**$**), (**f**), (**i**), (**$'**), (**v**), (**s**), (**t**), (**d**), (**V**), (**p**), (**c**) and (**Q**) are known, then its Quick or Acid Test Ratio Planned is:

$q = c - Vp/\{360[(U+\{\$[1-f-i]-\$'v[1+s]\}$
$[1-t][1-d])-Q]\}$

Steve Asikin ISBN 14: 978-1515158691, ISBN 10: 1515158691

Rule-20587:

If both (**/**), (**U**), (**\$**), (**f**), (**i**), (**\$'**), (**v**), (**s**), (**t**), (**d**), (**p**), (**c**) and (**q**) are known, then its Quoted Longterm Debt Planned is:

$$Q= /(U+\{\$[1-f-i]-\$'v[1+s]\}[1-t][1-d])$$
$$-\$vp/\{360[c-q]\}$$

Rule-20588:

If both (**Q**), (**U**), (**\$**), (**f**), (**i**), (**\$'**), (**v**), (**s**), (**t**), (**d**), (**p**), (**c**) and (**q**) are known, then its Leverage or Gearing Ratio Planned is:

$$/= (Q+\$vp/\{360[c-q]\})$$
$$/(U+\{\$[1-f-i]-\$'v[1+s]\}[1-t][1-d])$$

Rule-20589:

If both (**/**), (**Q**), (**\$**), (**f**), (**i**), (**\$'**), (**v**), (**s**), (**t**), (**d**), (**p**), (**c**) and (**q**) are known, then its Utilized or Starting Capital must be:

$$U=(Q+\$vp/\{360[c-q]\})//$$
$$-\{\$[1-f-i]-\$'v[1+s]\}[1-t][1-d]$$

Rule-20590:

If both (**/**), (**U**), (**Q**), (**f**), (**i**), (**\$'**), (**v**), (**s**), (**t**), (**d**), (**p**), (**c**) and (**q**) are known, then its Sales or Revenue Planned is:

$$\$= (Q-/\{U-\$'v[1+s][1-t][1-d]\}+\$vp/\{360[c-q]\})$$
$$/\{/[1-f-i][1-t][1-d]\})$$

Steve Asikin ISBN 14: 978-1515158691, ISBN 10: 1515158691

<u>Rule-20591</u>:

If both (**/**), (**U**), (**$**), (**Q**), (**i**), (**$'**), (**v**), (**s**), (**t**), (**d**), (**p**), (**c**) and (**q**) are known, then its Fixed Portion Planned is:

$$f= 1\text{-}i\text{-}\{\$'v[1+s]+[(Q+\$vp/\{360[c\text{-}q]\})/\!\!/U]$$
$$/\{[1\text{-}t][1\text{-}d]\}\}/\$$$

<u>Rule-20592</u>:

If both (**/**), (**U**), (**$**), (**f**), (**Q**), (**$'**), (**v**), (**s**), (**t**), (**d**), (**p**), (**c**) and (**q**) are known, then its Interest Portion Planned is:

$$i= 1\text{-}f\text{-}\{\$'v[1+s]+[(Q+\$vp/\{360[c\text{-}q]\})/\!\!/U]$$
$$/\{[1\text{-}t][1\text{-}d]\}\}/\$$$

<u>Rule-20593</u>:

If both (**/**), (**U**), (**$**), (**f**), (**i**), (**Q**), (**v**), (**s**), (**t**), (**d**), (**p**), (**c**) and (**q**) are known, then its Sales Past must be:

$$\$'= \{\$[1\text{-}f\text{-}i]\text{-}[(Q+\$vp/\{360[c\text{-}q]\})/\!\!/U]$$
$$/\{[1\text{-}t][1\text{-}d]\}\}/\{v[1+s]\}$$

<u>Rule-20594</u>:

If both (**/**), (**U**), (**$**), (**f**), (**i**), (**$'**), (**Q**), (**s**), (**t**), (**d**), (**p**), (**c**) and (**q**) are known, then its Variable Portion Planned is:

$$v= (/\{U+\$[1\text{-}f\text{-}i][1\text{-}t][1\text{-}d]\}\text{-}Q)$$
$$/(\$p/\{360[c\text{-}q]\}+\$'/[1+s]\,[1\text{-}t][1\text{-}d])$$

Steve Asikin ISBN 14: 978-1515158691, ISBN 10: 1515158691

Rule-20595:

If both (**/**), (**U**), (**$**), (**f**), (**i**), (**$'**), (**v**), (**Q**), (**t**), (**d**), (**p**), (**c**) and (**q**) are known, then its Sales Growth Planned is:

$s = \$/\$' - 1$

or it can also be found as

$s = \{\$[1\text{-}f\text{-}i]\text{-}[(Q+\$vp/\{360[c\text{-}q]\})/\text{\textit{f}}U]$
$/\{[1\text{-}t][1\text{-}d]\}\}/[\$'v]\text{-}1$

Rule-20596:

If both (**/**), (**U**), (**$**), (**f**), (**i**), (**$'**), (**v**), (**s**), (**Q**), (**d**), (**p**), (**c**) and (**q**) are known, then its Tax Rate Planned is:

$t = 1\text{-}\{[(Q+\$vp/\{360[c\text{-}q]\})/\text{\textit{f}}U]$
$/(\{\$[1\text{-}f\text{-}i]\text{-}\$'v[1+s]\}[1\text{-}t]\}$

Rule-20597:

If both (**/**), (**U**), (**$**), (**f**), (**i**), (**$'**), (**v**), (**s**), (**t**), (**Q**), (**p**), (**c**) and (**q**) are known, then its Dividend Payout Planned is:

$d = 1\text{-}\{[(Q+\$vp/\{360[c\text{-}q]\})/\text{\textit{f}}U]$
$/(\{\$[1\text{-}f\text{-}i]\text{-}\$'v[1+s]\}[1\text{-}t]\}$

Rule-20598:

If both (**/**), (**U**), (**$**), (**f**), (**i**), (**$'**), (**v**), (**s**), (**t**), (**d**), (**Q**), (**c**) and (**q**) are known, then its Procured Inventory Days Planned:

$p = 360[c\text{-}q][\text{\textit{f}}U+\{\$[1\text{-}f\text{-}i]\text{-}\$'v[1+s]\}$
$[1\text{-}t][1\text{-}d])\text{-}Q]/[\$v]$

Steve Asikin ISBN 14: 978-1515158691, ISBN 10: 1515158691

Rule-20599:

   If both (I), (U), (S), (f), (i), (S'), (v), (s), (t), (d), (p), (Q) and (q) are known, then its Current Ratio Planned is:

$$c = q - Svp/\{360[I U + \{S[1-f-i]-S'v[1+s]\}$$
$$[1-t][1-d])-Q]\}$$

Rule-20600:

   If both (I), (U), (S), (f), (i), (S'), (v), (s), (t), (d), (p), (c) and (Q) are known, then its Quick or Acid Test Ratio Planned is:

$$q = c - Svp/\{360[I U + \{S[1-f-i]-S'v[1+s]\}$$
$$[1-t][1-d])-Q]\}$$

Rule-20601:

   If both (I), (U), (S), (f), (i), (S'), (v), (s), (t), (d), (p), (c) and (q) are known, then its Quoted Longterm Debt Planned is:

$$Q = I U + \{S[1-f-i]-S'v[1+s]\}[1-t][1-d])$$
$$-S'vp[1+s]/\{360[c-q]\}$$

Rule-20602:

   If both (Q), (U), (S), (f), (i), (S'), (v), (s), (t), (d), (p), (c) and (q) are known, then its Leverage or Gearing Ratio Planned is:

$$I = (Q+S'vp[1+s]/\{360[c-q]\})$$
$$/(U+\{S[1-f-i]-S'v[1+s]\}[1-t][1-d])$$

Steve Asikin ISBN 14: 978-1515158691, ISBN 10: 1515158691

Rule-20603:
> If both (I), (Q), (S), (f), (i), (S'), (v), (s), (t), (d), (p), (c) and (q) are known, then its Utilized or Starting Capital must be:
>
> $$U=(Q+S'vp[1+s]/\{360[c-q]\})/I$$
> $$-\{S[1-f-i]-S'v[1+s]\}[1-t][1-d]$$

Rule-20604:
> If both (I), (U), (S), (f), (i), (S'), (v), (s), (t), (d), (p), (c) and (q) are known, then its Quoted Longterm Debt Planned is:
>
> $$S= (Q= IU+\{S[1-f-i]-S'v[1+s]\}[1-t][1-d])$$
> $$-S'vp[1+s]/\{360[c-q]\}$$

Rule-20605:
> If both (I), (U), (S), (Q), (i), (S'), (v), (s), (t), (d), (p), (c) and (q) are known, then its Fixed Portion Planned is:
>
> $$f= 1-i-\{S'v[1+s]+[(Q+S'vp[1+s]/\{360[c-q]\})/IU]$$
> $$/\{[1-t][1-d]\}\}/S$$

Rule-20606:
> If both (I), (U), (S), (f), (Q), (S'), (v), (s), (t), (d), (p), (c) and (q) are known, then its Interest Portion Planned is:
>
> $$i= 1-f-\{S'v[1+s]+[(Q+S'vp[1+s]/\{360[c-q]\})/IU]$$
> $$/\{[1-t][1-d]\}\}/S$$

Steve Asikin ISBN 14: 978-1515158691, ISBN 10: 1515158691

Rule-20607:

If both (**∫**), (**U**), (**$**), (**f**), (**i**), (**Q**), (**v**), (**s**), (**t**), (**d**), (**p**), (**c**) and (**q**) are known, then its Sales Past must be:

$$\text{\$'}= (\int\{U+\$[1\text{-}f\text{-}i][1\text{-}t][1\text{-}d]\}\text{-}Q)$$
$$/[v[1+s](p/\{360[c\text{-}q]\}+\int[1\text{-}t][1\text{-}d])]$$

Rule-20608:

If both (**∫**), (**U**), (**$**), (**f**), (**i**), (**$'**), (**Q**), (**s**), (**t**), (**d**), (**p**), (**c**) and (**q**) are known, then its Variable Portion Planned is:

$$v= (\int\{U+\$[1\text{-}f\text{-}i][1\text{-}t][1\text{-}d]\}\text{-}Q)$$
$$/[\text{\$'}[1+s](p/\{360[c\text{-}q]\}+\int[1\text{-}t][1\text{-}d])]$$

Rule-20609:

If both (**∫**), (**U**), (**$**), (**f**), (**i**), (**$'**), (**v**), (**Q**), (**t**), (**d**), (**p**), (**c**) and (**q**) are known, then its Sales Growth Planned is:

$$s= \$/\text{\$'}\text{-}1$$

or it can also be found as

$$s= (\int\{U+\$[1\text{-}f\text{-}i][1\text{-}t][1\text{-}d]\}\text{-}Q)$$
$$/[\text{\$'}v(p/\{360[c\text{-}q]\}+\int[1\text{-}t][1\text{-}d])]\text{-}1$$

Rule-20610:

If both (**∫**), (**U**), (**$**), (**f**), (**i**), (**$'**), (**v**), (**s**), (**Q**), (**d**), (**p**), (**c**) and (**q**) are known, then its Tax Rate Planned is:

$$t= 1\text{-}\{[(Q+\text{\$'}vp[1+s]/\{360[c\text{-}q]\})/\int U]$$
$$/(\{\$[1\text{-}f\text{-}i]\text{-}\text{\$'}v[1+s]\}[1\text{-}t]\}$$

Steve Asikin ISBN 14: 978-1515158691, ISBN 10: 1515158691

Rule-20611:

If both (f), (**U**), (**$**), (**f**), (**i**), (**$'**), (**v**), (**s**), (**t**), (**Q**), (**p**), (**c**) and (**q**) are known, then its Dividend Payout Planned is:

$$d = 1 - \{[(Q + \$'vp[1+s]/\{360[c-q]\})/fU]$$
$$/(\{\$[1-f-i]-\$'v[1+s]\}[1-t]\}$$

Rule-20612:

If both (f), (**U**), (**$**), (**f**), (**i**), (**$'**), (**v**), (**s**), (**t**), (**d**), (**Q**), (**c**) and (**q**) are known, then its Procured Inventory Days Planned:

$$p = 360[c-q][fU + \{\$[1-f-i]-\$'v[1+s]\}[1-t][1-d])-Q]$$
$$/\{\$'v[1+s]\}$$

Rule-20613:

If both (f), (**U**), (**$**), (**f**), (**i**), (**$'**), (**v**), (**s**), (**t**), (**d**), (**p**), (**Q**) and (**q**) are known, then its Current Ratio Planned is:

$$c = q - \$'vp[1+s]/\{360[fU + \{\$[1-f-i]-\$'v[1+s]\}$$
$$[1-t][1-d])-Q]\}$$

Rule-20614:

If both (f), (**U**), (**$**), (**f**), (**i**), (**$'**), (**v**), (**s**), (**t**), (**d**), (**p**), (**c**) and (**Q**) are known, then its Quick or Acid Test Ratio Planned is:

$$q = c - \$'vp[1+s]/\{360[fU + \{\$[1-f-i]-\$'v[1+s]\}$$
$$[1-t][1-d])-Q]\}$$

Steve Asikin ISBN 14: 978-1515158691, ISBN 10: 1515158691

Rule-20615:

If both (I), (U), (S), (f), (S'), (s), (v), (i), (T), (D) and (X) are known, then its Quoted Longterm Debt Planned is:

$$Q= I\{U+S[1-f]-S'[1+s][v+i]-T-D\}-X$$

Rule-20616:

If both (Q), (U), (S), (f), (S'), (s), (v), (i), (T), (D) and (X) are known, then its Leverage or Gearing Ratio Planned is:

$$I= [Q+X]/\{U+S[1-f]-S'[1+s][v+i]-T-D\}$$

Rule-20617:

If both (I), (Q), (S), (f), (S'), (s), (v), (i), (T), (D) and (X) are known, then its Utilized or Starting Capital must be:

$$U= T+D+[Q+X]/I-S[1-f]-S'[1+s][v+i]$$

Rule-20618:

If both (I), (U), (Q), (f), (S'), (s), (v), (i), (T), (D) and (X) are known, then its Sales or Revenue Planned is:

$$S= \{S'[1+s][v+i]+T+D+[Q+X]/I-U\}/[1-f]$$

Rule-20619:

If both (I), (U), (S), (Q), (S'), (s), (v), (i), (T), (D) and (X) are known, then its Fixed Portion Planned is:

$$f= 1-\{S'[1+s][v+i]+T+D+[Q+X]/I-U\}/S$$

Steve Asikin ISBN 14: 978-1515158691, ISBN 10: 1515158691

Rule-20620:

If both (**/**), (**U**), (**$**), (**f**), (**Q**), (**s**), (**v**), (**i**), (**T**), (**D**) and (**X**) are known, then its Sales Past must be:

$$\text{S'}= \{U+\$[1-f]-T-D-[Q+X]/\text{/}/\{[1+s][v+i]\}$$

Rule-20621:

If both (**/**), (**U**), (**$**), (**f**), (**$'**), (**Q**), (**v**), (**i**), (**T**), (**D**) and (**X**) are known, then its Sales Growth Planned is:

$$s= \$/\$'-1$$

or it can also be found as

$$s= \{U+\$[1-f]-T-D-[Q+X]/\text{/}/\{\$'[v+i]\}-1$$

Rule-20622:

If both (**/**), (**U**), (**$**), (**f**), (**$'**), (**s**), (**Q**), (**i**), (**T**), (**D**) and (**X**) are known, then its Variable Portion Planned is:

$$v= \{U+\$[1-f]-T-D-[Q+X]/\text{/}/\{\$'[1+s]\}-i$$

Rule-20623:

If both (**/**), (**U**), (**$**), (**f**), (**$'**), (**s**), (**v**), (**Q**), (**T**), (**D**) and (**X**) are known, then its Interest Portion Planned is:

$$i= \{U+\$[1-f]-T-D-[Q+X]/\text{/}/\{\$'[1+s]\}-v$$

Rule-20624:

If both (**/**), (**U**), (**$**), (**f**), (**$'**), (**s**), (**v**), (**i**), (**Q**), (**D**) and (**X**) are known, then its Tax Planned is:

$$T= U+\$[1-f]-\$'[1+s][v+i]-D-[Q+X]/\text{/}$$

Rule-20625:

If both (**/**), (**U**), (**$**), (**f**), (**$'**), (**s**), (**v**), (**i**), (**T**), (**Q**) and (**X**) are known, then its Dividend Planned is:

$$D= U+\$[1-f]-\$'[1+s][v+i]-T-[Q+X]/\text{/}$$

Steve Asikin ISBN 14: 978-1515158691, ISBN 10: 1515158691

Rule-20626:

If both (I), (U), (S), (f), (S'), (s), (v), (i), (T), (D) and (Q) are known, then its Xpress or Current Debt Planned is:

$$X= I\{U+S[1-f]-S'[1+s][v+i]-T-D\}-Q$$

Rule-20627:

If both (I), (U), (S), (f), (S'), (s), (v), (i), (T), (D), (P), (c) and (q) are known, then its Quoted Longterm Debt Planned is:

$$Q= I\{U+S[1-f]-S'[1+s][v+i]-T-D\}-P/[c-q]$$

Rule-20628:

If both (Q), (U), (S), (f), (S'), (s), (v), (i), (T), (D), (P), (c) and (q) are known, then its Leverage or Gearing Ratio Planned is:

$$I= \{Q+P/[c-q]\}/\{U+S[1-f]-S'[1+s][v+i]-T-D\}$$

Rule-20629:

If both (I), (Q), (S), (f), (S'), (s), (v), (i), (T), (D), (P), (c) and (q) are known, then its Utilized or Starting Capital must be:

$$U= T+D+\{Q+P/[c-q]\}/I-S[1-f]-S'[1+s][v+i]$$

Rule-20630:

If both (I), (U), (Q), (f), (S'), (s), (v), (i), (T), (D), (P), (c) and (q) are known, then its Sales or Revenue Planned is:

$$S= (S'[1+s][v+i]+T+D+\{Q+P/[c-q]\}/I-U)/[1-f]$$

Steve Asikin ISBN 14: 978-1515158691, ISBN 10: 1515158691

<u>Rule-20631</u>:

If both (I), (**U**), (**$**), (**Q**), (**$'**), (**s**), (**v**), (**i**), (**T**), (**D**), (**P**), (**c**) and (**q**) are known, then its Fixed Portion Planned is:

$$f= 1-(\$'[1+s][v+i]+T+D+\{Q+P/[c-q]\}/\mathit{I}\text{-}U)/\$$$

<u>Rule-20632</u>:

If both (I), (**U**), (**$**), (**f**), (**Q**), (**s**), (**v**), (**i**), (**T**), (**D**), (**P**), (**c**) and (**q**) are known, then its Sales Past must be:

$$\$'= (U+\$[1\text{-}f]\text{-}T\text{-}D\text{-}\{Q+P/[c\text{-}q]\}/\mathit{I})/\{[1+s][v+i]\}$$

<u>Rule-20633</u>:

If both (I), (**U**), (**$**), (**f**), (**$'**), (**Q**), (**v**), (**i**), (**T**), (**D**), (**P**), (**c**) and (**q**) are known, then its Sales Growth Planned is:

$$s= \$/\$'\text{-}1$$

or it can also be found as

$$s= (U+\$[1\text{-}f]\text{-}T\text{-}D\text{-}\{Q+P/[c\text{-}q]\}/\mathit{I})/\{\$'[v+i]\}\text{-}1$$

<u>Rule-20634</u>:

If both (I), (**U**), (**$**), (**f**), (**$'**), (**s**), (**Q**), (**i**), (**T**), (**D**), (**P**), (**c**) and (**q**) are known, then its Variable Portion Planned is:

$$v= (U+\$[1\text{-}f]\text{-}T\text{-}D\text{-}\{Q+P/[c\text{-}q]\}/\mathit{I})/\{\$'[1+s]\}\text{-}i$$

<u>Rule-20635</u>:

If both (I), (**U**), (**$**), (**f**), (**$'**), (**s**), (**v**), (**Q**), (**T**), (**D**), (**P**), (**c**) and (**q**) are known, then its Interest Portion Planned is:

$$i= (U+\$[1\text{-}f]\text{-}T\text{-}D\text{-}\{Q+P/[c\text{-}q]\}/\mathit{I})/\{\$'[1+s]\}\text{-}v$$

Steve Asikin ISBN 14: 978-1515158691, ISBN 10: 1515158691

Rule-20636:

If both (I, (**U**), (**$**), (**f**), (**$'**), (**s**), (**v**), (**i**), (**Q**), (**D**), (**P**), (**c**) and (**q**) are known, then its Tax Planned is:

$$T= U+\$[1-f]-\$'[1+s][v+i]-D-\{Q+P/[c-q]\}/I$$

Rule-20637:

If both (I, (**U**), (**$**), (**f**), (**$'**), (**s**), (**v**), (**i**), (**T**), (**Q**), (**P**), (**c**) and (**q**) are known, then its Dividend Planned is:

$$D= U+\$[1-f]-\$'[1+s][v+i]-T-\{Q+P/[c-q]\}/I$$

Rule-20638:

If both (I, (**U**), (**$**), (**f**), (**$'**), (**s**), (**v**), (**i**), (**T**), (**D**), (**Q**), (**c**) and (**q**) are known, then its Procured Inventory Planned is:

$$P= [c-q](I\{U+\$[1-f]-\$'[1+s][v+i]-T-D\}-Q)$$

Rule-20639:

If both (I, (**U**), (**$**), (**f**), (**$'**), (**s**), (**v**), (**i**), (**T**), (**D**), (**P**), (**Q**) and (**q**) are known, then its Current Ratio Planned is:

$$c= q+P/(I\{U+\$[1-f]-\$'[1+s][v+i]-T-D\}-Q)$$

Rule-20640:

If both (I, (**U**), (**$**), (**f**), (**$'**), (**s**), (**v**), (**i**), (**T**), (**D**), (**P**), (**c**) and (**Q**) are known, then its Quick or Acid Test Ratio Planned is:

$$q= c-P/(I\{U+\$[1-f]-\$'[1+s][v+i]-T-D\}-Q)$$

Steve Asikin ISBN 14: 978-1515158691, ISBN 10: 1515158691

Rule-20641:

If both (**/**), (**U**), (**\$**), (**f**), (**\$'**), (**s**), (**v**), (**i**), (**T**), (**D**), (**V**), (**p**), (**c**) and (**q**) are known, then its Quoted Longterm Debt Planned is:

$$Q = /\{U + \$[1-f] - \$'[1+s][v+i] - T - D\} - Vp/\{360[c-q]\}$$

Rule-20642:

If both (**Q**), (**U**), (**\$**), (**f**), (**\$'**), (**s**), (**v**), (**i**), (**T**), (**D**), (**V**), (**p**), (**c**) and (**q**) are known, then its Leverage or Gearing Ratio Planned is:

$$/= (Q + Vp/\{360[c-q]\})$$
$$/\{U + \$[1-f] - \$'[1+s][v+i] - T - D\}$$

Rule-20643:

If both (**/**), (**Q**), (**\$**), (**f**), (**\$'**), (**s**), (**v**), (**i**), (**T**), (**D**), (**V**), (**p**), (**c**) and (**q**) are known, then its Utilized or Starting Capital must be:

$$U = T + D + (Q + Vp/\{360[c-q]\})//\$[1-f] - \$'[1+s][v+i]$$

Rule-20644:

If both (**/**), (**U**), (**Q**), (**f**), (**\$'**), (**s**), (**v**), (**i**), (**T**), (**D**), (**V**), (**p**), (**c**) and (**q**) are known, then its Sales or Revenue Planned is:

$$S = [\$'[1+s][v+i] + T + D$$
$$+ (Q + Vp/\{360[c-q]\})//U]/[1-f]$$

Steve Asikin ISBN 14: 978-1515158691, ISBN 10: 1515158691

Rule-20645:

If both **(/)**, **(U)**, **($)**, **(Q)**, **($')**, **(s)**, **(v)**, **(i)**, **(T)**, **(D)**, **(V)**, **(p)**, **(c)** and **(q)** are known, then its Fixed Portion Planned is:

$$f= 1-[\mathbf{\$'}[1+\mathbf{s}][\mathbf{v+i}]+\mathbf{T+D}$$
$$+(\mathbf{Q+Vp}/\{360[\mathbf{c-q}]\})/\mathbf{/U}]/\mathbf{\$}$$

Rule-20646:

If both **(/)**, **(U)**, **($)**, **(f)**, **(Q)**, **(s)**, **(v)**, **(i)**, **(T)**, **(D)**, **(V)**, **(p)**, **(c)** and **(q)** are known, then its Sales Past must be:

$$\mathbf{\$'}= [\mathbf{U+\$}[1-\mathbf{f}]-\mathbf{T-D}-(\mathbf{Q+Vp}/\{360[\mathbf{c-q}]\})/\mathbf{/}]$$
$$/\{[1+\mathbf{s}][\mathbf{v+i}]\}$$

Rule-20647:

If both **(/)**, **(U)**, **($)**, **(f)**, **($')**, **(Q)**, **(v)**, **(i)**, **(T)**, **(D)**, **(V)**, **(p)**, **(c)** and **(q)** are known, then its Sales Growth Planned is:

$$s= \mathbf{\$}/\mathbf{\$'}-1$$

or it can also be found as

$$s= [\mathbf{U+\$}[1-\mathbf{f}]-\mathbf{T-D}-(\mathbf{Q+Vp}/\{360[\mathbf{c-q}]\})/\mathbf{/}]$$
$$/\{\mathbf{\$'}[\mathbf{v+i}]\}-1$$

Rule-20648:

If both **(/)**, **(U)**, **($)**, **(f)**, **($')**, **(s)**, **(Q)**, **(i)**, **(T)**, **(D)**, **(V)**, **(p)**, **(c)** and **(q)** are known, then its Variable Portion Planned is:

$$v= \mathbf{V}/\mathbf{\$}$$

or it can also be found as

$$v= [\mathbf{U+\$}[1-\mathbf{f}]-\mathbf{T-D}-(\mathbf{Q+Vp}/\{360[\mathbf{c-q}]\})/\mathbf{/}]$$
$$/\{\mathbf{\$'}[1+\mathbf{s}]\}-\mathbf{i}$$

Steve Asikin ISBN 14: 978-1515158691, ISBN 10: 1515158691

Rule-20649:

If both (I), (U), (S), (f), (S'), (s), (v), (Q), (T), (D), (V), (p), (c) and (q) are known, then its Interest Portion Planned is:

$$i = [U+S[1-f]-T-D-(Q+Vp/\{360[c-q]\})/I]$$
$$/\{S'[1+s]\}-v$$

Rule-20650:

If both (I), (U), (S), (f), (S'), (s), (v), (i), (Q), (D), (V), (p), (c) and (q) are known, then its Tax Planned is:

$$T = U+S[1-f]-S'[1+s][v+i]-D-(Q+Vp/\{360[c-q]\})/I$$

Rule-20651:

If both (I), (U), (S), (f), (S'), (s), (v), (i), (T), (Q), (V), (p), (c) and (q) are known, then its Dividend Planned is:

$$D = U+S[1-f]-S'[1+s][v+i]-T-(Q+Vp/\{360[c-q]\})/I$$

Rule-20652:

If both (I), (U), (S), (f), (S'), (s), (v), (i), (T), (D), (Q), (p), (c) and (q) are known, then its Variable Cost Planned is:

$$V = 360[c-q](I\{U+S[1-f]-S'[1+s][v+i]-T-D\}-Q)/p$$

Rule-20653:

If both (I), (U), (S), (f), (S'), (s), (v), (i), (T), (D), (V), (Q), (c) and (q) are known, then its Procured Inventory Days Planned:

$$p = 360[c-q](I\{U+S[1-f]-S'[1+s][v+i]-T-D\}-Q)/V$$

Steve Asikin ISBN 14: 978-1515158691, ISBN 10: 1515158691

Rule-20654:

If both (I), (U), (S), (f), (S'), (s), (v), (i), (T), (D), (V), (p), (Q) and (q) are known, then its Current Ratio Planned is:

$$c = q + Vp/[360(I\{U + S[1-f] - S'[1+s][v+i] - T - D\} - Q)]$$

Rule-20655:

If both (I), (U), (S), (f), (S'), (s), (v), (i), (T), (D), (V), (p), (c) and (Q) are known, then its Quick or Acid Test Ratio Planned is:

$$q = c - Vp/[360(I\{U + S[1-f] - S'[1+s][v+i] - T - D\} - Q)]$$

Rule-20656:

If both (I), (U), (S), (f), (S'), (s), (v), (i), (T), (D), (p), (c) and (q) are known, then its Quoted Longterm Debt Planned is:

$$Q = I\{U + S[1-f] - S'[1+s][v+i] - T - D\} - Svp/\{360[c-q]\}$$

Rule-20657:

If both (Q), (U), (S), (f), (S'), (s), (v), (i), (T), (D), (p), (c) and (q) are known, then its Leverage or Gearing Ratio Planned is:

$$I = (Q + Svp/\{360[c-q]\})$$
$$/\{U + S[1-f] - S'[1+s][v+i] - T - D\}$$

Rule-20658:

If both (I), (Q), (S), (f), (S'), (s), (v), (i), (T), (D), (p), (c) and (q) are known, then its Utilized or Starting Capital must be:

$$U = T + D + (Q + Svp/\{360[c-q]\})/I - S[1-f] - S'[1+s][v+i]$$

Steve Asikin ISBN 14: 978-1515158691, ISBN 10: 1515158691

Rule-20659:
> If both (**/**), (**U**), (**Q**), (**f**), (**S'**), (**s**), (**v**), (**i**), (**T**), (**D**), (**p**), (**c**) and (**q**) are known, then its Sales or Revenue Planned is:
>
> $S= (\ /\{U-S'[1+s][v+i]-T-D\}-Q)$
> $/(vp/\{360[c-q]\}-/[1-f])$

Rule-20660:
> If both (**/**), (**U**), (**S**), (**Q**), (**S'**), (**s**), (**v**), (**i**), (**T**), (**D**), (**p**), (**c**) and (**q**) are known, then its Fixed Portion Planned is:
>
> $f= 1-[S'[1+s][v+i]+T+D$
> $+(Q+Svp/\{360[c-q]\})/\textbf{/}U]/S$

Rule-20661:
> If both (**/**), (**U**), (**S**), (**f**), (**Q**), (**s**), (**v**), (**i**), (**T**), (**D**), (**p**), (**c**) and (**q**) are known, then its Sales Past must be:
>
> $S'= [U+S[1-f]-T-D-(Q+Svp/\{360[c-q]\})/\textbf{/}]$
> $/\{[1+s][v+i]\}$

Rule-20662:
> If both (**/**), (**U**), (**S**), (**f**), (**S'**), (**Q**), (**v**), (**i**), (**T**), (**D**), (**p**), (**c**) and (**q**) are known, then its Sales Growth Planned is:
>
> $s= S/S'-1$
>
> or it can also be found as
>
> $s= [U+S[1-f]-T-D-(Q+Svp/\{360[c-q]\})/\textbf{/}]$
> $/\{S'[v+i]\}-1$

Steve Asikin ISBN 14: 978-1515158691, ISBN 10: 1515158691

Rule-20663:

   If both (**/**), (**U**), (**$**), (**f**), (**$'**), (**s**), (**Q**), (**i**), (**T**), (**D**), (**p**), (**c**) and (**q**) are known, then its Variable Portion Planned is:

$$v = (\text{/}\{U+\$[1-f]-\$'i[1+s]-T-D\}-Q)$$
$$/(\$p/\{360[c-q]\}+\$'[1+s])$$

Rule-20664:

   If both (**/**), (**U**), (**$**), (**f**), (**$'**), (**s**), (**v**), (**Q**), (**T**), (**D**), (**p**), (**c**) and (**q**) are known, then its Interest Portion Planned is:

$$i = [U+\$[1-f]-T-D-(Q+\$vp/\{360[c-q]\})/\text{/}]$$
$$/\{\$'[1+s]\}-v$$

Rule-20665:

   If both (**/**), (**U**), (**$**), (**f**), (**$'**), (**s**), (**v**), (**i**), (**Q**), (**D**), (**p**), (**c**) and (**q**) are known, then its Tax Planned is:

$$T = U+\$[1-f]-\$'[1+s][v+i]-D-(Q+\$vp/\{360[c-q]\})/\text{/}$$

Rule-20666:

   If both (**/**), (**U**), (**$**), (**f**), (**$'**), (**s**), (**v**), (**i**), (**T**), (**Q**), (**p**), (**c**) and (**q**) are known, then its Dividend Planned is:

$$D = U+\$[1-f]-\$'[1+s][v+i]-T-(Q+\$vp/\{360[c-q]\})/\text{/}$$

Rule-20667:

   If both (**/**), (**U**), (**$**), (**f**), (**$'**), (**s**), (**v**), (**i**), (**T**), (**D**), (**Q**), (**c**) and (**q**) are known, then its Procured Inventory Days Planned:

$$p = 360[c-q](\text{/}\{U+\$[1-f]-\$'[1+s][v+i]-T-D\}-Q)/[\$v]$$

Steve Asikin ISBN 14: 978-1515158691, ISBN 10: 1515158691

Rule-20668:

If both (l), (U), (S), (f), (S'), (s), (v), (i), (T), (D), (p), (Q) and (q) are known, then its Current Ratio Planned is:

$$c = q + Svp/[360(l\{U+S[1-f]-S'[1+s][v+i]-T-D\}-Q)]$$

Rule-20669:

If both (l), (U), (S), (f), (S'), (s), (v), (i), (T), (D), (p), (c) and (Q) are known, then its Quick or Acid Test Ratio Planned is:

$$q = c - Svp/[360(l\{U+S[1-f]-S'[1+s][v+i]-T-D\}-Q)]$$

Rule-20670:

If both (l), (U), (S), (f), (S'), (s), (v), (i), (T), (D), (p), (c) and (q) are known, then its Quoted Longterm Debt Planned is:

$$Q = l\{U+S[1-f]-S'[1+s][v+i]-T-D\}$$
$$-S'vp[1+s]/\{360[c-q]\}$$

Rule-20671:

If both (Q), (U), (S), (f), (S'), (s), (v), (i), (T), (D), (p), (c) and (q) are known, then its Leverage or Gearing Ratio Planned is:

$$l = (Q + S'vp[1+s]/\{360[c-q]\})$$
$$/\{U+S[1-f]-S'[1+s][v+i]-T-D\}$$

Steve Asikin ISBN 14: 978-1515158691, ISBN 10: 1515158691

Rule-20672:

If both (I), (Q), (S), (f), (S'), (s), (v), (i), (T), (D), (p), (c) and (q) are known, then its Utilized or Starting Capital must be:

$$U = T+D+(Q+S'vp[1+s]/\{360[c-q]\})/I$$
$$-S[1-f]-S'[1+s][v+i]$$

Rule-20673:

If both (I), (U), (Q), (f), (S'), (s), (v), (i), (T), (D), (p), (c) and (q) are known, then its Sales or Revenue Planned is:

$$S = (Q-I\{U-S'[1+s][v+i]-T-D\}$$
$$+S'vp[1+s]/\{360[c-q]\})/\{I[1-f]\}$$

Rule-20674:

If both (I), (U), (S), (Q), (S'), (s), (v), (i), (T), (D), (p), (c) and (q) are known, then its Fixed Portion Planned is:

$$f = 1-[S'[1+s][v+i]+T+D$$
$$+(Q+S'vp[1+s]/\{360[c-q]\})/IU]/S$$

Rule-20675:

If both (I), (U), (S), (f), (Q), (s), (v), (i), (T), (D), (p), (c) and (q) are known, then its Sales Past must be:

$$S' = (I\{U+S[1-f]-S'i[1+s]-T-D\}-Q)$$
$$/[v[1+s](p/\{360[c-q]\}+I)]$$

Steve Asikin ISBN 14: 978-1515158691, ISBN 10: 1515158691

Rule-20676:

If both (I), (U), $(\$)$, (f), $(\$')$, (Q), (v), (i), (T), (D), (p), (c) and (q) are known, then its Sales Growth Planned is:

$$v = \$/\$' - 1$$

or it can also be found as

$$s = (I\{U + \$[1-f] - \$'i[1+s] - T - D\} - Q)$$
$$/[\$'v(p/\{360[c-q]\} + I)] - 1$$

Rule-20677:

If both (I), (U), $(\$)$, (f), $(\$')$, (s), (Q), (i), (T), (D), (p), (c) and (q) are known, then its Variable Portion Planned is:

$$v = (I\{U + \$[1-f] - \$'i[1+s] - T - D\} - Q)$$
$$/[\$'[1+s](p/\{360[c-q]\} + I)]$$

Rule-20678:

If both (I), (U), $(\$)$, (f), $(\$')$, (s), (v), (Q), (T), (D), (p), (c) and (q) are known, then its Interest Portion Planned is:

$$i = [U + \$[1-f] - T - D - (Q + \$'vp[1+s]/\{360[c-q]\})/I]$$
$$/\{\$'[1+s]\} - v$$

Rule-20679:

If both (I), (U), $(\$)$, (f), $(\$')$, (s), (v), (i), (Q), (D), (p), (c) and (q) are known, then its Tax Planned is:

$$T = U + \$[1-f] - \$'[1+s][v+i] - D$$
$$-(Q + \$'vp[1+s]/\{360[c-q]\})/I$$

Steve Asikin ISBN 14: 978-1515158691, ISBN 10: 1515158691

Rule-20680:
 If both (I), (U), $(\$)$, (f), $(\$')$, (s), (v), (i), (T), (Q), (p),
 (c) and (q) are known, then its Dividend Planned is:
 $$D= U+\$[1-f]-\$'[1+s][v+i]-T$$
 $$-(Q+\$'vp[1+s]/\{360[c-q]\})/I$$

Rule-20681:
 If both (I), (U), $(\$)$, (f), $(\$')$, (s), (v), (i), (T), (D), (Q),
 (c) and (q) are known, then its Procured Inventory
 Days Planned:
 $$p= 360[c-q](I\{U+\$[1-f]-\$'[1+s][v+i]-T-D\}-Q)$$
 $$/\{\$'v[1+s]\}$$

Rule-20682:
 If both (I), (U), $(\$)$, (f), $(\$')$, (s), (v), (i), (T), (D), (p),
 (Q) and (q) are known, then its Current Ratio Planned
 is:
 $$c= q+\$'v[1+s]$$
 $$/[360(I\{U+\$[1-f]-\$'[1+s][v+i]-T-D\}-Q)]$$

Rule-20683:
 If both (I), (U), $(\$)$, (f), $(\$')$, (s), (v), (i), (T), (D), (p),
 (c) and (Q) are known, then its Quick or Acid Test
 Ratio Planned is:
 $$q= c-\$'v[1+s]$$
 $$/[360(I\{U+\$[1-f]-\$'[1+s][v+i]-T-D\}-Q)]$$

Steve Asikin ISBN 14: 978-1515158691, ISBN 10: 1515158691

Rule-20684:

If both (I), (**U**), (**$**), (**f**), (**$'**), (**s**), (**v**), (**i**), (**T**), (**d**) and (**X**) are known, then its Quoted Longterm Debt Planned is:

$Q= I(U+\{\$[1-f]-\$'[1+s][v+i]-T\}[1-d])-X$

Rule-20685:

If both (**Q**), (**U**), (**$**), (**f**), (**$'**), (**s**), (**v**), (**i**), (**T**), (**d**) and (**X**) are known, then its Leverage or Gearing Ratio Planned is:

$I= [Q+X]/(U+\{\$[1-f]-\$'[1+s][v+i]-T\}[1-d])$

Rule-20686:

If both (I), (**Q**), (**$**), (**f**), (**$'**), (**s**), (**v**), (**i**), (**T**), (**d**) and (**X**) are known, then its Utilized or Starting Capital must be:

$U= [Q+X]/I\{\$[1-f]-\$'[1+s][v+i]-T\}[1-d]$

Rule-20687:

If both (I), (**U**), (**Q**), (**f**), (**$'**), (**s**), (**v**), (**i**), (**T**), (**d**) and (**X**) are known, then its Sales or Revenue Planned is:

$\$= (\$'[1+s][v+i]+T+\{[Q+X]/I-U\}/[1-d])/[1-f]$

Rule-20688:

If both (I), (**U**), (**$**), (**Q**), (**$'**), (**s**), (**v**), (**i**), (**T**), (**d**) and (**X**) are known, then its Fixed Portion Planned is:

$f= 1-(\$'[1+s][v+i]+T+\{[Q+X]/I-U\}/[1-d])/\$$

Steve Asikin ISBN 14: 978-1515158691, ISBN 10: 1515158691

Rule-20689:

If both (**/**), (**U**), (**$**), (**f**), (**Q**), (**s**), (**v**), (**i**), (**T**), (**d**) and (**X**) are known, then its Sales Past must be:

$$\$' = (\$[1-f]-T-\{[Q+X]/\textit{/}-U\}/[1-d])/\{[1+s][v+i]\}$$

Rule-20690:

If both (**/**), (**U**), (**$**), (**f**), (**$'**), (**Q**), (**v**), (**i**), (**T**), (**d**) and (**X**) are known, then its Sales Growth Planned is:

$$s = \$/\$' - 1$$

or it can also be found as

$$s = (\$[1-f]-T-\{[Q+X]/\textit{/}-U\}/[1-d])/\{\$'[v+i]\} - 1$$

Rule-20691:

If both (**/**), (**U**), (**$**), (**f**), (**$'**), (**s**), (**Q**), (**i**), (**T**), (**d**) and (**X**) are known, then its Variable Portion Planned is:

$$v = (\$[1-f]-T-\{[Q+X]/\textit{/}-U\}/[1-d])/\{\$'[1+s]\} - i$$

Rule-20692:

If both (**/**), (**U**), (**$**), (**f**), (**$'**), (**s**), (**v**), (**Q**), (**T**), (**d**) and (**X**) are known, then its Interest Portion Planned is:

$$i = (\$[1-f]-T-\{[Q+X]/\textit{/}-U\}/[1-d])/\{ \$'[1+s]-v$$

Rule-20693:

If both (**/**), (**U**), (**$**), (**f**), (**$'**), (**s**), (**v**), (**i**), (**Q**), (**d**) and (**X**) are known, then its Tax Planned is:

$$T = \$[1-f]-\$'[1+s][v+i]-\{[Q+X]/\textit{/}-U\}/[1-d]$$

Rule-20694:

If both (**/**), (**U**), (**$**), (**f**), (**$'**), (**s**), (**v**), (**i**), (**T**), (**Q**) and (**X**) are known, then its Dividend Payout Planned is:

$$d = 1-\{[Q+X]/\textit{/}-U\}/\{\$[1-f]-\$'[1+s][v+i]-T\}$$

Steve Asikin ISBN 14: 978-1515158691, ISBN 10: 1515158691

Rule-20695:

If both (I), (U), (S), (f), (S'), (s), (v), (i), (T), (d) and (Q) are known, then its Xpress or Current Debt Planned is:

$$X = I(U + \{S[1-f]-S'[1+s][v+i]-T\}[1-d]) - Q$$

Rule-20696:

If both (I), (U), (S), (f), (S'), (s), (v), (i), (T), (d), (P), (c) and (q) are known, then its Quoted Longterm Debt Planned is:

$$Q = I(U + \{S[1-f]-S'[1+s][v+i]-T\}[1-d]) - P/[c-q]$$

Rule-20697:

If both (Q), (U), (S), (f), (S'), (s), (v), (i), (T), (d), (P), (c) and (q) are known, then its Leverage or Gearing Ratio Planned is:

$$I = \{Q+P/[c-q]\}/(U + \{S[1-f]-S'[1+s][v+i]-T\}[1-d])$$

Rule-20698:

If both (I), (Q), (S), (f), (S'), (s), (v), (i), (T), (d), (P), (c) and (q) are known, then its Utilized or Starting Capital must be:

$$U = \{Q+P/[c-q]\}/I - \{S[1-f]-S'[1+s][v+i]-T\}[1-d]$$

Rule-20699:

If both (I), (U), (Q), (f), (S'), (s), (v), (i), (T), (d), (P), (c) and (q) are known, then its Sales or Revenue Planned is:

$$S = [S'[1+s][v+i]+T+(\{Q+P/[c-q]\}/I - U)/[1-d]]/[1-f]$$

Steve Asikin ISBN 14: 978-1515158691, ISBN 10: 1515158691

Rule-20700:

If both (f), (U), (S), (Q), (S'), (s), (v), (i), (T), (d), (P), (c) and (q) are known, then its Fixed Portion Planned is:

$$f= 1-[S'[1+s][v+i]+T+(\{Q+P/[c-q]\}/f-U)/[1-d]]/S$$

Rule-20701:

If both (f), (U), (S), (f), (Q), (s), (v), (i), (T), (d), (P), (c) and (q) are known, then its Sales Past must be:

$$S'= [S[1-f]-T-(\{Q+P/[c-q]\}/f-U)/[1-d]]$$
$$/\{[1+s][v+i]\}$$

Rule-20702:

If both (f), (U), (S), (f), (S'), (Q), (v), (i), (T), (d), (P), (c) and (q) are known, then its Sales Growth Planned is:

$$s= S/S'-1$$

or it can also be found as

$$s= [S[1-f]-T-(\{Q+P/[c-q]\}/f-U)/[1-d]]/\{S'[v+i]\}-1$$

Rule-20703:

If both (f), (U), (S), (f), (S'), (s), (Q), (i), (T), (d), (P), (c) and (q) are known, then its Variable Portion Planned is:

$$v= [S[1-f]-T-(\{Q+P/[c-q]\}/f-U)/[1-d]]/\{S'[1+s]\}-i$$

Rule-20704:

If both (f), (U), (S), (f), (S'), (s), (v), (Q), (T), (d), (P), (c) and (q) are known, then its Interest Portion Planned is:

$$i= [S[1-f]-T-(\{Q+P/[c-q]\}/f-U)/[1-d]]/\{ S'[1+s]-v$$

Steve Asikin ISBN 14: 978-1515158691, ISBN 10: 1515158691

Rule-20705:

If both (I), (U), $(\$)$, (f), $(\$')$, (s), (v), (i), (Q), (d), (P), (c) and (q) are known, then its Tax Planned is:

$$T = \$[1-f] - \$'[1+s][v+i] - (\{Q+P/[c-q]\}/I-U)/[1-d]$$

Rule-20706:

If both (I), (U), $(\$)$, (f), $(\$')$, (s), (v), (i), (T), (Q), (P), (c) and (q) are known, then its Dividend Payout Planned is:

$$d = 1 - (\{Q+P/[c-q]\}/I-U)/\{\$[1-f] - \$'[1+s][v+i] - T\}$$

Rule-20707:

If both (I), (U), $(\$)$, (f), $(\$')$, (s), (v), (i), (T), (d), (Q), (c) and (q) are known, then its Procured Inventory Planned is:

$$P = [c-q][I(U + \{\$[1-f] - \$'[1+s][v+i] - T\}[1-d]) - Q]$$

Rule-20708:

If both (I), (U), $(\$)$, (f), $(\$')$, (s), (v), (i), (T), (d), (P), (Q) and (q) are known, then its Current Ratio Planned is:

$$c = q + P/[I(U + \{\$[1-f] - \$'[1+s][v+i] - T\}[1-d]) - Q]$$

Rule-20709:

If both (I), (U), $(\$)$, (f), $(\$')$, (s), (v), (i), (T), (d), (P), (c) and (Q) are known, then its Quick or Acid Test Ratio Planned is:

$$q = c - P/[I(U + \{\$[1-f] - \$'[1+s][v+i] - T\}[1-d]) - Q]$$

Steve Asikin ISBN 14: 978-1515158691, ISBN 10: 1515158691

<u>Rule-20710</u>:

If both (I), (U), (S), (f), (S'), (s), (v), (i), (T), (d), (V), (p), (c) and (q) are known, then its Quoted Longterm Debt Planned is:

$$Q= I(U+\{S[1-f]-S'[1+s][v+i]-T\}[1-d])$$
$$-Vp/\{360[c-q]\}$$

<u>Rule-20711</u>:

If both (Q), (U), (S), (f), (S'), (s), (v), (i), (T), (d), (V), (p), (c) and (q) are known, then its Leverage or Gearing Ratio Planned is:

$$I= (Q+Vp/\{360[c-q]\})$$
$$/(U+\{S[1-f]-S'[1+s][v+i]-T\}[1-d])$$

<u>Rule-20712</u>:

If both (I), (Q), (S), (f), (S'), (s), (v), (i), (T), (d), (V), (p), (c) and (q) are known, then its Utilized or Starting Capital must be:

$$U= (Q+Vp/\{360[c-q]\})/I$$
$$-\{S[1-f]-S'[1+s][v+i]-T\}[1-d]$$

<u>Rule-20713</u>:

If both (I), (U), (Q), (f), (S'), (s), (v), (i), (T), (d), (V), (p), (c) and (q) are known, then its Sales or Revenue Planned is:

$$S= \{S'[1+s][v+i]+T+[(Q+Vp/\{360[c-q]\})/I-U]$$
$$/[1-d]\}/[1-f]$$

Steve Asikin ISBN 14: 978-1515158691, ISBN 10: 1515158691

Rule-20714:

    If both (**/**), (**U**), (**$**), (**Q**), (**$'**), (**s**), (**v**), (**i**), (**T**), (**d**), (**V**), (**p**), (**c**) and (**q**) are known, then its Fixed Portion Planned is:

$$f= 1-\{\$'[1+s][v+i]+T+[(Q+Vp/\{360[c-q]\})/I-U] /[1-d]\}/\$$$

Rule-20715:

    If both (**/**), (**U**), (**$**), (**f**), (**Q**), (**s**), (**v**), (**i**), (**T**), (**d**), (**V**), (**p**), (**c**) and (**q**) are known, then its Sales Past must be:

$$\$'= \{\$[1-f]-T-[(Q+Vp/\{360[c-q]\})/I-U]/[1-d]\} /\{[1+s][v+i]\}$$

Rule-20716:

    If both (**/**), (**U**), (**$**), (**f**), (**$'**), (**Q**), (**v**), (**i**), (**T**), (**d**), (**V**), (**p**), (**c**) and (**q**) are known, then its Sales Growth Planned is:

$$s= \$/\$'-1$$

        or it can also be found as

$$s=\{\$[1-f]-T-[(Q+Vp/\{360[c-q]\})/I-U]/[1-d]\} /\{\$'[v+i]\}-1$$

Rule-20717:

    If both (**/**), (**U**), (**$**), (**f**), (**$'**), (**s**), (**Q**), (**i**), (**T**), (**d**), (**V**), (**p**), (**c**) and (**q**) are known, then its Variable Portion Planned is:

$$v= V/\$$$

        or it can also be found as

$$v= \{\$[1-f]-T-[(Q+Vp/\{360[c-q]\})/I-U]/[1-d]\} /\{\$'[1+s]\}-i$$

Steve Asikin ISBN 14: 978-1515158691, ISBN 10: 1515158691

Rule-20718:

If both (**∫**), (**U**), (**\$**), (**f**), (**\$'**), (**s**), (**v**), (**Q**), (**T**), (**d**), (**V**), (**p**), (**c**) and (**q**) are known, then its Interest Portion Planned is:

$$i = \{\$[1\text{-}f]\text{-}T\text{-}[(Q+Vp/\{360[c\text{-}q]\})/\text{-}U]/[1\text{-}d]\}$$
$$/\{\ \$'[1+s]\text{-}v$$

Rule-20719:

If both (**∫**), (**U**), (**\$**), (**f**), (**\$'**), (**s**), (**v**), (**i**), (**Q**), (**d**), (**V**), (**p**), (**c**) and (**q**) are known, then its Tax Planned is:

$$T = \$[1\text{-}f]\text{-}\$'[1+s][v+i]$$
$$-[(Q+Vp/\{360[c\text{-}q]\})/\text{-}U]/[1\text{-}d]$$

Rule-20720:

If both (**∫**), (**U**), (**\$**), (**f**), (**\$'**), (**s**), (**v**), (**i**), (**T**), (**Q**), (**V**), (**p**), (**c**) and (**q**) are known, then its Dividend Payout Planned is:

$$d = 1\text{-}[(Q+Vp/\{360[c\text{-}q]\})/\text{-}U]$$
$$/\{\$[1\text{-}f]\text{-}\$'[1+s][v+i]\text{-}T\}$$

Rule-20721:

If both (**∫**), (**U**), (**\$**), (**f**), (**\$'**), (**s**), (**v**), (**i**), (**T**), (**d**), (**Q**), (**p**), (**c**) and (**q**) are known, then its Variable Cost Planned is:

$$V = 360[c\text{-}q]$$
$$[\text{∫}U+\{\$[1\text{-}f]\text{-}\$'[1+s][v+i]\text{-}T\}[1\text{-}d])\text{-}Q]/p$$

Steve Asikin ISBN 14: 978-1515158691, ISBN 10: 1515158691

Rule-20722:

If both (**/**), (**U**), (**$**), (**f**), (**$'**), (**s**), (**v**), (**i**), (**T**), (**d**), (**V**), (**Q**), (**c**) and (**q**) are known, then its Procured Inventory Days Planned:

$$p= 360[c\text{-}q]$$
$$[/U+\{\$[1\text{-}f]\text{-}\$'[1+s][v+i]\text{-}T\}[1\text{-}d])\text{-}Q]/V$$

Rule-20723:

If both (**/**), (**U**), (**$**), (**f**), (**$'**), (**s**), (**v**), (**i**), (**T**), (**d**), (**V**), (**p**), (**Q**) and (**q**) are known, then its Current Ratio Planned is:

$$c= q+Vp$$
$$/\{360[/U+\{\$[1\text{-}f]\text{-}\$'[1+s][v+i]\text{-}T\}[1\text{-}d])\text{-}Q]\}$$

Rule-20724:

If both (**/**), (**U**), (**$**), (**f**), (**$'**), (**s**), (**v**), (**i**), (**T**), (**d**), (**V**), (**p**), (**c**) and (**Q**) are known, then its Quick or Acid Test Ratio Planned is:

$$q= c\text{-}Vp$$
$$/\{360[/U+\{\$[1\text{-}f]\text{-}\$'[1+s][v+i]\text{-}T\}[1\text{-}d])\text{-}Q]\}$$

Rule-20725:

If both (**/**), (**U**), (**$**), (**f**), (**$'**), (**s**), (**v**), (**i**), (**T**), (**d**), (**p**), (**c**) and (**q**) are known, then its Quoted Longterm Debt Planned is:

$$Q= /U+\{\$[1\text{-}f]\text{-}\$'[1+s][v+i]\text{-}T\}[1\text{-}d])$$
$$-\$vp/\{360[c\text{-}q]\}$$

Steve Asikin ISBN 14: 978-1515158691, ISBN 10: 1515158691

Rule-20726:

If both $(\mathbf{Q})$, $(\mathbf{U})$, $(\mathbf{S})$, $(\mathbf{f})$, $(\mathbf{S'})$, $(\mathbf{s})$, $(\mathbf{v})$, $(\mathbf{i})$, $(\mathbf{T})$, $(\mathbf{d})$, $(\mathbf{p})$, $(\mathbf{c})$ and $(\mathbf{q})$ are known, then its Leverage or Gearing Ratio Planned is:

$$I = (\mathbf{Q} + \mathbf{Svp}/\{360[\mathbf{c} \text{-} \mathbf{q}]\})$$
$$/(\mathbf{U} + \{\mathbf{S}[1 \text{-} \mathbf{f}] \text{-} \mathbf{S'}[1 + \mathbf{s}][\mathbf{v} + \mathbf{i}] \text{-} \mathbf{T}\}[1 \text{-} \mathbf{d}])$$

Rule-20727:

If both (I), $(\mathbf{Q})$, $(\mathbf{S})$, $(\mathbf{f})$, $(\mathbf{S'})$, $(\mathbf{s})$, $(\mathbf{v})$, $(\mathbf{i})$, $(\mathbf{T})$, $(\mathbf{d})$, $(\mathbf{p})$, $(\mathbf{c})$ and $(\mathbf{q})$ are known, then its Utilized or Starting Capital must be:

$$\mathbf{U} = (\mathbf{Q} + \mathbf{Svp}/\{360[\mathbf{c} \text{-} \mathbf{q}]\})/I$$
$$\text{-}\{\mathbf{S}[1 \text{-} \mathbf{f}] \text{-} \mathbf{S'}[1 + \mathbf{s}][\mathbf{v} + \mathbf{i}] \text{-} \mathbf{T}\}[1 \text{-} \mathbf{d}]$$

Rule-20728:

If both (I), $(\mathbf{U})$, $(\mathbf{Q})$, $(\mathbf{f})$, $(\mathbf{S'})$, $(\mathbf{s})$, $(\mathbf{v})$, $(\mathbf{i})$, $(\mathbf{T})$, $(\mathbf{d})$, $(\mathbf{p})$, $(\mathbf{c})$ and $(\mathbf{q})$ are known, then its Sales or Revenue Planned is:

$$\mathbf{S} = [I\mathbf{U} \text{-} \{\mathbf{S'}[1 + \mathbf{s}][\mathbf{v} + \mathbf{i}] + \mathbf{T}\}[1 \text{-} \mathbf{d}]) \text{-} \mathbf{S}]$$
$$/(\mathbf{vp}/\{360[\mathbf{c} \text{-} \mathbf{q}]\} \text{-} I[1 \text{-} \mathbf{f}][1 \text{-} \mathbf{d}])$$

Rule-20729:

If both (I), $(\mathbf{U})$, $(\mathbf{S})$, $(\mathbf{Q})$, $(\mathbf{S'})$, $(\mathbf{s})$, $(\mathbf{v})$, $(\mathbf{i})$, $(\mathbf{T})$, $(\mathbf{d})$, $(\mathbf{p})$, $(\mathbf{c})$ and $(\mathbf{q})$ are known, then its Fixed Portion Planned is:

$$\mathbf{f} = 1 \text{-} \{\mathbf{S'}[1 + \mathbf{s}][\mathbf{v} + \mathbf{i}] + \mathbf{T}$$
$$+ [(\mathbf{Q} + \mathbf{Svp}/\{360[\mathbf{c} \text{-} \mathbf{q}]\})/I \text{-} \mathbf{U}]/[1 \text{-} \mathbf{d}]\}/\mathbf{S}$$

Steve Asikin ISBN 14: 978-1515158691, ISBN 10: 1515158691

Rule-20730:

If both (**/**), (**U**), (**$**), (**f**), (**Q**), (**s**), (**v**), (**i**), (**T**), (**d**), (**p**),
(**c**) and (**q**) are known, then its Sales Past must be:

$$\$' = \{\$[1\text{-}f]\text{-}T\text{-}[(Q+\$vp/\{360[c\text{-}q]\})/I\text{-}U]/[1\text{-}d]\}$$
$$/\{[1+s][v+i]\}$$

Rule-20731:

If both (**/**), (**U**), (**$**), (**f**), (**$'**), (**Q**), (**v**), (**i**), (**T**), (**d**), (**p**),
(**c**) and (**q**) are known, then its Sales Growth Planned
is:

$$s = \$/\$'\text{-}1$$

or it can also be found as

$$s = \{\$[1\text{-}f]\text{-}T\text{-}[(Q+\$vp/\{360[c\text{-}q]\})/I\text{-}U]/[1\text{-}d]\}$$
$$/\{\$'[v+i]\}\text{-}1$$

Rule-20732:

If both (**/**), (**U**), (**$**), (**f**), (**$'**), (**s**), (**Q**), (**i**), (**T**), (**d**), (**p**),
(**c**) and (**q**) are known, then its Variable Portion
Planned is:

$$v = [I U + \{\$[1\text{-}f]\text{-}\$'i[1+s]\text{-}T\}[1\text{-}d])\text{-}Q]$$
$$/(\$p/\{360[c\text{-}q]\}+\$'I[1+s][1\text{-}d])$$

Rule-20733:

If both (**/**), (**U**), (**$**), (**f**), (**$'**), (**s**), (**v**), (**Q**), (**T**), (**d**), (**p**),
(**c**) and (**q**) are known, then its Interest Portion
Planned is:

$$i = \{\$[1\text{-}f]\text{-}T\text{-}[(Q+\$vp/\{360[c\text{-}q]\})/I\text{-}U]/[1\text{-}d]\}$$
$$/\{\ \$'[1+s]\text{-}v$$

Steve Asikin ISBN 14: 978-1515158691, ISBN 10: 1515158691

Rule-20734:
  If both (*l*), (**U**), (**$**), (**f**), (**$'**), (**s**), (**v**), (**i**), (**Q**), (**d**), (**p**), (**c**) and (**q**) are known, then its Tax Planned is:

$$T= \$[1\text{-}f]\text{-}\$'[1+s][v+i]\text{-}[(Q+\$vp/\{360[c\text{-}q]\})/l\text{-}U]$$
$$/[1\text{-}d]$$

Rule-20735:
  If both (*l*), (**U**), (**$**), (**f**), (**$'**), (**s**), (**v**), (**i**), (**T**), (**Q**), (**p**), (**c**) and (**q**) are known, then its Dividend Payout Planned is:

$$d= 1\text{-}[(Q+\$vp/\{360[c\text{-}q]\})/l\text{-}U]$$
$$/\{\$[1\text{-}f]\text{-}\$'[1+s][v+i]\text{-}T\}$$

Rule-20736:
  If both (*l*), (**U**), (**$**), (**f**), (**$'**), (**s**), (**v**), (**i**), (**T**), (**d**), (**Q**), (**c**) and (**q**) are known, then its Procured Inventory Days Planned:

$$p= 360[c\text{-}q]$$
$$[l U+\{\$[1\text{-}f]\text{-}\$'[1+s][v+i]\text{-}T\}[1\text{-}d])\text{-}Q]/[\$v]$$

Rule-20737:
  If both (*l*), (**U**), (**$**), (**f**), (**$'**), (**s**), (**v**), (**i**), (**T**), (**d**), (**p**), (**Q**) and (**q**) are known, then its Current Ratio Planned is:

$$c= q+\$vp$$
$$/\{360[l U+\{\$[1\text{-}f]\text{-}\$'[1+s][v+i]\text{-}T\}[1\text{-}d])\text{-}Q]\}$$

Steve Asikin ISBN 14: 978-1515158691, ISBN 10: 1515158691

Rule-20738:

If both (I), (U), $(\$)$, (f), $(\$')$, (s), (v), (i), (T), (d), (p), (c) and (Q) are known, then its Quick or Acid Test Ratio Planned is:

$$q = c\text{-}\$vp$$
$$/\{360[I(U+\{\$[1\text{-}f]\text{-}\$'[1+s][v+i]\text{-}T\}[1\text{-}d])\text{-}Q]\}$$

Rule-20739:

If both (I), (U), $(\$)$, (f), $(\$')$, (s), (v), (i), (T), (d), (p), (c) and (q) are known, then its Quoted Longterm Debt Planned is:

$$Q = I(U+\{\$[1\text{-}f]\text{-}\$'[1+s][v+i]\text{-}T\}[1\text{-}d])$$
$$-\$'vp[1+s]/\{360[c\text{-}q]\}$$

Rule-20740:

If both (Q), (U), $(\$)$, (f), $(\$')$, (s), (v), (i), (T), (d), (p), (c) and (q) are known, then its Leverage or Geraing Ratio Planned is:

$$I = (Q+\$'vp[1+s]/\{360[c\text{-}q]\})$$
$$/(U+\{\$[1\text{-}f]\text{-}\$'[1+s][v+i]\text{-}T\}[1\text{-}d])$$

Rule-20741:

If both (I), (Q), $(\$)$, (f), $(\$')$, (s), (v), (i), (T), (d), (p), (c) and (q) are known, then its Utilized or Starting Capital must be:

$$U = (Q+\$'vp[1+s]/\{360[c\text{-}q]\})/I$$
$$-\{\$[1\text{-}f]\text{-}\$'[1+s][v+i]\text{-}T\}[1\text{-}d]$$

Steve Asikin ISBN 14: 978-1515158691, ISBN 10: 1515158691

Rule-20742:

If both ($\boldsymbol{I}$), (**U**), (**Q**), (**f**), (**S'**), (**s**), (**v**), (**i**), (**T**), (**d**), (**p**), (**c**) and (**q**) are known, then its Sales or Revenue Planned is:

$$S= [Q-I(U-\{S'[1+s][v+i]+T\}[1-d])$$
$$+S'vp[1+s]/\{360[c-q]\}]/\{I[1-f]\}$$

Rule-20743:

If both ($\boldsymbol{I}$), (**U**), (**S**), (**Q**), (**S'**), (**s**), (**v**), (**i**), (**T**), (**d**), (**p**), (**c**) and (**q**) are known, then its Fixed Portion Planned is:

$$f= 1-\{S'[1+s][v+i]+T+[(Q+S'vp[1+s]/\{360[c-q]\})/I$$
$$-U]/[1-d]\}/S$$

Rule-20744:

If both ($\boldsymbol{I}$), (**U**), (**S**), (**f**), (**Q**), (**s**), (**v**), (**i**), (**T**), (**d**), (**p**), (**c**) and (**q**) are known, then its Sales Past must be:

$$S'= [I(U+\{S[1-f]-S'i[1+s]-T\}[1-d])-Q]$$
$$/[v[1+s](p/\{360[c-q]\}+I[1-d])]$$

Rule-20745:

If both ($\boldsymbol{I}$), (**U**), (**S**), (**f**), (**S'**), (**Q**), (**v**), (**i**), (**T**), (**d**), (**p**), (**c**) and (**q**) are known, then its Sales Growth Planned is:

$$s= S/S'-1$$

or it can also be found as

$$s=[I(U+\{S[1-f]-S'i[1+s]-T\}[1-d])-Q]$$
$$/[S'v(p/\{360[c-q]\}+I[1-d])]-1$$

Steve Asikin ISBN 14: 978-1515158691, ISBN 10: 1515158691

Rule-20746:

If both (**/**), (**U**), (**$**), (**f**), (**$'**), (**s**), (**Q**), (**i**), (**T**), (**d**), (**p**), (**c**) and (**q**) are known, then its Variable Portion Planned is:

$$v= [/U+\{\$[1-f]-\$'i[1+s]-T\}[1-d])-Q]$$
$$/[\$'[1+s](p/\{360[c-q]\}+/[1-d])]$$

Rule-20747:

If both (**/**), (**U**), (**$**), (**f**), (**$'**), (**s**), (**v**), (**Q**), (**T**), (**d**), (**p**), (**c**) and (**q**) are known, then its Interest Portion Planned is:

$$i= \{\$[1-f]-T-[(Q+\$'vp[1+s]/\{360[c-q]\})//-U]$$
$$/[1-d]\}/\{ \$'[1+s]-v$$

Rule-20748:

If both (**/**), (**U**), (**$**), (**f**), (**$'**), (**s**), (**v**), (**i**), (**Q**), (**d**), (**p**), (**c**) and (**q**) are known, then its Tax Planned is:

$$T= \$[1-f]-\$'[1+s][v+i]$$
$$-[(Q+\$'vp[1+s]/\{360[c-q]\})//-U]/[1-d]$$

Rule-20749:

If both (**/**), (**U**), (**$**), (**f**), (**$'**), (**s**), (**v**), (**i**), (**T**), (**Q**), (**p**), (**c**) and (**q**) are known, then its Dividend Payout Planned is:

$$d= 1-[(Q+\$'vp[1+s]/\{360[c-q]\})//-U]$$
$$/\{\$[1-f]-\$'[1+s][v+i]-T\}$$

Steve Asikin ISBN 14: 978-1515158691, ISBN 10: 1515158691

Rule-20750:

If both (I), (U), (S), (f), (S'), (s), (v), (i), (T), (d), (Q), (c) and (q) are known, then its Procured Inventory Days Planned:

$$p= 360[c-q][I\{U+\{S[1-f]-S'[1+s][v+i]-T\}[1-d])-Q]/\{S'v[1+s]\}$$

Rule-20751:

If both (I), (U), (S), (f), (S'), (s), (v), (i), (T), (d), (p), (Q) and (q) are known, then its Current Ratio Planned is:

$$c= q+S'vp[1+s]/\{360[I\{U+\{S[1-f]-S'[1+s][v+i]-T\}[1-d])-Q]\}$$

Rule-20752:

If both (I), (U), (S), (f), (S'), (s), (v), (i), (T), (d), (p), (c) and (q) are known, then its Quoted Longterm Debt Planned is:

$$q= c-S'vp[1+s]/\{360[I\{U+\{S[1-f]-S'[1+s][v+i]-T\}[1-d])-Q]\}$$

Rule-20753:

If both (I), (U), (S), (f), (S'), (s), (v), (i), (T), (A), (d) and (X) are known, then its Quoted Longterm Debt Planned is:

$$Q= I\{U+S[1-f]-S'[1+s][v+i]-T-Ad\}-X$$

Steve Asikin ISBN 14: 978-1515158691, ISBN 10: 1515158691

Rule-20754:

If both $(\mathbf{Q})$, $(\mathbf{U})$, $(\mathbf{S})$, $(\mathbf{f})$, $(\mathbf{S'})$, $(\mathbf{s})$, $(\mathbf{v})$, $(\mathbf{i})$, $(\mathbf{T})$, $(\mathbf{A})$, $(\mathbf{d})$ and $(\mathbf{X})$ are known, then its Leverage or Gearing Ratio Planned is:

$$I = [Q+X]/\{U+S[1-f]-S'[1+s][v+i]-T-Ad\}$$

Rule-20755:

If both (I), $(\mathbf{Q})$, $(\mathbf{S})$, $(\mathbf{f})$, $(\mathbf{S'})$, $(\mathbf{s})$, $(\mathbf{v})$, $(\mathbf{i})$, $(\mathbf{T})$, $(\mathbf{A})$, $(\mathbf{d})$ and $(\mathbf{X})$ are known, then its Utilized or Starting Capital must be:

$$U = S'[1+s][v+i]+T+Ad+[Q+X]/I-S[1-f]$$

Rule-20756:

If both (I), $(\mathbf{U})$, $(\mathbf{Q})$, $(\mathbf{f})$, $(\mathbf{S'})$, $(\mathbf{s})$, $(\mathbf{v})$, $(\mathbf{i})$, $(\mathbf{T})$, $(\mathbf{A})$, $(\mathbf{d})$ and $(\mathbf{X})$ are known, then its Sales or Revenue Planned is:

$$S = \{S'[1+s][v+i]+T+Ad+[Q+X]/I-U\}/[1-f]$$

Rule-20757:

If both (I), $(\mathbf{U})$, $(\mathbf{S})$, $(\mathbf{Q})$, $(\mathbf{S'})$, $(\mathbf{s})$, $(\mathbf{v})$, $(\mathbf{i})$, $(\mathbf{T})$, $(\mathbf{A})$, $(\mathbf{d})$ and $(\mathbf{X})$ are known, then its Fixed Portion Planned is:

$$f = 1-\{S'[1+s][v+i]+T+Ad+[Q+X]/I-U\}/S$$

Rule-20758:

If both (I), $(\mathbf{U})$, $(\mathbf{S})$, $(\mathbf{f})$, $(\mathbf{Q})$, $(\mathbf{s})$, $(\mathbf{v})$, $(\mathbf{i})$, $(\mathbf{T})$, $(\mathbf{A})$, $(\mathbf{d})$ and $(\mathbf{X})$ are known, then its Sales or Revenue Planned is:

$$S' = \{U+S[1-f]-T-Ad-[Q+X]/I\}/\{[1+s][v+i]\}$$

Steve Asikin ISBN 14: 978-1515158691, ISBN 10: 1515158691

Rule-20759:

   If both (**/**), (**U**), (**$**), (**f**), (**$'**), (**Q**), (**v**), (**i**), (**T**), (**A**), (**d**) and (**X**) are known, then its Sales Growth Planned is:
   $s = \$/\$' - 1$

      or it can also be found as
   $s = \{U + \$[1-f] - T - Ad - [Q+X]/\text{/}\}/\{\$'[v+i]\} - 1$

Rule-20760:

   If both (**/**), (**U**), (**$**), (**f**), (**$'**), (**s**), (**Q**), (**i**), (**T**), (**A**), (**d**) and (**X**) are known, then its Variable Portion Planned is:
   $v = \{U + \$[1-f] - T - Ad - [Q+X]/\text{/}\}/\{\$'[1+s]\} - i$

Rule-20761:

   If both (**/**), (**U**), (**$**), (**f**), (**$'**), (**s**), (**v**), (**Q**), (**T**), (**A**), (**d**) and (**X**) are known, then its Interest Portion Planned is:
   $i = \{U + \$[1-f] - T - Ad - [Q+X]/\text{/}\}/\{\$'[1+s]\} - i$

Rule-20762:

   If both (**/**), (**U**), (**$**), (**f**), (**$'**), (**s**), (**v**), (**i**), (**Q**), (**A**), (**d**) and (**X**) are known, then its Tax Planned is:
   $T = U + \$[1-f] - \$'[1+s][v+i] - Ad - [Q+X]/\text{/}$

Rule-20763:

   If both (**/**), (**U**), (**$**), (**f**), (**$'**), (**s**), (**v**), (**i**), (**T**), (**Q**), (**d**) and (**X**) are known, then its After Tax Income Planned is:
   $A = \{U + \$[1-f] - \$'[1+s][v+i] - T - [Q+X]/\text{/}\}/d$

Steve Asikin ISBN 14: 978-1515158691, ISBN 10: 1515158691

Rule-20764:
> If both (**/**), (**U**), (**$**), (**f**), (**$'**), (**s**), (**v**), (**i**), (**T**), (**A**), (**Q**)
> and (**X**) are known, then its Dividend Payout Planned
> is:
> $$d= \{U+\$[1-f]-\$'[1+s][v+i]-T-[Q+X]//\}/A$$

Rule-20765:
> If both (**/**), (**U**), (**$**), (**f**), (**$'**), (**s**), (**v**), (**i**), (**T**), (**A**), (**d**)
> and (**Q**) are known, then its Xpress or Current Debt
> Planned is:
> $$X= /\{U+\$[1-f]-\$'[1+s][v+i]-T-Ad\}-Q$$

Rule-20766:
> If both (**/**), (**U**), (**$**), (**f**), (**$'**), (**s**), (**v**), (**i**), (**T**), (**A**), (**d**),
> (**P**), (**c**) and (**q**) are known, then its Quoted Longterm
> Debt Planned is:
> $$Q= /\{U+\$[1-f]-\$'[1+s][v+i]-T-Ad\}-P/[c-q]$$

Rule-20767:
> If both (**Q**), (**U**), (**$**), (**f**), (**$'**), (**s**), (**v**), (**i**), (**T**), (**A**), (**d**),
> (**P**), (**c**) and (**q**) are known, then its Leverage or
> Gearing Ratio Planned is:
> $$/= \{Q+P/[c-q]\}/\{U+\$[1-f]-\$'[1+s][v+i]-T-Ad\}$$

Rule-20768:
> If both (**/**), (**Q**), (**$**), (**f**), (**$'**), (**s**), (**v**), (**i**), (**T**), (**A**), (**d**),
> (**P**), (**c**) and (**q**) are known, then its Utilized or
> Starting Capital must be:
> $$U= \$'[1+s][v+i]+T+Ad+\{Q+P/[c-q]\}//-\$[1-f]$$

Steve Asikin ISBN 14: 978-1515158691, ISBN 10: 1515158691

Rule-20769:

If both (**/**), (**U**), (**Q**), (**f**), (**S'**), (**s**), (**v**), (**i**), (**T**), (**A**), (**d**), (**P**), (**c**) and (**q**) are known, then its Sales or Revenue Planned is:

$S = (S'[1+s][v+i]+T+Ad+\{Q+P/[c-q]\}/I\!\!-U)/[1-f]$

Rule-20770:

If both (**/**), (**U**), (**S**), (**Q**), (**S'**), (**s**), (**v**), (**i**), (**T**), (**A**), (**d**), (**P**), (**c**) and (**q**) are known, then its Fixed Portion Planned is:

$f = 1-(S'[1+s][v+i]+T+Ad+\{Q+P/[c-q]\}/I\!\!-U)/S$

Rule-20771:

If both (**/**), (**U**), (**S**), (**f**), (**Q**), (**s**), (**v**), (**i**), (**T**), (**A**), (**d**), (**P**), (**c**) and (**q**) are known, then its Sales Past must be:

$S' = (U+S[1-f]-T-Ad-\{Q+P/[c-q]\}/I\!\!)/\{[1+s][v+i]\}$

Rule-20772:

If both (**/**), (**U**), (**S**), (**f**), (**S'**), (**Q**), (**v**), (**i**), (**T**), (**A**), (**d**), (**P**), (**c**) and (**q**) are known, then its Sales Growth Planned is:

$s = S/S'-1$

or it can also be found as

$s = (U+S[1-f]-T-Ad-\{Q+P/[c-q]\}/I\!\!)/\{S'[v+i]\}-1$

Rule-20773:

If both (**/**), (**U**), (**S**), (**f**), (**S'**), (**s**), (**Q**), (**i**), (**T**), (**A**), (**d**), (**P**), (**c**) and (**q**) are known, then its Variable Portion Planned is:

$v = (U+S[1-f]-T-Ad-\{Q+P/[c-q]\}/I\!\!)/\{S'[1+s]\}-i$

Steve Asikin ISBN 14: 978-1515158691, ISBN 10: 1515158691

Rule-20774:

If both $(\textbf{\textit{I}})$, $(\textbf{U})$, $(\textbf{\$})$, $(\textbf{f})$, $(\textbf{\$'})$, $(\textbf{s})$, $(\textbf{v})$, $(\textbf{Q})$, $(\textbf{T})$, $(\textbf{A})$, $(\textbf{d})$, $(\textbf{P})$, $(\textbf{c})$ and $(\textbf{q})$ are known, then its Interest Portion Planned is:

$$i= (U+\$[1\text{-}f]\text{-}T\text{-}Ad\text{-}\{Q+P/[c\text{-}q]\}/I)/\{\$'[1+s]\}\text{-}v$$

Rule-20775:

If both $(\textbf{\textit{I}})$, $(\textbf{U})$, $(\textbf{\$})$, $(\textbf{f})$, $(\textbf{\$'})$, $(\textbf{s})$, $(\textbf{v})$, $(\textbf{i})$, $(\textbf{Q})$, $(\textbf{A})$, $(\textbf{d})$, $(\textbf{P})$, $(\textbf{c})$ and $(\textbf{q})$ are known, then its Tax Planned is:

$$T= U+\$[1\text{-}f]\text{-}\$'[1+s][v+i]\text{-}Ad\text{-}\{Q+P/[c\text{-}q]\}/I$$

Rule-20776:

If both $(\textbf{\textit{I}})$, $(\textbf{U})$, $(\textbf{\$})$, $(\textbf{f})$, $(\textbf{\$'})$, $(\textbf{s})$, $(\textbf{v})$, $(\textbf{i})$, $(\textbf{T})$, $(\textbf{Q})$, $(\textbf{d})$, $(\textbf{P})$, $(\textbf{c})$ and $(\textbf{q})$ are known, then its After Tax Income Planned is:

$$A= (U+\$[1\text{-}f]\text{-}\$'[1+s][v+i]\text{-}T\text{-}\{Q+P/[c\text{-}q]\}/I)/d$$

Rule-20777:

If both $(\textbf{\textit{I}})$, $(\textbf{U})$, $(\textbf{\$})$, $(\textbf{f})$, $(\textbf{\$'})$, $(\textbf{s})$, $(\textbf{v})$, $(\textbf{i})$, $(\textbf{T})$, $(\textbf{A})$, $(\textbf{Q})$, $(\textbf{P})$, $(\textbf{c})$ and $(\textbf{q})$ are known, then its Dividend Payout Planned is:

$$d= (U+\$[1\text{-}f]\text{-}\$'[1+s][v+i]\text{-}T\text{-}\{Q+P/[c\text{-}q]\}/I)/A$$

Rule-20778:

If both $(\textbf{\textit{I}})$, $(\textbf{U})$, $(\textbf{\$})$, $(\textbf{f})$, $(\textbf{\$'})$, $(\textbf{s})$, $(\textbf{v})$, $(\textbf{i})$, $(\textbf{T})$, $(\textbf{A})$, $(\textbf{d})$, $(\textbf{Q})$, $(\textbf{c})$ and $(\textbf{q})$ are known, then its Procured Inventory Planned is:

$$P= [c\text{-}q](I\{U+\$[1\text{-}f]\text{-}\$'[1+s][v+i]\text{-}T\text{-}Ad\}\text{-}Q)$$

Steve Asikin ISBN 14: 978-1515158691, ISBN 10: 1515158691

Rule-20779:

 If both (I), (U), (S), (f), (S'), (s), (v), (i), (T), (A), (d), (P), (Q) and (q) are known, then its Current Ratio Planned is:

$$c = q + P/(I\{U + S[1-f] - S'[1+s][v+i] - T - Ad\} - Q)$$

Rule-20780:

 If both (I), (U), (S), (f), (S'), (s), (v), (i), (T), (A), (d), (P), (c) and (Q) are known, then its Quick or Acid Test Ratio Planned is:

$$q = c - P/(I\{U + S[1-f] - S'[1+s][v+i] - T - Ad\} - Q)$$

Rule-20781:

 If both (I), (U), (S), (f), (S'), (s), (v), (i), (T), (A), (d), (V), (p), (c) and (q) are known, then its Quoted Longterm Debt Planned is:

$$Q = I\{U + S[1-f] - S'[1+s][v+i] - T - Ad\} - Vp/\{360[c-q]\}$$

Rule-20782:

 If both (Q), (U), (S), (f), (S'), (s), (v), (i), (T), (A), (d), (V), (p), (c) and (q) are known, then its Leverage or Gearing Ratio Planned is:

$$I = (Q + Vp/\{360[c-q]\})$$
$$/\{U + S[1-f] - S'[1+s][v+i] - T - Ad\}$$

Rule-20783:

 If both (I), (Q), (S), (f), (S'), (s), (v), (i), (T), (A), (d), (V), (p), (c) and (q) are known, then its Utilized or Starting Capital must be:

$$U = S'[1+s][v+i] + T + Ad + (Q + Vp/\{360[c-q]\})/I$$
$$- S[1-f]$$

Steve Asikin ISBN 14: 978-1515158691, ISBN 10: 1515158691

Rule-20784:

If both (**/**), (**U**), (**Q**), (**f**), (**S'**), (**s**), (**v**), (**i**), (**T**), (**A**), (**d**), (**V**), (**p**), (**c**) and (**q**) are known, then its Sales or Revenue Planned is:

$$S= [S'[1+s][v+i]+T+Ad$$
$$+(Q+Vp/\{360[c-q]\})/\textbf{/}U]/[1-f]$$

Rule-20785:

If both (**/**), (**U**), (**S**), (**Q**), (**S'**), (**s**), (**v**), (**i**), (**T**), (**A**), (**d**), (**V**), (**p**), (**c**) and (**q**) are known, then its Fixed Portion Planned is:

$$f= 1-[S'[1+s][v+i]+T+Ad$$
$$+(Q+Vp/\{360[c-q]\})/\textbf{/}U]/S$$

Rule-20786:

If both (**/**), (**U**), (**S**), (**f**), (**Q**), (**s**), (**v**), (**i**), (**T**), (**A**), (**d**), (**V**), (**p**), (**c**) and (**q**) are known, then its Sales Past must be:

$$S'= [U+S[1-f]-T-Ad$$
$$-(Q+Vp/\{360[c-q]\})/\textbf{/}]/\{[1+s][v+i]\}$$

Rule-20787:

If both (**/**), (**U**), (**S**), (**f**), (**S'**), (**Q**), (**v**), (**i**), (**T**), (**A**), (**d**), (**V**), (**p**), (**c**) and (**q**) are known, then its Sales Growth Planned is:

$$s= S/S'-1$$

or it can also be found as

$$s= [U+S[1-f]-T-Ad-(Q+Vp/\{360[c-q]\})/\textbf{/}]$$
$$/\{S'[v+i]\}-1$$

Steve Asikin ISBN 14: 978-1515158691, ISBN 10: 1515158691

Rule-20788:

If both (l), (U), $(\$)$, (f), $(\$')$, (s), (Q), (i), (T), (A), (d), (V), (p), (c) and (q) are known, then its Variable Portion Planned is:

$$v = V/\$$$

or it can also be found as

$$v = [U+\$[1-f]-T-Ad-(Q+Vp/\{360[c-q]\})/l]$$
$$/\{\$'[1+s]\}-i$$

Rule-20789:

If both (l), (U), $(\$)$, (f), $(\$')$, (s), (v), (Q), (T), (A), (d), (V), (p), (c) and (q) are known, then its Interest Portion Planned is:

$$i = [U+\$[1-f]-T-Ad-(Q+Vp/\{360[c-q]\})/l]$$
$$/\{\$'[1+s]\}-v$$

Rule-20790:

If both (l), (U), $(\$)$, (f), $(\$')$, (s), (v), (i), (Q), (A), (d), (V), (p), (c) and (q) are known, then its Tax Planned is:

$$T = U+\$[1-f]-\$'[1+s][v+i]-Ad-(Q+Vp/\{360[c-q]\})/l$$

Rule-20791:

If both (l), (U), $(\$)$, (f), $(\$')$, (s), (v), (i), (T), (Q), (d), (V), (p), (c) and (q) are known, then its After Tax Income Planned is:

$$A = [U+\$[1-f]-\$'[1+s][v+i]-T$$
$$-(Q+Vp/\{360[c-q]\})/l]/d$$

Steve Asikin ISBN 14: 978-1515158691, ISBN 10: 1515158691

<u>Rule-20792</u>:

If both (I), (**U**), (**$**), (**f**), (**$'**), (**s**), (**v**), (**i**), (**T**), (**A**), (**Q**), (**V**), (**p**), (**c**) and (**q**) are known, then its Dividend Payout Planned is:

$$d= [U+\$[1-f]-\$'[1+s][v+i]-T$$
$$-(Q+Vp/\{360[c-q]\})/I]/A$$

<u>Rule-20793</u>:

If both (I), (**U**), (**$**), (**f**), (**$'**), (**s**), (**v**), (**i**), (**T**), (**A**), (**d**), (**Q**), (**p**), (**c**) and (**q**) are known, then its Variable Cost Planned is:

$$V= 360[c-q](I\{U+\$[1-f]-\$'[1+s][v+i]-T-Ad\}-Q)/p$$

<u>Rule-20794</u>:

If both (I), (**U**), (**$**), (**f**), (**$'**), (**s**), (**v**), (**i**), (**T**), (**A**), (**d**), (**V**), (**Q**), (**c**) and (**q**) are known, then its Procured Inventory Days Planned:

$$p= 360[c-q](I\{U+\$[1-f]-\$'[1+s][v+i]-T-Ad\}-Q)/V$$

<u>Rule-20795</u>:

If both (I), (**U**), (**$**), (**f**), (**$'**), (**s**), (**v**), (**i**), (**T**), (**A**), (**d**), (**V**), (**p**), (**Q**) and (**q**) are known, then its Current Ratio Planned is:

$$c= q+Vp/[360(I\{U+\$[1-f]-\$'[1+s][v+i]-T-Ad\}-Q)]$$

<u>Rule-20796</u>:

If both (I), (**U**), (**$**), (**f**), (**$'**), (**s**), (**v**), (**i**), (**T**), (**A**), (**d**), (**V**), (**p**), (**c**) and (**Q**) are known, then its Quick or Acid Test Ratio Planned is:

$$q= c-Vp/[360(I\{U+\$[1-f]-\$'[1+s][v+i]-T-Ad\}-Q)]$$

Steve Asikin ISBN 14: 978-1515158691, ISBN 10: 1515158691

Rule-20797:

If both (I), (U), (S), (f), (S'), (s), (v), (i), (T), (A), (d), (p), (c) and (q) are known, then its Quoted Longterm Debt Planned is:

$$Q= I\{U+S[1-f]-S'[1+s][v+i]-T-Ad\}-Svp/\{360[c-q]\}$$

Rule-20798:

If both (Q), (U), (S), (f), (S'), (s), (v), (i), (T), (A), (d), (p), (c) and (q) are known, then its Leverage or Gearing Ratio Planned is:

$$I= (Q+Svp/\{360[c-q]\})$$
$$/\{U+S[1-f]-S'[1+s][v+i]-T-Ad\}$$

Rule-20799:

If both (I), (Q), (S), (f), (S'), (s), (v), (i), (T), (A), (d), (p), (c) and (q) are known, then its Utilized or Starting Capital must be:

$$U= S'[1+s][v+i]+T+Ad$$
$$+(Q+Svp/\{360[c-q]\})/I-S[1-f]$$

Rule-20800:

If both (I), (U), (Q), (f), (S'), (s), (v), (i), (T), (A), (d), (p), (c) and (q) are known, then its Sales or Revenue Planned is:

$$S= (I\{U-S'[1+s][v+i]-T-Ad\}-S)$$
$$/(vp/\{360[c-q]\}-I[1-f])$$

Steve Asikin ISBN 14: 978-1515158691, ISBN 10: 1515158691

Rule-20801:

If both (I), (**U**), (**$**), (**Q**), (**$'**), (**s**), (**v**), (**i**), (**T**), (**A**), (**d**), (**p**), (**c**) and (**q**) are known, then its Fixed Portion Planned is:

$$f = 1-[\$'[1+s][v+i]+T+Ad$$
$$+(Q+\$vp/\{360[c-q]\})/I\cdot U]/\$$$

Rule-20802:

If both (I), (**U**), (**$**), (**f**), (**Q**), (**s**), (**v**), (**i**), (**T**), (**A**), (**d**), (**p**), (**c**) and (**q**) are known, then its Sales Past must be:

$$\$' = [U+\$[1-f]-T-Ad-(Q+\$vp/\{360[c-q]\})/I]$$
$$/\{[1+s][v+i]\}$$

Rule-20803:

If both (I), (**U**), (**$**), (**f**), (**$'**), (**Q**), (**v**), (**i**), (**T**), (**A**), (**d**), (**p**), (**c**) and (**q**) are known, then its Sales Growth Planned is:

$$s = \$/\$'-1$$

or it can also be found as

$$s = [U+\$[1-f]-T-Ad-(Q+\$vp/\{360[c-q]\})/I]$$
$$/\{\$'[v+i]\}-1$$

Rule-20804:

If both (I), (**U**), (**$**), (**f**), (**$'**), (**s**), (**Q**), (**i**), (**T**), (**A**), (**d**), (**p**), (**c**) and (**q**) are known, then its Variable Portion Planned is:

$$v = (I\{U+\$[1-f]-T-Ad\}-Q)$$
$$/(\$p/\{360[c-q]\}+\$'I[1+s])$$

Steve Asikin ISBN 14: 978-1515158691, ISBN 10: 1515158691

Rule-20805:

If both (I), (U), (S), (f), (S'), (s), (v), (Q), (T), (A), (d), (p), (c) and (q) are known, then its Interest Portion Planned is:

$$i= [U+S[1-f]-T-Ad-(Q+Svp/\{360[c-q]\})/I] /\{S'[1+s]\}-v$$

Rule-20806:

If both (I), (U), (S), (f), (S'), (s), (v), (i), (Q), (A), (d), (p), (c) and (q) are known, then its Tax Planned is:

$$T= U+S[1-f]-S'[1+s][v+i]-Ad -(Q+Svp/\{360[c-q]\})/I$$

Rule-20807:

If both (I), (U), (S), (f), (S'), (s), (v), (i), (T), (Q), (d), (p), (c) and (q) are known, then its After Tax Income Planned is:

$$A= [U+S[1-f]-S'[1+s][v+i]-T -(Q+Svp/\{360[c-q]\})/I]/d$$

Rule-20808:

If both (I), (U), (S), (f), (S'), (s), (v), (i), (T), (A), (Q), (p), (c) and (q) are known, then its Dividend Payout Planned is:

$$d= [U+S[1-f]-S'[1+s][v+i]-T -(Q+Svp/\{360[c-q]\})/I]/A$$

Steve Asikin ISBN 14: 978-1515158691, ISBN 10: 1515158691

Rule-20809:

If both (I), (**U**), (**$**), (**f**), (**$'**), (**s**), (**v**), (**i**), (**T**), (**A**), (**d**), (**Q**), (**c**) and (**q**) are known, then its Procured Inventory Days Planned:

$$p= 360[c\text{-}q](I\{U+\$[1\text{-}f]\text{-}\$'[1+s][v+i]\text{-}T\text{-}Ad\}\text{-}Q)/[\$v]$$

Rule-20810:

If both (I), (**U**), (**$**), (**f**), (**$'**), (**s**), (**v**), (**i**), (**T**), (**A**), (**d**), (**p**), (**Q**) and (**q**) are known, then its Current Ratio Planned is:

$$c= q+\$vp/[360(I\{U+\$[1\text{-}f]\text{-}\$'[1+s][v+i]\text{-}T\text{-}Ad\}\text{-}Q)]$$

Rule-20811:

If both (I), (**U**), (**$**), (**f**), (**$'**), (**s**), (**v**), (**i**), (**T**), (**A**), (**d**), (**p**), (**c**) and (**Q**) are known, then its Quick or Acid Test Ratio Planned is:

$$q= c\text{-}\$vp/[360(I\{U+\$[1\text{-}f]\text{-}\$'[1+s][v+i]\text{-}T\text{-}Ad\}\text{-}Q)]$$

Rule-20812:

If both (I), (**U**), (**$**), (**f**), (**$'**), (**s**), (**v**), (**i**), (**T**), (**A**), (**d**), (**p**), (**c**) and (**q**) are known, then its Quoted Longterm Debt Planned is:

$$Q= I\{U+\$[1\text{-}f]\text{-}\$'[1+s][v+i]\text{-}T\text{-}Ad\}-\$vp[1+s]/\{360[c\text{-}q]\}$$

Steve Asikin ISBN 14: 978-1515158691, ISBN 10: 1515158691

Rule-20813:

If both **(Q)**, **(U)**, **($)**, **(f)**, **($')**, **(s)**, **(v)**, **(i)**, **(T)**, **(A)**, **(d)**, **(p)**, **(c)** and **(q)** are known, then its Leverage or Gearing Ratio Planned is:

$$\textit{l} = (Q+\$'\textbf{vp}[1+\textbf{s}]/\{360[\textbf{c-q}]\})$$
$$/\{\textbf{U}+\$[1-\textbf{f}]-\$'[1+\textbf{s}][\textbf{v}+\textbf{i}]-\textbf{T-Ad}\}$$

Rule-20814:

If both **(l)**, **(Q)**, **($)**, **(f)**, **($')**, **(s)**, **(v)**, **(i)**, **(T)**, **(A)**, **(d)**, **(p)**, **(c)** and **(q)** are known, then its Utilized or Starting Capital must be:

$$\textbf{U} = \$'[1+\textbf{s}][\textbf{v}+\textbf{i}]+\textbf{T+Ad}$$
$$+(Q+\$'\textbf{vp}[1+\textbf{s}]/\{360[\textbf{c-q}]\})/\textit{l}\$[1-\textbf{f}]$$

Rule-20815:

If both **(l)**, **(U)**, **(Q)**, **(f)**, **($')**, **(s)**, **(v)**, **(i)**, **(T)**, **(A)**, **(d)**, **(p)**, **(c)** and **(q)** are known, then its Sales or Revenue Planned is:

$$\$ = (Q-\textit{l}\{\textbf{U}-\$'[1+\textbf{s}][\textbf{v}+\textbf{i}]-\textbf{T-Ad}\}$$
$$+\$'\textbf{vp}[1+\textbf{s}]/\{360[\textbf{c-q}]\})/\{\textit{l}[1-\textbf{f}]\}$$

Rule-20816:

If both **(l)**, **(U)**, **($)**, **(Q)**, **($')**, **(s)**, **(v)**, **(i)**, **(T)**, **(A)**, **(d)**, **(p)**, **(c)** and **(q)** are known, then its Fixed Portion Planned is:

$$\textbf{f} = 1-[\$'[1+\textbf{s}][\textbf{v}+\textbf{i}]+\textbf{T+Ad}$$
$$+(Q+\$'\textbf{vp}[1+\textbf{s}]/\{360[\textbf{c-q}]\})/\textit{l}\textbf{U}]/\$$$

Steve Asikin ISBN 14: 978-1515158691, ISBN 10: 1515158691

Rule-20817:

If both (I), (U), (S), (f), (Q), (s), (v), (i), (T), (A), (d), (p), (c) and (q) are known, then its Sales Past must be:

$$S' = (I\{U+S[1-f]-T-Ad\}-Q)$$
$$/[v[1+s](p/\{360[c-q]\}+I)]$$

Rule-20818:

If both (I), (U), (S), (f), (S'), (Q), (v), (i), (T), (A), (d), (p), (c) and (q) are known, then its Sales Growth Planned is:

$$s = S/S'-1$$

or it can also be found as

$$s = (I\{U+S[1-f]-T-Ad\}-Q)/[S'v(p/\{360[c-q]\}+I)]-1$$

Rule-20819:

If both (I), (U), (S), (f), (S'), (s), (Q), (i), (T), (A), (d), (p), (c) and (q) are known, then its Variable Portion Planned is:

$$v = (I\{U+S[1-f]-T-Ad\}-Q)$$
$$/[S'[1+s](p/\{360[c-q]\}+I)]$$

Rule-20820:

If both (I), (U), (S), (f), (S'), (s), (v), (Q), (T), (A), (d), (p), (c) and (q) are known, then its Qinterest Portion Planned is:

$$i = [U+S[1-f]-T-Ad-(Q+S'vp[1+s]/\{360[c-q]\})/I]$$
$$/\{S'[1+s]\}-v$$

Steve Asikin ISBN 14: 978-1515158691, ISBN 10: 1515158691

Rule-20821:

If both (**/**), (**U**), (**$**), (**f**), (**$'**), (**s**), (**v**), (**i**), (**Q**), (**A**), (**d**), (**p**), (**c**) and (**q**) are known, then its Tax Planned is:

$$T= U+\$[1-f]-\$'[1+s][v+i]-Ad$$
$$-(Q+\$'vp[1+s]/\{360[c-q]\})/\text{/}$$

Rule-20822:

If both (**/**), (**U**), (**$**), (**f**), (**$'**), (**s**), (**v**), (**i**), (**T**), (**Q**), (**d**), (**p**), (**c**) and (**q**) are known, then its After Tax Income Planned is:

$$A= [U+\$[1-f]-\$'[1+s][v+i]-T$$
$$-(Q+\$'vp[1+s]/\{360[c-q]\})/\text{/}/d$$

Rule-20823:

If both (**/**), (**U**), (**$**), (**f**), (**$'**), (**s**), (**v**), (**i**), (**T**), (**A**), (**Q**), (**p**), (**c**) and (**q**) are known, then its Dividend Payout Planned is:

$$d= [U+\$[1-f]-\$'[1+s][v+i]-T$$
$$-(Q+\$'vp[1+s]/\{360[c-q]\})/\text{/}/A$$

Rule-20824:

If both (**/**), (**U**), (**$**), (**f**), (**$'**), (**s**), (**v**), (**i**), (**T**), (**A**), (**d**), (**p**), (**c**) and (**q**) are known, then its Quoted Longterm Debt Planned is:

$$p= 360[c-q](\text{/}\{U+\$[1-f]-\$'[1+s][v+i]-T-Ad\}-Q)$$
$$/\{\$'v[1+s]\}$$

Steve Asikin ISBN 14: 978-1515158691, ISBN 10: 1515158691

Rule-20825:

If both $(\textit{f})$, $(\textbf{U})$, $(\textbf{\$})$, $(\textbf{f})$, $(\textbf{\$'})$, $(\textbf{s})$, $(\textbf{v})$, $(\textbf{i})$, $(\textbf{T})$, $(\textbf{A})$, $(\textbf{d})$, $(\textbf{p})$, $(\textbf{Q})$ and $(\textbf{q})$ are known, then its Current Ratio Planned is:

$$c = q + \$'vp[1+s]$$
$$/[360(\textit{f}\{U + \$[1-f] - \$'[1+s][v+i] - T - Ad\} - Q)]$$

Rule-20826:

If both $(\textit{f})$, $(\textbf{U})$, $(\textbf{\$})$, $(\textbf{f})$, $(\textbf{\$'})$, $(\textbf{s})$, $(\textbf{v})$, $(\textbf{i})$, $(\textbf{T})$, $(\textbf{A})$, $(\textbf{d})$, $(\textbf{p})$, $(\textbf{c})$ and $(\textbf{Q})$ are known, then its Quick or Acid Test Ratio Planned is:

$$q = c - \$'vp[1+s]$$
$$/[360(\textit{f}\{U + \$[1-f] - \$'[1+s][v+i] - T - Ad\} - Q)]$$

Rule-20827:

If both $(\textit{f})$, $(\textbf{U})$, $(\textbf{\$})$, $(\textbf{f})$, $(\textbf{\$'})$, $(\textbf{s})$, $(\textbf{v})$, $(\textbf{i})$, $(\textbf{t})$, $(\textbf{D})$ and $(\textbf{X})$ are known, then its Quoted Longterm Debt Planned is:

$$Q = \textit{f}\{U + \$[1-f] - \$'[1+s][v+i]\}[1-t] - D) - X$$

Rule-20828:

If both $(\textbf{Q})$, $(\textbf{U})$, $(\textbf{\$})$, $(\textbf{f})$, $(\textbf{\$'})$, $(\textbf{s})$, $(\textbf{v})$, $(\textbf{i})$, $(\textbf{t})$, $(\textbf{D})$ and $(\textbf{X})$ are known, then its Leverage or Gearing Ratio Planned is:

$$\textit{f} = [Q + X] / (U + \$[1-f] - \$'[1+s][v+i]\}[1-t] - D)$$

Rule-20829:

If both $(\textit{f})$, $(\textbf{Q})$, $(\textbf{\$})$, $(\textbf{f})$, $(\textbf{\$'})$, $(\textbf{s})$, $(\textbf{v})$, $(\textbf{i})$, $(\textbf{t})$, $(\textbf{D})$ and $(\textbf{X})$ are known, then its Utilized or Starting Capital must be:

$$U = \$'[1+s][v+i]\}[1-t] + D + [Q+X] / \textit{f} - \$[1-f]$$

Steve Asikin ISBN 14: 978-1515158691, ISBN 10: 1515158691

Rule-20830:

If both (I), (U), (Q), (f), (S'), (s), (v), (i), (t), (D) and (X) are known, then its Sales or Revenue Planned is:

$S= \{S'[1+s][v+i]\}[1-t]+D+[Q+X]/I-U\}/[1-f]$

Rule-20831:

If both (I), (U), (S), (Q), (S'), (s), (v), (i), (t), (D) and (X) are known, then its Fixed Portion Planned is:

$f= 1-\{S'[1+s][v+i]\}[1-t]+D+[Q+X]/I-U\}/S$

Rule-20832:

If both (I), (U), (S), (f), (Q), (s), (v), (i), (t), (D) and (X) are known, then its Sales Past must be:

$S' = \{U+S[1-f]-D-[Q+X]/I\}/\{[1+s][v+i][1-t]\}$

Rule-20833:

If both (I), (U), (S), (f), (S'), (Q), (v), (i), (t), (D) and (X) are known, then its Sales Growth Planned is:

$s= S/S'-1$

or it can also be found as

$s= \{U+S[1-f]-D-[Q+X]/I\}/\{S'[v+i][1-t]\}-1$

Rule-20834:

If both (I), (U), (S), (f), (S'), (s), (Q), (i), (t), (D) and (X) are known, then its Variable Portion Planned is:

$v= \{U+S[1-f]-D-[Q+X]/I\}/\{S'[1+s][1-t]\}-i$

Rule-20835:

If both (I), (U), (S), (f), (S'), (s), (v), (Q), (t), (D) and (X) are known, then its Interest Portion Planned is:

$i= \{U+S[1-f]-D-[Q+X]/I\}/\{S'[1+s][1-t]\}-v$

Steve Asikin ISBN 14: 978-1515158691, ISBN 10: 1515158691

Rule-20836:

    If both (I), (**U**), (**$**), (**f**), (**$'**), (**s**), (**v**), (**i**), (**Q**), (**D**) and
(**X**) are known, then its Tax Rate Planned is:

      t= 1-{**U**+**$**[1-**f**]-**D**-[**Q**+**X**]/I/{**$'**[1+**s**][**v**+**i**]}

Rule-20837:

    If both (I), (**U**), (**$**), (**f**), (**$'**), (**s**), (**v**), (**i**), (**t**), (**Q**) and
(**X**) are known, then its Dividend Planned is:

      D= **U**+**$**[1-**f**]-**$'**[1+**s**][**v**+**i**]}[1-**t**]-[**Q**+**X**]/I

Rule-20838:

    If both (I), (**U**), (**$**), (**f**), (**$'**), (**s**), (**v**), (**i**), (**t**), (**D**) and
(**Q**) are known, then its Xpress or Current Debt
Planned is:

      X= I{**U**+**$**[1-**f**]-**$'**[1+**s**][**v**+**i**]}[1-**t**]-**D**)-**Q**

Rule-20839:

    If both (I), (**U**), (**$**), (**f**), (**$'**), (**s**), (**v**), (**i**), (**t**), (**D**), (**P**),
(**c**) and (**q**) are known, then its Quoted Longterm Debt
Planned is:

      Q= I{**U**+**$**[1-**f**]-**$'**[1+**s**][**v**+**i**]}[1-**t**]-**D**)-**P**/[**c**-**q**]

Rule-20840:

    If both (**Q**), (**U**), (**$**), (**f**), (**$'**), (**s**), (**v**), (**i**), (**t**), (**D**), (**P**),
(**c**) and (**q**) are known, then its Leverage or Gearing
Ratio Planned is:

      I= {**Q**+**P**/[**c**-**q**]}/(**U**+**$**[1-**f**]-**$'**[1+**s**][**v**+**i**]}[1-**t**]-**D**)

Steve Asikin ISBN 14: 978-1515158691, ISBN 10: 1515158691

Rule-20841:

If both (f), (Q), (S), (f), (S'), (s), (v), (i), (t), (D), (P), (c) and (q) are known, then its Utilized or Starting Capital must be:

$$U= S'[1+s][v+i]\}[1-t]+D+\{Q+P/[c-q]\}//S[1-f]$$

Rule-20842:

If both (f), (U), (Q), (f), (S'), (s), (v), (i), (t), (D), (P), (c) and (q) are known, then its Sales or Revenue Planned is:

$$S= (S'[1+s][v+i]\}[1-t]+D+\{Q+P/[c-q]\}//U)/[1-f]$$

Rule-20843:

If both (f), (U), (S), (Q), (S'), (s), (v), (i), (t), (D), (P), (c) and (q) are known, then its Fixed Portion Planned is:

$$f= 1-(S'[1+s][v+i]\}[1-t]+D+\{Q+P/[c-q]\}//U)/S$$

Rule-20844:

If both (f), (U), (S), (f), (Q), (s), (v), (i), (t), (D), (P), (c) and (q) are known, then its Sales Past must be:

$$S' = (U+S[1-f]-D-\{Q+P/[c-q]\}//)/\{[1+s][v+i][1-t]\}$$

Rule-20845:

If both (f), (U), (S), (f), (S'), (Q), (v), (i), (t), (D), (P), (c) and (q) are known, then its Sales Growth Planned is:

$$s= S/S'-1$$

or it can also be found as

$$s= (U+S[1-f]-D-\{Q+P/[c-q]\}//)/\{S'[v+i][1-t]\}-1$$

Steve Asikin ISBN 14: 978-1515158691, ISBN 10: 1515158691

Rule-20846:

If both (Λ), $(\mathbf{U})$, $(\mathbf{S})$, $(\mathbf{f})$, $(\mathbf{S'})$, $(\mathbf{s})$, $(\mathbf{Q})$, $(\mathbf{i})$, $(\mathbf{t})$, $(\mathbf{D})$, $(\mathbf{P})$, $(\mathbf{c})$ and $(\mathbf{q})$ are known, then its Variable Portion Planned is:

$$\mathbf{v} = (\mathbf{U} + \mathbf{S}[1-\mathbf{f}] - \mathbf{D} - \{\mathbf{Q} + \mathbf{P}/[\mathbf{c} - \mathbf{q}]\} / \Lambda) / \{\mathbf{S'}[1+\mathbf{s}][1-\mathbf{t}]\} - \mathbf{i}$$

Rule-20847:

If both (Λ), $(\mathbf{U})$, $(\mathbf{S})$, $(\mathbf{f})$, $(\mathbf{S'})$, $(\mathbf{s})$, $(\mathbf{v})$, $(\mathbf{Q})$, $(\mathbf{t})$, $(\mathbf{D})$, $(\mathbf{P})$, $(\mathbf{c})$ and $(\mathbf{q})$ are known, then its Interest Portion Planned is:

$$\mathbf{i} = (\mathbf{U} + \mathbf{S}[1-\mathbf{f}] - \mathbf{D} - \{\mathbf{Q} + \mathbf{P}/[\mathbf{c} - \mathbf{q}]\} / \Lambda) / \{\mathbf{S'}[1+\mathbf{s}][1-\mathbf{t}]\} - \mathbf{v}$$

Rule-20848:

If both (Λ), $(\mathbf{U})$, $(\mathbf{S})$, $(\mathbf{f})$, $(\mathbf{S'})$, $(\mathbf{s})$, $(\mathbf{v})$, $(\mathbf{i})$, $(\mathbf{Q})$, $(\mathbf{D})$, $(\mathbf{P})$, $(\mathbf{c})$ and $(\mathbf{q})$ are known, then its Tax Rate Planned is:

$$\mathbf{t} = 1 - (\mathbf{U} + \mathbf{S}[1-\mathbf{f}] - \mathbf{D} - \{\mathbf{Q} + \mathbf{P}/[\mathbf{c} - \mathbf{q}]\} / \Lambda) / \{\mathbf{S'}[1+\mathbf{s}][\mathbf{v} + \mathbf{i}]\}$$

Rule-20849:

If both (Λ), $(\mathbf{U})$, $(\mathbf{S})$, $(\mathbf{f})$, $(\mathbf{S'})$, $(\mathbf{s})$, $(\mathbf{v})$, $(\mathbf{i})$, $(\mathbf{t})$, $(\mathbf{Q})$, $(\mathbf{P})$, $(\mathbf{c})$ and $(\mathbf{q})$ are known, then its Dividend Planned is:

$$\mathbf{D} = \mathbf{U} + \mathbf{S}[1-\mathbf{f}] - \mathbf{S'}[1+\mathbf{s}][\mathbf{v} + \mathbf{i}]\} [1-\mathbf{t}] - \{\mathbf{Q} + \mathbf{P}/[\mathbf{c} - \mathbf{q}]\} / \Lambda$$

Rule-20850:

If both (Λ), $(\mathbf{U})$, $(\mathbf{S})$, $(\mathbf{f})$, $(\mathbf{S'})$, $(\mathbf{s})$, $(\mathbf{v})$, $(\mathbf{i})$, $(\mathbf{t})$, $(\mathbf{D})$, $(\mathbf{Q})$, $(\mathbf{c})$ and $(\mathbf{q})$ are known, then its Procured Inventory Planned is:

$$\mathbf{P} = [\mathbf{c} - \mathbf{q}][\Lambda(\mathbf{U} + \mathbf{S}[1-\mathbf{f}] - \mathbf{S'}[1+\mathbf{s}][\mathbf{v} + \mathbf{i}]\} [1-\mathbf{t}] - \mathbf{D}) - \mathbf{Q}]$$

Steve Asikin ISBN 14: 978-1515158691, ISBN 10: 1515158691

Rule-20851:

If both (*f*), (**U**), (**$**), (**f**), (**$'**), (**s**), (**v**), (**i**), (**t**), (**D**), (**P**), (**Q**) and (**q**) are known, then its Current Ratio Planned is:

$$c = q + P/[\{f U + \$[1-f] - \$'[1+s][v+i]\}[1-t] - D) - Q]$$

Rule-20852:

If both (*f*), (**U**), (**$**), (**f**), (**$'**), (**s**), (**v**), (**i**), (**t**), (**D**), (**P**), (**c**) and (**Q**) are known, then its Quick or Acid Test Ratio Planned is:

$$q = c - P/[\{f U + \$[1-f] - \$'[1+s][v+i]\}[1-t] - D) - Q]$$

Rule-20853:

If both (*f*), (**U**), (**$**), (**f**), (**$'**), (**s**), (**v**), (**i**), (**t**), (**D**), (**V**), (**p**), (**c**) and (**q**) are known, then its Quoted Longterm Debt Planned is:

$$Q = \{f U + \$[1-f] - \$'[1+s][v+i]\}[1-t] - D)$$
$$-Vp/\{360[c-q]\}$$

Rule-20854:

If both (**Q**), (**U**), (**$**), (**f**), (**$'**), (**s**), (**v**), (**i**), (**t**), (**D**), (**V**), (**p**), (**c**) and (**q**) are known, then its Leverage or Gearing Ratio Planned is:

$$f = (Q + Vp/\{360[c-q]\})$$
$$/(U + \$[1-f] - \$'[1+s][v+i]\}[1-t] - D)$$

Steve Asikin ISBN 14: 978-1515158691, ISBN 10: 1515158691

Rule-20855:

If both (**/**), (**Q**), (**$**), (**f**), (**$'**), (**s**), (**v**), (**i**), (**t**), (**D**), (**V**), (**p**), (**c**) and (**q**) are known, then its Utilized or Starting Capital must be:

$$U = \$'[1+s][v+i]\}[1\text{-}t]+D$$
$$+(Q+Vp/\{360[c\text{-}q]\})/\text{/-}\$[1\text{-}f]$$

Rule-20856:

If both (**/**), (**U**), (**Q**), (**f**), (**$'**), (**s**), (**v**), (**i**), (**t**), (**D**), (**V**), (**p**), (**c**) and (**q**) are known, then its Sales or Revenue Planned is:

$$\$ = [\$'[1+s][v+i]\}[1\text{-}t]+D+(Q+Vp/\{360[c\text{-}q]\})/\text{/-}U]$$
$$/[1\text{-}f]$$

Rule-20857:

If both (**/**), (**U**), (**$**), (**Q**), (**$'**), (**s**), (**v**), (**i**), (**t**), (**D**), (**V**), (**p**), (**c**) and (**q**) are known, then its Fixed Portion Planned is:

$$f = 1\text{-}[\$'[1+s][v+i]\}[1\text{-}t]+D$$
$$+(Q+Vp/\{360[c\text{-}q]\})/\text{/-}U]/\$$$

Rule-20858:

If both (**/**), (**U**), (**$**), (**f**), (**Q**), (**s**), (**v**), (**i**), (**t**), (**D**), (**V**), (**p**), (**c**) and (**q**) are known, then its Sales Past must be:

$$\$' = [U+\$[1\text{-}f]\text{-}D\text{-}(Q+Vp/\{360[c\text{-}q]\})/\text{/}]$$
$$/\{[1+s][v+i][1\text{-}t]\}$$

Steve Asikin ISBN 14: 978-1515158691, ISBN 10: 1515158691

<u>Rule-20859</u>:

If both $(\textit{l})$, $(\textbf{U})$, $(\textbf{\$})$, $(\textbf{f})$, $(\textbf{\$'})$, $(\textbf{Q})$, $(\textbf{v})$, $(\textbf{i})$, $(\textbf{t})$, $(\textbf{D})$, $(\textbf{V})$, $(\textbf{p})$, $(\textbf{c})$ and $(\textbf{q})$ are known, then its Sales Growth Planned is:

$$s= \$/\$'-1$$

or it can also be found as

$$s= [U+\$[1-f]-D-(Q+Vp/\{360[c-q]\})/\textit{l}]$$
$$/\{\$'[v+i][1-t]\}-1$$

<u>Rule-20860</u>:

If both $(\textit{l})$, $(\textbf{U})$, $(\textbf{\$})$, $(\textbf{f})$, $(\textbf{\$'})$, $(\textbf{s})$, $(\textbf{Q})$, $(\textbf{i})$, $(\textbf{t})$, $(\textbf{D})$, $(\textbf{V})$, $(\textbf{p})$, $(\textbf{c})$ and $(\textbf{q})$ are known, then its Variable Portion Planned is:

$$v= V/\$$$

or it can also be found as

$$v=[U+\$[1-f]-D-(Q+Vp/\{360[c-q]\})/\textit{l}]$$
$$/\{\$'[1+s][1-t]\}-i$$

<u>Rule-20861</u>:

If both $(\textit{l})$, $(\textbf{U})$, $(\textbf{\$})$, $(\textbf{f})$, $(\textbf{\$'})$, $(\textbf{s})$, $(\textbf{v})$, $(\textbf{Q})$, $(\textbf{t})$, $(\textbf{D})$, $(\textbf{V})$, $(\textbf{p})$, $(\textbf{c})$ and $(\textbf{q})$ are known, then its Interest Portion Planned is:

$$i= [U+\$[1-f]-D-(Q+Vp/\{360[c-q]\})/\textit{l}]$$
$$/\{\$'[1+s][1-t]\}-v$$

<u>Rule-20862</u>:

If both $(\textit{l})$, $(\textbf{U})$, $(\textbf{\$})$, $(\textbf{f})$, $(\textbf{\$'})$, $(\textbf{s})$, $(\textbf{v})$, $(\textbf{i})$, $(\textbf{Q})$, $(\textbf{D})$, $(\textbf{V})$, $(\textbf{p})$, $(\textbf{c})$ and $(\textbf{q})$ are known, then its Tax Rate Planned is:

$$t= 1-[U+\$[1-f]-D-(Q+Vp/\{360[c-q]\})/\textit{l}]$$
$$/\{\$'[1+s][v+i]\}$$

Steve Asikin ISBN 14: 978-1515158691, ISBN 10: 1515158691

Rule-20863:

If both (**/**), (**U**), (**$**), (**f**), (**$'**), (**s**), (**v**), (**i**), (**t**), (**Q**), (**V**), (**p**), (**c**) and (**q**) are known, then its Dividend Planned is:

$$D= U+\$[1-f]-\$'[1+s][v+i]\}[1-t]$$
$$-(Q+Vp/\{360[c-q]\})//$$

Rule-20864:

If both (**/**), (**U**), (**$**), (**f**), (**$'**), (**s**), (**v**), (**i**), (**t**), (**D**), (**Q**), (**p**), (**c**) and (**q**) are known, then its Variable Cost Planned is:

$$V= 36-[c-q][\{U+\$[1-f]-\$'[1+s][v+i]\}[1-t]-D)-Q]/p$$

Rule-20865:

If both (**/**), (**U**), (**$**), (**f**), (**$'**), (**s**), (**v**), (**i**), (**t**), (**D**), (**V**), (**Q**), (**c**) and (**q**) are known, then its Procured Inventory Days Planned:

$$p= 36-[c-q][\{U+\$[1-f]-\$'[1+s][v+i]\}[1-t]-D)-Q]/V$$

Rule-20866:

If both (**/**), (**U**), (**$**), (**f**), (**$'**), (**s**), (**v**), (**i**), (**t**), (**D**), (**V**), (**p**), (**Q**) and (**q**) are known, then its Current Ratio Planned is:

$$c= q+Vp$$
$$/\{360[\{U+\$[1-f]-\$'[1+s][v+i]\}[1-t]-D)-Q]\}$$

Rule-20867:

If both (**/**), (**U**), (**$**), (**f**), (**$'**), (**s**), (**v**), (**i**), (**t**), (**D**), (**V**), (**p**), (**c**) and (**q**) are known, then its Quoted Longterm Debt Planned is:

$$q= c-Vp$$
$$/\{360[\{U+\$[1-f]-\$'[1+s][v+i]\}[1-t]-D)-Q]\}$$

Steve Asikin ISBN 14: 978-1515158691, ISBN 10: 1515158691

Rule-20868:
  If both (**/**), (**U**), (**$**), (**f**), (**$'**), (**s**), (**v**), (**i**), (**t**), (**D**), (**p**),
  (**c**) and (**q**) are known, then its Quoted Longterm Debt
  Planned is:
  $$Q= /(U+\$[1-f]-\$'[1+s][v+i]\}[1-t]-D)$$
  $$-\$vp/\{360[c-q]\}$$

Rule-20869:
  If both (**Q**), (**U**), (**$**), (**f**), (**$'**), (**s**), (**v**), (**i**), (**t**), (**D**), (**p**),
  (**c**) and (**q**) are known, then its Leverage or Gearing
  Ratio Planned is:
  $$/= (Q+\$vp/\{360[c-q]\})$$
  $$/(U+\$[1-f]-\$'[1+s][v+i]\}[1-t]-D)$$

Rule-20870:
  If both (**/**), (**Q**), (**$**), (**f**), (**$'**), (**s**), (**v**), (**i**), (**t**), (**D**), (**p**),
  (**c**) and (**q**) are known, then its Utilized or Starting
  Capital must be:
  $$U= \$'[1+s][v+i]\}[1-t]+D$$
  $$+(Q+\$vp/\{360[c-q]\})//\$[1-f]$$

Rule-20871:
  If both (**/**), (**U**), (**Q**), (**f**), (**$'**), (**s**), (**v**), (**i**), (**t**), (**D**), (**p**),
  (**c**) and (**q**) are known, then its Sales or Revenue
  Planned is:
  $$S= [/(U-\$'[1+s][v+i]\}[1-t]-D)-Q]$$
  $$/(vp/\{360[c-q]\}-/[1-f][1-t])$$

Steve Asikin ISBN 14: 978-1515158691, ISBN 10: 1515158691

Rule-20872:

If both $(\textit{\textbf{l}})$, $(\textbf{U})$, $(\textbf{\$})$, $(\textbf{Q})$, $(\textbf{\$'})$, $(\textbf{s})$, $(\textbf{v})$, $(\textbf{i})$, $(\textbf{t})$, $(\textbf{D})$, $(\textbf{p})$, $(\textbf{c})$ and $(\textbf{q})$ are known, then its Fixed Portion Planned is:

$$f= 1-[\$'[1+s][v+i]\}[1-t]+D$$
$$+(Q+\$vp/\{360[c-q]\})/\textit{l}U]/\$$$

Rule-20873:

If both $(\textit{\textbf{l}})$, $(\textbf{U})$, $(\textbf{\$})$, $(\textbf{f})$, $(\textbf{Q})$, $(\textbf{s})$, $(\textbf{v})$, $(\textbf{i})$, $(\textbf{t})$, $(\textbf{D})$, $(\textbf{p})$, $(\textbf{c})$ and $(\textbf{q})$ are known, then its Sales Past must be:

$$\$' = [U+\$[1-f]-D-(Q+\$vp/\{360[c-q]\})/\textit{l}]$$
$$/\{[1+s][v+i][1-t]\}$$

Rule-20874:

If both $(\textit{\textbf{l}})$, $(\textbf{U})$, $(\textbf{\$})$, $(\textbf{f})$, $(\textbf{\$'})$, $(\textbf{Q})$, $(\textbf{v})$, $(\textbf{i})$, $(\textbf{t})$, $(\textbf{D})$, $(\textbf{p})$, $(\textbf{c})$ and $(\textbf{q})$ are known, then its Sales Growth Planned is:

$$s= \$/\$'-1$$

or it can also be found as

$$s= [U+\$[1-f]-D-(Q+\$vp/\{360[c-q]\})/\textit{l}]$$
$$/\{\$'[v+i][1-t]\}-1$$

Rule-20875:

If both $(\textit{\textbf{l}})$, $(\textbf{U})$, $(\textbf{\$})$, $(\textbf{f})$, $(\textbf{\$'})$, $(\textbf{s})$, $(\textbf{Q})$, $(\textbf{i})$, $(\textbf{t})$, $(\textbf{D})$, $(\textbf{p})$, $(\textbf{c})$ and $(\textbf{q})$ are known, then its Variable Portion Planned is:

$$v= [\textit{l}U+\$[1-f]-\$'i[1+s]\}[1-t]-D)-Q]$$
$$/(\$p/\{360[c-q]\}+\$'\textit{l}[1+s][1-t])$$

Steve Asikin ISBN 14: 978-1515158691, ISBN 10: 1515158691

Rule-20876:

If both (**/**), (**U**), (**$**), (**f**), (**$'**), (**s**), (**v**), (**Q**), (**t**), (**D**), (**p**), (**c**) and (**q**) are known, then its Interest Portion Planned is:

$$i= [U+\$[1-f]-D-(Q+\$vp/\{360[c-q]\})//]$$
$$/\{\$'[1+s][1-t]\}-v$$

Rule-20877:

If both (**/**), (**U**), (**$**), (**f**), (**$'**), (**s**), (**v**), (**i**), (**Q**), (**D**), (**p**), (**c**) and (**q**) are known, then its Tax Rate Planned is:

$$t= 1-[U+\$[1-f]-D-(Q+\$vp/\{360[c-q]\})//]$$
$$/\{\$'[1+s][v+i]\}$$

Rule-20878:

If both (**/**), (**U**), (**$**), (**f**), (**$'**), (**s**), (**v**), (**i**), (**t**), (**Q**), (**p**), (**c**) and (**q**) are known, then its Dividend Planned is:

$$D= U+\$[1-f]-\$'[1+s][v+i]\}[1-t]$$
$$-(Q+\$vp/\{360[c-q]\})//$$

Rule-20879:

If both (**/**), (**U**), (**$**), (**f**), (**$'**), (**s**), (**v**), (**i**), (**t**), (**D**), (**Q**), (**c**) and (**q**) are known, then its Procured Inventory Days Planned:

$$p= 36-[c-q][/U+\$[1-f]-\$'[1+s][v+i]\}[1-t]-D)-Q]$$
$$/[\$v]$$

Rule-20880:

If both (**/**), (**U**), (**$**), (**f**), (**$'**), (**s**), (**v**), (**i**), (**t**), (**D**), (**p**), (**Q**) and (**q**) are known, then its Current Ratio Planned is:

$$c= q+\$vp$$
$$/\{360[/U+\$[1-f]-\$'[1+s][v+i]\}[1-t]-D)-Q]\}$$

Steve Asikin ISBN 14: 978-1515158691, ISBN 10: 1515158691

Rule-20881:

If both $(\textit{l})$, $(\textbf{U})$, $(\textbf{\$})$, $(\textbf{f})$, $(\textbf{\$'})$, $(\textbf{s})$, $(\textbf{v})$, $(\textbf{i})$, $(\textbf{t})$, $(\textbf{D})$, $(\textbf{p})$, $(\textbf{c})$ and $(\textbf{Q})$ are known, then its Quick or Acid Test Ratio Planned is:

$$q = c\text{-}\$vp$$
$$/\{360[\textit{l}\textbf{U}+\$[1\text{-}f]\text{-}\$'[1+s][v+i]\}[1\text{-}t]\text{-}\textbf{D})\text{-}\textbf{Q}]\}$$

Rule-20882:

If both $(\textit{l})$, $(\textbf{U})$, $(\textbf{\$})$, $(\textbf{f})$, $(\textbf{\$'})$, $(\textbf{s})$, $(\textbf{v})$, $(\textbf{i})$, $(\textbf{t})$, $(\textbf{D})$, $(\textbf{p})$, $(\textbf{c})$ and $(\textbf{q})$ are known, then its Quoted Longterm Debt Planned is:

$$Q = \textit{l}(\textbf{U}+\$[1\text{-}f]\text{-}\$'[1+s][v+i]\}[1\text{-}t]\text{-}\textbf{D})$$
$$-\$'vp[1+s]/\{360[c\text{-}q]\}$$

Rule-20883:

If both $(\textbf{Q})$, $(\textbf{U})$, $(\textbf{\$})$, $(\textbf{f})$, $(\textbf{\$'})$, $(\textbf{s})$, $(\textbf{v})$, $(\textbf{i})$, $(\textbf{t})$, $(\textbf{D})$, $(\textbf{p})$, $(\textbf{c})$ and $(\textbf{q})$ are known, then its Leverage or Gearing Ratio Planned is:

$$\textit{l} = (\textbf{Q}+\$'vp[1+s]/\{360[c\text{-}q]\})$$
$$/(\textbf{U}+\$[1\text{-}f]\text{-}\$'[1+s][v+i]\}[1\text{-}t]\text{-}\textbf{D})$$

Rule-20884:

If both $(\textit{l})$, $(\textbf{Q})$, $(\textbf{\$})$, $(\textbf{f})$, $(\textbf{\$'})$, $(\textbf{s})$, $(\textbf{v})$, $(\textbf{i})$, $(\textbf{t})$, $(\textbf{D})$, $(\textbf{p})$, $(\textbf{c})$ and $(\textbf{q})$ are known, then its Utilized or Starting Capital must be:

$$U = \$'[1+s][v+i]\}[1\text{-}t]+\textbf{D}$$
$$+(\textbf{Q}+\$'vp[1+s]/\{360[c\text{-}q]\})/\textit{l}\text{-}\$[1\text{-}f]$$

Steve Asikin ISBN 14: 978-1515158691, ISBN 10: 1515158691

Rule-20885:

> If both (I), (U), (Q), (f), (S'), (s), (v), (i), (t), (D), (p), (c) and (q) are known, then its Sales or Revenue Planned is:
>
> $$S= [Q-I\{U-S'[1+s][v+i]\}[1-t]-D)$$
> $$+S'vp[1+s]/\{360[c-q]\}]/\{I[1-f][1-t]\}$$

Rule-20886:

> If both (I), (U), (S), (Q), (S'), (s), (v), (i), (t), (D), (p), (c) and (q) are known, then its Fixed Portion Planned is:
>
> $$f= 1-[S'[1+s][v+i]\}[1-t]+D$$
> $$+(Q+S'vp[1+s]/\{360[c-q]\})/IU]/S$$

Rule-20887:

> If both (I), (U), (S), (f), (Q), (s), (v), (i), (t), (D), (p), (c) and (q) are known, then its Sales Past must be:
>
> $$S' = [\{IU+S[1-f]-S'i[1+s]\}[1-t]-D)-Q]$$
> $$/[v[1+s](p/\{360[c-q]\}+I[1-t])$$

Rule-20888:

> If both (I), (U), (S), (f), (S'), (Q), (v), (i), (t), (D), (p), (c) and (q) are known, then its Sales Growth Planned is:
>
> $$s= S/S'-1$$
>
> or it can also be found as
>
> $$s= [\{IU+S[1-f]-S'i[1+s]\}[1-t]-D)-Q]$$
> $$/[S'v(p/\{360[c-q]\}+I[1-t])-1$$

Steve Asikin ISBN 14: 978-1515158691, ISBN 10: 1515158691

Rule-20889:

If both (I), $(\mathbf{U})$, $(\mathbf{S})$, $(\mathbf{f})$, $(\mathbf{S'})$, $(\mathbf{s})$, $(\mathbf{Q})$, $(\mathbf{i})$, $(\mathbf{t})$, $(\mathbf{D})$, $(\mathbf{p})$, $(\mathbf{c})$ and $(\mathbf{q})$ are known, then its Variable Portion Planned is:

$$v = [(\mathbf{I}\mathbf{U} + \mathbf{S}[1-\mathbf{f}] - \mathbf{S'i}[1+\mathbf{s}]\}[1-\mathbf{t}] - \mathbf{D}) - \mathbf{Q}]$$
$$/[\mathbf{S'}[1+\mathbf{s}](\mathbf{p}/\{360[\mathbf{c} - \mathbf{q}]\} + I[1-\mathbf{t}])$$

Rule-20890:

If both (I), $(\mathbf{U})$, $(\mathbf{S})$, $(\mathbf{f})$, $(\mathbf{S'})$, $(\mathbf{s})$, $(\mathbf{v})$, $(\mathbf{Q})$, $(\mathbf{t})$, $(\mathbf{D})$, $(\mathbf{p})$, $(\mathbf{c})$ and $(\mathbf{q})$ are known, then its Interest Portion Planned is:

$$i = [\mathbf{U} + \mathbf{S}[1-\mathbf{f}] - \mathbf{D} - (\mathbf{Q} + \mathbf{S'vp}[1+\mathbf{s}]/\{360[\mathbf{c} - \mathbf{q}]\})/I]$$
$$/\{\mathbf{S'}[1+\mathbf{s}][1-\mathbf{t}]\} - \mathbf{v}$$

Rule-20891:

If both (I), $(\mathbf{U})$, $(\mathbf{S})$, $(\mathbf{f})$, $(\mathbf{S'})$, $(\mathbf{s})$, $(\mathbf{v})$, $(\mathbf{i})$, $(\mathbf{Q})$, $(\mathbf{D})$, $(\mathbf{p})$, $(\mathbf{c})$ and $(\mathbf{q})$ are known, then its Tax Rate Planned is:

$$t = 1 - [\mathbf{U} + \mathbf{S}[1-\mathbf{f}] - \mathbf{D} - (\mathbf{Q} + \mathbf{S'vp}[1+\mathbf{s}]/\{360[\mathbf{c} - \mathbf{q}]\})/I]$$
$$/\{\mathbf{S'}[1+\mathbf{s}][\mathbf{v} + \mathbf{i}]\}$$

Rule-20892:

If both (I), $(\mathbf{U})$, $(\mathbf{S})$, $(\mathbf{f})$, $(\mathbf{S'})$, $(\mathbf{s})$, $(\mathbf{v})$, $(\mathbf{i})$, $(\mathbf{t})$, $(\mathbf{Q})$, $(\mathbf{p})$, $(\mathbf{c})$ and $(\mathbf{q})$ are known, then its Dividend Planned is:

$$D = \mathbf{U} + \mathbf{S}[1-\mathbf{f}] - \mathbf{S'}[1+\mathbf{s}][\mathbf{v} + \mathbf{i}]\}[1-\mathbf{t}]$$
$$- (\mathbf{Q} + \mathbf{S'vp}[1+\mathbf{s}]/\{360[\mathbf{c} - \mathbf{q}]\})/I$$

Steve Asikin ISBN 14: 978-1515158691, ISBN 10: 1515158691

Rule-20893:
   If both (**/**), (**U**), (**$**), (**f**), (**$'**), (**s**), (**v**), (**i**), (**t**), (**D**), (**Q**), (**c**) and (**q**) are known, then its Procured Inventory Days Planned:
   $$p= 36-[c-q][/(U+\$[1-f]-\$'[1+s][v+i]\}[1-t]-D)-Q]$$
   $$/\{\$'v[1+s]\}$$

Rule-20894:
   If both (**/**), (**U**), (**$**), (**f**), (**$'**), (**s**), (**v**), (**i**), (**t**), (**D**), (**p**), (**Q**) and (**q**) are known, then its Current Ratio Planned is:
   $$c= q+\$'vp[1+s]$$
   $$/\{360[/(U+\$[1-f]-\$'[1+s][v+i]\}[1-t]-D)-Q]\}$$

Rule-20895:
   If both (**/**), (**U**), (**$**), (**f**), (**$'**), (**s**), (**v**), (**i**), (**t**), (**D**), (**p**), (**c**) and (**q**) are known, then its Quoted Longterm Debt Planned is:
   $$q= c-\$'vp[1+s]$$
   $$/\{360[/(U+\$[1-f]-\$'[1+s][v+i]\}[1-t]-D)-Q]\}$$

Rule-20896:
   If both (**/**), (**U**), (**$**), (**f**), (**$'**), (**s**), (**v**), (**i**), (**t**), (**A**), (**d**) and (**X**) are known, then its Quoted Longterm Debt Planned is:
   $$Q= /(U+\{\$[1-f]-\$'[1+s][v+i]\}[1-t]-Ad)-X$$

Rule-20897:
   If both (**Q**), (**U**), (**$**), (**f**), (**$'**), (**s**), (**v**), (**i**), (**t**), (**A**), (**d**) and (**X**) are known, then its Leverage or Gearing Ratio Planned is:
   $$/= [Q+X]/(U+\{\$[1-f]-\$'[1+s][v+i]\}[1-t]-Ad)$$

Steve Asikin ISBN 14: 978-1515158691, ISBN 10: 1515158691

Rule-20898:

If both (I), (Q), $(\$)$, (f), $(\$')$, (s), (v), (i), (t), (A), (d) and (X) are known, then its Utilized or Starting Capital must be:

$U = Ad + [Q+X]/I - \{\$[1-f] - \$'[1+s][v+i]\}[1-t]$

Rule-20899:

If both (I), (U), (Q), (f), $(\$')$, (s), (v), (i), (t), (A), (d) and (X) are known, then its Sales or Revenue Planned is:

$\$ = (\$'[1+s][v+i] + \{Ad + [Q+X]/I - U\}/[1-t])/[1-f]$

Rule-20900:

If both (I), (U), $(\$)$, (Q), $(\$')$, (s), (v), (i), (t), (A), (d) and (X) are known, then its Fixed Portion Planned is:

$f = 1 - (\$'[1+s][v+i] + \{Ad + [Q+X]/I - U\}/[1-t])/\$$

Rule-20901:

If both (I), (U), $(\$)$, (f), (Q), (s), (v), (i), (t), (A), (d) and (X) are known, then its Sales Past must be:

$\$' = (\$[1-f] - \{Ad + [Q+X]/I - U\}/[1-t])/\{[1+s][v+i]\}$

Rule-20902:

If both (I), (U), $(\$)$, (f), $(\$')$, (Q), (v), (i), (t), (A), (d) and (X) are known, then its Sales Growth Planned is:

$s = \$/\$' - 1$

or it can also be found as

$s = (\$[1-f] - \{Ad + [Q+X]/I - U\}/[1-t])/\{\$'[v+i]\} - 1$

Steve Asikin ISBN 14: 978-1515158691, ISBN 10: 1515158691

Rule-20903:

If both (f), (U), (S), (f), (S'), (s), (Q), (i), (t), (A), (d) and (X) are known, then its Variable Portion Planned is:

$$v= (S[1\text{-}f]\text{-}\{Ad+[Q+X]/fU\}/[1\text{-}t])/\{S'[1+s]\}\text{-}i$$

Rule-20904:

If both (f), (U), (S), (f), (S'), (s), (v), (Q), (t), (A), (d) and (X) are known, then its Interest Portion Planned is:

$$i = (S[1\text{-}f]\text{-}\{Ad+[Q+X]/fU\}/[1\text{-}t])/\{S'[1+s]\}\text{-}v$$

Rule-20905:

If both (f), (U), (S), (f), (S'), (s), (v), (i), (Q), (A), (d) and (X) are known, then its Tax Rate Planned is:

$$t= 1\text{-}\{Ad+[Q+X]/fU\}/\{S[1\text{-}f]\text{-}S'[1+s][v+i]\}$$

Rule-20906:

If both (f), (U), (S), (f), (S'), (s), (v), (i), (t), (Q), (d) and (X) are known, then its After Tax Income Planned is:

$$A= (U+S[1\text{-}f]\text{-}S'[1+s][v+i]\}[1\text{-}t]\text{-}[Q+X]/f)/d$$

Rule-20907:

If both (f), (U), (S), (f), (S'), (s), (v), (i), (t), (A), (Q) and (X) are known, then its Dividend Payout Planned is:

$$d= (U+S[1\text{-}f]\text{-}S'[1+s][v+i]\}[1\text{-}t]\text{-}[Q+X]/f)/A$$

Steve Asikin ISBN 14: 978-1515158691, ISBN 10: 1515158691

Rule-20908:

If both (I), (**U**), (**$**), (**f**), (**$'**), (**s**), (**v**), (**i**), (**t**), (**A**), (**d**) and (**Q**) are known, then its Xpress or Current Debt Planned is:

$$X= I(U+\$[1-f]-\$'[1+s][v+i])[1-t]-Ad)-Q$$

Rule-20896:

If both (I), (**U**), (**$**), (**f**), (**$'**), (**s**), (**v**), (**i**), (**t**), (**A**), (**d**), (**P**), (**c**) and (**q**) are known, then its Quoted Longterm Debt Planned is:

$$Q= I(U+\$[1-f]-\$'[1+s][v+i])[1-t]-Ad)-P/[c-q]$$

Rule-20897:

If both (**Q**), (**U**), (**$**), (**f**), (**$'**), (**s**), (**v**), (**i**), (**t**), (**A**), (**d**), (**P**), (**c**) and (**q**) are known, then its Leverage or Gearing Ratio Planned is:

$$I= \{Q+P/[c-q]\}/(U+\{\$[1-f]-\$'[1+s][v+i]\}[1-t]-Ad)$$

Rule-20898:

If both (I), (**Q**), (**$**), (**f**), (**$'**), (**s**), (**v**), (**i**), (**t**), (**A**), (**d**), (**P**), (**c**) and (**q**) are known, then its Utilized or Starting Capital must be:

$$U= Ad+\{Q+P/[c-q]\}/I-\{\$[1-f]-\$'[1+s][v+i]\}[1-t]$$

Rule-20899:

If both (I), (**U**), (**Q**), (**f**), (**$'**), (**s**), (**v**), (**i**), (**t**), (**A**), (**d**), (**P**), (**c**) and (**q**) are known, then its Sales or Revenue Planned is:

$$\$= [\$'[1+s][v+i]+(Ad+\{Q+P/[c-q]\}/I-U)/[1-t]]/[1-f]$$

Steve Asikin ISBN 14: 978-1515158691, ISBN 10: 1515158691

Rule-20900:

    If both $(\boldsymbol{I})$, $(\mathbf{U})$, $(\mathbf{\$})$, $(\mathbf{Q})$, $(\mathbf{\$'})$, $(\mathbf{s})$, $(\mathbf{v})$, $(\mathbf{i})$, $(\mathbf{t})$, $(\mathbf{A})$, $(\mathbf{d})$, $(\mathbf{P})$, $(\mathbf{c})$ and $(\mathbf{q})$ are known, then its Fixed Portion Planned is:

$$f= 1-[\$'[1+s][v+i]+(Ad+\{Q+P/[c-q]\}/IU)/[1-t]]/\$$$

Rule-20901:

    If both $(\boldsymbol{I})$, $(\mathbf{U})$, $(\mathbf{\$})$, $(\mathbf{f})$, $(\mathbf{Q})$, $(\mathbf{s})$, $(\mathbf{v})$, $(\mathbf{i})$, $(\mathbf{t})$, $(\mathbf{A})$, $(\mathbf{d})$, $(\mathbf{P})$, $(\mathbf{c})$ and $(\mathbf{q})$ are known, then its Sales Past must be:

$$\$'= [\$[1-f]-(Ad+\{Q+P/[c-q]\}/IU)/[1-t]] /\{[1+s][v+i]\}$$

Rule-20902:

    If both $(\boldsymbol{I})$, $(\mathbf{U})$, $(\mathbf{\$})$, $(\mathbf{f})$, $(\mathbf{\$'})$, $(\mathbf{Q})$, $(\mathbf{v})$, $(\mathbf{i})$, $(\mathbf{t})$, $(\mathbf{A})$, $(\mathbf{d})$, $(\mathbf{P})$, $(\mathbf{c})$ and $(\mathbf{q})$ are known, then its Sales Growth Planned is:

$$s= \$/\$'-1$$

        or it can also be found as

$$s = [\$[1-f]-(Ad+\{Q+P/[c-q]\}/IU)/[1-t]] /\{\$'[v+i]\}-1$$

Rule-20903:

    If both $(\boldsymbol{I})$, $(\mathbf{U})$, $(\mathbf{\$})$, $(\mathbf{f})$, $(\mathbf{\$'})$, $(\mathbf{s})$, $(\mathbf{Q})$, $(\mathbf{i})$, $(\mathbf{t})$, $(\mathbf{A})$, $(\mathbf{d})$, $(\mathbf{P})$, $(\mathbf{c})$ and $(\mathbf{q})$ are known, then its Variable Portion Planned is:

$$v= [\$[1-f]-(Ad+\{Q+P/[c-q]\}/IU)/[1-t]] /\{\$'[1+s]\}-i$$

Steve Asikin ISBN 14: 978-1515158691, ISBN 10: 1515158691

Rule-20904:

If both (**/**), (**U**), (**$**), (**f**), (**$'**), (**s**), (**v**), (**Q**), (**t**), (**A**), (**d**), (**P**), (**c**) and (**q**) are known, then its Interest Portion Planned is:

$$i = [\$[1\text{-}f]\text{-}(Ad+\{Q+P/[c\text{-}q]\}//U)/[1\text{-}t]]$$
$$/\{\$'[1+s]\}\text{-}v$$

Rule-20905:

If both (**/**), (**U**), (**$**), (**f**), (**$'**), (**s**), (**v**), (**i**), (**Q**), (**A**), (**d**), (**P**), (**c**) and (**q**) are known, then its Tax Rate Planned is:

$$t = 1\text{-}\{Ad+[Q+X]//U\}/\{\$[1\text{-}f]\text{-}\$'[1+s][v+i]\}$$

Rule-20906:

If both (**/**), (**U**), (**$**), (**f**), (**$'**), (**s**), (**v**), (**i**), (**t**), (**Q**), (**d**), (**P**), (**c**) and (**q**) are known, then its After Tax Income Planned is:

$$A = [U+\$[1\text{-}f]\text{-}\$'[1+s][v+i]\}[1\text{-}t]\text{-}\{Q+P/[c\text{-}q]\}//]/d$$

Rule-20907:

If both (**/**), (**U**), (**$**), (**f**), (**$'**), (**s**), (**v**), (**i**), (**t**), (**A**), (**Q**), (**P**), (**c**) and (**q**) are known, then its Dividend Payout Planned is:

$$d = [U+\$[1\text{-}f]\text{-}\$'[1+s][v+i]\}[1\text{-}t]\text{-}\{Q+P/[c\text{-}q]\}//]/A$$

Rule-20908:

If both (**/**), (**U**), (**$**), (**f**), (**$'**), (**s**), (**v**), (**i**), (**t**), (**A**), (**d**), (**Q**), (**c**) and (**q**) are known, then its Qprocured Inventory Planned is:

$$P = [c\text{-}q][/U+\$[1\text{-}f]\text{-}\$'[1+s][v+i]\}[1\text{-}t]\text{-}Ad)\text{-}Q]$$

Steve Asikin ISBN 14: 978-1515158691, ISBN 10: 1515158691

Rule-20909:

If both (I), (U), $(\$)$, (f), $(\$')$, (s), (v), (i), (t), (A), (d), (P), (Q) and (q) are known, then its Current Ratio Planned is:

$$c = q + P/[\{I(U + \$[1-f] - \$'[1+s][v+i]\}[1-t] - Ad) - Q]$$

Rule-20910:

If both (I), (U), $(\$)$, (f), $(\$')$, (s), (v), (i), (t), (A), (d), (P), (c) and (Q) are known, then its Quick or Acid Test Ratio Planned is:

$$q = c - P/[\{I(U + \$[1-f] - \$'[1+s][v+i]\}[1-t] - Ad) - Q]$$

Rule-20911:

If both (I), (U), $(\$)$, (f), $(\$')$, (s), (v), (i), (t), (A), (d), (V), (p), (c) and (q) are known, then its Quoted Longterm Debt Planned is:

$$Q = I(U + \$[1-f] - \$'[1+s][v+i]\}[1-t] - Ad)$$
$$-Vp/\{360[c-q]\}$$

Rule-20912:

If both (Q), (U), $(\$)$, (f), $(\$')$, (s), (v), (i), (t), (A), (d), (V), (p), (c) and (q) are known, then its Leverage or Gearing Ratio Planned is:

$$I = (Q + Vp/\{360[c-q]\})$$
$$/(U + \{\$[1-f] - \$'[1+s][v+i]\}[1-t] - Ad)$$

Steve Asikin ISBN 14: 978-1515158691, ISBN 10: 1515158691

Rule-20913:

 If both (**/**), (**Q**), (**S**), (**f**), (**S'**), (**s**), (**v**), (**i**), (**t**), (**A**), (**d**), (**V**), (**p**), (**c**) and (**q**) are known, then its Utilized or Starting Capital must be:

$$U= Ad+(Q+Vp/\{360[c-q]\})/I$$
$$-\{S[1-f]-S'[1+s][v+i]\}[1-t]$$

Rule-20914:

 If both (**/**), (**U**), (**Q**), (**f**), (**S'**), (**s**), (**v**), (**i**), (**t**), (**A**), (**d**), (**V**), (**p**), (**c**) and (**q**) are known, then its Sales or Revenue Planned is:

$$S= \{S'[1+s][v+i]+[Ad+(Q+Vp/\{360[c-q]\})/I-U]$$
$$/[1-t]\}/[1-f]$$

Rule-20915:

 If both (**/**), (**U**), (**S**), (**Q**), (**S'**), (**s**), (**v**), (**i**), (**t**), (**A**), (**d**), (**V**), (**p**), (**c**) and (**q**) are known, then its Fixed Portion Planned is:

$$f= 1-\{S'[1+s][v+i]+[Ad+(Q+Vp/\{360[c-q]\})/I-U]$$
$$/[1-t]\}/S$$

Rule-20916:

 If both (**/**), (**U**), (**S**), (**f**), (**Q**), (**s**), (**v**), (**i**), (**t**), (**A**), (**d**), (**V**), (**p**), (**c**) and (**q**) are known, then its Sales Past must be:

$$S'= \{S[1-f]-[Ad+(Q+Vp/\{360[c-q]\})/I-U]/[1-t]\}$$
$$/\{[1+s][v+i]\}$$

Steve Asikin ISBN 14: 978-1515158691, ISBN 10: 1515158691

Rule-20917:

If both $(\textbf{\textit{I}})$, $(\textbf{U})$, $(\textbf{\$})$, $(\textbf{f})$, $(\textbf{\$'})$, $(\textbf{Q})$, $(\textbf{v})$, $(\textbf{i})$, $(\textbf{t})$, $(\textbf{A})$, $(\textbf{d})$, $(\textbf{V})$, $(\textbf{p})$, $(\textbf{c})$ and $(\textbf{q})$ are known, then its Sales Growth Planned is:

$s = \$/\$' - 1$

　　or it can also be found as

$s = \{\$[1\text{-}f]\text{-}[Ad\text{+}(Q\text{+}Vp/\{360[c\text{-}q]\})/\textit{I}U]/[1\text{-}t]\}$
$/\{\$'[v\text{+}i]\} - 1$

Rule-20918:

If both $(\textbf{\textit{I}})$, $(\textbf{U})$, $(\textbf{\$})$, $(\textbf{f})$, $(\textbf{\$'})$, $(\textbf{s})$, $(\textbf{Q})$, $(\textbf{i})$, $(\textbf{t})$, $(\textbf{A})$, $(\textbf{d})$, $(\textbf{V})$, $(\textbf{p})$, $(\textbf{c})$ and $(\textbf{q})$ are known, then its Variable Portion Planned is:

$v = V/\$$

　　or it can also be found as

$v = \{\$[1\text{-}f]\text{-}[Ad\text{+}(Q\text{+}Vp/\{360[c\text{-}q]\})/\textit{I}U]/[1\text{-}t]\}$
$/\{\$'[1\text{+}s]\} - i$

Rule-20919:

If both $(\textbf{\textit{I}})$, $(\textbf{U})$, $(\textbf{\$})$, $(\textbf{f})$, $(\textbf{\$'})$, $(\textbf{s})$, $(\textbf{v})$, $(\textbf{Q})$, $(\textbf{t})$, $(\textbf{A})$, $(\textbf{d})$, $(\textbf{V})$, $(\textbf{p})$, $(\textbf{c})$ and $(\textbf{q})$ are known, then its Interest Portion Planned is:

$i = \{\$[1\text{-}f]\text{-}[Ad\text{+}(Q\text{+}Vp/\{360[c\text{-}q]\})/\textit{I}U]/[1\text{-}t]\}$
$/\{\$'[1\text{+}s]\} - v$

Rule-20920:

If both $(\textbf{\textit{I}})$, $(\textbf{U})$, $(\textbf{\$})$, $(\textbf{f})$, $(\textbf{\$'})$, $(\textbf{s})$, $(\textbf{v})$, $(\textbf{i})$, $(\textbf{Q})$, $(\textbf{A})$, $(\textbf{d})$, $(\textbf{V})$, $(\textbf{p})$, $(\textbf{c})$ and $(\textbf{q})$ are known, then its Tax Rate Planned is:

$t = 1\text{-}[Ad\text{+}(Q\text{+}Vp/\{360[c\text{-}q]\})/\textit{I}U]$
$/\{\$[1\text{-}f]\text{-}\$'[1\text{+}s][v\text{+}i]\}$

Steve Asikin ISBN 14: 978-1515158691, ISBN 10: 1515158691

Rule-20921:

If both (*I*), (**U**), (**$**), (**f**), (**$'**), (**s**), (**v**), (**i**), (**t**), (**Q**), (**d**), (**V**), (**p**), (**c**) and (**q**) are known, then its After Tax Income Planned is:

$$A= \{U+\$[1\text{-}f]\text{-}\$'[1+s][v+i]\}[1\text{-}t]$$
$$\quad -[Ad+(Q+Vp/\{360[c\text{-}q]\})/\textit{I}U]/\textit{I}/d$$

Rule-20922:

If both (*I*), (**U**), (**$**), (**f**), (**$'**), (**s**), (**v**), (**i**), (**t**), (**A**), (**d**), (**V**), (**p**), (**c**) and (**q**) are known, then its Quoted Longterm Debt Planned is:

$$d=\{U+\$[1\text{-}f]\text{-}\$'[1+s][v+i]\}[1\text{-}t]$$
$$\quad -[Ad+(Q+Vp/\{360[c\text{-}q]\})/\textit{I}U]/\textit{I}/A$$

Rule-20923:

If both (*I*), (**U**), (**$**), (**f**), (**$'**), (**s**), (**v**), (**i**), (**t**), (**A**), (**d**), (**Q**), (**p**), (**c**) and (**q**) are known, then its Variable Cost Planned is:

$$V= 360[c\text{-}q]$$
$$\quad [\textit{I}(U+\$[1\text{-}f]\text{-}\$'[1+s][v+i]\}[1\text{-}t]\text{-}Ad)\text{-}Q]/p$$

Rule-20924:

If both (*I*), (**U**), (**$**), (**f**), (**$'**), (**s**), (**v**), (**i**), (**t**), (**A**), (**d**), (**V**), (**Q**), (**c**) and (**q**) are known, then its Procured Inventory Days Planned:

$$p= 360[c\text{-}q]$$
$$\quad [\textit{I}(U+\$[1\text{-}f]\text{-}\$'[1+s][v+i]\}[1\text{-}t]\text{-}Ad)\text{-}Q]/V$$

Steve Asikin ISBN 14: 978-1515158691, ISBN 10: 1515158691

Rule-20925:

If both (I), (U), (S), (f), (S'), (s), (v), (i), (t), (A), (d), (V), (p), (Q) and (q) are known, then its Current Ratio Planned is:

$$c = q + Vp$$
$$/\{360[I(U+S[1-f]-S'[1+s][v+i]\}[1-t]-Ad)-Q]\}$$

Rule-20926:

If both (I), (U), (S), (f), (S'), (s), (v), (i), (t), (A), (d), (V), (p), (c) and (Q) are known, then its Quick or Acid Test Ratio Planned is:

$$q = c - Vp$$
$$/\{360[I(U+S[1-f]-S'[1+s][v+i]\}[1-t]-Ad)-Q]\}$$

Rule-20927:

If both (I), (U), (S), (f), (S'), (s), (v), (i), (t), (A), (d), (p), (c) and (q) are known, then its Quoted Longterm Debt Planned is:

$$Q = I(U+S[1-f]-S'[1+s][v+i]\}[1-t]-Ad)$$
$$-Svp/\{360[c-q]\}$$

Rule-20928:

If both (Q), (U), (S), (f), (S'), (s), (v), (i), (t), (A), (d), (p), (c) and (q) are known, then its Leverage or Gearing Ratio Planned is:

$$I = (Q+Svp/\{360[c-q]\})$$
$$/(U+\{S[1-f]-S'[1+s][v+i]\}[1-t]-Ad)$$

Steve Asikin ISBN 14: 978-1515158691, ISBN 10: 1515158691

Rule-20929:

If both (**/**), (**Q**), (**$**), (**f**), (**$'**), (**s**), (**v**), (**i**), (**t**), (**A**), (**d**), (**p**), (**c**) and (**q**) are known, then its Utilized or Starting Capital must be:

$$U= Ad+(Q+\$vp/\{360[c\text{-}q]\})//$$
$$-\{\$[1\text{-}f]\text{-}\$'[1+s][v+i]\}[1\text{-}t]$$

Rule-20930:

If both (**/**), (**U**), (**Q**), (**f**), (**$'**), (**s**), (**v**), (**i**), (**t**), (**A**), (**d**), (**p**), (**c**) and (**q**) are known, then its Sales or Revenue Planned is:

$$\$= [/\{U\text{-}\$'[1+s][v+i]\}[1\text{-}t]\text{-}Ad)\text{-}\$]$$
$$/(vp/\{360[c\text{-}q]\}\text{-}/[1\text{-}f][1\text{-}t])$$

Rule-20931:

If both (**/**), (**U**), (**$**), (**Q**), (**$'**), (**s**), (**v**), (**i**), (**t**), (**A**), (**d**), (**p**), (**c**) and (**q**) are known, then its Fixed Portion Planned is:

$$f= 1\text{-}\{\$'[1+s][v+i]$$
$$+[Ad+(Q+\$vp/\{360[c\text{-}q]\})//U]/[1\text{-}t]\}/\$$$

Rule-20932:

If both (**/**), (**U**), (**$**), (**f**), (**Q**), (**s**), (**v**), (**i**), (**t**), (**A**), (**d**), (**p**), (**c**) and (**q**) are known, then its Sales Past must be:

$$\$'= \{\$[1\text{-}f]\text{-}[Ad+(Q+\$vp/\{360[c\text{-}q]\})//U]/[1\text{-}t]\}$$
$$/\{[1+s][v+i]\}$$

Steve Asikin ISBN 14: 978-1515158691, ISBN 10: 1515158691

Rule-20933:

If both (I), (U), (S), (f), (S'), (Q), (v), (i), (t), (A), (d), (p), (c) and (q) are known, then its Sales Growth Planned is:

$$s= S/S'-1$$

or it can also be found as

$$s = \{S[1\text{-}f]\text{-}[Ad+(Q+Svp/\{360[c\text{-}q]\})/I\text{-}U]/[1\text{-}t]\}/\{S'[v+i]\}\text{-}1$$

Rule-20934:

If both (I), (U), (S), (f), (S'), (s), (Q), (i), (t), (A), (d), (p), (c) and (q) are known, then its Variable Portion Planned is:

$$v= [I(U+S[1\text{-}f]\text{-}S'i[1+s]\}[1\text{-}t]\text{-}Ad)\text{-}Q]/(Sp/\{360[c\text{-}q]\}+S'I[1+s][1\text{-}t])$$

Rule-20935:

If both (I), (U), (S), (f), (S'), (s), (v), (Q), (t), (A), (d), (p), (c) and (q) are known, then its Interest Portion Planned is:

$$i = \{S[1\text{-}f]\text{-}[Ad+(Q+Svp/\{360[c\text{-}q]\})/I\text{-}U]/[1\text{-}t]\}/\{S'[1+s]\}\text{-}v$$

Rule-20936:

If both (I), (U), (S), (f), (S'), (s), (v), (i), (Q), (A), (d), (p), (c) and (q) are known, then its Tax Rate Planned is:

$$t= 1\text{-}[Ad+(Q+Svp/\{360[c\text{-}q]\})/I\text{-}U]/\{S[1\text{-}f]\text{-}S'[1+s][v+i]\}$$

Steve Asikin ISBN 14: 978-1515158691, ISBN 10: 1515158691

Rule-20937:

If both (**/**), (**U**), (**$**), (**f**), (**$'**), (**s**), (**v**), (**i**), (**t**), (**Q**), (**d**), (**p**), (**c**) and (**q**) are known, then its After Tax Income Planned is:

$$A = \{U+\$[1-f]-\$'[1+s][v+i]\}[1-t]$$
$$-[Ad+(Q+\$vp/\{360[c-q]\})/\!-U]/\!/\!/d$$

Rule-20938:

If both (**/**), (**U**), (**$**), (**f**), (**$'**), (**s**), (**v**), (**i**), (**t**), (**A**), (**Q**), (**p**), (**c**) and (**q**) are known, then its Dividend Payout Planned is:

$$d = \{U+\$[1-f]-\$'[1+s][v+i]\}[1-t]$$
$$-[Ad+(Q+\$vp/\{360[c-q]\})/\!-U]/\!/\!/A$$

Rule-20939:

If both (**/**), (**U**), (**$**), (**f**), (**$'**), (**s**), (**v**), (**i**), (**t**), (**A**), (**d**), (**Q**), (**c**) and (**q**) are known, then its Procured Inventory Days Planned:

$$p = 360[c-q]$$
$$[/\!(U+\$[1-f]-\$'[1+s][v+i]\}[1-t]-Ad)-Q]/[\$v]$$

Rule-20940:

If both (**/**), (**U**), (**$**), (**f**), (**$'**), (**s**), (**v**), (**i**), (**t**), (**A**), (**d**), (**p**), (**Q**) and (**q**) are known, then its Current Ratio Planned is:

$$c = q+\$vp$$
$$/\{360[/\!(U+\$[1-f]-\$'[1+s][v+i]\}[1-t]-Ad)-Q]\}$$

Steve Asikin ISBN 14: 978-1515158691, ISBN 10: 1515158691

Rule-20941:
  If both (I, (**U**), (**$**), (**f**), (**$'**), (**s**), (**v**), (**i**), (**t**), (**A**), (**d**), (**p**), (**c**) and (**Q**) are known, then its Quick or Acid Test Ratio Planned is:

  q= c-$vp
  $$/\{360[I(\mathbf{U}+\$[1\text{-}f]\text{-}\$'[1+s][v+i]\}[1\text{-}t]\text{-}\mathbf{Ad})\text{-}\mathbf{Q}]\}$$

Rule-20942:
  If both (I, (**U**), (**$**), (**f**), (**$'**), (**s**), (**v**), (**i**), (**t**), (**A**), (**d**), (**p**), (**c**) and (**q**) are known, then its Quoted Longterm Debt Planned is:

  $$\mathbf{Q}= I(\mathbf{U}+\$[1\text{-}f]\text{-}\$'[1+s][v+i]\}[1\text{-}t]\text{-}\mathbf{Ad})$$
  $$\text{-}\$'vp[1+s]/\{360[\mathbf{c}\text{-}\mathbf{q}]\}$$

Rule-20943:
  If both (**Q**), (**U**), (**$**), (**f**), (**$'**), (**s**), (**v**), (**i**), (**t**), (**A**), (**d**), (**p**), (**c**) and (**q**) are known, then its Leverage or Gearing Ratio Planned is:

  $$I\text{=} (\mathbf{Q}+\$'vp[1+s]/\{360[\mathbf{c}\text{-}\mathbf{q}]\})$$
  $$/(\mathbf{U}+\{\$[1\text{-}f]\text{-}\$'[1+s][v+i]\}[1\text{-}t]\text{-}\mathbf{Ad})$$

Rule-20944:
  If both (I, (**Q**), (**$**), (**f**), (**$'**), (**s**), (**v**), (**i**), (**t**), (**A**), (**d**), (**p**), (**c**) and (**q**) are known, then its Utilized or Starting Capital must be:

  $$\mathbf{U}= \mathbf{Ad}+(\mathbf{Q}+\$'vp[1+s]/\{360[\mathbf{c}\text{-}\mathbf{q}]\})/I$$
  $$\text{-}\{\$[1\text{-}f]\text{-}\$'[1+s][v+i]\}[1\text{-}t]$$

Steve Asikin ISBN 14: 978-1515158691, ISBN 10: 1515158691

Rule-20945:

If both (I), $(\mathbf{U})$, $(\mathbf{Q})$, $(\mathbf{f})$, $(\mathbf{S'})$, $(\mathbf{s})$, $(\mathbf{v})$, $(\mathbf{i})$, $(\mathbf{t})$, $(\mathbf{A})$, $(\mathbf{d})$, $(\mathbf{p})$, $(\mathbf{c})$ and $(\mathbf{q})$ are known, then its Sales or Revemue Planned is:

$$\mathbf{S} = [\mathbf{Q} - I\{\mathbf{U} - \mathbf{S'}[1+\mathbf{s}][\mathbf{v}+\mathbf{i}]\}[1-\mathbf{t}] - \mathbf{Ad})$$
$$+ \mathbf{S'vp}[1+\mathbf{s}]/\{360[\mathbf{c}-\mathbf{q}]\}]/\{I[1-\mathbf{f}][1-\mathbf{t}])$$

Rule-20946:

If both (I), $(\mathbf{U})$, $(\mathbf{S})$, $(\mathbf{Q})$, $(\mathbf{S'})$, $(\mathbf{s})$, $(\mathbf{v})$, $(\mathbf{i})$, $(\mathbf{t})$, $(\mathbf{A})$, $(\mathbf{d})$, $(\mathbf{p})$, $(\mathbf{c})$ and $(\mathbf{q})$ are known, then its Fixed Portion Planned is:

$$\mathbf{f} = 1 - \{\mathbf{S'}[1+\mathbf{s}][\mathbf{v}+\mathbf{i}] + [\mathbf{Ad} + (\mathbf{Q} + \mathbf{S'vp}[1+\mathbf{s}]$$
$$/\{360[\mathbf{c}-\mathbf{q}]\})/I\mathbf{U}]/[1-\mathbf{t}]\}/\mathbf{S}$$

Rule-20947:

If both (I), $(\mathbf{U})$, $(\mathbf{S})$, $(\mathbf{f})$, $(\mathbf{Q})$, $(\mathbf{s})$, $(\mathbf{v})$, $(\mathbf{i})$, $(\mathbf{t})$, $(\mathbf{A})$, $(\mathbf{d})$, $(\mathbf{p})$, $(\mathbf{c})$ and $(\mathbf{q})$ are known, then its Sales Past must be:

$$\mathbf{S'} = [\{I\mathbf{U} + \mathbf{S}[1-\mathbf{f}] - \mathbf{S'i}[1+\mathbf{s}]\}[1-\mathbf{t}] - \mathbf{Ad}) - \mathbf{Q}]$$
$$/[\mathbf{v}[1+\mathbf{s}](\mathbf{p}/\{360[\mathbf{c}-\mathbf{q}]\} + I[1-\mathbf{t}])]$$

Rule-20948:

If both (I), $(\mathbf{U})$, $(\mathbf{S})$, $(\mathbf{f})$, $(\mathbf{S'})$, $(\mathbf{Q})$, $(\mathbf{v})$, $(\mathbf{i})$, $(\mathbf{t})$, $(\mathbf{A})$, $(\mathbf{d})$, $(\mathbf{p})$, $(\mathbf{c})$ and $(\mathbf{q})$ are known, then its Sales Growth Planned is:

$$\mathbf{s} = [\{I\mathbf{U} + \mathbf{S}[1-\mathbf{f}] - \mathbf{S'i}[1+\mathbf{s}]\}[1-\mathbf{t}] - \mathbf{Ad}) - \mathbf{Q}]$$
$$/[\mathbf{S'v}(\mathbf{p}/\{360[\mathbf{c}-\mathbf{q}]\} + I[1-\mathbf{t}])] - 1$$

Steve Asikin ISBN 14: 978-1515158691, ISBN 10: 1515158691

Rule-20949:

If both (f), (U), (S), (f), (S'), (s), (Q), (i), (t), (A), (d), (p), (c) and (q) are known, then its Variable Portion Planned is:

$$v= [\{U+S[1-f]-S'i[1+s]\}[1-t]-Ad)$$
$$-Q]/[S'[1+s](p/\{360[c-q]\}+\{1-t])]$$

Rule-20950:

If both (f), (U), (S), (f), (S'), (s), (v), (Q), (t), (A), (d), (p), (c) and (q) are known, then its Interest Portion Planned is:

$$i = \{S[1-f]-[Ad+(Q+S'vp[1+s]/\{360[c-q]\})/\textit{f}U]$$
$$/[1-t]\}/\{S'[1+s]\}-v$$

Rule-20951:

If both (f), (U), (S), (f), (S'), (s), (v), (i), (Q), (A), (d), (p), (c) and (q) are known, then its Tax Rate Planned is:

$$t= 1-[Ad+(Q+S'vp[1+s]/\{360[c-q]\})/\textit{f}U]$$
$$/\{S[1-f]-S'[1+s][v+i]\}$$

Rule-20952:

If both (f), (U), (S), (f), (S'), (s), (v), (i), (t), (Q), (d), (p), (c) and (q) are known, then its After Tax Income Planned is:

$$A= \{U+S[1-f]-S'[1+s][v+i]\}[1-t]$$
$$-[Ad+(Q+S'vp[1+s]/\{360[c-q]\})/\textit{f}U]/\textit{f}/d$$

Steve Asikin ISBN 14: 978-1515158691, ISBN 10: 1515158691

Rule-20953:

If both (*l*), (**U**), (**$**), (**f**), (**$'**), (**s**), (**v**), (**i**), (**t**), (**A**), (**Q**), (**p**), (**c**) and (**q**) are known, then its Dividend Payout Planned is:

$$d=\{U+\$[1-f]-\$'[1+s][v+i]\}[1-t]$$
$$-[Ad+(Q+\$'vp[1+s]/\{360[c-q]\})/lU]/l/A$$

Rule-20954:

If both (*l*), (**U**), (**$**), (**f**), (**$'**), (**s**), (**v**), (**i**), (**t**), (**A**), (**d**), (**Q**), (**c**) and (**q**) are known, then its Procured Inventory Days Planned:

$$p=360[c-q][lU+\$[1-f]-\$'[1+s][v+i]\}[1-t]-Ad)-Q]$$
$$/\{\$'v[1+s]\}$$

Rule-20955:

If both (*l*), (**U**), (**$**), (**f**), (**$'**), (**s**), (**v**), (**i**), (**t**), (**A**), (**d**), (**p**), (**Q**) and (**q**) are known, then its Current Ratio Planned is:

$$c=q+\$'vp[1+s]$$
$$/\{360[lU+\$[1-f]-\$'[1+s][v+i]\}[1-t]-Ad)-Q]\}$$

Rule-20956:

If both (*l*), (**U**), (**$**), (**f**), (**$'**), (**s**), (**v**), (**i**), (**t**), (**A**), (**d**), (**p**), (**c**) and (**Q**) are known, then its Quick or Acid test Ratio Planned is:

$$q=c-\$'vp[1+s]$$
$$/\{360[lU+\$[1-f]-\$'[1+s][v+i]\}[1-t]-Ad)-Q]\}$$

Steve Asikin ISBN 14: 978-1515158691, ISBN 10: 1515158691

Rule-20957:

If both (**/**), (**U**), (**$**), (**f**), (**$'**), (**s**), (**v**), (**i**), (**t**), (**d**) and (**X**) are known, then its Quoted Longterm Debt Planned is:

$$Q= /(U+\$[1-f]-\$'[1+s][v+i])[1-t][1-d])-X$$

Rule-20958:

If both (**Q**), (**U**), (**$**), (**f**), (**$'**), (**s**), (**v**), (**i**), (**t**), (**d**) and (**X**) are known, then its Leverage or Gearing Ratio Planned is:

$$/= [Q+X]/(U+\$[1-f]-\$'[1+s][v+i])[1-t][1-d])$$

Rule-20959:

If both (**/**), (**Q**), (**$**), (**f**), (**$'**), (**s**), (**v**), (**i**), (**t**), (**d**) and (**X**) are known, then its Utilized or Starting Capital must be:

$$U= [Q+X]//\{\$[1-f]= \$'[1+s][v+i]\}[1-t][1-d]$$

Rule-20960:

If both (**/**), (**U**), (**Q**), (**f**), (**$'**), (**s**), (**v**), (**i**), (**t**), (**d**) and (**X**) are known, then its Sales or Revenue Planned is:

$$\$= (\$'[1+s][v+i]+\{[Q+X]//U\}/\{[1-t][1-d]\})/[1-f]$$

Rule-20961:

If both (**/**), (**U**), (**$**), (**Q**), (**$'**), (**s**), (**v**), (**i**), (**t**), (**d**) and (**X**) are known, then its Fixed Portion Planned is:

$$f= 1-(\$'[1+s][v+i]+\{[Q+X]//U\}/\{[1-t][1-d]\})/\$$$

Rule-20962:

If both (**/**), (**U**), (**$**), (**f**), (**Q**), (**s**), (**v**), (**i**), (**t**), (**d**) and (**X**) are known, then its Sales Past must be:

$$\$'= (\$[1-f]-\{[Q+X]//U\}/\{[1-t][1-d]\})/\{[1+s][v+i]\}$$

Steve Asikin ISBN 14: 978-1515158691, ISBN 10: 1515158691

Rule-20963:

If both (I), (U), $(\$)$, (f), $(\$')$, (Q), (v), (i), (t), (d) and (X) are known, then its Sales Growth Planned is:

$s = \$/\$' - 1$

or it can also be found as

$s = (\$[1-f] - \{[Q+X]/I\text{-}U\}/\{[1-t][1-d]\})/\{\$'[v+i]\} - 1$

Rule-20964:

If both (I), (U), $(\$)$, (f), $(\$')$, (s), (Q), (i), (t), (d) and (X) are known, then its Variable Portion Planned is:

$v = (\$[1-f] - \{[Q+X]/I\text{-}U\}/\{[1-t][1-d]\})/\{\$'[1+s]\} - i$

Rule-20965:

If both (I), (U), $(\$)$, (f), $(\$')$, (s), (v), (Q), (t), (d) and (X) are known, then its Interest Portion Planned is:

$i = (\$[1-f] - \{[Q+X]/I\text{-}U\}/\{[1-t][1-d]\})/\{\$'[1+s]\} - v$

Rule-20966:

If both (I), (U), $(\$)$, (f), $(\$')$, (s), (v), (i), (Q), (d) and (X) are known, then its Tax Rate Planned is:

$t = 1 - \{[Q+X]/I\text{-}U\}/(\{\$[1-f] - \$'[1+s][v+i]\}[1-d])$

Rule-20967:

If both (I), (U), $(\$)$, (f), $(\$')$, (s), (v), (i), (t), (Q) and (X) are known, then its Dividend Payout Planned is:

$d = 1 - \{[Q+X]/I\text{-}U\}/(\{\$[1-f] - \$'[1+s][v+i]\}[1-t])$

Steve Asikin ISBN 14: 978-1515158691, ISBN 10: 1515158691

Rule-20968:
  If both (**∫**), (**U**), (**$**), (**f**), (**$'**), (**s**), (**v**), (**i**), (**t**), (**d**) and
  (**Q**) are known, then its Xpress or Current Debt
  Planned is:
  $$X= \text{∫}U+\{\$[1\text{-}f]\text{-}\$'[1+s][v+i]\}[1\text{-}t][1\text{-}d])\text{-}Q$$

Rule-20969:
  If both (**∫**), (**U**), (**$**), (**f**), (**$'**), (**s**), (**v**), (**i**), (**t**), (**d**), (**P**),
  (**c**) and (**q**) are known, then its Quoted Longterm Debt
  Planned is:
  $$Q= \text{∫}U+\{\$[1\text{-}f]\text{-}\$'[1+s][v+i]\}[1\text{-}t][1\text{-}d])\text{-}P/[c\text{-}q]$$

Rule-20970:
  If both (**Q**), (**U**), (**$**), (**f**), (**$'**), (**s**), (**v**), (**i**), (**t**), (**d**), (**P**),
  (**c**) and (**q**) are known, then its Leverage or Gearing
  Ratio Planned is:
  $$\text{∫}= \{Q+P/[c\text{-}q]\}/(U+\$[1\text{-}f]\text{-}\$'[1+s][v+i]\}[1\text{-}t][1\text{-}d])$$

Rule-20971:
  If both (**∫**), (**Q**), (**$**), (**f**), (**$'**), (**s**), (**v**), (**i**), (**t**), (**d**), (**P**),
  (**c**) and (**q**) are known, then its Utilized or Starting
  Capital must be:
  $$U= \{Q+P/[c\text{-}q]\}/\text{∫}\{\$[1\text{-}f]= \$'[1+s][v+i]\}[1\text{-}t][1\text{-}d]$$

Rule-20972:
  If both (**∫**), (**U**), (**Q**), (**f**), (**$'**), (**s**), (**v**), (**i**), (**t**), (**d**), (**P**),
  (**c**) and (**q**) are known, then its Sales or Revenue
  Planned is:
  $$\$= [\$'[1+s][v+i]+(\{Q+P/[c\text{-}q]\}/\text{∫}U)/\{[1\text{-}t][1\text{-}d]\}]$$
  $$/[1\text{-}f]$$

Steve Asikin ISBN 14: 978-1515158691, ISBN 10: 1515158691

Rule-20973:

If both (**/**), (**U**), (**$**), (**Q**), (**$'**), (**s**), (**v**), (**i**), (**t**), (**d**), (**P**), (**c**) and (**q**) are known, then its Fixed Portion Planned is:

$$f = 1 - [\$'[1+s][v+i] + (\{Q+P/[c-q]\}//U) /\{[1-t][1-d]\}]/\$$$

Rule-20974:

If both (**/**), (**U**), (**$**), (**f**), (**Q**), (**s**), (**v**), (**i**), (**t**), (**d**), (**P**), (**c**) and (**q**) are known, then its Sales Past must be:

$$\$' = [\$[1-f] - (\{Q+P/[c-q]\}//U)/\{[1-t][1-d]\}] /\{[1+s][v+i]\}$$

Rule-20975:

If both (**/**), (**U**), (**$**), (**f**), (**$'**), (**Q**), (**v**), (**i**), (**t**), (**d**), (**P**), (**c**) and (**q**) are known, then its Sales Growth Planned is:

$$s = \$/\$' - 1$$

or it can also be found as

$$s = [\$[1-f] - (\{Q+P/[c-q]\}//U)/\{[1-t][1-d]\}] /\{\$'[v+i]\} - 1$$

Rule-20976:

If both (**/**), (**U**), (**$**), (**f**), (**$'**), (**s**), (**Q**), (**i**), (**t**), (**d**), (**P**), (**c**) and (**q**) are known, then its Variable Portion Planned is:

$$v = [\$[1-f] - (\{Q+P/[c-q]\}//U)/\{[1-t][1-d]\}] /\{\$'[1+s]\} - i$$

Steve Asikin ISBN 14: 978-1515158691, ISBN 10: 1515158691

Rule-20977:

If both (I), (U), (S), (f), (S'), (s), (v), (Q), (t), (d), (P), (c) and (q) are known, then its Interest Portion Planned is:

$$i= [S[1-f]-(\{Q+P/[c-q]\}/IU)$$
$$/\{[1-t][1-d]\}]/\{S'[1+s]\}-v$$

Rule-20978:

If both (I), (U), (S), (f), (S'), (s), (v), (i), (Q), (d), (P), (c) and (q) are known, then its Tax Rate Planned is:

$$t= 1-(\{Q+P/[c-q]\}/IU)$$
$$/(\{S[1-f]-S'[1+s][v+i]\}[1-d])$$

Rule-20979:

If both (I), (U), (S), (f), (S'), (s), (v), (i), (t), (Q), (P), (c) and (q) are known, then its Dividend Payout Planned is:

$$d= 1-(\{Q+P/[c-q]\}/IU)$$
$$/(\{S[1-f]-S'[1+s][v+i]\}[1-t])$$

Rule-20980:

If both (I), (U), (S), (f), (S'), (s), (v), (i), (t), (d), (Q), (c) and (q) are known, then its Procured Inventory Planned is:

$$P= [c-q][IU+\{S[1-f]-S'[1+s][v+i]\}[1-t][1-d])-Q]$$

Rule-20981:

If both (I), (U), (S), (f), (S'), (s), (v), (i), (t), (d), (P), (Q) and (q) are known, then its Current Ratio Planned is:

$$c= q+P/[IU+\{S[1-f]-S'[1+s][v+i]\}[1-t][1-d])-Q]$$

Steve Asikin ISBN 14: 978-1515158691, ISBN 10: 1515158691

Rule-20982:

If both (**/**), (**U**), (**$**), (**f**), (**$'**), (**s**), (**v**), (**i**), (**t**), (**d**), (**P**), (**c**) and (**Q**) are known, then its Quick or Acid Test Ratio Planned is:

$$q = c\text{-}P/[\text{/}U + \{\$[1\text{-}f]\text{-}\$'[1+s][v+i]\}[1\text{-}t][1\text{-}d]) \text{-}Q]$$

Rule-20983:

If both (**/**), (**U**), (**$**), (**f**), (**$'**), (**s**), (**v**), (**i**), (**t**), (**d**), (**V**), (**p**), (**c**) and (**q**) are known, then its Quoted Longterm Debt Planned is:

$$Q = \text{/}U + \{\$[1\text{-}f]\text{-}\$'[1+s][v+i]\}[1\text{-}t][1\text{-}d])$$
$$\text{-}Vp/\{360[c\text{-}q]\}$$

Rule-20984:

If both (**Q**), (**U**), (**$**), (**f**), (**$'**), (**s**), (**v**), (**i**), (**t**), (**d**), (**V**), (**p**), (**c**) and (**q**) are known, then its Leverage or Gearing Ratio Planned is:

$$\text{/} = (Q + Vp/\{360[c\text{-}q]\})$$
$$/(U + \$[1\text{-}f]\text{-}\$'[1+s][v+i]\}[1\text{-}t][1\text{-}d])$$

Rule-20985:

If both (**/**), (**Q**), (**$**), (**f**), (**$'**), (**s**), (**v**), (**i**), (**t**), (**d**), (**V**), (**p**), (**c**) and (**q**) are known, then its Utilized or Starting Capital must be:

$$U = (Q + Vp/\{360[c\text{-}q]\})/\text{/}$$
$$\text{-}\{\$[1\text{-}f] = \$'[1+s][v+i]\}[1\text{-}t][1\text{-}d]$$

Steve Asikin ISBN 14: 978-1515158691, ISBN 10: 1515158691

Rule-20986:

If both (**/**), (**U**), (**Q**), (**f**), (**S'**), (**s**), (**v**), (**i**), (**t**), (**d**), (**V**), (**p**), (**c**) and (**q**) are known, then its Sales or Revenue Planned is:

$$S= \{S'[1+s][v+i]+[(Q+Vp/\{360[c-q]\})/I\text{-}U] /\{[1-t][1-d]\}\}/[1-f]$$

Rule-20987:

If both (**/**), (**U**), (**S**), (**Q**), (**S'**), (**s**), (**v**), (**i**), (**t**), (**d**), (**V**), (**p**), (**c**) and (**q**) are known, then its Fixed Portion Planned is:

$$f= 1-\{S'[1+s][v+i]+[(Q+Vp/\{360[c-q]\})/I\text{-}U] /\{[1-t][1-d]\}\}/S$$

Rule-20988:

If both (**/**), (**U**), (**S**), (**f**), (**Q**), (**s**), (**v**), (**i**), (**t**), (**d**), (**V**), (**p**), (**c**) and (**q**) are known, then its Sales Past must be:

$$S'= \{S[1-f]-[(Q+Vp/\{360[c-q]\})/I\text{-}U] /\{[1-t][1-d]\}\}/\{[1+s][v+i]\}$$

Rule-20989:

If both (**/**), (**U**), (**S**), (**f**), (**S'**), (**Q**), (**v**), (**i**), (**t**), (**d**), (**V**), (**p**), (**c**) and (**q**) are known, then its Sales Growth Planned is:

$$s= S/S'-1$$

or it can also be found as

$$s= \{S[1-f]-[(Q+Vp/\{360[c-q]\})/I\text{-}U]/\{[1-t][1-d]\}\} /\{S'[v+i]\}-1$$

Steve Asikin ISBN 14: 978-1515158691, ISBN 10: 1515158691

Rule-20990:

If both $(\textbf{I})$, $(\textbf{U})$, $(\textbf{S})$, $(\textbf{f})$, $(\textbf{S'})$, $(\textbf{s})$, $(\textbf{Q})$, $(\textbf{i})$, $(\textbf{t})$, $(\textbf{d})$, $(\textbf{V})$, $(\textbf{p})$, $(\textbf{c})$ and $(\textbf{q})$ are known, then its Variable Portion Planned is:

$$\textbf{v}= \textbf{V}/\textbf{S}$$

or it can also be found as

$$\textbf{v}=\{\textbf{S}[1\text{-}\textbf{f}]\text{-}[(\textbf{Q}+\textbf{Vp}/\{360[\textbf{c-q}]\})/\textbf{I-U}]/\{[1\text{-}\textbf{t}][1\text{-}\textbf{d}]\}\}$$
$$/\{\textbf{S'}[1+\textbf{s}]\}\text{-}\textbf{i}$$

Rule-20991:

If both $(\textbf{I})$, $(\textbf{U})$, $(\textbf{S})$, $(\textbf{f})$, $(\textbf{S'})$, $(\textbf{s})$, $(\textbf{v})$, $(\textbf{Q})$, $(\textbf{t})$, $(\textbf{d})$, $(\textbf{V})$, $(\textbf{p})$, $(\textbf{c})$ and $(\textbf{q})$ are known, then its Interest Portion Planned is:

$$\textbf{i}= \{\textbf{S}[1\text{-}\textbf{f}]\text{-}[(\textbf{Q}+\textbf{Vp}/\{360[\textbf{c-q}]\})/\textbf{I-U}]/\{[1\text{-}\textbf{t}][1\text{-}\textbf{d}]\}\}$$
$$/\{\textbf{S'}[1+\textbf{s}]\}\text{-}\textbf{v}$$

Rule-20992:

If both $(\textbf{I})$, $(\textbf{U})$, $(\textbf{S})$, $(\textbf{f})$, $(\textbf{S'})$, $(\textbf{s})$, $(\textbf{v})$, $(\textbf{i})$, $(\textbf{Q})$, $(\textbf{d})$, $(\textbf{V})$, $(\textbf{p})$, $(\textbf{c})$ and $(\textbf{q})$ are known, then its Tax Rate Planned is:

$$\textbf{t}= 1\text{-}[(\textbf{Q}+\textbf{Vp}/\{360[\textbf{c-q}]\})/\textbf{I-U}]$$
$$/(\{\textbf{S}[1\text{-}\textbf{f}]\text{-}\textbf{S'}[1+\textbf{s}][\textbf{v}+\textbf{i}]\}[1\text{-}\textbf{d}])$$

Rule-20993:

If both $(\textbf{I})$, $(\textbf{U})$, $(\textbf{S})$, $(\textbf{f})$, $(\textbf{S'})$, $(\textbf{s})$, $(\textbf{v})$, $(\textbf{i})$, $(\textbf{t})$, $(\textbf{Q})$, $(\textbf{V})$, $(\textbf{p})$, $(\textbf{c})$ and $(\textbf{q})$ are known, then its Dividend Payout Planned is:

$$\textbf{d}= 1\text{-}[(\textbf{Q}+\textbf{Vp}/\{360[\textbf{c-q}]\})/\textbf{I-U}]$$
$$/(\{\textbf{S}[1\text{-}\textbf{f}]\text{-}\textbf{S'}[1+\textbf{s}][\textbf{v}+\textbf{i}]\}[1\text{-}\textbf{t}])$$

Steve Asikin ISBN 14: 978-1515158691, ISBN 10: 1515158691

Rule-20994:

   If both (**/**), (**U**), (**$**), (**f**), (**$'**), (**s**), (**v**), (**i**), (**t**), (**d**), (**Q**),
   (**p**), (**c**) and (**q**) are known, then its Variable Cost
   Planned is:

   $V= 360[c-q]$
   $$[/U+\{\$[1-f]-\$'[1+s][v+i]\}[1-t][1-d])-Q]/p$$

Rule-20995:

   If both (**/**), (**U**), (**$**), (**f**), (**$'**), (**s**), (**v**), (**i**), (**t**), (**d**), (**V**),
   (**Q**), (**c**) and (**q**) are known, then its Procured
   Inventory Days Planned:

   $p= 360[c-q]$
   $$[/U+\{\$[1-f]-\$'[1+s][v+i]\}[1-t][1-d])-Q]/V$$

Rule-20996:

   If both (**/**), (**U**), (**$**), (**f**), (**$'**), (**s**), (**v**), (**i**), (**t**), (**d**), (**V**),
   (**p**), (**Q**) and (**q**) are known, then its Current Ratio
   Planned is:

   $$c= q+Vp/\{360[/U+\{\$[1-f]-\$'[1+s][v+i]\}$$
   $$[1-t][1-d])-Q]\}$$

Rule-20997:

   If both (**/**), (**U**), (**$**), (**f**), (**$'**), (**s**), (**v**), (**i**), (**t**), (**d**), (**V**),
   (**p**), (**c**) and (**Q**) are known, then its Quick or Acid
   Test Ratio Planned is:

   $$q= c-Vp/\{360[/U+\{\$[1-f]-\$'[1+s][v+i]\}$$
   $$[1-t][1-d])-Q]\}$$

Steve Asikin ISBN 14: 978-1515158691, ISBN 10: 1515158691

Rule-20998:
>If both (**/**), (**U**), (**$**), (**f**), (**$'**), (**s**), (**v**), (**i**), (**t**), (**d**), (**p**),
>(**c**) and (**q**) are known, then its Quoted Longterm Debt
>Planned is:
>$$Q= \textit{/}\!(U+\{\$[1\text{-}f]\text{-}\$'[1+s][v+i]\}[1\text{-}t][1\text{-}d])$$
>$$-\$vp/\{360[c\text{-}q]\}$$

Rule-20999:
>If both (**Q**), (**U**), (**$**), (**f**), (**$'**), (**s**), (**v**), (**i**), (**t**), (**d**), (**p**),
>(**c**) and (**q**) are known, then its Levergare or Gearing
>Ratio Planned is:
>$$\textit{/} = (Q+\$vp/\{360[c\text{-}q]\})$$
>$$/(U+\$[1\text{-}f]\text{-}\$'[1+s][v+i]\}[1\text{-}t][1\text{-}d])$$

Rule-21000:
>If both (**/**), (**Q**), (**$**), (**f**), (**$'**), (**s**), (**v**), (**i**), (**t**), (**d**), (**p**),
>(**c**) and (**q**) are known, then its Utilized or Starting
>Capital must be:
>$$U= (Q+\$vp/\{360[c\text{-}q]\})/\textit{/}\{\$[1\text{-}f]$$
>$$- \$'[1+s][v+i]\}[1\text{-}t][1\text{-}d]$$

Rule-21001:
>If both (**/**), (**U**), (**Q**), (**f**), (**$'**), (**s**), (**v**), (**i**), (**t**), (**d**), (**p**),
>(**c**) and (**q**) are known, then its Sales or Revenue
>Planned is:
>$$\$= (\textit{/}\!\{U\text{-}\$'[1+s][v+i][1\text{-}t][1\text{-}d]\}\text{-}Q)$$
>$$/(vp/\{360[c\text{-}q]\}\text{-}\textit{/}[1\text{-}f][1\text{-}t][1\text{-}d])$$

Steve Asikin ISBN 14: 978-1515158691, ISBN 10: 1515158691

Rule-21002:

If both (f), (U), (S), (Q), (S'), (s), (v), (i), (t), (d), (p), (c) and (q) are known, then its Fixed Portion Planned is:

$$f= 1-\{S'[1+s][v+i]+[(Q+Svp/\{360[c-q]\})/fU]$$
$$/\{[1-t][1-d]\}\}/S$$

Rule-21003:

If both (f), (U), (S), (f), (Q), (s), (v), (i), (t), (d), (p), (c) and (q) are known, then its Sales Past must be:
$$S'= \{S[1-f]-[(Q+Svp/\{360[c-q]\})/fU]$$
$$/\{[1-t][1-d]\}\}/\{[1+s][v+i]\}$$

Rule-21004:

If both (f), (U), (S), (f), (S'), (Q), (v), (i), (t), (d), (p), (c) and (q) are known, then its Sales Growth Planned is:

$$s= S/S'-1$$

or it can also be found
$$s= \{S[1-f]-[(Q+Svp/\{360[c-q]\})/fU]$$
$$/\{[1-t][1-d]\}\}/\{S'[v+i]\}-1$$

Rule-21005:

If both (f), (U), (S), (f), (S'), (s), (Q), (i), (t), (d), (p), (c) and (q) are known, then its Variable Portion Planned is:

$$v= [fU+\{S[1-f]-S'i[1+s]\}[1-t][1-d])-Q]$$
$$/(Sp/\{360[c-q]\} -S'f[1+s][1-t][1-d])$$

Steve Asikin ISBN 14: 978-1515158691, ISBN 10: 1515158691

Rule-21006:

If both (**l**), (**U**), (**$**), (**f**), (**$'**), (**s**), (**v**), (**Q**), (**t**), (**d**), (**p**), (**c**) and (**q**) are known, then its Interest Portion Planned is:

$$i = \{\$[1\text{-}f]\text{-}[(Q+\$vp/\{360[c\text{-}q]\})/\textit{l}U]$$
$$/\{[1\text{-}t][1\text{-}d]\}\}/\{\$'[1+s]\}\text{-}v$$

Rule-21007:

If both (**l**), (**U**), (**$**), (**f**), (**$'**), (**s**), (**v**), (**i**), (**Q**), (**d**), (**p**), (**c**) and (**q**) are known, then its Tax Rate Planned is:

$$t = 1\text{-}[(Q+\$vp/\{360[c\text{-}q]\})/\textit{l}U]$$
$$/(\{\$[1\text{-}f]\text{-}\$'[1+s][v+i]\}[1\text{-}d])$$

Rule-21008:

If both (**l**), (**U**), (**$**), (**f**), (**$'**), (**s**), (**v**), (**i**), (**t**), (**Q**), (**p**), (**c**) and (**q**) are known, then its Dividend Payout Planned is:

$$d = 1\text{-}[(Q+\$vp/\{360[c\text{-}q]\})/\textit{l}U]$$
$$/(\{\$[1\text{-}f]\text{-}\$'[1+s][v+i]\}[1\text{-}t])$$

Rule-21009:

If both (**l**), (**U**), (**$**), (**f**), (**$'**), (**s**), (**v**), (**i**), (**t**), (**d**), (**Q**), (**c**) and (**q**) are known, then its Procured Inventory Days Planned:

$$p = 360[c\text{-}q][\textit{l}U+\{\$[1\text{-}f]\text{-}\$'[1+s][v+i]\}$$
$$[1\text{-}t][1\text{-}d])\text{-}Q]/[\$v]$$

Steve Asikin ISBN 14: 978-1515158691, ISBN 10: 1515158691

Rule-21010:

If both $(\pounds)$, $(\mathbf{U})$, $(\$)$, $(\mathbf{f})$, $(\$')$, $(\mathbf{s})$, $(\mathbf{v})$, $(\mathbf{i})$, $(\mathbf{t})$, $(\mathbf{d})$, $(\mathbf{p})$, $(\mathbf{Q})$ and $(\mathbf{q})$ are known, then its Current Ratio Planned is:

$$c = q + \$vp/\{360[\pounds U + \{\$[1-f] - \$'[1+s][v+i]\}$$
$$[1-t][1-d]) - Q]\}$$

Rule-21011:

If both $(\pounds)$, $(\mathbf{U})$, $(\$)$, $(\mathbf{f})$, $(\$')$, $(\mathbf{s})$, $(\mathbf{v})$, $(\mathbf{i})$, $(\mathbf{t})$, $(\mathbf{d})$, $(\mathbf{p})$, $(\mathbf{c})$ and $(\mathbf{Q})$ are known, then its Quick or Acid Test Ratio Planned is:

$$q = c - \$vp/\{360[\pounds U + \{\$[1-f] - \$'[1+s][v+i]\}$$
$$[1-t][1-d]) - Q]\}$$

Rule-21012:

If both $(\pounds)$, $(\mathbf{U})$, $(\$)$, $(\mathbf{f})$, $(\$')$, $(\mathbf{s})$, $(\mathbf{v})$, $(\mathbf{i})$, $(\mathbf{t})$, $(\mathbf{d})$, $(\mathbf{p})$, $(\mathbf{c})$ and $(\mathbf{q})$ are known, then its Quoted Longterm Debt Planned is:

$$Q = \pounds U + \{\$[1-f] - \$'[1+s][v+i]\}[1-t][1-d])$$
$$- \$'vp[1+s]/\{360[c-q]\}$$

Rule-21013:

If both $(\mathbf{Q})$, $(\mathbf{U})$, $(\$)$, $(\mathbf{f})$, $(\$')$, $(\mathbf{s})$, $(\mathbf{v})$, $(\mathbf{i})$, $(\mathbf{t})$, $(\mathbf{d})$, $(\mathbf{p})$, $(\mathbf{c})$ and $(\mathbf{q})$ are known, then its Leverage or Gearing Ratio Planned is:

$$\pounds = (Q + \$'vp[1+s]/\{360[c-q]\})$$
$$/(U + \$[1-f] - \$'[1+s][v+i]\}[1-t][1-d])$$

Steve Asikin ISBN 14: 978-1515158691, ISBN 10: 1515158691

Rule-21014:

> If both (f), (Q), (S), (f), (S'), (s), (v), (i), (t), (d), (p), (c) and (q) are known, then its Utilized or Starting Capital must be:
>
> $$U = (Q + S'vp[1+s]/\{360[c-q]\})/f$$
> $$- \{S[1-f] - S'[1+s][v+i]\}[1-t][1-d]$$

Rule-21015:

> If both (f), (U), (Q), (f), (S'), (s), (v), (i), (t), (d), (p), (c) and (q) are known, then its Sales or Revenue Planned is:
>
> $$S = (Q - f\{U - S'[1+s][v+i][1-t][1-d]\}$$
> $$+ S'vp[1+s]/\{360[c-q]\})/\{f[1-f][1-t][1-d]\})$$

Rule-21016:

> If both (f), (U), (S), (Q), (S'), (s), (v), (i), (t), (d), (p), (c) and (q) are known, then its Fixed Portion Planned is:
>
> $$f = 1 - \{S'[1+s][v+i] + [(Q + S'vp[1+s]$$
> $$/\{360[c-q]\})/f U]/\{[1-t][1-d]\}\}/S$$

Rule-21017:

> If both (f), (U), (S), (f), (Q), (s), (v), (i), (t), (d), (p), (c) and (q) are known, then its Sales Past must be:
>
> $$S' = [f U + \{S[1-f] - S'i[1+s]\}[1-t][1-d]) - Q]$$
> $$/[v[1+s](p/\{360[c-q]\} - f[1-t][1-d])]$$

Steve Asikin ISBN 14: 978-1515158691, ISBN 10: 1515158691

<u>Rule-21018</u>:

If both (**/**), (**U**), (**S**), (**f**), (**S'**), (**Q**), (**v**), (**i**), (**t**), (**d**), (**p**), (**c**) and (**q**) are known, then its Sales Growth Planned is:

$s= S/S'-1$

or it can also be found as

$s= [/(U+\{S[1-f]-S'i[1+s]\}[1-t][1-d])-Q]$
$/[S'v[1+s](p/\{360[c-q]\}-/[1-t][1-d])]-1$

<u>Rule-21019</u>:

If both (**/**), (**U**), (**S**), (**f**), (**S'**), (**s**), (**Q**), (**i**), (**t**), (**d**), (**p**), (**c**) and (**q**) are known, then its Variable Portion Planned is:

$v= [/(U+\{S[1-f]-S'i[1+s]\}[1-t][1-d])-Q]$
$/[S'[1+s](p/\{360[c-q]\}-/[1-t][1-d])]$

<u>Rule-21020</u>:

If both (**/**), (**U**), (**S**), (**f**), (**S'**), (**s**), (**v**), (**Q**), (**t**), (**d**), (**p**), (**c**) and (**q**) are known, then its Interest Portion Planned is:

$i= \{S[1-f]-[(Q+S'vp[1+s]/\{360[c-q]\})//U]$
$/\{[1-t][1-d]\}\}/\{S'[1+s]\}-v$

<u>Rule-21021</u>:

If both (**/**), (**U**), (**S**), (**f**), (**S'**), (**s**), (**v**), (**i**), (**Q**), (**d**), (**p**), (**c**) and (**q**) are known, then its Tax Rate Planned is:

$t= 1-[(Q+S'vp[1+s]/\{360[c-q]\})//U]$
$/(\{S[1-f]-S'[1+s][v+i]\}[1-d])$

Steve Asikin ISBN 14: 978-1515158691, ISBN 10: 1515158691

Rule-21022:

If both (**/**), (**U**), (**$**), (**f**), (**$'**), (**s**), (**v**), (**i**), (**t**), (**Q**), (**p**), (**c**) and (**q**) are known, then its Dividend Planned is:

$$d = 1 - [(Q + S'vp[1+s]/\{360[c-q]\})/\mathit{l}U]$$
$$/(\{S[1-f]-S'[1+s][v+i]\}[1-t])$$

Rule-21023:

If both (**/**), (**U**), (**$**), (**f**), (**$'**), (**s**), (**v**), (**i**), (**t**), (**d**), (**Q**), (**c**) and (**q**) are known, then its Procured Inventory Days Planned:

$$p = 360[c-q][\mathit{l}U + \{S[1-f]-S'[1+s][v+i]\}$$
$$[1-t][1-d])-Q]/\{S'v[1+s]\}$$

Rule-21024:

If both (**/**), (**U**), (**$**), (**f**), (**$'**), (**s**), (**v**), (**i**), (**t**), (**d**), (**p**), (**Q**) and (**q**) are known, then its Current Ratio Planned is:

$$c = q + S'vp[1+s]/\{360[\mathit{l}U + \{S[1-f]-S'[1+s][v+i]\}$$
$$[1-t][1-d])-Q]\}$$

Rule-21025:

If both (**/**), (**U**), (**$**), (**f**), (**$'**), (**s**), (**v**), (**i**), (**t**), (**d**), (**p**), (**c**) and (**Q**) are known, then its Quick or Acid Test Ratio Planned is:

$$q = c - S'vp[1+s]/\{360[\mathit{l}U + \{S[1-f]-S'[1+s][v+i]\}$$
$$[1-t][1-d])-Q]\}$$

Steve Asikin ISBN 14: 978-1515158691, ISBN 10: 1515158691

Rule-21026:

If both (I), (U), (S), (S'), (s), (v), (f), (I), (T), (D) and (X) are known, then its Quoted Longterm Debt Planned is:

$Q = I\{U+S-S'[1+s][v+f]-I-T-D\}-X$

Rule-21027:

If both (Q), (U), (S), (S'), (s), (v), (f), (I), (T), (D) and (X) are known, then its Leverage or Gearing Ratio Planned is:

$I = [Q+X]/\{U+S-S'[1+s][v+f]-I-T-D\}$

Rule-21028:

If both (I), (Q), (S), (S'), (s), (v), (f), (I), (T), (D) and (X) are known, then its Utilized or Starting Capital must be:

$U = S'[1+s][v+f]+I+T+D+[Q+X]/I-S$

Rule-21029:

If both (I), (U), (Q), (S'), (s), (v), (f), (I), (T), (D) and (X) are known, then its Sales or Revenue Planned is:

$S = S'[1+s][v+f] = I+T+D+[Q+X]/I-U$

Rule-21030:

If both (I), (U), (S), (Q), (s), (v), (f), (I), (T), (D) and (X) are known, then its Sales Past must be:

$S' = \{U+S-I-T-D-[Q+X]/I\}/\{[1+s][v+f]\}$

Steve Asikin ISBN 14: 978-1515158691, ISBN 10: 1515158691

Rule-21031:

If both (I), (**U**), (**$**), (**$'**), (**Q**), (**v**), (**f**), (**I**), (**T**), (**D**) and (**X**) are known, then its Sales Growth Planned is:

$s= \$/\$'-1$

or it can also be found as

$s= \{U+\$-I-T-D-[Q+X]/I\}/\{\$'[v+f]\}-1$

Rule-21032:

If both (I), (**U**), (**$**), (**$'**), (**s**), (**Q**), (**f**), (**I**), (**T**), (**D**) and (**X**) are known, then its Variable Portion Planned is:

$v= \{U+\$-I-T-D-[Q+X]/I\}/\{\$'[1+s]\}-f$

Rule-21033:

If both (I), (**U**), (**$**), (**$'**), (**s**), (**v**), (**Q**), (**I**), (**T**), (**D**) and (**X**) are known, then its Fixed Portion Planned is:

$f= \{U+\$-I-T-D-[Q+X]/I\}/\{\$'[1+s]\}-v$

Rule-21034:

If both (I), (**U**), (**$**), (**$'**), (**s**), (**v**), (**f**), (**Q**), (**T**), (**D**) and (**X**) are known, then its Interest Expense Planned is:

$I= U+\$-\$'[1+s][v+f]-T-D-[Q+X]/I$

Rule-21035:

If both (I), (**U**), (**$**), (**$'**), (**s**), (**v**), (**f**), (**I**), (**Q**), (**D**) and (**X**) are known, then its Tax Planned is:

$T= U+\$-\$'[1+s][v+f]-I-D-[Q+X]/I$

Rule-21036:

If both (I), (**U**), (**$**), (**$'**), (**s**), (**v**), (**f**), (**I**), (**T**), (**Q**) and (**X**) are known, then its Dividend Planned is:

$D= U+\$-\$'[1+s][v+f]-I-T-[Q+X]/I$

Steve Asikin ISBN 14: 978-1515158691, ISBN 10: 1515158691

Rule-21037:

If both $(\textit{I})$, $(\mathbf{U})$, $(\mathbf{\$})$, $(\mathbf{\$'})$, $(\mathbf{s})$, $(\mathbf{v})$, $(\mathbf{f})$, $(\mathbf{I})$, $(\mathbf{T})$, $(\mathbf{D})$ and $(\mathbf{Q})$ are known, then its Xpress or Current Debt Planned is:

$$X = \textit{I}\{U+\$-\$'[1+s][v+f]-I-T-D\}-Q$$

Rule-21038:

If both $(\textit{I})$, $(\mathbf{U})$, $(\mathbf{\$})$, $(\mathbf{\$'})$, $(\mathbf{s})$, $(\mathbf{v})$, $(\mathbf{f})$, $(\mathbf{I})$, $(\mathbf{T})$, $(\mathbf{D})$, $(\mathbf{P})$, $(\mathbf{c})$ and $(\mathbf{q})$ are known, then its Quoted Longterm Debt Planned is:

$$Q = \textit{I}\{U+\$-\$'[1+s][v+f]-I-T-D\}-P/[c-q]$$

Rule-21039:

If both $(\mathbf{Q})$, $(\mathbf{U})$, $(\mathbf{\$})$, $(\mathbf{\$'})$, $(\mathbf{s})$, $(\mathbf{v})$, $(\mathbf{f})$, $(\mathbf{I})$, $(\mathbf{T})$, $(\mathbf{D})$, $(\mathbf{P})$, $(\mathbf{c})$ and $(\mathbf{q})$ are known, then its Leverage or Gearing Ratio Planned is:

$$\textit{I} = \{Q+P/[c-q]\}/\{U+\$-\$'[1+s][v+f]-I-T-D\}$$

Rule-21040:

If both $(\textit{I})$, $(\mathbf{Q})$, $(\mathbf{\$})$, $(\mathbf{\$'})$, $(\mathbf{s})$, $(\mathbf{v})$, $(\mathbf{f})$, $(\mathbf{I})$, $(\mathbf{T})$, $(\mathbf{D})$, $(\mathbf{P})$, $(\mathbf{c})$ and $(\mathbf{q})$ are known, then its Utilized or Starting Capital must be:

$$U = \$'[1+s][v+f]+I+T+D+\{Q+P/[c-q]\}/\textit{I}-\$$$

Rule-21141:

If both $(\textit{I})$, $(\mathbf{U})$, $(\mathbf{Q})$, $(\mathbf{\$'})$, $(\mathbf{s})$, $(\mathbf{v})$, $(\mathbf{f})$, $(\mathbf{I})$, $(\mathbf{T})$, $(\mathbf{D})$, $(\mathbf{P})$, $(\mathbf{c})$ and $(\mathbf{q})$ are known, then its Sales or Revenue Planned is:

$$\$ = \$'[1+s][v+f] = I+T+D+\{Q+P/[c-q]\}/\textit{I}-U$$

Steve Asikin ISBN 14: 978-1515158691, ISBN 10: 1515158691

Rule-21142:
>   If both (**/**), (**U**), (**$**), (**Q**), (**s**), (**v**), (**f**), (**I**), (**T**), (**D**), (**P**), (**c**) and (**q**) are known, then its Sales Past must be:
> $$S' = (U+S-I-T-D-\{Q+P/[c-q]\}//)/\{[1+s][v+f]\}$$

Rule-21043:
>   If both (**/**), (**U**), (**$**), (**$'**), (**Q**), (**v**), (**f**), (**I**), (**T**), (**D**), (**P**), (**c**) and (**q**) are known, then its Sales Growth Planned is:
> $$s = S/S' - 1$$
>   or it can also be found as
> $$s = (U+S-I-T-D-\{Q+P/[c-q]\}//)/\{S'[v+f]\} - 1$$

Rule-21044:
>   If both (**/**), (**U**), (**$**), (**$'**), (**s**), (**Q**), (**f**), (**I**), (**T**), (**D**), (**P**), (**c**) and (**q**) are known, then its Variable Portion Planned is:
> $$v = (U+S-I-T-D-\{Q+P/[c-q]\}//)/\{S'[1+s]\} - f$$

Rule-21045:
>   If both (**/**), (**U**), (**$**), (**$'**), (**s**), (**v**), (**Q**), (**I**), (**T**), (**D**), (**P**), (**c**) and (**q**) are known, then its Fixed Portion Planned is:
> $$f = (U+S-I-T-D-\{Q+P/[c-q]\}//)/\{S'[1+s]\} - v$$

Rule-21046:
>   If both (**/**), (**U**), (**$**), (**$'**), (**s**), (**v**), (**f**), (**Q**), (**T**), (**D**), (**P**), (**c**) and (**q**) are known, then its Interest Expense Planned is:
> $$I = U+S-S'[1+s][v+f]-T-D-\{Q+P/[c-q]\}//$$

Steve Asikin ISBN 14: 978-1515158691, ISBN 10: 1515158691

Rule-21047:
  If both (**/**), (**U**), (**$**), (**$'**), (**s**), (**v**), (**f**), (**I**), (**Q**), (**D**), (**P**),
  (**c**) and (**q**) are known, then its Tax Planned is:
  $$T= U+\$-\$'[1+s][v+f]-I-D-\{Q+P/[c-q]\}//$$

Rule-21048:
  If both (**/**), (**U**), (**$**), (**$'**), (**s**), (**v**), (**f**), (**I**), (**T**), (**Q**), (**P**),
  (**c**) and (**q**) are known, then its Dividend Planned is:
  $$D= U+\$-\$'[1+s][v+f]-I-T-\{Q+P/[c-q]\}//$$

Rule-21049:
  If both (**/**), (**U**), (**$**), (**$'**), (**s**), (**v**), (**f**), (**I**), (**T**), (**D**), (**Q**),
  (**c**) and (**q**) are known, then its Procured Inventory
  Planned is:
  $$P= [c-q](/\{U+\$-\$'[1+s][v+f]-I-T-D\}-Q)$$

Rule-21050:
  If both (**/**), (**U**), (**$**), (**$'**), (**s**), (**v**), (**f**), (**I**), (**T**), (**D**), (**P**),
  (**Q**) and (**q**) are known, then its Current Ratio Planned
  is:
  $$c= q+P/(/\{U+\$-\$'[1+s][v+f]-I-T-D\}-Q)$$

Rule-21051:
  If both (**/**), (**U**), (**$**), (**$'**), (**s**), (**v**), (**f**), (**I**), (**T**), (**D**), (**P**),
  (**c**) and (**Q**) are known, then its Quick or Acid Test
  Ratio Planned is:
  $$q= c-P/(/\{U+\$-\$'[1+s][v+f]-I-T-D\}-Q)$$

Steve Asikin ISBN 14: 978-1515158691, ISBN 10: 1515158691

Rule-21052:

If both (I), $(\mathbf{U})$, $(\mathbf{\$})$, $(\mathbf{\$'})$, $(\mathbf{s})$, $(\mathbf{v})$, $(\mathbf{f})$, $(\mathbf{I})$, $(\mathbf{T})$, $(\mathbf{D})$, $(\mathbf{V})$, $(\mathbf{p})$, $(\mathbf{c})$ and $(\mathbf{q})$ are known, then its Quoted Longterm Debt Planned is:

$$Q = I\{U+\$-\$'[1+s][v+f]-I-T-D\}-Vp/\{360[c-q]\}$$

Rule-21053:

If both $(\mathbf{Q})$, $(\mathbf{U})$, $(\mathbf{\$})$, $(\mathbf{\$'})$, $(\mathbf{s})$, $(\mathbf{v})$, $(\mathbf{f})$, $(\mathbf{I})$, $(\mathbf{T})$, $(\mathbf{D})$, $(\mathbf{V})$, $(\mathbf{p})$, $(\mathbf{c})$ and $(\mathbf{q})$ are known, then its Leverage or Gearing Ratio Planned is:

$$I = (Q+Vp/\{360[c-q]\})/\{U+\$-\$'[1+s][v+f]-I-T-D\}$$

Rule-21054:

If both (I), $(\mathbf{Q})$, $(\mathbf{\$})$, $(\mathbf{\$'})$, $(\mathbf{s})$, $(\mathbf{v})$, $(\mathbf{f})$, $(\mathbf{I})$, $(\mathbf{T})$, $(\mathbf{D})$, $(\mathbf{V})$, $(\mathbf{p})$, $(\mathbf{c})$ and $(\mathbf{q})$ are known, then its Utilized or Starting Capital must be:

$$U = \$'[1+s][v+f]+I+T+D+(Q+Vp/\{360[c-q]\})/I\$$$

Rule-21055:

If both (I), $(\mathbf{U})$, $(\mathbf{Q})$, $(\mathbf{\$'})$, $(\mathbf{s})$, $(\mathbf{v})$, $(\mathbf{f})$, $(\mathbf{I})$, $(\mathbf{T})$, $(\mathbf{D})$, $(\mathbf{V})$, $(\mathbf{p})$, $(\mathbf{c})$ and $(\mathbf{q})$ are known, then its Sales or Revenue Planned is:

$$\$ = \$'[1+s][v+f] = I+T+D+(Q+Vp/\{360[c-q]\})/I\,U$$

Rule-21056:

If both (I), $(\mathbf{U})$, $(\mathbf{\$})$, $(\mathbf{Q})$, $(\mathbf{s})$, $(\mathbf{v})$, $(\mathbf{f})$, $(\mathbf{I})$, $(\mathbf{T})$, $(\mathbf{D})$, $(\mathbf{V})$, $(\mathbf{p})$, $(\mathbf{c})$ and $(\mathbf{q})$ are known, then its Sales Past must be:

$$\$' = [U+\$-I-T-D-(Q+Vp/\{360[c-q]\})/I]$$
$$/\{[1+s][v+f]\}$$

Steve Asikin ISBN 14: 978-1515158691, ISBN 10: 1515158691

Rule-21057:

If both $(\textbf{/})$, $(\textbf{U})$, $(\textbf{S})$, $(\textbf{S'})$, $(\textbf{Q})$, $(\textbf{v})$, $(\textbf{f})$, $(\textbf{I})$, $(\textbf{T})$, $(\textbf{D})$, $(\textbf{V})$, $(\textbf{p})$, $(\textbf{c})$ and $(\textbf{q})$ are known, then its Sales Growth Planned is:

$s= S/S'-1$

or it can also be found as

$s= [U+S-I-T-D-(Q+Vp/\{360[c-q]\})//]/\{S'[v+f]\}-1$

Rule-21058:

If both $(\textbf{/})$, $(\textbf{U})$, $(\textbf{S})$, $(\textbf{S'})$, $(\textbf{s})$, $(\textbf{Q})$, $(\textbf{f})$, $(\textbf{I})$, $(\textbf{T})$, $(\textbf{D})$, $(\textbf{V})$, $(\textbf{p})$, $(\textbf{c})$ and $(\textbf{q})$ are known, then its Variable Portion Planned is:

$v= [U+S-I-T-D-(Q+Vp/\{360[c-q]\})//]/\{S'[1+s]\}-f$

Rule-21059:

If both $(\textbf{/})$, $(\textbf{U})$, $(\textbf{S})$, $(\textbf{S'})$, $(\textbf{s})$, $(\textbf{v})$, $(\textbf{Q})$, $(\textbf{I})$, $(\textbf{T})$, $(\textbf{D})$, $(\textbf{V})$, $(\textbf{p})$, $(\textbf{c})$ and $(\textbf{q})$ are known, then its Fixed Portion Planned is:

$f= [U+S-I-T-D-(Q+Vp/\{360[c-q]\})//]/\{S'[1+s]\}-v$

Rule-21060:

If both $(\textbf{/})$, $(\textbf{U})$, $(\textbf{S})$, $(\textbf{S'})$, $(\textbf{s})$, $(\textbf{v})$, $(\textbf{f})$, $(\textbf{Q})$, $(\textbf{T})$, $(\textbf{D})$, $(\textbf{V})$, $(\textbf{p})$, $(\textbf{c})$ and $(\textbf{q})$ are known, then its Interest Expense Planned is:

$I= U+S-S'[1+s][v+f]-T-D-(Q+Vp/\{360[c-q]\})//$

Rule-21061:

If both $(\textbf{/})$, $(\textbf{U})$, $(\textbf{S})$, $(\textbf{S'})$, $(\textbf{s})$, $(\textbf{v})$, $(\textbf{f})$, $(\textbf{I})$, $(\textbf{Q})$, $(\textbf{D})$, $(\textbf{V})$, $(\textbf{p})$, $(\textbf{c})$ and $(\textbf{q})$ are known, then its Tax Planned is:

$T= U+S-S'[1+s][v+f]-I-D-(Q+Vp/\{360[c-q]\})//$

Steve Asikin ISBN 14: 978-1515158691, ISBN 10: 1515158691

Rule-21062:

If both (I), (U), (S), (S'), (s), (v), (f), (I), (T), (Q), (V), (p), (c) and (q) are known, then its Dividend Planned is:

$$D= U+S-S'[1+s][v+f]-I-T-(Q+Vp/\{360[c-q]\})/I$$

Rule-21063:

If both (I), (U), (S), (S'), (s), (v), (f), (I), (T), (D), (Q), (p), (c) and (q) are known, then its Variable Cost Planned is:

$$V= 360[c-q](I\{U+S-S'[1+s][v+f]-I-T-D\}-Q)/p$$

Rule-21064:

If both (I), (U), (S), (S'), (s), (v), (f), (I), (T), (D), (V), (Q), (c) and (q) are known, then its Procured Inventory Days Planned:

$$p= 360[c-q](I\{U+S-S'[1+s][v+f]-I-T-D\}-Q)/V$$

Rule-21065:

If both (I), (U), (S), (S'), (s), (v), (f), (I), (T), (D), (V), (p), (Q) and (q) are known, then its Current Ratio Planned is:

$$c= q+Vp/[360(I\{U+S-S'[1+s][v+f]-I-T-D\}-Q)]$$

Rule-21066:

If both (I), (U), (S), (S'), (s), (v), (f), (I), (T), (D), (V), (p), (c) and (Q) are known, then its Quick or Acid Test Ratio Planned is:

$$q= c-Vp/[360(I\{U+S-S'[1+s][v+f]-I-T-D\}-Q)]$$

Steve Asikin ISBN 14: 978-1515158691, ISBN 10: 1515158691

Rule-21067:

If both (I), (**U**), (**S**), (**S'**), (**s**), (**v**), (**f**), (**I**), (**T**), (**D**), (**p**), (**c**) and (**q**) are known, then its Quoted Longterm Debt Planned is:

$$Q = I\{U+S-S'[1+s][v+f]-I-T-D\}-Svp/\{360[c-q]\}$$

Rule-21068:

If both (**Q**), (**U**), (**S**), (**S'**), (**s**), (**v**), (**f**), (**I**), (**T**), (**D**), (**p**), (**c**) and (**q**) are known, then its Leverage or Gearing Ratio Planned is:

$$I = (Q+Svp/\{360[c-q]\})/\{U+S-S'[1+s][v+f]-I-T-D\}$$

Rule-21069:

If both (I), (**Q**), (**S**), (**S'**), (**s**), (**v**), (**f**), (**I**), (**T**), (**D**), (**p**), (**c**) and (**q**) are known, then its Utilized or Starting Capital must be:

$$U = S'[1+s][v+f]+I+T+D+(Q+Svp/\{360[c-q]\})/I-S$$

Rule-21070:

If both (I), (**U**), (**Q**), (**S'**), (**s**), (**v**), (**f**), (**I**), (**T**), (**D**), (**p**), (**c**) and (**q**) are known, then its Sales or Revenue Planned is:

$$S = (I\{U-S'[1+s][v+f]-I-T-D\}-Q)/(vp/\{360[c-q]\}-I)$$

Rule-21071:

If both (I), (**U**), (**S**), (**Q**), (**s**), (**v**), (**f**), (**I**), (**T**), (**D**), (**p**), (**c**) and (**q**) are known, then its Sales Past must be:

$$S' = [U+S-I-T-D-(Q+Svp/\{360[c-q]\})/I] /\{[1+s][v+f]\}$$

Steve Asikin ISBN 14: 978-1515158691, ISBN 10: 1515158691

Rule-21072:

If both $(\textit{\textbf{I}})$, $(\textbf{U})$, $(\textbf{S})$, $(\textbf{S'})$, $(\textbf{Q})$, $(\textbf{v})$, $(\textbf{f})$, $(\textbf{I})$, $(\textbf{T})$, $(\textbf{D})$, $(\textbf{p})$, $(\textbf{c})$ and $(\textbf{q})$ are known, then its Sales Growth Planned is:

$$s = S/S'-1$$

or it can also be found as

$$s = [U+S-I-T-D-(Q+Svp/\{360[c-q]\})/\textit{\textbf{I}}]/\{S'[v+f]\}-1$$

Rule-21073:

If both $(\textit{\textbf{I}})$, $(\textbf{U})$, $(\textbf{S})$, $(\textbf{S'})$, $(\textbf{s})$, $(\textbf{Q})$, $(\textbf{f})$, $(\textbf{I})$, $(\textbf{T})$, $(\textbf{D})$, $(\textbf{p})$, $(\textbf{c})$ and $(\textbf{q})$ are known, then its Variable Portion Planned is:

$$v = (\textit{\textbf{I}}\{U+S-S'f[1+s]-I-T-D\}-Q)$$
$$/(Sp/\{360[c-q]\}+S'\textit{\textbf{I}}[1+s])$$

Rule-21074:

If both $(\textit{\textbf{I}})$, $(\textbf{U})$, $(\textbf{S})$, $(\textbf{S'})$, $(\textbf{s})$, $(\textbf{v})$, $(\textbf{Q})$, $(\textbf{I})$, $(\textbf{T})$, $(\textbf{D})$, $(\textbf{p})$, $(\textbf{c})$ and $(\textbf{q})$ are known, then its Fixed Portion Planned is:

$$f = [U+S-I-T-D-(Q+Svp/\{360[c-q]\})/\textit{\textbf{I}}]/\{S'[1+s]\}-v$$

Rule-21075:

If both $(\textit{\textbf{I}})$, $(\textbf{U})$, $(\textbf{S})$, $(\textbf{S'})$, $(\textbf{s})$, $(\textbf{v})$, $(\textbf{f})$, $(\textbf{Q})$, $(\textbf{T})$, $(\textbf{D})$, $(\textbf{p})$, $(\textbf{c})$ and $(\textbf{q})$ are known, then its Interest Expense Planned is:

$$I = U+S-S'[1+s][v+f]-T-D-(Q+Svp/\{360[c-q]\})/\textit{\textbf{I}}$$

Rule-21076:

If both $(\textit{\textbf{I}})$, $(\textbf{U})$, $(\textbf{S})$, $(\textbf{S'})$, $(\textbf{s})$, $(\textbf{v})$, $(\textbf{f})$, $(\textbf{I})$, $(\textbf{Q})$, $(\textbf{D})$, $(\textbf{p})$, $(\textbf{c})$ and $(\textbf{q})$ are known, then its Tax Rate Planned is:

$$T = U+S-S'[1+s][v+f]-I-D-(Q+Svp/\{360[c-q]\})/\textit{\textbf{I}}$$

Steve Asikin ISBN 14: 978-1515158691, ISBN 10: 1515158691

Rule-21077:

If both (**/**), (**U**), (**$**), (**$'**), (**s**), (**v**), (**f**), (**I**), (**T**), (**Q**), (**p**), (**c**) and (**q**) are known, then its Dividend Planned is:

$$D= U+\$-\$'[1+s][v+f]-I-T-(Q+\$vp/\{360[c-q]\})/I$$

Rule-21078:

If both (**/**), (**U**), (**$**), (**$'**), (**s**), (**v**), (**f**), (**I**), (**T**), (**D**), (**Q**), (**c**) and (**q**) are known, then its Procured Inventory Days Planned:

$$p= 360[c-q](I\{U+\$-\$'[1+s][v+f]-I-T-D\}-Q)/[\$v]$$

Rule-21079:

If both (**/**), (**U**), (**$**), (**$'**), (**s**), (**v**), (**f**), (**I**), (**T**), (**D**), (**p**), (**Q**) and (**q**) are known, then its Current Ratio Planned is:

$$c= q+\$vp/[360(I\{U+\$-\$'[1+s][v+f]-I-T-D\}-Q)]$$

Rule-21080:

If both (**/**), (**U**), (**$**), (**$'**), (**s**), (**v**), (**f**), (**I**), (**T**), (**D**), (**p**), (**c**) and (**Q**) are known, then its Quick or Acid Test Ratio Planned is:

$$q= c-\$vp/[360(I\{U+\$-\$'[1+s][v+f]-I-T-D\}-Q)]$$

Rule-21081:

If both (**/**), (**U**), (**$**), (**$'**), (**s**), (**v**), (**f**), (**I**), (**T**), (**D**), (**p**), (**c**) and (**q**) are known, then its Quoted Longterm Debt Planned is:

$$Q= I\{U+\$-\$'[1+s][v+f]-I-T-D\}$$
$$-\$'vp[1+s]/\{360[c-q]\}$$

Steve Asikin ISBN 14: 978-1515158691, ISBN 10: 1515158691

Rule-21082:

If both (**Q**), (**U**), (**$**), (**$'**), (**s**), (**v**), (**f**), (**I**), (**T**), (**D**), (**p**), (**c**) and (**q**) are known, then its Leverage or Gearing Ratio Planned is:

$$\text{\textit{I}} = (Q + \text{\$'}vp[1+s]/\{360[c\text{-}q]\})$$
$$/\{U + \$ - \$'[1+s][v+f] - I - T - D\}$$

Rule-21083:

If both (**I**), (**Q**), (**$**), (**$'**), (**s**), (**v**), (**f**), (**I**), (**T**), (**D**), (**p**), (**c**) and (**q**) are known, then its Utilized or Starting Capital must be:

$$U = \$'[1+s][v+f] + I + T + D$$
$$+ (Q + \text{\$'}vp[1+s]/\{360[c\text{-}q]\})/\text{\textit{I}}\$$$

Rule-21084:

If both (**I**), (**U**), (**Q**), (**$'**), (**s**), (**v**), (**f**), (**I**), (**T**), (**D**), (**p**), (**c**) and (**q**) are known, then its Sales or Revenue Planned is:

$$\$ = (Q - \text{\textit{I}}\{U - \$'[1+s][v+f] - I - T - D\}$$
$$+ \$'vp[1+s]/\{360[c\text{-}q]\})/\text{\textit{I}}$$

Rule-21085:

If both (**I**), (**U**), (**$**), (**Q**), (**s**), (**v**), (**f**), (**I**), (**T**), (**D**), (**p**), (**c**) and (**q**) are known, then its Sales Past must be:

$$\$' = (\text{\textit{I}}\{U + \$ - \$'f[1+s] - I - T - D\} - Q)$$
$$/[v[1+s](p/\{360[c\text{-}q]\} + \text{\textit{I}})]$$

Steve Asikin ISBN 14: 978-1515158691, ISBN 10: 1515158691

Rule-21086:

If both (I), (U), (S), (S'), (Q), (v), (f), (I), (T), (D), (p), (c) and (q) are known, then its Sales Growth Planned is:

$s = S/S' - 1$

or it can also be found as

$s = (I\{U+S-S'f[1+s]-I-T-D\}-Q)$
$/[S'v(p/\{360[c-q]\}+I)-1$

Rule-21087:

If both (I), (U), (S), (S'), (s), (Q), (f), (I), (T), (D), (p), (c) and (q) are known, then its Variable Portion Planned is:

$v = (I\{U+S-S'f[1+s]-I-T-D\}-Q)$
$/[S'[1+s](p/\{360[c-q]\}+I)$

Rule-21088:

If both (I), (U), (S), (S'), (s), (v), (Q), (I), (T), (D), (p), (c) and (q) are known, then its Fixed Portion Planned is:

$f = [U+S-I-T-D-(Q+S'vp[1+s]/\{360[c-q]\})/I]$
$/\{S'[1+s]\}-v$

Rule-21089:

If both (I), (U), (S), (S'), (s), (v), (f), (Q), (T), (D), (p), (c) and (q) are known, then its Interest Expense Planned is:

$I = U+S-S'[1+s][v+f]-T-D$
$-(Q+S'vp[1+s]/\{360[c-q]\})/I$

Steve Asikin ISBN 14: 978-1515158691, ISBN 10: 1515158691

Rule-21090:

If both (**/**), (**U**), (**$**), (**$'**), (**s**), (**v**), (**f**), (**I**), (**Q**), (**D**), (**p**), (**c**) and (**q**) are known, then its Tax Planned is:

$$T= U+\$-\$'[1+s][v+f]-I-D$$
$$-(Q+\$'vp[1+s]/\{360[c-q]\})/I$$

Rule-21091:

If both (**/**), (**U**), (**$**), (**$'**), (**s**), (**v**), (**f**), (**I**), (**T**), (**Q**), (**p**), (**c**) and (**q**) are known, then its Dividend Planned is:

$$D= U+\$-\$'[1+s][v+f]-I-T$$
$$-(Q+\$'vp[1+s]/\{360[c-q]\})/I$$

Rule-21092:

If both (**/**), (**U**), (**$**), (**$'**), (**s**), (**v**), (**f**), (**I**), (**T**), (**D**), (**Q**), (**c**) and (**q**) are known, then its Procured Inventory Days Planned:

$$p= 360[c-q](I\{U+\$-\$'[1+s][v+f]-I-T-D\}-Q)$$
$$/\{\$'v[1+s]\}$$

Rule-21093:

If both (**/**), (**U**), (**$**), (**$'**), (**s**), (**v**), (**f**), (**I**), (**T**), (**D**), (**p**), (**Q**) and (**q**) are known, then its Current Ratio Planned is:

$$c= q+\$'vp[1+s]$$
$$/[360(I\{U+\$-\$'[1+s][v+f]-I-T-D\}-Q)]$$

Steve Asikin ISBN 14: 978-1515158691, ISBN 10: 1515158691

Rule-21094:

If both (f), (U), (S), (S'), (s), (v), (f), (I), (T), (D), (p), (c) and (Q) are known, then its Quick or Acid Test Ratio Planned is:

$q= c\text{-}S'vp[1+s]$
$/[360(f\{U+S\text{-}S'[1+s][v+f]\text{-}I\text{-}T\text{-}D\}\text{-}Q)]$

Rule-21095:

If both (f), (U), (S), (S'), (s), (v), (f), (I), (T), (d) and (X) are known, then its Quoted Longterm Debt Planned is:

$Q= f(U+\{S\text{-}S'[1+s][v+f]\text{-}I\text{-}T\}[1\text{-}d])\text{-}X$

Rule-21096:

If both (Q), (U), (S), (S'), (s), (v), (f), (I), (T), (d) and (X) are known, then its Leverage or Gearing Ratio Planned is:

$f= [Q+X]/(U+\{S\text{-}S'[1+s][v+f]\text{-}I\text{-}T\}[1\text{-}d])$

Rule-21097:

If both (f), (Q), (S), (S'), (s), (v), (f), (I), (T), (d) and (X) are known, then its Utilized or Starting Capital must be:

$U= [Q+X]/f\{S\text{-}S'[1+s][v+f]\text{-}I\text{-}T\}[1\text{-}d]$

Rule-21098:

If both (f), (U), (Q), (S'), (s), (v), (f), (I), (T), (d) and (X) are known, then its Sales or Revenue Planned is:

$S= S'[1+s][v+f]+I+T+\{[Q+X]/f\text{-}U\}/[1\text{-}d]$

Steve Asikin ISBN 14: 978-1515158691, ISBN 10: 1515158691

Rule-21099:
> If both (**ƒ**), (**U**), (**$**), (**Q**), (**s**), (**v**), (**f**), (**I**), (**T**), (**d**) and
> (**X**) are known, then its Sales Past must be:
> $'= ($-I-T-{[Q+X]/ƒ-U}/[1-d])/{[1+s][v+f]}

Rule-21100:
> If both (**ƒ**), (**U**), (**$**), (**$'**), (**Q**), (**v**), (**f**), (**I**), (**T**), (**d**) and
> (**X**) are known, then its Sales Growth Planned is:
> s= $/$'-1
> > or it can also be found as
> s= ($-I-T-{[Q+X]/ƒ-U}/[1-d])/{$'[v+f]}-1

Rule-21101:
> If both (**ƒ**), (**U**), (**$**), (**$'**), (**s**), (**Q**), (**f**), (**I**), (**T**), (**d**) and
> (**X**) are known, then its Variable Portion Planned is:
> v= ($-I-T-{[Q+X]/ƒ-U}/[1-d])/{$'[1+s]}-f

Rule-21102:
> If both (**ƒ**), (**U**), (**$**), (**$'**), (**s**), (**v**), (**Q**), (**I**), (**T**), (**d**) and
> (**X**) are known, then its Fixed Portion Planned is:
> f= ($-I-T-{[Q+X]/ƒ-U}/[1-d])/{$'[1+s]}-v

Rule-21103:
> If both (**ƒ**), (**U**), (**$**), (**$'**), (**s**), (**v**), (**f**), (**Q**), (**T**), (**d**) and
> (**X**) are known, then its Interest Expense Planned is:
> I= $-$'[1+s][v+f]-T-{[Q+X]/ƒ-U}/[1-d]

Rule-21104:
> If both (**ƒ**), (**U**), (**$**), (**$'**), (**s**), (**v**), (**f**), (**I**), (**Q**), (**d**) and
> (**X**) are known, then its Tax Planned is:
> T= $-$'[1+s][v+f]-I-{[Q+X]/ƒ-U}/[1-d]

Steve Asikin ISBN 14: 978-1515158691, ISBN 10: 1515158691

Rule-21105:

If both $(/)$, $(\mathbf{U})$, $(\$)$, $(\$')$, (s), (v), (f), $(\mathbf{I})$, $(\mathbf{T})$, $(\mathbf{Q})$ and $(\mathbf{X})$ are known, then its Dividend Payout Planned is:
$\mathbf{d}= 1-\{[\mathbf{Q}+\mathbf{X}]//-\mathbf{U}\}/\{\$-\$'[1+s][v+f]-\mathbf{I}-\mathbf{T}\}$

Rule-21106:

If both $(/)$, $(\mathbf{U})$, $(\$)$, $(\$')$, (s), (v), (f), $(\mathbf{I})$, $(\mathbf{T})$, $(\mathbf{d})$ and $(\mathbf{Q})$ are known, then its Xpress or Current Debt Planned is:
$\mathbf{X}= /(\mathbf{U}+\{\$-\$'[1+s][v+f]-\mathbf{I}-\mathbf{T}\}[1-\mathbf{d}])-\mathbf{Q}$

Rule-21107:

If both $(/)$, $(\mathbf{U})$, $(\$)$, $(\$')$, (s), (v), (f), $(\mathbf{I})$, $(\mathbf{T})$, $(\mathbf{d})$, $(\mathbf{P})$, $(\mathbf{c})$ and $(\mathbf{q})$ are known, then its Quoted Longterm Debt Planned is:
$\mathbf{Q}= /(\mathbf{U}+\{\$-\$'[1+s][v+f]-\mathbf{I}-\mathbf{T}\}[1-\mathbf{d}])-\mathbf{P}/[\mathbf{c}-\mathbf{q}]$

Rule-21108:

If both $(\mathbf{Q})$, $(\mathbf{U})$, $(\$)$, $(\$')$, (s), (v), (f), $(\mathbf{I})$, $(\mathbf{T})$, $(\mathbf{d})$, $(\mathbf{P})$, $(\mathbf{c})$ and $(\mathbf{q})$ are known, then its Leverage or Gearing Ratio Planned is:
$/= \{\mathbf{Q}+\mathbf{P}/[\mathbf{c}-\mathbf{q}]\}/(\mathbf{U}+\{\$-\$'[1+s][v+f]-\mathbf{I}-\mathbf{T}\}[1-\mathbf{d}])$

Rule-21109:

If both $(/)$, $(\mathbf{Q})$, $(\$)$, $(\$')$, (s), (v), (f), $(\mathbf{I})$, $(\mathbf{T})$, $(\mathbf{d})$, $(\mathbf{P})$, $(\mathbf{c})$ and $(\mathbf{q})$ are known, then its Utilized or Starting Capital must be:
$\mathbf{U}= \{\mathbf{Q}+\mathbf{P}/[\mathbf{c}-\mathbf{q}]\}//-\{\$-\$'[1+s][v+f]-\mathbf{I}-\mathbf{T}\}[1-\mathbf{d}]$

Steve Asikin ISBN 14: 978-1515158691, ISBN 10: 1515158691

Rule-21110:

If both (I), $(\mathbf{U})$, $(\mathbf{Q})$, $(\mathbf{S'})$, $(\mathbf{s})$, $(\mathbf{v})$, $(\mathbf{f})$, $(\mathbf{I})$, $(\mathbf{T})$, $(\mathbf{d})$, $(\mathbf{P})$, $(\mathbf{c})$ and $(\mathbf{q})$ are known, then its Sales or Revenue Planned is:

$$S = S'[1+s][v+f] + I + T + (\{Q + P/[c-q]\}/I - U)/[1-d]$$

Rule-21111:

If both (I), $(\mathbf{U})$, $(\mathbf{S})$, $(\mathbf{Q})$, $(\mathbf{s})$, $(\mathbf{v})$, $(\mathbf{f})$, $(\mathbf{I})$, $(\mathbf{T})$, $(\mathbf{d})$, $(\mathbf{P})$, $(\mathbf{c})$ and $(\mathbf{q})$ are known, then its Sales Past must be:

$$S' = [S - I - T - (\{Q + P/[c-q]\}/I - U)/[1-d]]/\{[1+s][v+f]\}$$

Rule-21112:

If both (I), $(\mathbf{U})$, $(\mathbf{S})$, $(\mathbf{S'})$, $(\mathbf{Q})$, $(\mathbf{v})$, $(\mathbf{f})$, $(\mathbf{I})$, $(\mathbf{T})$, $(\mathbf{d})$, $(\mathbf{P})$, $(\mathbf{c})$ and $(\mathbf{q})$ are known, then its Sales Growth Planned is:

$$s = S/S' - 1$$

or it can also be found as

$$s = [S - I - T - (\{Q + P/[c-q]\}/I - U)/[1-d]]/\{S'[v+f]\} - 1$$

Rule-21113:

If both (I), $(\mathbf{U})$, $(\mathbf{S})$, $(\mathbf{S'})$, $(\mathbf{s})$, $(\mathbf{Q})$, $(\mathbf{f})$, $(\mathbf{I})$, $(\mathbf{T})$, $(\mathbf{d})$, $(\mathbf{P})$, $(\mathbf{c})$ and $(\mathbf{q})$ are known, then its Variable Portion Planned is:

$$v = [S - I - T - (\{Q + P/[c-q]\}/I - U)/[1-d]]/\{S'[1+s]\} - f$$

Rule-21114:

If both (I), $(\mathbf{U})$, $(\mathbf{S})$, $(\mathbf{S'})$, $(\mathbf{s})$, $(\mathbf{v})$, $(\mathbf{Q})$, $(\mathbf{I})$, $(\mathbf{T})$, $(\mathbf{d})$, $(\mathbf{P})$, $(\mathbf{c})$ and $(\mathbf{q})$ are known, then its Fixed Portion Planned is:

$$f = [S - I - T - (\{Q + P/[c-q]\}/I - U)/[1-d]]/\{S'[1+s]\} - v$$

Steve Asikin ISBN 14: 978-1515158691, ISBN 10: 1515158691

Rule-21115:

If both (I), (U), (S), (S'), (s), (v), (f), (Q), (T), (d), (P), (c) and (q) are known, then its Interest Expense Planned is:

$I = S-S'[1+s][v+f]-T-(\{Q+P/[c-q]\}/I-U)/[1-d]$

Rule-21116:

If both (I), (U), (S), (S'), (s), (v), (f), (I), (Q), (d), (P), (c) and (q) are known, then its TaxPlanned is:

$T = S-S'[1+s][v+f]-I-(\{Q+P/[c-q]\}/I-U)/[1-d]$

Rule-21117:

If both (I), (U), (S), (S'), (s), (v), (f), (I), (T), (Q), (P), (c) and (q) are known, then its Dividend Payout Planned is

$d = 1-(\{Q+P/[c-q]\}/I-U)/\{S-S'[1+s][v+f]-I-T\}$

Rule-21118:

If both (I), (U), (S), (S'), (s), (v), (f), (I), (T), (d). (Q), (c) and (q) are known, then its Procured Inventory Planned is:

$P = [c-q][I(U+\{S-S'[1+s][v+f]-I-T\}[1-d])-Q]$

Rule-21119:

If both (I), (U), (S), (S'), (s), (v), (f), (I), (T), (d). (P), (Q) and (q) are known, then its Current Ratio Planned is:

$c = q+P/[I(U+\{S-S'[1+s][v+f]-I-T\}[1-d])-Q]$

Steve Asikin ISBN 14: 978-1515158691, ISBN 10: 1515158691

Rule-21120:

If both (**/**), (**U**), (**$**), (**$'**), (**s**), (**v**), (**f**), (**I**), (**T**), (**d**). (**P**), (**c**) and (**Q**) are known, then its Quick or Acid Test Ratio Planned is:

$$q = c - P/[/(U + \{\$-\$'[1+s][v+f]-I-T\}[1-d])-Q]$$

Rule-21121:

If both (**/**), (**U**), (**$**), (**$'**), (**s**), (**v**), (**f**), (**I**), (**T**), (**d**). (**V**), (**p**), (**c**) and (**q**) are known, then its Quoted Longterm Debt Planned is:

$$Q = /(U + \{\$-\$'[1+s][v+f]-I-T\}[1-d])-Vp/\{360[c-q]\}$$

Rule-21122:

If both (**Q**), (**U**), (**$**), (**$'**), (**s**), (**v**), (**f**), (**I**), (**T**), (**d**). (**V**), (**p**), (**c**) and (**q**) are known, then its Leverage or Gearing Ratio Planned is:

$$/= (Q + Vp/\{360[c-q]\})$$
$$/(U + \{\$-\$'[1+s][v+f]-I-T\}[1-d])$$

Rule-21123:

If both (**/**), (**Q**), (**$**), (**$'**), (**s**), (**v**), (**f**), (**I**), (**T**), (**d**). (**V**), (**p**), (**c**) and (**q**) are known, then its Utilized or Starting Capital must be:

$$U = (Q + Vp/\{360[c-q]\})//\{\$-\$'[1+s][v+f]-I-T\}[1-d]$$

Rule-21124:

If both (**/**), (**U**), (**Q**), (**$'**), (**s**), (**v**), (**f**), (**I**), (**T**), (**d**). (**V**), (**p**), (**c**) and (**q**) are known, then its Sales or Revenue Planned is:

$$\$ = \$'[1+s][v+f]+I+T+[(Q+Vp/\{360[c-q]\})//-U]$$
$$/[1-d]$$

Steve Asikin ISBN 14: 978-1515158691, ISBN 10: 1515158691

Rule-21125:

    If both (**/**), (**U**), (**S**), (**Q**), (**s**), (**v**), (**f**), (**I**), (**T**), (**d**). (**V**), (**p**), (**c**) and (**q**) are known, then its Sales Past must be:

$$S' = \{S\text{-}I\text{-}T\text{-}[(Q+Vp/\{360[c\text{-}q]\})/\text{/-}U]/[1\text{-}d]\} / \{[1+s][v+f]\}$$

Rule-21126:

    If both (**/**), (**U**), (**S**), (**S'**), (**Q**), (**v**), (**f**), (**I**), (**T**), (**d**). (**V**), (**p**), (**c**) and (**q**) are known, then its Sales Growth Planned is:

$$s = S/S' - 1$$

        or it can also be found as

$$s = \{S\text{-}I\text{-}T\text{-}[(Q+Vp/\{360[c\text{-}q]\})/\text{/-}U]/[1\text{-}d]\} / \{S'[v+f]\} - 1$$

Rule-21127:

    If both (**/**), (**U**), (**S**), (**S'**), (**s**), (**Q**), (**f**), (**I**), (**T**), (**d**). (**V**), (**p**), (**c**) and (**q**) are known, then its Variable Portion Planned is:

$$v = V/S$$

        or it can also be found as

$$v = \{S\text{-}I\text{-}T\text{-}[(Q+Vp/\{360[c\text{-}q]\})/\text{/-}U]/[1\text{-}d]\} / \{S'[1+s]\} - f$$

Rule-21128:

    If both (**/**), (**U**), (**S**), (**S'**), (**s**), (**v**), (**Q**), (**I**), (**T**), (**d**). (**V**), (**p**), (**c**) and (**q**) are known, then its Fixed Portion Planned is:

$$f = \{S\text{-}I\text{-}T\text{-}[(Q+Vp/\{360[c\text{-}q]\})/\text{/-}U]/[1\text{-}d]\} / \{S'[1+s]\} - v$$

Steve Asikin ISBN 14: 978-1515158691, ISBN 10: 1515158691

Rule-21129:

If both (*I*), (**U**), (**$**), (**$'**), (**s**), (**v**), (**f**), (**Q**), (**T**), (**d**). (**V**), (**p**), (**c**) and (**q**) are known, then its Interest Expense Planned is:

$$I= \$-\$'[1+s][v+f]-T-[(Q+Vp/\{360[c-q]\})/I-U] /[1-d]$$

Rule-21130:

If both (*I*), (**U**), (**$**), (**$'**), (**s**), (**v**), (**f**), (**I**), (**Q**), (**d**). (**V**), (**p**), (**c**) and (**q**) are known, then its Tax Planned is:

$$T= \$-\$'[1+s][v+f]-I-[(Q+Vp/\{360[c-q]\})/I-U] /[1-d]$$

Rule-21131:

If both (*I*), (**U**), (**$**), (**$'**), (**s**), (**v**), (**f**), (**I**), (**T**), (**Q**). (**V**), (**p**), (**c**) and (**q**) are known, then its Dividend Payout Planned is:

$$d= 1-[(Q+Vp/\{360[c-q]\})/I-U] /\{\$-\$'[1+s][v+f]-I-T\}$$

Rule-21132:

If both (*I*), (**U**), (**$**), (**$'**), (**s**), (**v**), (**f**), (**I**), (**T**), (**d**). (**Q**), (**p**), (**c**) and (**q**) are known, then its Variable Cost Planned is:

$$V= 360[c-q][(I U+\{\$-\$'[1+s][v+f]-I-T\}[1-d])-Q]/p$$

Rule-21133:

If both (*I*), (**U**), (**$**), (**$'**), (**s**), (**v**), (**f**), (**I**), (**T**), (**d**). (**V**), (**Q**), (**c**) and (**q**) are known, then its Procured Inventory Days Planned:

$$p= 360[c-q][(I U+\{\$-\$'[1+s][v+f]-I-T\}[1-d])-Q]/V$$

Steve Asikin ISBN 14: 978-1515158691, ISBN 10: 1515158691

Rule-21134:

If both (**/**), (**U**), (**$**), (**$'**), (**s**), (**v**), (**f**), (**I**), (**T**), (**d**). (**V**), (**p**), (**Q**) and (**q**) are known, then its Current Ratio Planned is:

$$c= q+Vp/\{360[/(U+\{\$-\$'[1+s][v+f]-I-T\}[1-d])-Q]\}$$

Rule-21135:

If both (**/**), (**U**), (**$**), (**$'**), (**s**), (**v**), (**f**), (**I**), (**T**), (**d**). (**V**), (**p**), (**c**) and (**Q**) are known, then its Quick or Acid Test Ratio Planned is:

$$q= c-Vp/\{360[/(U+\{\$-\$'[1+s][v+f]-I-T\}[1-d])-Q]\}$$

Rule-21136:

If both (**/**), (**U**), (**$**), (**$'**), (**s**), (**v**), (**f**), (**I**), (**T**), (**d**). (**p**), (**c**) and (**q**) are known, then its Quoted Longterm Debt Planned is:

$$Q= /(U+\{\$-\$'[1+s][v+f]-I-T\}[1-d])-\$vp/\{360[c-q]\}$$

Rule-21137:

If both (**Q**), (**U**), (**$**), (**$'**), (**s**), (**v**), (**f**), (**I**), (**T**), (**d**). (**p**), (**c**) and (**q**) are known, then its Leverage or Gearing Ratio Planned is:

$$/= (Q+\$vp/\{360[c-q]\})$$
$$/(U+\{\$-\$'[1+s][v+f]-I-T\}[1-d])$$

Rule-21138:

If both (**/**), (**Q**), (**$**), (**$'**), (**s**), (**v**), (**f**), (**I**), (**T**), (**d**). (**p**), (**c**) and (**q**) are known, then its Utilized or Starting Capital must be:

$$U= (Q+\$vp/\{360[c-q]\})//$$
$$-\{\$-\$'[1+s][v+f]-I-T\}[1-d]$$

Steve Asikin ISBN 14: 978-1515158691, ISBN 10: 1515158691

<u>Rule-21139</u>:

If both (**/**), (**U**), (**Q**), (**S'**), (**s**), (**v**), (**f**), (**I**), (**T**), (**d**). (**p**), (**c**) and (**q**) are known, then its Sales or Revenue Planned is:

$$S= [\text{/(U-}\{S'[1+s][v+f]+I+T\}[1-d])-S]$$
$$/(vp/\{360[c-q]\}-\text{/}[1-d])$$

<u>Rule-21140</u>:

If both (**/**), (**U**), (**S**), (**Q**), (**s**), (**v**), (**f**), (**I**), (**T**), (**d**). (**p**), (**c**) and (**q**) are known, then its Sales Past must be:

$$S'= \{S-I-T-[(Q+Svp/\{360[c-q]\})/\text{/}-U]/[1-d]\}$$
$$/\{[1+s][v+f]\}$$

<u>Rule-21141</u>:

If both (**/**), (**U**), (**S**), (**S'**), (**Q**), (**v**), (**f**), (**I**), (**T**), (**d**). (**p**), (**c**) and (**q**) are known, then its Sales Growth Planned is:

$$s= S/S'-1$$

or it can also be found as

$$s=\{S-I-T-[(Q+Svp/\{360[c-q]\})/\text{/}-U]$$
$$/[1-d]\}/\{S'[v+f]\}-1$$

<u>Rule-21142</u>:

If both (**/**), (**U**), (**S**), (**S'**), (**s**), (**Q**), (**f**), (**I**), (**T**), (**d**). (**p**), (**c**) and (**q**) are known, then its Variable Portion Planned is:

$$v= [\text{/(U+}\{S-S'f[1+s]-I-T\}[1-d])-Q]$$
$$/(Sp/\{360[c-q]\}+S'\text{/}[1+s][1-d])$$

Steve Asikin ISBN 14: 978-1515158691, ISBN 10: 1515158691

Rule-21143:

If both $(\textit{l})$, $(\textbf{U})$, $(\textbf{S})$, $(\textbf{S'})$, $(\textbf{s})$, $(\textbf{v})$, $(\textbf{Q})$, $(\textbf{I})$, $(\textbf{T})$, $(\textbf{d})$. $(\textbf{p})$, $(\textbf{c})$ and $(\textbf{q})$ are known, then its Fixed Portion Planned is:

$$f = \{S-I-T-[(Q+Svp/\{360[c-q]\})/I-U]/[1-d]\}$$
$$/\{S'[1+s]\}-v$$

Rule-21144:

If both $(\textit{l})$, $(\textbf{U})$, $(\textbf{S})$, $(\textbf{S'})$, $(\textbf{s})$, $(\textbf{v})$, $(\textbf{f})$, $(\textbf{Q})$, $(\textbf{T})$, $(\textbf{d})$. $(\textbf{p})$, $(\textbf{c})$ and $(\textbf{q})$ are known, then its Interest Expense Planned is:

$$I = S-S'[1+s][v+f]-T-[(Q+Svp/\{360[c-q]\})/I-U]$$
$$/[1-d]$$

Rule-21145:

If both $(\textit{l})$, $(\textbf{U})$, $(\textbf{S})$, $(\textbf{S'})$, $(\textbf{s})$, $(\textbf{v})$, $(\textbf{f})$, $(\textbf{I})$, $(\textbf{Q})$, $(\textbf{d})$. $(\textbf{p})$, $(\textbf{c})$ and $(\textbf{q})$ are known, then its Tax Planned is:

$$T = S-S'[1+s][v+f]-I$$
$$-[(Q+Svp/\{360[c-q]\})/I-U]/[1-d]$$

Rule-21146:

If both $(\textit{l})$, $(\textbf{U})$, $(\textbf{S})$, $(\textbf{S'})$, $(\textbf{s})$, $(\textbf{v})$, $(\textbf{f})$, $(\textbf{I})$, $(\textbf{T})$, $(\textbf{Q})$. $(\textbf{p})$, $(\textbf{c})$ and $(\textbf{q})$ are known, then its Dividend Payout Planned is:

$$d = 1-[(Q+Svp/\{360[c-q]\})/I-U]$$
$$/\{S-S'[1+s][v+f]-I-T\}$$

Steve Asikin ISBN 14: 978-1515158691, ISBN 10: 1515158691

Rule-21147:

If both ($\jmath$), (**U**), (**S**), (**S'**), (**s**), (**v**), (**f**), (**I**), (**T**), (**d**). (**p**), (**c**) and (**q**) are known, then its Quoted Longterm Debt Planned is:

$$p= 360[c\text{-}q][\jmath U+\{S\text{-}S'[1+s][v+f]\text{-}I\text{-}T\}[1\text{-}d])\text{-}Q]$$
$$/[Sv]$$

Rule-21148:

If both ($\jmath$), (**U**), (**S**), (**S'**), (**s**), (**v**), (**f**), (**I**), (**T**), (**d**). (**p**), (**Q**) and (**q**) are known, then its Current Ratio Planned is:

$$c= q+Svp$$
$$/\{360[\jmath U+\{S\text{-}S'[1+s][v+f]\text{-}I\text{-}T\}[1\text{-}d])\text{-}Q]\}$$

Rule-21149:

If both ($\jmath$), (**U**), (**S**), (**S'**), (**s**), (**v**), (**f**), (**I**), (**T**), (**d**). (**p**), (**c**) and (**Q**) are known, then its Quick or Acid Test Ratio Planned is:

$$q= c\text{-}Svp/\{360[\jmath U+\{S\text{-}S'[1+s][v+f]\text{-}I\text{-}T\}[1\text{-}d])\text{-}Q]\}$$

Rule-21150:

If both ($\jmath$), (**U**), (**S**), (**S'**), (**s**), (**v**), (**f**), (**I**), (**T**), (**d**). (**p**), (**c**) and (**q**) are known, then its Quoted Longterm Debt Planned is:

$$Q= \jmath U+\{S\text{-}S'[1+s][v+f]\text{-}I\text{-}T\}[1\text{-}d])$$
$$-S'vp[1+s]/\{360[c\text{-}q]\}$$

Steve Asikin ISBN 14: 978-1515158691, ISBN 10: 1515158691

Rule-21151:

If both $(\mathbf{Q})$, $(\mathbf{U})$, $(\mathbf{S})$, $(\mathbf{S'})$, $(\mathbf{s})$, $(\mathbf{v})$, $(\mathbf{f})$, $(\mathbf{I})$, $(\mathbf{T})$, $(\mathbf{d})$. $(\mathbf{p})$, $(\mathbf{c})$ and $(\mathbf{q})$ are known, then its Leverage or Gearing Ratio Planned is:

$$\mathbf{l} = (\mathbf{Q} + \mathbf{S'vp}[1+\mathbf{s}]/\{360[\mathbf{c}-\mathbf{q}]\})$$
$$/(\mathbf{U} + \{\mathbf{S} - \mathbf{S'}[1+\mathbf{s}][\mathbf{v}+\mathbf{f}] - \mathbf{I} - \mathbf{T}\}[1-\mathbf{d}])$$

Rule-21152:

If both $(\mathbf{l})$, $(\mathbf{Q})$, $(\mathbf{S})$, $(\mathbf{S'})$, $(\mathbf{s})$, $(\mathbf{v})$, $(\mathbf{f})$, $(\mathbf{I})$, $(\mathbf{T})$, $(\mathbf{d})$. $(\mathbf{p})$, $(\mathbf{c})$ and $(\mathbf{q})$ are known, then its Utilized or Starting Capital must be:

$$\mathbf{U} = (\mathbf{Q} + \mathbf{S'vp}[1+\mathbf{s}]/\{360[\mathbf{c}-\mathbf{q}]\})/\mathbf{l}$$
$$-\{\mathbf{S} - \mathbf{S'}[1+\mathbf{s}][\mathbf{v}+\mathbf{f}] - \mathbf{I} - \mathbf{T}\}[1-\mathbf{d}]$$

Rule-21153:

If both $(\mathbf{l})$, $(\mathbf{U})$, $(\mathbf{Q})$, $(\mathbf{S'})$, $(\mathbf{s})$, $(\mathbf{v})$, $(\mathbf{f})$, $(\mathbf{I})$, $(\mathbf{T})$, $(\mathbf{d})$. $(\mathbf{p})$, $(\mathbf{c})$ and $(\mathbf{q})$ are known, then its Sales or Revenue Planned is:

$$\mathbf{S} = [\mathbf{Q} - \mathbf{l}(\mathbf{U} - \{\mathbf{S'}[1+\mathbf{s}][\mathbf{v}+\mathbf{f}] + \mathbf{I} + \mathbf{T}\}[1-\mathbf{d}])$$
$$+\mathbf{S'vp}[1+\mathbf{s}]/\{360[\mathbf{c}-\mathbf{q}]\}]/\{\mathbf{l}[1-\mathbf{d}]\}$$

Rule-21154:

If both $(\mathbf{l})$, $(\mathbf{U})$, $(\mathbf{S})$, $(\mathbf{Q})$, $(\mathbf{s})$, $(\mathbf{v})$, $(\mathbf{f})$, $(\mathbf{I})$, $(\mathbf{T})$, $(\mathbf{d})$. $(\mathbf{p})$, $(\mathbf{c})$ and $(\mathbf{q})$ are known, then its Sales Past must be:

$$\mathbf{S'} = [\mathbf{l}\mathbf{U} + \{\mathbf{S} - \mathbf{S'f}[1+\mathbf{s}] - \mathbf{I} - \mathbf{T}\}[1-\mathbf{d}]) - \mathbf{Q}]$$
$$/[\mathbf{v}[1+\mathbf{s}](\mathbf{p}/\{360[\mathbf{c}-\mathbf{q}]\} + \mathbf{l}[1-\mathbf{d}])]$$

Steve Asikin ISBN 14: 978-1515158691, ISBN 10: 1515158691

Rule-21155:

If both (f), $(\mathbf{U})$, $(\mathbf{S})$, $(\mathbf{S'})$, $(\mathbf{Q})$, $(\mathbf{v})$, $(\mathbf{f})$, $(\mathbf{I})$, $(\mathbf{T})$, $(\mathbf{d})$. $(\mathbf{p})$, $(\mathbf{c})$ and $(\mathbf{q})$ are known, then its Sales Growth Planned is:

$s = S/S' - 1$

> or it can also be found as

$s = [\mathit{f}(U + \{S - S'f[1+s] - I - T\}[1-d]) - Q]$
$/[S'v[1+s](p/\{360[c-q]\} + \mathit{f}[1-d])] - 1$

Rule-21156:

If both (f), $(\mathbf{U})$, $(\mathbf{S})$, $(\mathbf{S'})$, $(\mathbf{s})$, $(\mathbf{Q})$, $(\mathbf{f})$, $(\mathbf{I})$, $(\mathbf{T})$, $(\mathbf{d})$. $(\mathbf{p})$, $(\mathbf{c})$ and $(\mathbf{q})$ are known, then its Variable Portion Planned is:

$v = [\mathit{f}(U + \{S - S'f[1+s] - I - T\}[1-d]) - Q]$
$/[S'[1+s](p/\{360[c-q]\} + \mathit{f}[1-d])]$

Rule-21157:

If both (f), $(\mathbf{U})$, $(\mathbf{S})$, $(\mathbf{S'})$, $(\mathbf{s})$, $(\mathbf{v})$, $(\mathbf{Q})$, $(\mathbf{I})$, $(\mathbf{T})$, $(\mathbf{d})$. $(\mathbf{p})$, $(\mathbf{c})$ and $(\mathbf{q})$ are known, then its Fixed Portion Planned is:

$f = \{S - I - T - [(Q + S'vp[1+s]/\{360[c-q]\})/\mathit{f} - U]/[1-d]\}$
$/\{S'[1+s]\} - v$

Rule-21158:

If both (f), $(\mathbf{U})$, $(\mathbf{S})$, $(\mathbf{S'})$, $(\mathbf{s})$, $(\mathbf{v})$, $(\mathbf{f})$, $(\mathbf{Q})$, $(\mathbf{T})$, $(\mathbf{d})$. $(\mathbf{p})$, $(\mathbf{c})$ and $(\mathbf{q})$ are known, then its Interest Expense Planned is:

$I = S - S'[1+s][v+f] - T$
$-[(Q + S'vp[1+s]/\{360[c-q]\})/\mathit{f} - U]/[1-d]$

Steve Asikin ISBN 14: 978-1515158691, ISBN 10: 1515158691

Rule-21159:

 If both (I), (U), $(\$)$, $(\$')$, (s), (v), (f), (I), (Q), (d). (p), (c) and (q) are known, then its Tax Planned is:

$$T= \$-\$'[1+s][v+f]-I$$
$$-[(Q+\$'vp[1+s]/\{360[c-q]\})/I-U]/[1-d]$$

Rule-21160:

 If both (I), (U), $(\$)$, $(\$')$, (s), (v), (f), (I), (T), (Q). (p), (c) and (q) are known, then its Dividend Payout Planned is:

$$d= 1-[(Q+\$'vp[1+s]/\{360[c-q]\})/I-U]$$
$$/\{\$-\$'[1+s][v+f]-I-T\}$$

Rule-21161:

 If both (I), (U), $(\$)$, $(\$')$, (s), (v), (f), (I), (T), (d). (Q), (c) and (q) are known, then its Procured Inventory Days Planned:

$$p= 360[c-q][I(U+\{\$-\$'[1+s][v+f]-I-T\}[1-d])-Q]$$
$$/\{\$'v[1+s]\}$$

Rule-21162:

 If both (I), (U), $(\$)$, $(\$')$, (s), (v), (f), (I), (T), (d). (p), (Q) and (q) are known, then its Current Ratio Planned is:

$$c= q+\$'vp[1+s]$$
$$/\{360[I(U+\{\$-\$'[1+s][v+f]-I-T\}[1-d])-Q]\}$$

Steve Asikin ISBN 14: 978-1515158691, ISBN 10: 1515158691

Rule-21163:

    If both (**/**), (**U**), (**\$**), (**\$'**), (**s**), (**v**), (**f**), (**I**), (**T**), (**d**). (**p**), (**c**) and (**Q**) are known, then its Quick or Acid Test Ratio Planned is:

$$q= c\text{-}\$'vp[1+s]$$
$$/\{360[/U+\{\$\text{-}\$'[1+s][v+f]\text{-}I\text{-}T\}[1\text{-}d])\text{-}Q]\}$$

Rule-21164:

    If both (**/**), (**U**), (**\$**), (**\$'**), (**s**), (**v**), (**f**), (**I**), (**T**), (**A**). (**d**) and (**X**) are known, then its Quoted Longterm Debt Planned is:

$$Q= /\{U+\$\text{-}\$'[1+s][v+f]\text{-}I\text{-}T\text{-}Ad\}\text{-}X$$

Rule-21165:

    If both (**Q**), (**U**), (**\$**), (**\$'**), (**s**), (**v**), (**f**), (**I**), (**T**), (**A**). (**d**) and (**X**) are known, then its Leverage or Gearing Ratio Planned is:

$$/= [Q+X]/\{U+\$\text{-}\$'[1+s][v+f]\text{-}I\text{-}T\text{-}Ad\}$$

Rule-21166:

    If both (**/**), (**Q**), (**\$**), (**\$'**), (**s**), (**v**), (**f**), (**I**), (**T**), (**A**). (**d**) and (**X**) are known, then its Utilized or Starting Capital must be:

$$U = \$'[1+s][v+f]+I+T+Ad+[Q+X]//\$$$

Rule-21167:

    If both (**/**), (**U**), (**Q**), (**\$'**), (**s**), (**v**), (**f**), (**I**), (**T**), (**A**). (**d**) and (**X**) are known, then its Sales or Revenue Planned is:

$$\$= \$'[1+s][v+f]+I+T+Ad+[Q+X]//U$$

Steve Asikin ISBN 14: 978-1515158691, ISBN 10: 1515158691

Rule-21168:

   If both (I), $(\mathbf{U})$, $(\mathbf{S})$, $(\mathbf{Q})$, (s), (v), $(\mathbf{f})$, $(\mathbf{I})$, $(\mathbf{T})$, $(\mathbf{A})$. $(\mathbf{d})$ and $(\mathbf{X})$ are known, then its Sales Past must be:

$$S'= \{U+S-I-T-Ad-[Q+X]/I\}/\{[1+s][v+f]\}$$

Rule-21169:

   If both (I), $(\mathbf{U})$, $(\mathbf{S})$, (S'), $(\mathbf{Q})$, (v), $(\mathbf{f})$, $(\mathbf{I})$, $(\mathbf{T})$, $(\mathbf{A})$. $(\mathbf{d})$ and $(\mathbf{X})$ are known, then its Sales Growth Planned is:

$$s= S/S'-1$$

       or it can also be found as

$$s= \{U+S-I-T-Ad-[Q+X]/I\}/\{S'[v+f]\}-1$$

Rule-21170:

   If both (I), $(\mathbf{U})$, $(\mathbf{S})$, (S'), (s), $(\mathbf{Q})$, $(\mathbf{f})$, $(\mathbf{I})$, $(\mathbf{T})$, $(\mathbf{A})$. $(\mathbf{d})$ and $(\mathbf{X})$ are known, then its Variable Portion Planned is:

$$v= \{U+S-I-T-Ad-[Q+X]/I\}/\{S'[1+s]\}-f$$

Rule-21171:

   If both (I), $(\mathbf{U})$, $(\mathbf{S})$, (S'), (s), (v), $(\mathbf{Q})$, $(\mathbf{I})$, $(\mathbf{T})$, $(\mathbf{A})$. $(\mathbf{d})$ and $(\mathbf{X})$ are known, then its Fixed Portion Planned is:

$$f= \{U+S-I-T-Ad-[Q+X]/I\}/\{S'[1+s]\}-v$$

Rule-21172:

   If both (I), $(\mathbf{U})$, $(\mathbf{S})$, (S'), (s), (v), $(\mathbf{f})$, $(\mathbf{Q})$, $(\mathbf{T})$, $(\mathbf{A})$. $(\mathbf{d})$ and $(\mathbf{X})$ are known, then its Interest Expense Planned is:

$$I= U+S-S'[1+s][v+f]-T-Ad-[Q+X]/I$$

Steve Asikin ISBN 14: 978-1515158691, ISBN 10: 1515158691

Rule-21173:

If both (**/**), (**U**), (**$**), (**$'**), (**s**), (**v**), (**f**), (**I**), (**Q**), (**A**). (**d**) and (**X**) are known, then its Tax Planned is:

$$T= U+\text{$}-\text{$}'[1+s][v+f]-I-Ad-[Q+X]/\text{/}$$

Rule-21174:

If both (**/**), (**U**), (**$**), (**$'**), (**s**), (**v**), (**f**), (**I**), (**T**), (**Q**). (**d**) and (**X**) are known, then its After Tax Income Planned is:

$$A= \{U+\text{$}-\text{$}'[1+s][v+f]-I-T-[Q+X]/\text{/}\}/d$$

Rule-21175:

If both (**/**), (**U**), (**$**), (**$'**), (**s**), (**v**), (**f**), (**I**), (**T**), (**A**). (**Q**) and (**X**) are known, then its Dividend Portion Planned is:

$$d= \{U+\text{$}-\text{$}'[1+s][v+f]-I-T-[Q+X]/\text{/}\}/A$$

Rule-21176:

If both (**/**), (**U**), (**$**), (**$'**), (**s**), (**v**), (**f**), (**I**), (**T**), (**A**). (**d**) and (**Q**) are known, then its Xpress or Current Debt Planned is:

$$X= \text{/}\{U+\text{$}-\text{$}'[1+s][v+f]-I-T-Ad\}-Q$$

Rule-21177:

If both (**/**), (**U**), (**$**), (**$'**), (**s**), (**v**), (**f**), (**I**), (**T**), (**A**). (**d**), (**P**), (**c**) and (**X**) are known, then its Quoted Longterm Debt Planned is:

$$Q= \text{/}\{U+\text{$}-\text{$}'[1+s][v+f]-I-T-Ad\}-P/[c-q]$$

Steve Asikin ISBN 14: 978-1515158691, ISBN 10: 1515158691

Rule-21178:

If both $(\mathbf{Q})$, $(\mathbf{U})$, $(\mathbf{S})$, $(\mathbf{S'})$, $(\mathbf{s})$, $(\mathbf{v})$, $(\mathbf{f})$, $(\mathbf{I})$, $(\mathbf{T})$, $(\mathbf{A})$. $(\mathbf{d})$, $(\mathbf{P})$, $(\mathbf{c})$ and $(\mathbf{X})$ are known, then its Leverage or Gearing Ratio Planned is:

$$l = \{Q+P/[c-q]\}/\{U+S-S'[1+s][v+f]-I-T-Ad\}$$

Rule-21179:

If both $(\mathbf{l})$, $(\mathbf{Q})$, $(\mathbf{S})$, $(\mathbf{S'})$, $(\mathbf{s})$, $(\mathbf{v})$, $(\mathbf{f})$, $(\mathbf{I})$, $(\mathbf{T})$, $(\mathbf{A})$. $(\mathbf{d})$, $(\mathbf{P})$, $(\mathbf{c})$ and $(\mathbf{X})$ are known, then its Utilized or Starting Capital must be:

$$U = S'[1+s][v+f]+I+T+Ad+\{Q+P/[c-q]\}/l-S$$

Rule-21180:

If both $(\mathbf{l})$, $(\mathbf{U})$, $(\mathbf{Q})$, $(\mathbf{S'})$, $(\mathbf{s})$, $(\mathbf{v})$, $(\mathbf{f})$, $(\mathbf{I})$, $(\mathbf{T})$, $(\mathbf{A})$. $(\mathbf{d})$, $(\mathbf{P})$, $(\mathbf{c})$ and $(\mathbf{X})$ are known, then its Sales or Revenue Planned is:

$$S = S'[1+s][v+f]+I+T+Ad+\{Q+P/[c-q]\}/l-U$$

Rule-21181:

If both $(\mathbf{l})$, $(\mathbf{U})$, $(\mathbf{S})$, $(\mathbf{Q})$, $(\mathbf{s})$, $(\mathbf{v})$, $(\mathbf{f})$, $(\mathbf{I})$, $(\mathbf{T})$, $(\mathbf{A})$. $(\mathbf{d})$, $(\mathbf{P})$, $(\mathbf{c})$ and $(\mathbf{X})$ are known, then its Sales Past must be:

$$S' = (U+S-I-T-Ad-\{Q+P/[c-q]\}/l)/\{[1+s][v+f]\}$$

Rule-21182:

If both $(\mathbf{l})$, $(\mathbf{U})$, $(\mathbf{S})$, $(\mathbf{S'})$, $(\mathbf{Q})$, $(\mathbf{v})$, $(\mathbf{f})$, $(\mathbf{I})$, $(\mathbf{T})$, $(\mathbf{A})$. $(\mathbf{d})$, $(\mathbf{P})$, $(\mathbf{c})$ and $(\mathbf{X})$ are known, then its Sales Growth Planned is:

$$s = S/S'-1$$

or it can also be found as

$$s = (U+S-I-T-Ad-\{Q+P/[c-q]\}/l)/\{S'[v+f]\}-1$$

Steve Asikin ISBN 14: 978-1515158691, ISBN 10: 1515158691

<u>Rule-21183</u>:

If both (I), $(\mathbf{U})$, $(\$)$, $(\$')$, (s), $(\mathbf{Q})$, $(\mathbf{f})$, $(\mathbf{I})$, $(\mathbf{T})$, $(\mathbf{A})$. $(\mathbf{d})$, $(\mathbf{P})$, $(\mathbf{c})$ and $(\mathbf{X})$ are known, then its Variable Portion Planned is:

$$v = (\mathbf{U} + \$ - \mathbf{I} - \mathbf{T} - \mathbf{Ad} - \{\mathbf{Q} + \mathbf{P}/[\mathbf{c} - \mathbf{q}]\}/I)/\{\$'[1 + s]\} - \mathbf{f}$$

<u>Rule-21184</u>:

If both (I), $(\mathbf{U})$, $(\$)$, $(\$')$, (s), (v), $(\mathbf{Q})$, $(\mathbf{I})$, $(\mathbf{T})$, $(\mathbf{A})$. $(\mathbf{d})$, $(\mathbf{P})$, $(\mathbf{c})$ and $(\mathbf{X})$ are known, then its Fixed Portion Planned is:

$$f = (\mathbf{U} + \$ - \mathbf{I} - \mathbf{T} - \mathbf{Ad} - \{\mathbf{Q} + \mathbf{P}/[\mathbf{c} - \mathbf{q}]\}/I)/\{\$'[1 + s]\} - v$$

<u>Rule-21185</u>:

If both (I), $(\mathbf{U})$, $(\$)$, $(\$')$, (s), (v), $(\mathbf{f})$, $(\mathbf{Q})$, $(\mathbf{T})$, $(\mathbf{A})$. $(\mathbf{d})$, $(\mathbf{P})$, $(\mathbf{c})$ and $(\mathbf{X})$ are known, then its Interest Expense Planned is:

$$\mathbf{I} = \mathbf{U} + \$ - \$'[1 + s][v + \mathbf{f}] - \mathbf{T} - \mathbf{Ad} - \{\mathbf{Q} + \mathbf{P}/[\mathbf{c} - \mathbf{q}]\}/I$$

<u>Rule-21186</u>:

If both (I), $(\mathbf{U})$, $(\$)$, $(\$')$, (s), (v), $(\mathbf{f})$, $(\mathbf{I})$, $(\mathbf{Q})$, $(\mathbf{A})$. $(\mathbf{d})$, $(\mathbf{P})$, $(\mathbf{c})$ and $(\mathbf{X})$ are known, then its Tax Planned is:

$$\mathbf{T} = \mathbf{U} + \$ - \$'[1 + s][v + \mathbf{f}] - \mathbf{I} - \mathbf{Ad} - \{\mathbf{Q} + \mathbf{P}/[\mathbf{c} - \mathbf{q}]\}/I$$

<u>Rule-21187</u>:

If both (I), $(\mathbf{U})$, $(\$)$, $(\$')$, (s), (v), $(\mathbf{f})$, $(\mathbf{I})$, $(\mathbf{T})$, $(\mathbf{Q})$. $(\mathbf{d})$, $(\mathbf{P})$, $(\mathbf{c})$ and $(\mathbf{X})$ are known, then its After Tax Income Planned is:

$$\mathbf{A} = (\mathbf{U} + \$ - \$'[1 + s][v + \mathbf{f}] - \mathbf{I} - \mathbf{T} - \{\mathbf{Q} + \mathbf{P}/[\mathbf{c} - \mathbf{q}]\}/I)/\mathbf{d}$$

Steve Asikin ISBN 14: 978-1515158691, ISBN 10: 1515158691

Rule-21188:

If both (**∫**), (**U**), (**S**), (**S'**), (**s**), (**v**), (**f**), (**I**), (**T**), (**A**). (**Q**), (**P**), (**c**) and (**X**) are known, then its Dividend Payout Planned is:

$$d = (U+S-S'[1+s][v+f]-I-T-\{Q+P/[c-q]\}/∫)/A$$

Rule-21189:

If both (**∫**), (**U**), (**S**), (**S'**), (**s**), (**v**), (**f**), (**I**), (**T**), (**A**). (**d**), (**Q**), (**c**) and (**X**) are known, then its Procured Inventory Planned is:

$$P = [c-q](∫\{U+S-S'[1+s][v+f]-I-T-Ad\}-Q)$$

Rule-21190:

If both (**∫**), (**U**), (**S**), (**S'**), (**s**), (**v**), (**f**), (**I**), (**T**), (**A**). (**d**), (**P**), (**Q**) and (**X**) are known, then its Current Ratio Planned is:

$$c = q+P/(∫\{U+S-S'[1+s][v+f]-I-T-Ad\}-Q)$$

Rule-21191:

If both (**∫**), (**U**), (**S**), (**S'**), (**s**), (**v**), (**f**), (**I**), (**T**), (**A**). (**d**), (**P**), (**c**) and (**X**) are known, then its Quoted Longterm Debt Planned is:

$$q = c-P/(∫\{U+S-S'[1+s][v+f]-I-T-Ad\}-Q)$$

Rule-21192:

If both (**∫**), (**U**), (**S**), (**S'**), (**s**), (**v**), (**f**), (**I**), (**T**), (**A**). (**d**), (**V**), (**p**), (**c**) and (**q**) are known, then its Quoted Longterm Debt Planned is:

$$Q = ∫\{U+S-S'[1+s][v+f]-I-T-Ad\}-Vp/\{360[c-q]\}$$

Steve Asikin ISBN 14: 978-1515158691, ISBN 10: 1515158691

Rule-21193:

If both $(\mathbf{Q})$, $(\mathbf{U})$, $(\mathbf{S})$, $(\mathbf{S'})$, $(\mathbf{s})$, $(\mathbf{v})$, $(\mathbf{f})$, $(\mathbf{I})$, $(\mathbf{T})$, $(\mathbf{A})$. $(\mathbf{d})$, $(\mathbf{V})$, $(\mathbf{p})$, $(\mathbf{c})$ and $(\mathbf{q})$ are known, then its Leverage or Gearing Ratio Planned is:

$$I = (Q+Vp/\{360[c-q]\})/\{U+S-S'[1+s][v+f]-I-T-Ad\}$$

Rule-21194:

If both $(\mathbf{I})$, $(\mathbf{Q})$, $(\mathbf{S})$, $(\mathbf{S'})$, $(\mathbf{s})$, $(\mathbf{v})$, $(\mathbf{f})$, $(\mathbf{I})$, $(\mathbf{T})$, $(\mathbf{A})$. $(\mathbf{d})$, $(\mathbf{V})$, $(\mathbf{p})$, $(\mathbf{c})$ and $(\mathbf{q})$ are known, then its Utilized or Starting Capital must be:

$$U = S'[1+s][v+f]+I+T+Ad+(Q+Vp/\{360[c-q]\})/I-S$$

Rule-21195:

If both $(\mathbf{I})$, $(\mathbf{U})$, $(\mathbf{Q})$, $(\mathbf{S'})$, $(\mathbf{s})$, $(\mathbf{v})$, $(\mathbf{f})$, $(\mathbf{I})$, $(\mathbf{T})$, $(\mathbf{A})$. $(\mathbf{d})$, $(\mathbf{V})$, $(\mathbf{p})$, $(\mathbf{c})$ and $(\mathbf{q})$ are known, then its Sales or Revenue Planned is:

$$S = S'[1+s][v+f]+I+T+Ad+(Q+Vp/\{360[c-q]\})/I-U$$

Rule-21196:

If both $(\mathbf{I})$, $(\mathbf{U})$, $(\mathbf{S})$, $(\mathbf{Q})$, $(\mathbf{s})$, $(\mathbf{v})$, $(\mathbf{f})$, $(\mathbf{I})$, $(\mathbf{T})$, $(\mathbf{A})$. $(\mathbf{d})$, $(\mathbf{V})$, $(\mathbf{p})$, $(\mathbf{c})$ and $(\mathbf{q})$ are known, then its Sales Past must be:

$$S' = [U+S-I-T-Ad-(Q+Vp/\{360[c-q]\})/I]$$
$$/\{[1+s][v+f]\}$$

Rule-21197:

If both $(\mathbf{I})$, $(\mathbf{U})$, $(\mathbf{S})$, $(\mathbf{S'})$, $(\mathbf{Q})$, $(\mathbf{v})$, $(\mathbf{f})$, $(\mathbf{I})$, $(\mathbf{T})$, $(\mathbf{A})$. $(\mathbf{d})$, $(\mathbf{V})$, $(\mathbf{p})$, $(\mathbf{c})$ and $(\mathbf{q})$ are known, then its Sales Growth Planned is:

$$s = S/S'-1$$

or it can also be found as

$$s = [U+S-I-T-Ad-(Q+Vp/\{360[c-q]\})/I]/\{S'[v+f]\}-1$$

Steve Asikin ISBN 14: 978-1515158691, ISBN 10: 1515158691

Rule-21198:

If both (f), (U), (S), (S'), (s), (Q), (f), (I), (T), (A). (d), (V), (p), (c) and (q) are known, then its Variable Portion Planned is:

$$s = V/S$$

or it can also be found as

$$v = [U+S-I-T-Ad-(Q+Vp/\{360[c-q]\})/f]/\{S'[1+s]\}-f$$

Rule-21199:

If both (f), (U), (S), (S'), (s), (v), (Q), (I), (T), (A). (d), (V), (p), (c) and (q) are known, then its Fixed Portion Planned is:

$$f = [U+S-I-T-Ad-(Q+Vp/\{360[c-q]\})/f]/\{S'[1+s]\}-v$$

Rule-21200:

If both (f), (U), (S), (S'), (s), (v), (f), (Q), (T), (A). (d), (V), (p), (c) and (q) are known, then its Interest Expense Planned is:

$$I = U+S-S'[1+s][v+f]-T-Ad-(Q+Vp/\{360[c-q]\})/f$$

Rule-21201:

If both (f), (U), (S), (S'), (s), (v), (f), (I), (Q), (A). (d), (V), (p), (c) and (2) are known, then its Tax Planned is:

$$T = U+S-S'[1+s][v+f]-I-Ad-(Q+Vp/\{360[c-q]\})/f$$

Rule-21202:

If both (f), (U), (S), (S'), (s), (v), (f), (I), (T), (Q). (d), (V), (p), (c) and (q) are known, then its After Tax Income Planned is:

$$A = [U+S-S'[1+s][v+f]-I-T-(Q+Vp/\{360[c-q]\})/f]/d$$

Steve Asikin ISBN 14: 978-1515158691, ISBN 10: 1515158691

Rule-21203:

If both (I), (**U**), (**$**), (**$'**), (**s**), (**v**), (**f**), (**I**), (**T**), (**A**). (**Q**), (**V**), (**p**), (**c**) and (**q**) are known, then its Dividend Payout Planned is:

$$d= [U+\$-\$'[1+s][v+f]-I-T-(Q+Vp/\{360[c-q]\})/I]/A$$

Rule-21204:

If both (I), (**U**), (**$**), (**$'**), (**s**), (**v**), (**f**), (**I**), (**T**), (**A**). (**d**), (**Q**), (**p**), (**c**) and (**q**) are known, then its Variable Cost Planned is:

$$V= 360[c-q](I\{U+\$-\$'[1+s][v+f]-I-T-Ad\}-Q)/p$$

Rule-21205:

If both (I), (**U**), (**$**), (**$'**), (**s**), (**v**), (**f**), (**I**), (**T**), (**A**). (**d**), (**V**), (**Q**), (**c**) and (**q**) are known, then its Procured Inventory Days Planned:

$$p= 360[c-q](I\{U+\$-\$'[1+s][v+f]-I-T-Ad\}-Q)/V$$

Rule-21206:

If both (I), (**U**), (**$**), (**$'**), (**s**), (**v**), (**f**), (**I**), (**T**), (**A**). (**d**), (**V**), (**p**), (**Q**) and (**X**) are known, then its Current Ratio Planned is:

$$c= q+Vp/[360(I\{U+\$-\$'[1+s][v+f]-I-T-Ad\}-Q)]$$

Rule-21207:

If both (I), (**U**), (**$**), (**$'**), (**s**), (**v**), (**f**), (**I**), (**T**), (**A**). (**d**), (**V**), (**p**), (**c**) and (**Q**) are known, then its Quick or Acid Test Ratio Planned is:

$$q= c-Vp/[360(I\{U+\$-\$'[1+s][v+f]-I-T-Ad\}-Q)]$$

Steve Asikin ISBN 14: 978-1515158691, ISBN 10: 1515158691

Rule-21208:

If both (I), (U), (S), (S'), (s), (v), (f), (I), (T), (A). (d), (p), (c) and (q) are known, then its Quoted Longterm Debt Planned is:

$$Q = I\{U+S-S'[1+s][v+f]-I-T-Ad\}-Svp/\{360[c-q]\}$$

Rule-21209:

If both (Q), (U), (S), (S'), (s), (v), (f), (I), (T), (A). (d), (p), (c) and (q) are known, then its Leverage or Gearing Ratio Planned is:

$$I = (Q+Svp/\{360[c-q]\})/\{U+S-S'[1+s][v+f]-I-T-Ad\}$$

Rule-21210:

If both (I), (Q), (S), (S'), (s), (v), (f), (I), (T), (A). (d), (p), (c) and (q) are known, then its Utilized or Starting Capital must be:

$$U = S'[1+s][v+f]+I+T+Ad+(Q+Svp/\{360[c-q]\})/I-S$$

Rule-21211:

If both (I), (U), (Q), (S'), (s), (v), (f), (I), (T), (A). (d), (p), (c) and (q) are known, then its Sales or Revenue Planned is:

$$S = [I\{U-S'[1+s][v+f]-I-T-Ad\}-Q]$$
$$/(vp/\{360[c-q]\}-I)$$

Rule-21212:

If both (I), (U), (S), (Q), (s), (v), (f), (I), (T), (A). (d), (p), (c) and (q) are known, then its Sales Past must be:

$$S' = [U+S-I-T-Ad-(Q+Svp/\{360[c-q]\})/I]$$
$$/\{[1+s][v+f]\}$$

Steve Asikin ISBN 14: 978-1515158691, ISBN 10: 1515158691

Rule-21213:

If both (**/**), (**U**), (**\$**), (**\$'**), (**Q**), (**v**), (**f**), (**I**), (**T**), (**A**). (**d**), (**p**), (**c**) and (**q**) are known, then its Sales Growth Planned is:

$$s = [U+\$-I-T-Ad-(Q+\$vp/\{360[c-q]\})/\textbf{\textit{I}}]$$
$$/\{\$'[v+f]\}-1$$

Rule-21214:

If both (**/**), (**U**), (**\$**), (**\$'**), (**s**), (**Q**), (**f**), (**I**), (**T**), (**A**). (**d**), (**p**), (**c**) and (**q**) are known, then its Variable Portion Planned is:

$$v = [\textbf{\textit{I}}\{U+\$-\$'f[1+s]-I-T-Ad\}-Q]$$
$$/(\$p/\{360[c-q]\}+\$'\textbf{\textit{I}}[1+s])$$

Rule-21215:

If both (**/**), (**U**), (**\$**), (**\$'**), (**s**), (**v**), (**Q**), (**I**), (**T**), (**A**). (**d**), (**p**), (**c**) and (**q**) are known, then its Fixed Cost Planned is:

$$f = [U+\$-I-T-Ad-(Q+\$vp/\{360[c-q]\})/\textbf{\textit{I}}]$$
$$/\{\$'[1+s]\}-v$$

Rule-21216:

If both (**/**), (**U**), (**\$**), (**\$'**), (**s**), (**v**), (**f**), (**Q**), (**T**), (**A**). (**d**), (**p**), (**c**) and (**q**) are known, then its Interest Expense Planned is:

$$I = U+\$-\$'[1+s][v+f]-T-Ad-(Q+\$vp/\{360[c-q]\})/\textbf{\textit{I}}$$

Rule-21217:

If both (**/**), (**U**), (**\$**), (**\$'**), (**s**), (**v**), (**f**), (**I**), (**Q**), (**A**). (**d**), (**p**), (**c**) and (**q**) are known, then its Tax Planned is:

$$T = U+\$-\$'[1+s][v+f]-I-Ad-(Q+\$vp/\{360[c-q]\})/\textbf{\textit{I}}$$

Steve Asikin ISBN 14: 978-1515158691, ISBN 10: 1515158691

Rule-21218:

If both (Λ), $(\mathbf{U})$, $(\mathbf{S})$, $(\mathbf{S'})$, $(\mathbf{s})$, $(\mathbf{v})$, $(\mathbf{f})$, $(\mathbf{I})$, $(\mathbf{T})$, $(\mathbf{Q})$. $(\mathbf{d})$, $(\mathbf{p})$, $(\mathbf{c})$ and $(\mathbf{q})$ are known, then its After Tax Income Planned is:

$$A= [U+S-S'[1+s][v+f]-I-T$$
$$-(Q+Svp/\{360[c-q]\})/\Lambda/d$$

Rule-21219:

If both (Λ), $(\mathbf{U})$, $(\mathbf{S})$, $(\mathbf{S'})$, $(\mathbf{s})$, $(\mathbf{v})$, $(\mathbf{f})$, $(\mathbf{I})$, $(\mathbf{T})$, $(\mathbf{A})$. $(\mathbf{Q})$, $(\mathbf{p})$, $(\mathbf{c})$ and $(\mathbf{q})$ are known, then its Dividend Payout Planned is:

$$d= [U+S-S'[1+s][v+f]-I-T$$
$$-(Q+Svp/\{360[c-q]\})/\Lambda/A$$

Rule-21220:

If both (Λ), $(\mathbf{U})$, $(\mathbf{S})$, $(\mathbf{S'})$, $(\mathbf{s})$, $(\mathbf{v})$, $(\mathbf{f})$, $(\mathbf{I})$, $(\mathbf{T})$, $(\mathbf{A})$. $(\mathbf{d})$, $(\mathbf{Q})$, $(\mathbf{c})$ and $(\mathbf{q})$ are known, then its Procured Inventory Days Planned:

$$p= 360[c-q](\Lambda\{U+S-S'[1+s][v+f]-I-T-Ad\}-Q)/[Sv]$$

Rule-21221:

If both (Λ), $(\mathbf{U})$, $(\mathbf{S})$, $(\mathbf{S'})$, $(\mathbf{s})$, $(\mathbf{v})$, $(\mathbf{f})$, $(\mathbf{I})$, $(\mathbf{T})$, $(\mathbf{A})$. $(\mathbf{d})$, $(\mathbf{p})$, $(\mathbf{Q})$ and $(\mathbf{q})$ are known, then its Current Ratio Planned is:

$$c= q+Svp/[360(\Lambda\{U+S-S'[1+s][v+f]-I-T-Ad\}-Q)]$$

Rule-21222:

If both (Λ), $(\mathbf{U})$, $(\mathbf{S})$, $(\mathbf{S'})$, $(\mathbf{s})$, $(\mathbf{v})$, $(\mathbf{f})$, $(\mathbf{I})$, $(\mathbf{T})$, $(\mathbf{A})$. $(\mathbf{d})$, $(\mathbf{p})$, $(\mathbf{c})$ and $(\mathbf{Q})$ are known, then its Quick or Acid Test Ratio Planned is:

$$q= c-Svp/[360(\Lambda\{U+S-S'[1+s][v+f]-I-T-Ad\}-Q)]$$

Steve Asikin ISBN 14: 978-1515158691, ISBN 10: 1515158691

Rule-21223:

If both (I), (U), (S), (S'), (s), (v), (f), (I), (T), (A). (d), (p), (c) and (q) are known, then its Quoted Longterm Debt Planned is:

$$Q= I\{U+S-S'[1+s][v+f]-I-T-Ad\}$$
$$-S'vp[1+s]/\{360[c-q]\}$$

Rule-21224:

If both (Q), (U), (S), (S'), (s), (v), (f), (I), (T), (A). (d), (p), (c) and (q) are known, then its Leverage or Gearing Ratio Planned is:

$$I= (Q+S'vp[1+s]/\{360[c-q]\})$$
$$/\{U+S-S'[1+s][v+f]-I-T-Ad\}$$

Rule-21225:

If both (I), (Q), (S), (S'), (s), (v), (f), (I), (T), (A). (d), (p), (c) and (q) are known, then its Utilized or Starting Capital must be:

$$U = S'[1+s][v+f]+I+T+Ad$$
$$+(Q+S'vp[1+s]/\{360[c-q]\})/I-S$$

Rule-21226:

If both (I), (U), (Q), (S'), (s), (v), (f), (I), (T), (A). (d), (p), (c) and (q) are known, then its Sales or Revenue Planned is:

$$S= (Q-I\{U-S'[1+s][v+f]-I-T-Ad\}$$
$$+S'vp[1+s]/\{360[c-q]\})/I$$

Steve Asikin ISBN 14: 978-1515158691, ISBN 10: 1515158691

Rule-21227:
> If both (**/**), (**U**), (**$**), (**Q**), (**s**), (**v**), (**f**), (**I**), (**T**), (**A**). (**d**),
> (**p**), (**c**) and (**q**) are known, then its Sales Past must be:
> $$\$' = [/\{U+\$-\$'f[1+s]-I-T-Ad\}-Q]$$
> $$/[v[1+s](p/\{360[c-q]\}+/)]$$

Rule-21228:
> If both (**/**), (**U**), (**$**), (**$'**), (**Q**), (**v**), (**f**), (**I**), (**T**), (**A**). (**d**),
> (**p**), (**c**) and (**q**) are known, then its Sales Growth
> Planned is:
> $$s = \$/\$'-1$$
>
> or it can also be found as
> $$s = [/\{U+\$-\$'f[1+s]-I-T-Ad\}-Q]$$
> $$/[\$'v(p/\{360[c-q]\}+/)]-1$$

Rule-21229:
> If both (**/**), (**U**), (**$**), (**$'**), (**s**), (**Q**), (**f**), (**I**), (**T**), (**A**). (**d**),
> (**p**), (**c**) and (**q**) are known, then its Variable Portion
> Planned is:
> $$v = [/\{U+\$-\$'f[1+s]-I-T-Ad\}-Q]$$
> $$/[\$'[1+s](p/\{360[c-q]\}+/)]$$

Rule-21230:
> If both (**/**), (**U**), (**$**), (**$'**), (**s**), (**v**), (**Q**), (**I**), (**T**), (**A**). (**d**),
> (**p**), (**c**) and (**q**) are known, then its Fixed Portion
> Planned is:
> $$f = [U+\$-I-T-Ad-(Q+\$'vp[1+s]/\{360[c-q]\})//]$$
> $$/\{\$'[1+s]\}-v$$

Steve Asikin ISBN 14: 978-1515158691, ISBN 10: 1515158691

Rule-21231:

If both (*I*), (**U**), (**S**), (**S'**), (**s**), (**v**), (**f**), (**Q**), (**T**), (**A**). (**d**), (**p**), (**c**) and (**q**) are known, then its Interest Expense Planned is:

$$I= U+S-S'[1+s][v+f]-T-Ad$$
$$-(Q+S'vp[1+s]/\{360[c-q]\})/I$$

Rule-21232:

If both (*I*), (**U**), (**S**), (**S'**), (**s**), (**v**), (**f**), (**I**), (**Q**), (**A**). (**d**), (**p**), (**c**) and (**q**) are known, then its Tax Planned is:

$$T= U+S-S'[1+s][v+f]-I-Ad$$
$$-(Q+S'vp[1+s]/\{360[c-q]\})/I$$

Rule-21233:

If both (*I*), (**U**), (**S**), (**S'**), (**s**), (**v**), (**f**), (**I**), (**T**), (**Q**). (**d**), (**p**), (**c**) and (**q**) are known, then its After Tax Income Planned is:

$$A= [U+S-S'[1+s][v+f]-I-T$$
$$-(Q+S'vp[1+s]/\{360[c-q]\})/I]/d$$

Rule-21234:

If both (*I*), (**U**), (**S**), (**S'**), (**s**), (**v**), (**f**), (**I**), (**T**), (**A**). (**Q**), (**p**), (**c**) and (**q**) are known, then its Dividend Payout Planned is:

$$d= [U+S-S'[1+s][v+f]-I-T$$
$$-(Q+S'vp[1+s]/\{360[c-q]\})/I]/A$$

Steve Asikin ISBN 14: 978-1515158691, ISBN 10: 1515158691

Rule-21235:

If both (**/**), (**U**), (**$**), (**$'**), (**s**), (**v**), (**f**), (**I**), (**T**), (**A**). (**d**), (**Q**), (**c**) and (**q**) are known, then its Procured Inventory Days Planned:

$$\mathbf{p}= 360[\mathbf{c\text{-}q}](\textit{/}\{\mathbf{U+-'}[1+\mathbf{s}][\mathbf{v+f}]\mathbf{\text{-}I\text{-}T\text{-}Ad}\}\mathbf{\text{-}Q})$$
$$/\{\mathbf{$'v}[1+\mathbf{s}]\}$$

Rule-21236:

If both (**/**), (**U**), (**$**), (**$'**), (**s**), (**v**), (**f**), (**I**), (**T**), (**A**). (**d**), (**p**), (**Q**) and (**q**) are known, then its Current Ratio Planned is:

$$\mathbf{c}= \mathbf{q+$'vp}[1+\mathbf{s}]$$
$$/[360(\textit{/}\{\mathbf{U+-'}[1+\mathbf{s}][\mathbf{v+f}]\mathbf{\text{-}I\text{-}T\text{-}Ad}\}\mathbf{\text{-}Q})]$$

Rule-21237:

If both (**/**), (**U**), (**$**), (**$'**), (**s**), (**v**), (**f**), (**I**), (**T**), (**A**). (**d**), (**p**), (**c**) and (**q**) are known, then its Quoted Longterm Debt Planned is:

$$\mathbf{q}= \mathbf{c\text{-}$'vp}[1+\mathbf{s}]$$
$$/[360(\textit{/}\{\mathbf{U+-'}[1+\mathbf{s}][\mathbf{v+f}]\mathbf{\text{-}I\text{-}T\text{-}Ad}\}\mathbf{\text{-}Q})]$$

Rule-21238:

If both (**/**), (**U**), (**$**), (**$'**), (**s**), (**v**), (**f**), (**I**), (**t**), (**D**) and (**X**) are known, then its Quoted Longterm Debt Planned is:

$$\mathbf{Q}= \textit{/}\{\mathbf{U+}\{\mathbf{-'}[1+\mathbf{s}][\mathbf{v+f}]\mathbf{\text{-}I}\}[1\mathbf{\text{-}t}]\mathbf{\text{-}D})\mathbf{\text{-}X}$$

Steve Asikin ISBN 14: 978-1515158691, ISBN 10: 1515158691

Rule-21239:

    If both **(Q)**, **(U)**, **($)**, **($')**, **(s)**, **(v)**, **(f)**, **(I)**, **(t)**, **(D)** and **(X)** are known, then its Leverage or Gearing Ratio Planned is:

$$I= [Q+X]/(U+\{\$-\$'[1+s][v+f]-I\}[1-t]-D)$$

Rule-21240:

    If both **(I)**, **(Q)**, **($)**, **($')**, **(s)**, **(v)**, **(f)**, **(I)**, **(t)**, **(D)** and **(X)** are known, then its Utilized or Starting Capital must be:

$$U= D+[Q+X]/I\{\$-\$'[1+s][v+f]-I\}[1-t]$$

Rule-21241:

    If both **(I)**, **(U)**, **(Q)**, **($')**, **(s)**, **(v)**, **(f)**, **(I)**, **(t)**, **(D)** and **(X)** are known, then its Sales or Revenue Planned is:

$$\$= \$'[1+s][v+f]+I+\{D+[Q+X]/I-U\}/[1-t]$$

Rule-21242:

    If both **(I)**, **(U)**, **($)**, **(Q)**, **(s)**, **(v)**, **(f)**, **(I)**, **(t)**, **(D)** and **(X)** are known, then its Sales Past must be:

$$\$'= (\$-I-\{D+[Q+X]/I-U\}/[1-t])/\{[1+s][v+f]\}$$

Rule-21243:

    If both **(I)**, **(U)**, **($)**, **($')**, **(Q)**, **(v)**, **(f)**, **(I)**, **(t)**, **(D)** and **(X)** are known, then its Sales Growth Planned is:

$$s= \$/\$'-1$$

        or it can also be found as

$$s= (\$-I-\{D+[Q+X]/I-U\}/[1-t])/\{\$'[v+f]\}-1$$

Steve Asikin ISBN 14: 978-1515158691, ISBN 10: 1515158691

Rule-21244:

If both (I), (**U**), (**$**), (**$'**), (**s**), (**Q**), (**f**), (**I**), (**t**), (**D**) and (**X**) are known, then its Variable Portion Planned is:

$v= (\$-I-\{D+[Q+X]/I-U\}/[1-t])/\{\$'[1+s]\}-f$

Rule-21245:

If both (I), (**U**), (**$**), (**$'**), (**s**), (**v**), (**Q**), (**I**), (**t**), (**D**) and (**X**) are known, then its Fixed Portion Planned is:

$f= (\$-I-\{D+[Q+X]/I-U\}/[1-t])/\{\$'[1+s]\}-v$

Rule-21246:

If both (I), (**U**), (**$**), (**$'**), (**s**), (**v**), (**f**), (**Q**), (**t**), (**D**) and (**X**) are known, then its Interest Portion Planned is:

$I= \$-\$'[1+s][v+f]-\{D+[Q+X]/I-U\}/[1-t]$

Rule-21247:

If both (I), (**U**), (**$**), (**$'**), (**s**), (**v**), (**f**), (**I**), (**Q**), (**D**) and (**X**) are known, then its Tax Rate Planned is:

$t= 1-\{D+[Q+X]/I-U\}/\{\$-\$'[1+s][v+f]-I\}$

Rule-21248:

If both (I), (**U**), (**$**), (**$'**), (**s**), (**v**), (**f**), (**I**), (**t**), (**Q**) and (**X**) are known, then its Dividend Payout Planned is:

$D= U+\{\$-\$'[1+s][v+f]-I\}[1-t]-[Q+X]/I$

Rule-21249:

If both (I), (**U**), (**$**), (**$'**), (**s**), (**v**), (**f**), (**I**), (**t**), (**D**) and (**Q**) are known, then its Xpress or Current Debt Planned is:

$X= I(U+\{\$-\$'[1+s][v+f]-I\}[1-t]-D)-Q$

Steve Asikin ISBN 14: 978-1515158691, ISBN 10: 1515158691

Rule-21250:

If both (l), (**U**), (**$**), (**$'**), (**s**), (**v**), (**f**), (**I**), (**t**), (**D**), (**P**), (**c**) and (**q**) are known, then its Quoted Longterm Debt Planned is:

$$Q= l(U+\{\$-\$'[1+s][v+f]-I\}[1-t]-D)-P/[c-q]$$

Rule-21251:

If both (**Q**), (**U**), (**$**), (**$'**), (**s**), (**v**), (**f**), (**I**), (**t**), (**D**), (**P**), (**c**) and (**q**) are known, then its Leverage or Gearing Ratio Planned is:

$$l= \{Q+P/[c-q]\}/(U+\{\$-\$'[1+s][v+f]-I\}[1-t]-D)$$

Rule-21252:

If both (l), (**Q**), (**$**), (**$'**), (**s**), (**v**), (**f**), (**I**), (**t**), (**D**), (**P**), (**c**) and (**q**) are known, then its Utilized or Starting Capital must be:

$$U= D+\{Q+P/[c-q]\}/l\{\$-\$'[1+s][v+f]-I\}[1-t]$$

Rule-21253:

If both (l), (**U**), (**Q**), (**$'**), (**s**), (**v**), (**f**), (**I**), (**t**), (**D**), (**P**), (**c**) and (**q**) are known, then its Sales or Revenue Planned is:

$$\$= \$'[1+s][v+f]+I+(D+\{Q+P/[c-q]\}/l-U)/[1-t]$$

Rule-21254:

If both (l), (**U**), (**$**), (**Q**), (**s**), (**v**), (**f**), (**I**), (**t**), (**D**), (**P**), (**c**) and (**q**) are known, then its Sales Past must be:

$$\$'= [\$-I-(D+\{Q+P/[c-q]\}/l-U)/ [1-t]]/\{[1+s][v+f]\}$$

Steve Asikin ISBN 14: 978-1515158691, ISBN 10: 1515158691

Rule-21255:

If both (I), (U), $(\$)$, $(\$')$, (Q), (v), (f), (I), (t), (D), (P), (c) and (q) are known, then its Sales Growth Planned is:

$$s= \$/\$'-1$$

or it can also be found as

$$s= [\$-I-(D+\{Q+P/[c-q]\}/I-U)/ [1-t]]/\{\$'[v+f]\}-1$$

Rule-21256:

If both (I), (U), $(\$)$, $(\$')$, (s), (Q), (f), (I), (t), (D), (P), (c) and (q) are known, then its Variable Portion Planned is:

$$v= [\$-I-(D+\{Q+P/[c-q]\}/I-U)/[1-t]]/\{\$'[1+s]\}-f$$

Rule-21257:

If both (I), (U), $(\$)$, $(\$')$, (s), (v), (Q), (I), (t), (D), (P), (c) and (q) are known, then its Fixed Portion Planned is:

$$f= [\$-I-(D+\{Q+P/[c-q]\}/I-U)/[1-t]]/\{\$'[1+s]\}-v$$

Rule-21258:

If both (I), (U), $(\$)$, $(\$')$, (s), (v), (f), (Q), (t), (D), (P), (c) and (q) are known, then its Interest Expense Planned is:

$$I= \$-\$'[1+s][v+f]-(D+\{Q+P/[c-q]\}/I-U)/[1-t]$$

Rule-21259:

If both (I), (U), $(\$)$, $(\$')$, (s), (v), (f), (I), (Q), (D), (P), (c) and (q) are known, then its Tax Rate Planned is:

$$t= 1-(D+\{Q+P/[c-q]\}/I-U)/\{\$-\$'[1+s][v+f]-I\}$$

Steve Asikin ISBN 14: 978-1515158691, ISBN 10: 1515158691

Rule-21260:

If both (**/**), (**U**), (**S**), (**S'**), (**s**), (**v**), (**f**), (**I**), (**t**), (**Q**), (**P**), (**c**) and (**q**) are known, then its Dividend Planned is:

$$D= U+\{S\text{-}S'[1+s][v+f]\text{-}I\}[1\text{-}t]\text{-}\{Q+P/[c\text{-}q]\}//$$

Rule-21261:

If both (**/**), (**U**), (**S**), (**S'**), (**s**), (**v**), (**f**), (**I**), (**t**), (**D**), (**Q**), (**c**) and (**q**) are known, then its Procured Inventory Planned is:

$$P= [c\text{-}q][/U+\{S\text{-}S'[1+s][v+f]\text{-}I\}[1\text{-}t]\text{-}D)\text{-}Q]$$

Rule-21262:

If both (**/**), (**U**), (**S**), (**S'**), (**s**), (**v**), (**f**), (**I**), (**t**), (**D**), (**P**), (**Q**) and (**q**) are known, then its Current Ratio Planned is:

$$c= q+P/[/U+\{S\text{-}S'[1+s][v+f]\text{-}I\}[1\text{-}t]\text{-}D)\text{-}Q]$$

Rule-21263:

If both (**/**), (**U**), (**S**), (**S'**), (**s**), (**v**), (**f**), (**I**), (**t**), (**D**), (**P**), (**c**) and (**Q**) are known, then its Quick or Acid test Ratio Planned is:

$$q= c\text{-}P/[/U+\{S\text{-}S'[1+s][v+f]\text{-}I\}[1\text{-}t]\text{-}D)\text{-}Q]$$

Rule-21264:

If both (**/**), (**U**), (**S**), (**S'**), (**s**), (**v**), (**f**), (**I**), (**t**), (**D**), (**V**), (**p**), (**c**) and (**q**) are known, then its Quoted Longterm Debt Planned is:

$$Q= /U+\{S\text{-}S'[1+s][v+f]\text{-}I\}[1\text{-}t]\text{-}D)\text{-}Vp/\{360[c\text{-}q]\}$$

Steve Asikin ISBN 14: 978-1515158691, ISBN 10: 1515158691

Rule-21265:

If both (**Q**), (**U**), (**$**), (**$'**), (**s**), (**v**), (**f**), (**I**), (**t**), (**D**), (**V**), (**p**), (**c**) and (**q**) are known, then its Leverage or Gearing Ratio Planned is:

$$I\!= (Q+Vp/\{360[c\text{-}q]\})$$
$$/(U+\{\$\text{-}\$'[1+s][v+f]\text{-}I\}[1\text{-}t]\text{-}D)$$

Rule-21266:

If both (**I**), (**Q**), (**$**), (**$'**), (**s**), (**v**), (**f**), (**I**), (**t**), (**D**), (**V**), (**p**), (**c**) and (**q**) are known, then its Utilized or Starting Capital must be:

$$U= D+(Q+Vp/\{360[c\text{-}q]\})/I$$
$$-\{\$\text{-}\$'[1+s][v+f]\text{-}I\}[1\text{-}t]$$

Rule-21267:

If both (**I**), (**U**), (**Q**), (**$'**), (**s**), (**v**), (**f**), (**I**), (**t**), (**D**), (**V**), (**p**), (**c**) and (**q**) are known, then its Sales or Revenue Planned is:

$$\$= \$'[1+s][v+f]+I$$
$$+[D+(Q+Vp/\{360[c\text{-}q]\})/I\text{-}U]/[1\text{-}t]$$

Rule-21268:

If both (**I**), (**U**), (**$**), (**Q**), (**s**), (**v**), (**f**), (**I**), (**t**), (**D**), (**V**), (**p**), (**c**) and (**q**) are known, then its Sales Past must be:

$$\$'= \{\$\text{-}I\text{-}[D+(Q+Vp/\{360[c\text{-}q]\})/I\text{-}U]/[1\text{-}t]\}$$
$$/\{[1+s][v+f]\}$$

Steve Asikin ISBN 14: 978-1515158691, ISBN 10: 1515158691

<u>Rule-21269</u>:

If both (**/**), (**U**), (**$**), (**$'**), (**Q**), (**v**), (**f**), (**I**), (**t**), (**D**), (**V**), (**p**), (**c**) and (**q**) are known, then its Sales Growth Planned is:

$s = \$/\$' - 1$

or itcan also be found as

$s = \{\$-I-[D+(Q+Vp/\{360[c-q]\})/I-U]/[1-t]\}$
$/\{\$'[v+f]\}-1$

<u>Rule-21270</u>:

If both (**/**), (**U**), (**$**), (**$'**), (**s**), (**Q**), (**f**), (**I**), (**t**), (**D**), (**V**), (**p**), (**c**) and (**q**) are known, then its Variable Portion Planned is:

$v = V/\$$

or it can also be found as

$v = \{\$-I-[D+(Q+Vp/\{360[c-q]\})/I-U]/[1-t]\}$
$/\{\$'[1+s]\}-f$

<u>Rule-21271</u>:

If both (**/**), (**U**), (**$**), (**$'**), (**s**), (**v**), (**Q**), (**I**), (**t**), (**D**), (**V**), (**p**), (**c**) and (**q**) are known, then its Fixed Portion Planned is:

$f = \{\$-I-[D+(Q+Vp/\{360[c-q]\})/I-U]/[1-t]\}$
$/\{\$'[1+s]\}-v$

<u>Rule-21272</u>:

If both (**/**), (**U**), (**$**), (**$'**), (**s**), (**v**), (**f**), (**Q**), (**t**), (**D**), (**V**), (**p**), (**c**) and (**q**) are known, then its Interest Expense Planned is:

$I = \$-\$'[1+s][v+f]-[D+(Q+Vp/\{360[c-q]\})/I-U]$
$/[1-t]$

Steve Asikin ISBN 14: 978-1515158691, ISBN 10: 1515158691

Rule-21273:

If both (I), (**U**), (**S**), (**S'**), (**s**), (**v**), (**f**), (**I**), (**Q**), (**D**), (**V**), (**p**), (**c**) and (**q**) are known, then its Tax Rate Planned is:

$$t= 1-[D+(Q+Vp/\{360[c\text{-}q]\})/I\text{-}U]$$
$$/\{S\text{-}S'[1+s][v+f]\text{-}I\}$$

Rule-21274:

If both (I), (**U**), (**S**), (**S'**), (**s**), (**v**), (**f**), (**I**), (**t**), (**Q**), (**V**), (**p**), (**c**) and (**q**) are known, then its Dividend Planned is:

$$D= U+\{S\text{-}S'[1+s][v+f]\text{-}I\}[1\text{-}t]$$
$$-(Q+Vp/\{360[c\text{-}q]\})/I$$

Rule-21275:

If both (I), (**U**), (**S**), (**S'**), (**s**), (**v**), (**f**), (**I**), (**t**), (**D**), (**Q**), (**p**), (**c**) and (**q**) are known, then its Variable Cost Planned is:

$$V= 360[c\text{-}q][I(U+\{S\text{-}S'[1+s][v+f]\text{-}I\}[1\text{-}t]\text{-}D)\text{-}Q]/p$$

Rule-21276:

If both (I), (**U**), (**S**), (**S'**), (**s**), (**v**), (**f**), (**I**), (**t**), (**D**), (**V**), (**Q**), (**c**) and (**q**) are known, then its Procured Inventory Days Planned:

$$p= 360[c\text{-}q][I(U+\{S\text{-}S'[1+s][v+f]\text{-}I\}[1\text{-}t]\text{-}D)\text{-}Q]/V$$

Rule-21277:

If both (I), (**U**), (**S**), (**S'**), (**s**), (**v**), (**f**), (**I**), (**t**), (**D**), (**V**), (**p**), (**Q**) and (**q**) are known, then its Current Ratio Planned is:

$$c= q+Vp/\{360[I(U+\{S\text{-}S'[1+s][v+f]\text{-}I\}[1\text{-}t]\text{-}D)\text{-}Q]\}$$

Steve Asikin ISBN 14: 978-1515158691, ISBN 10: 1515158691

Rule-21278:

If both (*I*), (**U**), (**$**), (**$'**), (**s**), (**v**), (**f**), (**I**), (**t**), (**D**), (**V**), (**p**), (**c**) and (**Q**) are known, then its Quick or Acid test Ratio Planned is:

$$q = c - Vp / \{360[\textit{I}U + \{\$-\$'[1+s][v+f]-I\}[1-t]-D)-Q]\}$$

Rule-21279:

If both (*I*), (**U**), (**$**), (**$'**), (**s**), (**v**), (**f**), (**I**), (**t**), (**D**), (**p**), (**c**) and (**q**) are known, then its Quoted Longterm Debt Planned is:

$$Q = \textit{I}U + \{\$-\$'[1+s][v+f]-I\}[1-t]-D)-\$vp / \{360[c-q]\}$$

Rule-21280:

If both (**Q**), (**U**), (**$**), (**$'**), (**s**), (**v**), (**f**), (**I**), (**t**), (**D**), (**p**), (**c**) and (**q**) are known, then its Leverage or Gearing Ratio Planned is:

$$\textit{I} = (Q + \$vp / \{360[c-q]\})$$
$$/(U + \{\$-\$'[1+s][v+f]-I\}[1-t]-D)$$

Rule-21281:

If both (*I*), (**Q**), (**$**), (**$'**), (**s**), (**v**), (**f**), (**I**), (**t**), (**D**), (**p**), (**c**) and (**q**) are known, then its Utilized or Starting Capital must be:

$$U = D + (Q + \$vp / \{360[c-q]\}) / \textit{I}$$
$$-\{\$-\$'[1+s][v+f]-I\}[1-t]$$

Steve Asikin ISBN 14: 978-1515158691, ISBN 10: 1515158691

Rule-21282:

If both (f), (U), (Q), (S'), (s), (v), (f), (I), (t), (D), (p), (c) and (q) are known, then its Sales or Revenue Planned is:

$$S= [f(U-\{S'[1+s][v+f]+I\}[1-t]-D)-Q]$$
$$/(vp/\{360[c-q]\}+f[1-t])$$

Rule-21283:

If both (f), (U), (S), (Q), (s), (v), (f), (I), (t), (D), (p), (c) and (q) are known, then its Sales Past must be:

$$S'= \{S-I-[D+(Q+Svp/\{360[c-q]\})/f-U]/[1-t]\}$$
$$/\{[1+s][v+f]\}$$

Rule-21284:

If both (f), (U), (S), (S'), (Q), (v), (f), (I), (t), (D), (p), (c) and (q) are known, then its Sales Growth Planned is:

$$s= S/S'-1$$

or it can also be found as

$$s= \{S-I-[D+(Q+Svp/\{360[c-q]\})/f-U]$$
$$/[1-t]\}/\{S'[v+f]\}-1$$

Rule-21285:

If both (f), (U), (S), (S'), (s), (Q), (f), (I), (t), (D), (p), (c) and (q) are known, then its Variable Portion Planned is:

$$v= [f(U+\{S-S'f[1+s]-I\}[1-t]-D)-Q]$$
$$/(Sp/\{360[c-q]\}-S'f[1+s][1-t])$$

Steve Asikin ISBN 14: 978-1515158691, ISBN 10: 1515158691

Rule-21286:

If both $(\textit{l})$, $(\mathbf{U})$, $(\mathbf{S})$, $(\mathbf{S'})$, $(\mathbf{s})$, $(\mathbf{v})$, $(\mathbf{Q})$, $(\mathbf{I})$, $(\mathbf{t})$, $(\mathbf{D})$, $(\mathbf{p})$, $(\mathbf{c})$ and $(\mathbf{q})$ are known, then its Fixed Portion Planned is:

$$f = \{S\text{-}I\text{-}[D+(Q+Svp/\{360[c\text{-}q]\})/\textit{l}\text{-}U]/[1\text{-}t]\} /\{S'[1+s]\}\text{-}v$$

Rule-21287:

If both $(\textit{l})$, $(\mathbf{U})$, $(\mathbf{S})$, $(\mathbf{S'})$, $(\mathbf{s})$, $(\mathbf{v})$, $(\mathbf{f})$, $(\mathbf{Q})$, $(\mathbf{t})$, $(\mathbf{D})$, $(\mathbf{p})$, $(\mathbf{c})$ and $(\mathbf{q})$ are known, then its Interest Expense Planned is:

$$I = S\text{-}S'[1+s][v+f] -[D+(Q+Svp/\{360[c\text{-}q]\})/\textit{l}\text{-}U]/[1\text{-}t]$$

Rule-21288:

If both $(\textit{l})$, $(\mathbf{U})$, $(\mathbf{S})$, $(\mathbf{S'})$, $(\mathbf{s})$, $(\mathbf{v})$, $(\mathbf{f})$, $(\mathbf{I})$, $(\mathbf{Q})$, $(\mathbf{D})$, $(\mathbf{p})$, $(\mathbf{c})$ and $(\mathbf{q})$ are known, then its Tax Rate Planned is:

$$t = 1\text{-}[D+(Q+Svp/\{360[c\text{-}q]\})/\textit{l}\text{-}U] /\{S\text{-}S'[1+s][v+f]\text{-}I\}$$

Rule-21289:

If both $(\textit{l})$, $(\mathbf{U})$, $(\mathbf{S})$, $(\mathbf{S'})$, $(\mathbf{s})$, $(\mathbf{v})$, $(\mathbf{f})$, $(\mathbf{I})$, $(\mathbf{t})$, $(\mathbf{Q})$, $(\mathbf{p})$, $(\mathbf{c})$ and $(\mathbf{q})$ are known, then its Dividend Planned is:

$$D = U+\{S\text{-}S'[1+s][v+f]\text{-}I\}[1\text{-}t] -(Q+Svp/\{360[c\text{-}q]\})/\textit{l}$$

Steve Asikin ISBN 14: 978-1515158691, ISBN 10: 1515158691

Rule-21290:

If both (I), (U), (S), (S'), (s), (v), (f), (I), (t), (D), (Q), (c) and (q) are known, then its Procured Inventory Days Planned:

$$p = 360[c-q]$$
$$[I(U+\{S-S'[1+s][v+f]-I\}[1-t]-D)-Q]/[Sv]$$

Rule-21291:

If both (I), (U), (S), (S'), (s), (v), (f), (I), (t), (D), (p), (Q) and (q) are known, then its Current Ratio Planned is:

$$c = q+Svp$$
$$/\{360[I(U+\{S-S'[1+s][v+f]-I\}[1-t]-D)-Q]\}$$

Rule-21292:

If both (I), (U), (S), (S'), (s), (v), (f), (I), (t), (D), (p), (c) and (Q) are known, then its Quick or Acid Test Ratio Planned is:

$$q = c-Svp/\{360[I(U+\{S-S'[1+s][v+f]-I\}[1-t]-D)-Q]\}$$

Rule-21293:

If both (I), (U), (S), (S'), (s), (v), (f), (I), (t), (D), (p), (c) and (q) are known, then its Quoted Longterm Debt Planned is:

$$Q = I(U+\{S-S'[1+s][v+f]-I\}[1-t]-D)$$
$$-S'vp[1+s]/\{360[c-q]\}$$

Steve Asikin ISBN 14: 978-1515158691, ISBN 10: 1515158691

Rule-21294:

> If both $(\mathbf{Q})$, $(\mathbf{U})$, $(\mathbf{S})$, $(\mathbf{S'})$, $(\mathbf{s})$, $(\mathbf{v})$, $(\mathbf{f})$, $(\mathbf{I})$, $(\mathbf{t})$, $(\mathbf{D})$, $(\mathbf{p})$, $(\mathbf{c})$ and $(\mathbf{q})$ are known, then its Leverage or Gearing Ratio Planned is:
>
> $$\mathit{l} = (\mathbf{Q} + \mathbf{S'vp}[1+\mathbf{s}]/\{360[\mathbf{c}-\mathbf{q}]\})$$
> $$/(\mathbf{U} + \{\mathbf{S} - \mathbf{S'}[1+\mathbf{s}][\mathbf{v}+\mathbf{f}] - \mathbf{I}\}[1-\mathbf{t}] - \mathbf{D})$$

Rule-21295:

> If both (l), $(\mathbf{Q})$, $(\mathbf{S})$, $(\mathbf{S'})$, $(\mathbf{s})$, $(\mathbf{v})$, $(\mathbf{f})$, $(\mathbf{I})$, $(\mathbf{t})$, $(\mathbf{D})$, $(\mathbf{p})$, $(\mathbf{c})$ and $(\mathbf{q})$ are known, then its Utilized or Starting Equity must be:
>
> $$\mathbf{U} = \mathbf{D} + (\mathbf{Q} + \mathbf{S'vp}[1+\mathbf{s}]/\{360[\mathbf{c}-\mathbf{q}]\})/\mathit{l}$$
> $$-\{\mathbf{S} - \mathbf{S'}[1+\mathbf{s}][\mathbf{v}+\mathbf{f}] - \mathbf{I}\}[1-\mathbf{t}]$$

Rule-21296:

> If both (l), $(\mathbf{U})$, $(\mathbf{Q})$, $(\mathbf{S'})$, $(\mathbf{s})$, $(\mathbf{v})$, $(\mathbf{f})$, $(\mathbf{I})$, $(\mathbf{t})$, $(\mathbf{D})$, $(\mathbf{p})$, $(\mathbf{c})$ and $(\mathbf{q})$ are known, then its Sales or Revenue Planned is:
>
> $$\mathbf{S} = [\mathbf{Q} - \mathit{l}(\mathbf{U} - \{\mathbf{S'}[1+\mathbf{s}][\mathbf{v}+\mathbf{f}] + \mathbf{I}\}[1-\mathbf{t}] - \mathbf{D})$$
> $$+ \mathbf{S'vp}[1+\mathbf{s}]/\{360[\mathbf{c}-\mathbf{q}]\}]/\{\mathit{l}[1-\mathbf{t}]\}$$

Rule-21297:

> If both (l), $(\mathbf{U})$, $(\mathbf{S})$, $(\mathbf{Q})$, $(\mathbf{s})$, $(\mathbf{v})$, $(\mathbf{f})$, $(\mathbf{I})$, $(\mathbf{t})$, $(\mathbf{D})$, $(\mathbf{p})$, $(\mathbf{c})$ and $(\mathbf{q})$ are known, then its Sales Past must be:
>
> $$\mathbf{S'} = [\mathit{l}(\mathbf{U} + \{\mathbf{S} - \mathbf{S'f}[1+\mathbf{s}] - \mathbf{I}\}[1-\mathbf{t}] - \mathbf{D}) - \mathbf{Q}]$$
> $$/[\mathbf{v}[1+\mathbf{s}](\mathbf{p}/\{360[\mathbf{c}-\mathbf{q}]\} - \mathit{l}[1-\mathbf{t}])]$$

Steve Asikin ISBN 14: 978-1515158691, ISBN 10: 1515158691

Rule-21298:

If both (**/**), (**U**), (**$**), (**$'**), (**Q**), (**v**), (**f**), (**I**), (**t**), (**D**), (**p**), (**c**) and (**q**) are known, then its Sales Growth Planned is:

$s= \$/\$'-1$

or it can also be found as

$s= [\text{/}U+\{\$-\$'f[1+s]-I\}[1-t]-D)-Q]$
$/[\$'v(p/\{360[c-q]\}-\text{/}[1-t])]-1$

Rule-21299:

If both (**/**), (**U**), (**$**), (**$'**), (**s**), (**Q**), (**f**), (**I**), (**t**), (**D**), (**p**), (**c**) and (**q**) are known, then its Variable Portion Planned is:

$v= [\text{/}U+\{\$-\$'f[1+s]-I\}[1-t]-D)-Q]$
$/[\$'[1+s](p/\{360[c-q]\}-\text{/}[1-t])]$

Rule-21300:

If both (**/**), (**U**), (**$**), (**$'**), (**s**), (**v**), (**Q**), (**I**), (**t**), (**D**), (**p**), (**c**) and (**q**) are known, then its Fixed Portion Planned is:

$f= \{\$-I-[D+(Q+\$'vp[1+s]/\{360[c-q]\})/\text{/-}U]/[1-t]\}$
$/\{\$'[1+s]\}-v$

Rule-21301:

If both (**/**), (**U**), (**$**), (**$'**), (**s**), (**v**), (**f**), (**Q**), (**t**), (**D**), (**p**), (**c**) and (**q**) are known, then its Interest Expense Planned is:

$I= \$-\$'[1+s][v+f]$
$-[D+(Q+\$'vp[1+s]/\{360[c-q]\})/\text{/-}U]/[1-t]$

Steve Asikin ISBN 14: 978-1515158691, ISBN 10: 1515158691

Rule-21302:

If both (**/**), (**U**), (**$**), (**$'**), (**s**), (**v**), (**f**), (**I**), (**Q**), (**D**), (**p**), (**c**) and (**q**) are known, then its Tax Rate Planned is:

$$t= 1-[D+(Q+$'vp[1+s]/\{360[c-q]\})/I-U]$$
$$/\{$-$'[1+s][v+f]-I\}$$

Rule-21303:

If both (**/**), (**U**), (**$**), (**$'**), (**s**), (**v**), (**f**), (**I**), (**t**), (**Q**), (**p**), (**c**) and (**q**) are known, then its Dividend Planned is:

$$D= U+\{$-$'[1+s][v+f]-I\}[1-t]$$
$$-(Q+$'vp[1+s]/\{360[c-q]\})/I$$

Rule-21304:

If both (**/**), (**U**), (**$**), (**$'**), (**s**), (**v**), (**f**), (**I**), (**t**), (**D**), (**Q**), (**c**) and (**q**) are known, then its Procured Inventory Days Planned:

$$p= 360[c-q][(U+\{$-$'[1+s][v+f]-I\}[1-t]-D)-Q]$$
$$/\{$'v[1+s]\}$$

Rule-21305:

If both (**/**), (**U**), (**$**), (**$'**), (**s**), (**v**), (**f**), (**I**), (**t**), (**D**), (**p**), (**Q**) and (**q**) are known, then its Current Ratio Planned is:

$$c= q+$'vp[1+s]$$
$$/\{360[(U+\{$-$'[1+s][v+f]-I\}[1-t]-D)-Q]\}$$

Steve Asikin ISBN 14: 978-1515158691, ISBN 10: 1515158691

Rule-21306:

If both (I), (U), (S), (S'), (s), (v), (f), (I), (t), (D), (p), (c) and (Q) are known, then its Quick or Acid Test Ratio Planned is:

$q= c-S'vp[1+s]$
$/\{360[I(U+\{S-S'[1+s][v+f]-I\}[1-t]-D)-Q]\}$

Rule-21307:

If both (I), (U), (S), (S'), (s), (v), (f), (I), (t), (A), (d) and (X) are known, then its Quoted Longterm Debt Planned is:

$Q= I(U+\{S-S'[1+s][v+f]-I\}[1-t]-Ad)-X$

Rule-21308:

If both (Q), (U), (S), (S'), (s), (v), (f), (I), (t), (A), (d) and (X) are known, then its Leverage or Gearing Ratio Planned is:

$I= [Q+X]/(U+\{S-S'[1+s][v+f]-I\}[1-t]-Ad)$

Rule-21309:

If both (I), (Q), (S), (S'), (s), (v), (f), (I), (t), (A), (d) and (X) are known, then its Utilized or Starting Capital must be:

$U= Ad+[Q+X]/I-\{S-S'[1+s][v+f]-I\}[1-t]$

Rule-21310:

If both (I), (U), (Q), (S'), (s), (v), (f), (I), (t), (A), (d) and (X) are known, then its Sales or Revenue Planned is:

$S= S'[1+s][v+f]+I+\{Ad+[Q+X]/I-U\}/[1-t]$

Steve Asikin ISBN 14: 978-1515158691, ISBN 10: 1515158691

Rule-21311:

If both (I), (U), (S), (Q), (s), (v), (f), (I), (t), (A), (d) and (X) are known, then its Sales Past must be:
$$S' = (S\text{-}I\text{-}Ad + \{[Q+X]/I\text{-}U\}/[1\text{-}t])/\{[1+s][v+f]\}$$

Rule-21312:

If both (I), (U), (S), (S'), (Q), (v), (f), (I), (t), (A), (d) and (X) are known, then its Sales Growth Planned is:
$$s = S/S'\text{-}1$$
or it can also be found as
$$s = (S\text{-}I\text{-}Ad + \{[Q+X]/I\text{-}U\}/[1\text{-}t])/\{S'[v+f]\}\text{-}1$$

Rule-21313:

If both (I), (U), (S), (S'), (s), (Q), (f), (I), (t), (A), (d) and (X) are known, then its Variable Portion Planned is:
$$v = (S\text{-}I\text{-}Ad + \{[Q+X]/I\text{-}U\}/[1\text{-}t])/\{S'[1+s]\}\text{-}f$$

Rule-21314:

If both (I), (U), (S), (S'), (s), (v), (Q), (I), (t), (A), (d) and (X) are known, then its Fixed Portion Planned is:
$$f = (S\text{-}I\text{-}Ad + \{[Q+X]/I\text{-}U\}/[1\text{-}t])/\{S'[1+s]\}\text{-}v$$

Rule-21315:

If both (I), (U), (S), (S'), (s), (v), (f), (Q), (t), (A), (d) and (X) are known, then its Interest Expense Planned is:
$$I = S\text{-}S'[1+s][v+f]\text{-}\{Ad+[Q+X]/I\text{-}U\}/[1\text{-}t]$$

Steve Asikin ISBN 14: 978-1515158691, ISBN 10: 1515158691

Rule-21316:

If both (I), (U), (S), (S'), (s), (v), (f), (I), (Q), (A), (d) and (X) are known, then its Tax Rate Planned is:

$$t = 1 - \{Ad + [Q+X]/I \cdot U\} / \{S - S'[1+s][v+f] - I\}$$

Rule-21317:

If both (I), (U), (S), (S'), (s), (v), (f), (I), (t), (Q), (d) and (X) are known, then its After Tax Income Planned is:

$$A = (U + \{S - S'[1+s][v+f] - I\}[1-t] - [Q+X]/I)/d$$

Rule-21318:

If both (I), (U), (S), (S'), (s), (v), (f), (I), (t), (A), (Q) and (X) are known, then its Dividend Payout Planned is:

$$d = (U + \{S - S'[1+s][v+f] - I\}[1-t] - [Q+X]/I)/A$$

Rule-21319:

If both (I), (U), (S), (S'), (s), (v), (f), (I), (t), (A), (d) and (Q) are known, then its Xpress or Current Debt Planned is:

$$X = I(U + \{S - S'[1+s][v+f] - I\}[1-t] - Ad) - Q$$

Rule-21320:

If both (I), (U), (S), (S'), (s), (v), (f), (I), (t), (A), (d), (P), (c) and (q) are known, then its Quoted Longterm Debt Planned is:

$$Q = I(U + \{S - S'[1+s][v+f] - I\}[1-t] - Ad) - P/[c-q]$$

Steve Asikin ISBN 14: 978-1515158691, ISBN 10: 1515158691

Rule-21321:

If both $(\mathbf{Q})$, $(\mathbf{U})$, $(\mathbf{S})$, $(\mathbf{S'})$, $(\mathbf{s})$, $(\mathbf{v})$, $(\mathbf{f})$, $(\mathbf{I})$, $(\mathbf{t})$, $(\mathbf{A})$, $(\mathbf{d})$, $(\mathbf{P})$, $(\mathbf{c})$ and $(\mathbf{q})$ are known, then its Leverage or Gearing Ratio Planned is:

$$I = \{Q+P/[c-q]\}/(U+\{S-S'[1+s][v+f]-I\}[1-t]-Ad)$$

Rule-21322:

If both (l), $(\mathbf{Q})$, $(\mathbf{S})$, $(\mathbf{S'})$, $(\mathbf{s})$, $(\mathbf{v})$, $(\mathbf{f})$, $(\mathbf{I})$, $(\mathbf{t})$, $(\mathbf{A})$, $(\mathbf{d})$, $(\mathbf{P})$, $(\mathbf{c})$ and $(\mathbf{q})$ are known, then its Utilized or Starting Capital must be:

$$U = Ad+\{Q+P/[c-q]\}/I\{S-S'[1+s][v+f]-I\}[1-t]$$

Rule-21323:

If both (l), $(\mathbf{U})$, $(\mathbf{Q})$, $(\mathbf{S'})$, $(\mathbf{s})$, $(\mathbf{v})$, $(\mathbf{f})$, $(\mathbf{I})$, $(\mathbf{t})$, $(\mathbf{A})$, $(\mathbf{d})$, $(\mathbf{P})$, $(\mathbf{c})$ and $(\mathbf{q})$ are known, then its Sales or Revenue Planned is:

$$S = S'[1+s][v+f]+I+(Ad+\{Q+P/[c-q]\}/IU)/[1-t]$$

Rule-21324:

If both (l), $(\mathbf{U})$, $(\mathbf{S})$, $(\mathbf{Q})$, $(\mathbf{s})$, $(\mathbf{v})$, $(\mathbf{f})$, $(\mathbf{I})$, $(\mathbf{t})$, $(\mathbf{A})$, $(\mathbf{d})$, $(\mathbf{P})$, $(\mathbf{c})$ and $(\mathbf{q})$ are known, then its Sales Past must be:

$$S' = [S-I-Ad+(\{Q+P/[c-q]\}/IU)/[1-t]]/\{[1+s][v+f]\}$$

Rule-21325:

If both (l), $(\mathbf{U})$, $(\mathbf{S})$, $(\mathbf{S'})$, $(\mathbf{Q})$, $(\mathbf{v})$, $(\mathbf{f})$, $(\mathbf{I})$, $(\mathbf{t})$, $(\mathbf{A})$, $(\mathbf{d})$, $(\mathbf{P})$, $(\mathbf{c})$ and $(\mathbf{q})$ are known, then its Sales Growth Planned is:

$$s = S/S'-1$$

or it can also be found as

$$s = [S-I-Ad+(\{Q+P/[c-q]\}/IU)/[1-t]]/\{S'[v+f]\}-1$$

Steve Asikin ISBN 14: 978-1515158691, ISBN 10: 1515158691

Rule-21326:

If both (I), (U), (S), (S'), (s), (Q), (f), (I), (t), (A), (d), (P), (c) and (q) are known, then its Vartiable Portion Planned is:

$$v= [S\text{-}I\text{-}Ad+(\{Q+P/[c\text{-}q]\}/I\text{-}U)/[1\text{-}t]]/\{S'[1+s]\}\text{-}f$$

Rule-21327:

If both (I), (U), (S), (S'), (s), (v), (Q), (I), (t), (A), (d), (P), (c) and (q) are known, then its Fixed Portion Planned is:

$$f= [S\text{-}I\text{-}Ad+(\{Q+P/[c\text{-}q]\}/I\text{-}U)/[1\text{-}t]]/\{S'[1+s]\}\text{-}v$$

Rule-21328:

If both (I), (U), (S), (S'), (s), (v), (f), (Q), (t), (A), (d), (P), (c) and (q) are known, then its Interest Expense Planned is:

$$I= S\text{-}S'[1+s][v+f]\text{-}(Ad+\{Q+P/[c\text{-}q]\}/I\text{-}U)/[1\text{-}t]$$

Rule-21329:

If both (I), (U), (S), (S'), (s), (v), (f), (I), (Q), (A), (d), (P), (c) and (q) are known, then its Tax Rate Planned is:

$$t= 1\text{-}(Ad+\{Q+P/[c\text{-}q]\}/I\text{-}U)/\{S\text{-}S'[1+s][v+f]\text{-}I\}$$

Rule-21330:

If both (I), (U), (S), (S'), (s), (v), (f), (I), (t), (Q), (d), (P), (c) and (q) are known, then its After Tax Income Planned is:

$$A= (U+\{S\text{-}S'[1+s][v+f]\text{-}I\}[1\text{-}t]\text{-}\{Q+P/[c\text{-}q]\}/I)/d$$

Steve Asikin ISBN 14: 978-1515158691, ISBN 10: 1515158691

Rule-21331:
   If both (**/**), (**U**), (**$**), (**$'**), (**s**), (**v**), (**f**), (**I**), (**t**), (**A**), (**Q**), (**P**), (**c**) and (**q**) are known, then its Dividend Payout Planned is:
   $$d= (U+\{S-S'[1+s][v+f]-I\}[1-t]-\{Q+P/[c-q]\}//)/A$$

Rule-21332:
   If both (**/**), (**U**), (**$**), (**$'**), (**s**), (**v**), (**f**), (**I**), (**t**), (**A**), (**d**), (**Q**), (**c**) and (**q**) are known, then its Procured Inventory Planned is:
   $$P= [c-q][/U+\{S-S'[1+s][v+f]-I\}[1-t]-Ad)-Q]$$

Rule-21333:
   If both (**/**), (**U**), (**$**), (**$'**), (**s**), (**v**), (**f**), (**I**), (**t**), (**A**), (**d**), (**P**), (**Q**) and (**q**) are known, then its Current Ratio Planned is:
   $$c= q+P/[/U+\{S-S'[1+s][v+f]-I\}[1-t]-Ad)-Q]$$

Rule-21334:
   If both (**/**), (**U**), (**$**), (**$'**), (**s**), (**v**), (**f**), (**I**), (**t**), (**A**), (**d**), (**P**), (**c**) and (**Q**) are known, then its Quick or Acid Test Ratio Planned is:
   $$q= c-P/[/U+\{S-S'[1+s][v+f]-I\}[1-t]-Ad)-Q]$$

Rule-21335:
   If both (**/**), (**U**), (**$**), (**$'**), (**s**), (**v**), (**f**), (**I**), (**t**), (**A**), (**d**), (**V**), (**p**), (**c**) and (**q**) are known, then its Quoted Longterm Debt Planned is:
   $$Q= /U+\{S-S'[1+s][v+f]-I\}[1-t]-Ad)$$
   $$-Vp/\{360[c-q]\}$$

Steve Asikin ISBN 14: 978-1515158691, ISBN 10: 1515158691

Rule-21336:

If both $(\mathbf{Q})$, $(\mathbf{U})$, $(\mathbf{S})$, $(\mathbf{S'})$, $(\mathbf{s})$, $(\mathbf{v})$, $(\mathbf{f})$, $(\mathbf{I})$, $(\mathbf{t})$, $(\mathbf{A})$, $(\mathbf{d})$, $(\mathbf{V})$, $(\mathbf{p})$, $(\mathbf{c})$ and $(\mathbf{q})$ are known, then its Leverage or Gearing Ratio Planned is:

$$\mathit{I} = \{\mathbf{Q}+\mathbf{Vp}/\{360[\mathbf{c}\text{-}\mathbf{q}]\})$$
$$/ (\mathbf{U}+\{\mathbf{S}\text{-}\mathbf{S'}[1+\mathbf{s}][\mathbf{v}+\mathbf{f}]\text{-}\mathbf{I}\}[1\text{-}\mathbf{t}]\text{-}\mathbf{Ad})$$

Rule-21337:

If both (I), $(\mathbf{Q})$, $(\mathbf{S})$, $(\mathbf{S'})$, $(\mathbf{s})$, $(\mathbf{v})$, $(\mathbf{f})$, $(\mathbf{I})$, $(\mathbf{t})$, $(\mathbf{A})$, $(\mathbf{d})$, $(\mathbf{V})$, $(\mathbf{p})$, $(\mathbf{c})$ and $(\mathbf{q})$ are known, then its Utilized or Starting Capital must be:

$$\mathbf{U}= \mathbf{Ad}+\{\mathbf{Q}+\mathbf{Vp}/\{360[\mathbf{c}\text{-}\mathbf{q}]\})/\mathit{I}$$
$$-\{\mathbf{S}\text{-}\mathbf{S'}[1+\mathbf{s}][\mathbf{v}+\mathbf{f}]\text{-}\mathbf{I}\}[1\text{-}\mathbf{t}]$$

Rule-21338:

If both (I), $(\mathbf{U})$, $(\mathbf{Q})$, $(\mathbf{S'})$, $(\mathbf{s})$, $(\mathbf{v})$, $(\mathbf{f})$, $(\mathbf{I})$, $(\mathbf{t})$, $(\mathbf{A})$, $(\mathbf{d})$, $(\mathbf{V})$, $(\mathbf{p})$, $(\mathbf{c})$ and $(\mathbf{q})$ are known, then its Sales or Revenue Planned is:

$$\mathbf{S}= \mathbf{S'}[1+\mathbf{s}][\mathbf{v}+\mathbf{f}]+\mathbf{I}+[\mathbf{Ad}+\{\mathbf{Q}+\mathbf{Vp}/\{360[\mathbf{c}\text{-}\mathbf{q}]\})/\mathit{I}\text{-}\mathbf{U}]$$
$$/[1\text{-}\mathbf{t}]$$

Rule-21339:

If both (I), $(\mathbf{U})$, $(\mathbf{S})$, $(\mathbf{Q})$, $(\mathbf{s})$, $(\mathbf{v})$, $(\mathbf{f})$, $(\mathbf{I})$, $(\mathbf{t})$, $(\mathbf{A})$, $(\mathbf{d})$, $(\mathbf{V})$, $(\mathbf{p})$, $(\mathbf{c})$ and $(\mathbf{q})$ are known, then its Sales Past must be:

$$\mathbf{S'}= [\mathbf{S}\text{-}\mathbf{I}\text{-}\mathbf{Ad}+[(\mathbf{Q}+\mathbf{Vp}/\{360[\mathbf{c}\text{-}\mathbf{q}]\})/\mathit{I}\text{-}\mathbf{U}]/[1\text{-}\mathbf{t}]\}$$
$$/\{[1+\mathbf{s}][\mathbf{v}+\mathbf{f}]\}$$

Steve Asikin ISBN 14: 978-1515158691, ISBN 10: 1515158691

Rule-21340:

If both (I), (U), (S), (S'), (Q), (v), (f), (I), (t), (A), (d), (V), (p), (c) and (q) are known, then its Sales Growth Planned is:

$s= S/S'-1$

or it can also be found as

$s= \{S\text{-}I\text{-}Ad+[(Q+Vp/\{360[c\text{-}q]\})/I\cdot U]/[1\text{-}t]\}$
$/\{S'[v+f]\}\text{-}1$

Rule-21341:

If both (I), (U), (S), (S'), (s), (Q), (f), (I), (t), (A), (d), (V), (p), (c) and (q) are known, then its Variable Portion Planned is:

$v= V/S$

or it can also be found as

$v= \{S\text{-}I\text{-}Ad+[(Q+Vp/\{360[c\text{-}q]\})/I\cdot U]/[1\text{-}t]\}$
$/\{S'[1+s]\}\text{-}f$

Rule-21342:

If both (I), (U), (S), (S'), (s), (v), (Q), (I), (t), (A), (d), (V), (p), (c) and (q) are known, then its Fixed Portion Planned is:

$f= \{S\text{-}I\text{-}Ad+[(Q+Vp/\{360[c\text{-}q]\})/I\cdot U]/[1\text{-}t]\}$
$/\{S'[1+s]\}\text{-}v$

Rule-21343:

If both (I), (U), (S), (S'), (s), (v), (f), (Q), (t), (A), (d), (V), (p), (c) and (q) are known, then its Interest Expense Planned is:

$I= S\text{-}S'[1+s][v+f]$
$-[Ad+(Q+Vp/\{360[c\text{-}q]\})/I\cdot U]/[1\text{-}t]$

Steve Asikin ISBN 14: 978-1515158691, ISBN 10: 1515158691

Rule-21344:

If both (*l*), (**U**), (**$**), (**$'**), (**s**), (**v**), (**f**), (**I**), (**Q**), (**A**), (**d**), (**V**), (**p**), (**c**) and (**q**) are known, then its Tax Rate Planned is:

$$t= 1-[\textbf{Ad}+(\textbf{Q}+\textbf{Vp}/\{360[\textbf{c-q}]\})/\textit{l}\textbf{U}]$$
$$/\{\textbf{\$-\$'}[1+\textbf{s}][\textbf{v+f}]-\textbf{I}\}$$

Rule-21345:

If both (*l*), (**U**), (**$**), (**$'**), (**s**), (**v**), (**f**), (**I**), (**t**), (**Q**), (**d**), (**V**), (**p**), (**c**) and (**q**) are known, then its After Tax Income Planned is:

$$A= [\textbf{U}+\{\textbf{\$-\$'}[1+\textbf{s}][\textbf{v+f}]-\textbf{I}\}[1-\textbf{t}]$$
$$-(\textbf{Q}+\textbf{Vp}/\{360[\textbf{c-q}]\})/\textit{l}]/\textbf{d}$$

Rule-21346:

If both (*l*), (**U**), (**$**), (**$'**), (**s**), (**v**), (**f**), (**I**), (**t**), (**A**), (**Q**), (**V**), (**p**), (**c**) and (**q**) are known, then its Dividend Payout Planned is:

$$d= [\textbf{U}+\{\textbf{\$-\$'}[1+\textbf{s}][\textbf{v+f}]-\textbf{I}\}[1-\textbf{t}]$$
$$-(\textbf{Q}+\textbf{Vp}/\{360[\textbf{c-q}]\})/\textit{l}]/\textbf{A}$$

Rule-21347:

If both (*l*), (**U**), (**$**), (**$'**), (**s**), (**v**), (**f**), (**I**), (**t**), (**A**), (**d**), (**Q**), (**p**), (**c**) and (**q**) are known, then its Variable Cost Planned is:

$$V= 360[\textbf{c-q}]$$
$$[\textit{l}(\textbf{U}+\{\textbf{\$-\$'}[1+\textbf{s}][\textbf{v+f}]-\textbf{I}\}[1-\textbf{t}]-\textbf{Ad})-\textbf{Q}]/\textbf{p}$$

Steve Asikin ISBN 14: 978-1515158691, ISBN 10: 1515158691

Rule-21348:

If both $(\textit{I})$, $(\textbf{U})$, $(\textbf{S})$, $(\textbf{S'})$, $(\textbf{s})$, $(\textbf{v})$, $(\textbf{f})$, $(\textbf{I})$, $(\textbf{t})$, $(\textbf{A})$, $(\textbf{d})$, $(\textbf{V})$, $(\textbf{Q})$, $(\textbf{c})$ and $(\textbf{q})$ are known, then its Procured Inventory Days Planned:

$$p= 360[\textbf{c-q}]$$
$$[\textit{I}\textbf{U}+\{\textbf{S-S'}[1+\textbf{s}][\textbf{v+f}]\textbf{-I}\}[1\textbf{-t}]\textbf{-Ad})\textbf{-Q}]/\textbf{V}$$

Rule-21349:

If both $(\textit{I})$, $(\textbf{U})$, $(\textbf{S})$, $(\textbf{S'})$, $(\textbf{s})$, $(\textbf{v})$, $(\textbf{f})$, $(\textbf{I})$, $(\textbf{t})$, $(\textbf{A})$, $(\textbf{d})$, $(\textbf{V})$, $(\textbf{p})$, $(\textbf{Q})$ and $(\textbf{q})$ are known, then its Current Ratio Planned is:

$$c= q+\textbf{Vp}$$
$$/\{360[\textit{I}\textbf{U}+\{\textbf{S-S'}[1+\textbf{s}][\textbf{v+f}]\textbf{-I}\}[1\textbf{-t}]\textbf{-Ad})\textbf{-Q}]\}$$

Rule-21350:

If both $(\textit{I})$, $(\textbf{U})$, $(\textbf{S})$, $(\textbf{S'})$, $(\textbf{s})$, $(\textbf{v})$, $(\textbf{f})$, $(\textbf{I})$, $(\textbf{t})$, $(\textbf{A})$, $(\textbf{d})$, $(\textbf{V})$, $(\textbf{p})$, $(\textbf{c})$ and $(\textbf{Q})$ are known, then its Quick or Acid Test Ratio Planned is:

$$q= c\textbf{-Vp}$$
$$/\{360[\textit{I}\textbf{U}+\{\textbf{S-S'}[1+\textbf{s}][\textbf{v+f}]\textbf{-I}\}[1\textbf{-t}]\textbf{-Ad})\textbf{-Q}]\}$$

Rule-21351:

If both $(\textit{I})$, $(\textbf{U})$, $(\textbf{S})$, $(\textbf{S'})$, $(\textbf{s})$, $(\textbf{v})$, $(\textbf{f})$, $(\textbf{I})$, $(\textbf{t})$, $(\textbf{A})$, $(\textbf{d})$, $(\textbf{p})$, $(\textbf{c})$ and $(\textbf{q})$ are known, then its Quoted Longterm Debt Planned is:

$$Q= \textit{I}\textbf{U}+\{\textbf{S-S'}[1+\textbf{s}][\textbf{v+f}]\textbf{-I}\}[1\textbf{-t}]\textbf{-Ad})$$
$$\textbf{-Svp}/\{360[\textbf{c-q}]\}$$

Steve Asikin ISBN 14: 978-1515158691, ISBN 10: 1515158691

Rule-21352:

 If both $(\mathbf{Q})$, $(\mathbf{U})$, $(\mathbf{S})$, $(\mathbf{S'})$, $(\mathbf{s})$, $(\mathbf{v})$, $(\mathbf{f})$, $(\mathbf{I})$, $(\mathbf{t})$, $(\mathbf{A})$, $(\mathbf{d})$, $(\mathbf{p})$, $(\mathbf{c})$ and $(\mathbf{q})$ are known, then its Leverage or Gearing Ratio Planned is:

$$\mathbf{I} = \{\mathbf{Q} + \mathbf{Svp}/\{360[\mathbf{c}\text{-}\mathbf{q}]\})$$
$$/(\mathbf{U} + \{\mathbf{S}\text{-}\mathbf{S'}[1+\mathbf{s}][\mathbf{v}+\mathbf{f}]\text{-}\mathbf{I}\}[1\text{-}\mathbf{t}]\text{-}\mathbf{Ad})$$

Rule-21353:

 If both $(\mathbf{I})$, $(\mathbf{Q})$, $(\mathbf{S})$, $(\mathbf{S'})$, $(\mathbf{s})$, $(\mathbf{v})$, $(\mathbf{f})$, $(\mathbf{I})$, $(\mathbf{t})$, $(\mathbf{A})$, $(\mathbf{d})$, $(\mathbf{p})$, $(\mathbf{c})$ and $(\mathbf{q})$ are known, then its Utilized or Starting Capital must be:

$$\mathbf{U} = \mathbf{Ad} + \{\mathbf{Q} + \mathbf{Svp}/\{360[\mathbf{c}\text{-}\mathbf{q}]\})/\mathbf{I}$$
$$-\{\mathbf{S}\text{-}\mathbf{S'}[1+\mathbf{s}][\mathbf{v}+\mathbf{f}]\text{-}\mathbf{I}\}[1\text{-}\mathbf{t}]$$

Rule-21354:

 If both $(\mathbf{I})$, $(\mathbf{U})$, $(\mathbf{Q})$, $(\mathbf{S'})$, $(\mathbf{s})$, $(\mathbf{v})$, $(\mathbf{f})$, $(\mathbf{I})$, $(\mathbf{t})$, $(\mathbf{A})$, $(\mathbf{d})$, $(\mathbf{p})$, $(\mathbf{c})$ and $(\mathbf{q})$ are known, then its Sales or Revenue Planned is:

$$\mathbf{S} = [\mathbf{I}(\mathbf{U} - \{\mathbf{S'}[1+\mathbf{s}][\mathbf{v}+\mathbf{f}]+\mathbf{I}\}[1\text{-}\mathbf{t}]\text{-}\mathbf{Ad})\text{-}\mathbf{Q}]$$
$$/(\mathbf{vp}/\{360[\mathbf{c}\text{-}\mathbf{q}]\}\text{-}\mathbf{I}[1\text{-}\mathbf{t}])$$

Rule-21355:

 If both $(\mathbf{I})$, $(\mathbf{U})$, $(\mathbf{S})$, $(\mathbf{Q})$, $(\mathbf{s})$, $(\mathbf{v})$, $(\mathbf{f})$, $(\mathbf{I})$, $(\mathbf{t})$, $(\mathbf{A})$, $(\mathbf{d})$, $(\mathbf{p})$, $(\mathbf{c})$ and $(\mathbf{q})$ are known, then its Sales Past must be:

$$\mathbf{S'} = [\mathbf{S}\text{-}\mathbf{I}\text{-}\mathbf{Ad} + [(\mathbf{Q} + \mathbf{Svp}/\{360[\mathbf{c}\text{-}\mathbf{q}]\})/\mathbf{I}\text{-}\mathbf{U}]/[1\text{-}\mathbf{t}]\}$$
$$/\{[1+\mathbf{s}][\mathbf{v}+\mathbf{f}]\}$$

Steve Asikin ISBN 14: 978-1515158691, ISBN 10: 1515158691

Rule-21356:

If both $(\textbf{\textit{l}})$, $(\textbf{U})$, $(\textbf{S})$, $(\textbf{S'})$, $(\textbf{Q})$, $(\textbf{v})$, $(\textbf{f})$, $(\textbf{I})$, $(\textbf{t})$, $(\textbf{A})$, $(\textbf{d})$, $(\textbf{p})$, $(\textbf{c})$ and $(\textbf{q})$ are known, then its Sales Growth Planned is:

$$s = S/S' - 1$$

or it can also be found as

$$s = \{S\text{-}I\text{-}Ad+[(Q+Svp/\{360[c\text{-}q]\})/\textbf{\textit{l}}U]/[1\text{-}t]\} /\{S'[v+f]\}-1$$

Rule-21357:

If both $(\textbf{\textit{l}})$, $(\textbf{U})$, $(\textbf{S})$, $(\textbf{S'})$, $(\textbf{s})$, $(\textbf{Q})$, $(\textbf{f})$, $(\textbf{I})$, $(\textbf{t})$, $(\textbf{A})$, $(\textbf{d})$, $(\textbf{p})$, $(\textbf{c})$ and $(\textbf{q})$ are known, then its Variable Portion Planned is:

$$v = [\textbf{\textit{l}}U+\{S\text{-}S'f[1+s]\text{-}I\}[1\text{-}t]\text{-}Ad)\text{-}Q] /(Sp/\{360[c\text{-}q]\}+S'\textbf{\textit{l}}[1+s][1\text{-}t])$$

Rule-21358:

If both $(\textbf{\textit{l}})$, $(\textbf{U})$, $(\textbf{S})$, $(\textbf{S'})$, $(\textbf{s})$, $(\textbf{v})$, $(\textbf{Q})$, $(\textbf{I})$, $(\textbf{t})$, $(\textbf{A})$, $(\textbf{d})$, $(\textbf{p})$, $(\textbf{c})$ and $(\textbf{q})$ are known, then its Fixed Portion Planned is:

$$f = \{S\text{-}I\text{-}Ad+[(Q+Svp/\{360[c\text{-}q]\})/\textbf{\textit{l}}U]/[1\text{-}t]\} /\{S'[1+s]\}-v$$

Rule-21359:

If both $(\textbf{\textit{l}})$, $(\textbf{U})$, $(\textbf{S})$, $(\textbf{S'})$, $(\textbf{s})$, $(\textbf{v})$, $(\textbf{f})$, $(\textbf{Q})$, $(\textbf{t})$, $(\textbf{A})$, $(\textbf{d})$, $(\textbf{p})$, $(\textbf{c})$ and $(\textbf{q})$ are known, then its Interest Expense Planned is:

$$I = S\text{-}S'[1+s][v+f]\text{-}[Ad+(Q+Svp/\{360[c\text{-}q]\})/\textbf{\textit{l}}U] /[1\text{-}t]$$

Steve Asikin ISBN 14: 978-1515158691, ISBN 10: 1515158691

Rule-21360:

> If both $(\textbf{\textit{l}})$, $(\textbf{U})$, $(\textbf{S})$, $(\textbf{S'})$, $(\textbf{s})$, $(\textbf{v})$, $(\textbf{f})$, $(\textbf{I})$, $(\textbf{Q})$, $(\textbf{A})$, $(\textbf{d})$, $(\textbf{p})$, $(\textbf{c})$ and $(\textbf{q})$ are known, then its Tax Rate Planned is:
>
> $$t= 1-[\textbf{Ad}+(\textbf{Q}+\textbf{Svp}/\{360[\textbf{c-q}]\})/\textbf{\textit{l}U}]$$
> $$/\{\textbf{S-S'}[1+\textbf{s}][\textbf{v+f}]-\textbf{I}\}$$

Rule-21361:

> If both $(\textbf{\textit{l}})$, $(\textbf{U})$, $(\textbf{S})$, $(\textbf{S'})$, $(\textbf{s})$, $(\textbf{v})$, $(\textbf{f})$, $(\textbf{I})$, $(\textbf{t})$, $(\textbf{Q})$, $(\textbf{d})$, $(\textbf{p})$, $(\textbf{c})$ and $(\textbf{q})$ are known, then its After Tax Income Planned is:
>
> $$A= [U+\{S-S'[1+s][v+f]-I\}[1-t]$$
> $$-(Q+Svp/\{360[c-q]\})/\textbf{\textit{l}}]/d$$

Rule-21362:

> If both $(\textbf{\textit{l}})$, $(\textbf{U})$, $(\textbf{S})$, $(\textbf{S'})$, $(\textbf{s})$, $(\textbf{v})$, $(\textbf{f})$, $(\textbf{I})$, $(\textbf{t})$, $(\textbf{A})$, $(\textbf{Q})$, $(\textbf{p})$, $(\textbf{c})$ and $(\textbf{q})$ are known, then its Dividend Payout Planned is:
>
> $$d= [U+\{S-S'[1+s][v+f]-I\}[1-t]$$
> $$-(Q+Svp/\{360[c-q]\})/\textbf{\textit{l}}]/A$$

Rule-21363:

> If both $(\textbf{\textit{l}})$, $(\textbf{U})$, $(\textbf{S})$, $(\textbf{S'})$, $(\textbf{s})$, $(\textbf{v})$, $(\textbf{f})$, $(\textbf{I})$, $(\textbf{t})$, $(\textbf{A})$, $(\textbf{d})$, $(\textbf{Q})$, $(\textbf{c})$ and $(\textbf{q})$ are known, then its Procured Inventory Days Planned:
>
> $$p= 360[c-q][\textbf{\textit{l}}(U+\{S-S'[1+s][v+f]-I\}[1-t]-Ad)-Q]$$
> $$/[Sv]$$

Steve Asikin ISBN 14: 978-1515158691, ISBN 10: 1515158691

Rule-21364:

If both (**/**), (**U**), (**$**), (**$'**), (**s**), (**v**), (**f**), (**I**), (**t**), (**A**), (**d**), (**p**), (**Q**) and (**q**) are known, then its Current Ratio Planned is:

$$c = q + \mathbf{\$vp}$$
$$/\{360[\mathbf{/U} + \{\mathbf{\$-\$'}[1+\mathbf{s}][\mathbf{v+f}]-\mathbf{I}\}[1-\mathbf{t}]-\mathbf{Ad})-\mathbf{Q}]\}$$

Rule-21365:

If both (**/**), (**U**), (**$**), (**$'**), (**s**), (**v**), (**f**), (**I**), (**t**), (**A**), (**d**), (**p**), (**c**) and (**Q**) are known, then its Quick or Acid Test Ratio Planned is:

$$q = c - \mathbf{\$vp}$$
$$/\{360[\mathbf{/U} + \{\mathbf{\$-\$'}[1+\mathbf{s}][\mathbf{v+f}]-\mathbf{I}\}[1-\mathbf{t}]-\mathbf{Ad})-\mathbf{Q}]\}$$

Rule-21366:

If both (**/**), (**U**), (**$**), (**$'**), (**s**), (**v**), (**f**), (**I**), (**t**), (**A**), (**d**), (**p**), (**c**) and (**q**) are known, then its Quoted Longterm Debt Planned is:

$$\mathbf{Q} = \mathbf{/U} + \{\mathbf{\$-\$'}[1+\mathbf{s}][\mathbf{v+f}]-\mathbf{I}\}[1-\mathbf{t}]-\mathbf{Ad})$$
$$-\mathbf{\$'vp}[1+\mathbf{s}]/\{360[\mathbf{c-q}]\}$$

Rule-21367:

If both (**Q**), (**U**), (**$**), (**$'**), (**s**), (**v**), (**f**), (**I**), (**t**), (**A**), (**d**), (**p**), (**c**) and (**q**) are known, then its Leverage or Gearing Ratio Planned is:

$$\mathbf{/} = \{\mathbf{Q} + \mathbf{\$'vp}[1+\mathbf{s}]/\{360[\mathbf{c-q}]\}\})$$
$$/ (\mathbf{U} + \{\mathbf{\$-\$'}[1+\mathbf{s}][\mathbf{v+f}]-\mathbf{I}\}[1-\mathbf{t}]-\mathbf{Ad})$$

Steve Asikin ISBN 14: 978-1515158691, ISBN 10: 1515158691

Rule-21368:

If both (I), $(\mathbf{Q})$, $(\mathbf{S})$, $(\mathbf{S'})$, $(\mathbf{s})$, $(\mathbf{v})$, $(\mathbf{f})$, $(\mathbf{I})$, $(\mathbf{t})$, $(\mathbf{A})$, $(\mathbf{d})$, $(\mathbf{p})$, $(\mathbf{c})$ and $(\mathbf{q})$ are known, then its Utilized or Starting Capital must be:

$$U= Ad+\{Q+S'vp[1+s]/\{360[c\text{-}q]\})/I$$
$$-\{S\text{-}S'[1+s][v+f]\text{-}I\}[1\text{-}t]$$

Rule-21369:

If both (I), $(\mathbf{U})$, $(\mathbf{S})$, $(\mathbf{Q})$, $(\mathbf{s})$, $(\mathbf{v})$, $(\mathbf{f})$, $(\mathbf{I})$, $(\mathbf{t})$, $(\mathbf{A})$, $(\mathbf{d})$, $(\mathbf{p})$, $(\mathbf{c})$ and $(\mathbf{q})$ are known, then its Sales or Revenue Planned is:

$$S= [Q\text{-}I(U\text{-}\{S'[1+s][v+f]+I\}[1\text{-}t]\text{-}Ad)$$
$$+S'vp[1+s]/\{360[c\text{-}q]\}]/\{I[1\text{-}t]\}$$

Rule-21370:

If both (I), $(\mathbf{U})$, $(\mathbf{S})$, $(\mathbf{Q})$, $(\mathbf{s})$, $(\mathbf{v})$, $(\mathbf{f})$, $(\mathbf{I})$, $(\mathbf{t})$, $(\mathbf{A})$, $(\mathbf{d})$, $(\mathbf{p})$, $(\mathbf{c})$ and $(\mathbf{q})$ are known, then its Sales Past must be:

$$S'= [I(U+\{S\text{-}S'f[1+s]\text{-}I\}[1\text{-}t]\text{-}Ad)\text{-}Q]$$
$$/[v[1+s](p/\{360[c\text{-}q]\}+I[1\text{-}t])]$$

Rule-21371:

If both (I), $(\mathbf{U})$, $(\mathbf{S})$, $(\mathbf{S'})$, $(\mathbf{Q})$, $(\mathbf{v})$, $(\mathbf{f})$, $(\mathbf{I})$, $(\mathbf{t})$, $(\mathbf{A})$, $(\mathbf{d})$, $(\mathbf{p})$, $(\mathbf{c})$ and $(\mathbf{q})$ are known, then its Sales Growth Planned is:

$$s= S/S'\text{-}1$$

or it can also be found as

$$s= [I(U+\{S\text{-}S'f[1+s]\text{-}I\}[1\text{-}t]\text{-}Ad)\text{-}Q]$$
$$/[S'v(p/\{360[c\text{-}q]\}+I[1\text{-}t])]\text{-}1$$

Steve Asikin ISBN 14: 978-1515158691, ISBN 10: 1515158691

Rule-21372:

   If both (I), $(\mathbf{U})$, $(\mathbf{S})$, $(\mathbf{S'})$, $(\mathbf{s})$, $(\mathbf{Q})$, $(\mathbf{f})$, $(\mathbf{I})$, $(\mathbf{t})$, $(\mathbf{A})$, $(\mathbf{d})$, $(\mathbf{p})$, $(\mathbf{c})$ and $(\mathbf{q})$ are known, then its Variable Portion Planned is:

$$v= [IU+\{S\text{-}S'f[1+s]\text{-}I\}[1\text{-}t]\text{-}Ad)\text{-}Q]$$
$$/[S'[1+s](p/\{360[c\text{-}q]\}+I[1\text{-}t])]$$

Rule-21373:

   If both (I), $(\mathbf{U})$, $(\mathbf{S})$, $(\mathbf{S'})$, $(\mathbf{s})$, $(\mathbf{v})$, $(\mathbf{Q})$, $(\mathbf{I})$, $(\mathbf{t})$, $(\mathbf{A})$, $(\mathbf{d})$, $(\mathbf{p})$, $(\mathbf{c})$ and $(\mathbf{q})$ are known, then its Fixed Portion Planned is:

$$f= \{S\text{-}I\text{-}Ad+[(Q+S'vp[1+s]/\{360[c\text{-}q]\})/IU]/[1\text{-}t]\}$$
$$/\{S'[1+s]\}\text{-}v$$

Rule-21374:

   If both (I), $(\mathbf{U})$, $(\mathbf{S})$, $(\mathbf{S'})$, $(\mathbf{s})$, $(\mathbf{v})$, $(\mathbf{f})$, $(\mathbf{Q})$, $(\mathbf{t})$, $(\mathbf{A})$, $(\mathbf{d})$, $(\mathbf{p})$, $(\mathbf{c})$ and $(\mathbf{q})$ are known, then its Interest Expense Planned is:

$$I= S\text{-}S'[1+s][v+f]\text{-}[Ad+(Q+S'vp[1+s]$$
$$/\{360[c\text{-}q]\})/IU]/[1\text{-}t]$$

Rule-21375:

   If both (I), $(\mathbf{U})$, $(\mathbf{S})$, $(\mathbf{S'})$, $(\mathbf{s})$, $(\mathbf{v})$, $(\mathbf{f})$, $(\mathbf{I})$, $(\mathbf{Q})$, $(\mathbf{A})$, $(\mathbf{d})$, $(\mathbf{p})$, $(\mathbf{c})$ and $(\mathbf{q})$ are known, then its Tax Planned is:

$$t= 1\text{-}[Ad+(Q+S'vp[1+s]/\{360[c\text{-}q]\})/IU]$$
$$/\{S\text{-}S'[1+s][v+f]\text{-}I\}$$

Steve Asikin ISBN 14: 978-1515158691, ISBN 10: 1515158691

Rule-21376:

If both (**/**), (**U**), (**$**), (**$'**), (**s**), (**v**), (**f**), (**I**), (**t**), (**Q**), (**d**), (**p**), (**c**) and (**q**) are known, then its After Tax Income Planned is:

$$A= [U+\{\$-\$'[1+s][v+f]-I\}[1-t]$$
$$-(Q+\$'vp[1+s]/\{360[c-q]\})/\text{/}/d$$

Rule-21377:

If both (**/**), (**U**), (**$**), (**$'**), (**s**), (**v**), (**f**), (**I**), (**t**), (**A**), (**Q**), (**p**), (**c**) and (**q**) are known, then its Dividend Payout Planned is:

$$d= [U+\{\$-\$'[1+s][v+f]-I\}[1-t]$$
$$-(Q+\$'vp[1+s]/\{360[c-q]\})/\text{/}/A$$

Rule-21378:

If both (**/**), (**U**), (**$**), (**$'**), (**s**), (**v**), (**f**), (**I**), (**t**), (**A**), (**d**), (**Q**), (**c**) and (**q**) are known, then its Procured Inventory Days Planned:

$$p= 360[c-q][\text{/}U+\{\$-\$'[1+s][v+f]-I\}[1-t]-Ad)-Q]$$
$$/\{\$'v[1+s]\}$$

Rule-21379:

If both (**/**), (**U**), (**$**), (**$'**), (**s**), (**v**), (**f**), (**I**), (**t**), (**A**), (**d**), (**p**), (**Q**) and (**q**) are known, then its Current Ratio Planned is:

$$c= q+\$'vp[1+s]$$
$$/\{360[\text{/}U+\{\$-\$'[1+s][v+f]-I\}[1-t]-Ad)-Q]\}$$

Steve Asikin ISBN 14: 978-1515158691, ISBN 10: 1515158691

Rule-21380:

If both (f), (U), $($)$, $($')$, (s), (v), (f), (I), (t), (A), (d), (p), (c) and (q) are known, then its Quoted Longterm Debt Planned is:

$$q = c\text{-}S'vp[1+s]$$
$$/\{360[f(U+\{S\text{-}S'[1+s][v+f]\text{-}I\}[1\text{-}t]\text{-}Ad)\text{-}Q]\}$$

Rule-21381:

If both (f), (U), $($)$, $($')$, (s), (v), (f), (I), (t), (d) and (X) are known, then its Quoted Longterm Debt Planned is:

$$Q = f(U+\{S\text{-}S'[1+s][v+f]\text{-}I\}[1\text{-}t][1\text{-}d])\text{-}X$$

Rule-21382:

If both (Q), (U), $($)$, $($')$, (s), (v), (f), (I), (t), (d) and (X) are known, then its Leverage or Gearing Ratio Planned is:

$$f = [Q+X]/(U+\{S\text{-}S'[1+s][v+f]\text{-}I\}[1\text{-}t][1\text{-}d])$$

Rule-21383:

If both (f), (Q), $($)$, $($')$, (s), (v), (f), (I), (t), (d) and (X) are known, then its Utilized or Starting Capital must be:

$$U = [Q+X]/f\{S\text{-}S'[1+s][v+f]\text{-}I\}[1\text{-}t][1\text{-}d]$$

Rule-21384:

If both (f), (U), (Q), $($')$, (s), (v), (f), (I), (t), (d) and (X) are known, then its Sales or Revenue Planned is:

$$S = S'[1+s][v+f]+I+\{[Q+X]/fU\}/\{[1\text{-}t][1\text{-}d]\}$$

Steve Asikin ISBN 14: 978-1515158691, ISBN 10: 1515158691

Rule-21385:

If both (I), (U), $(\$)$, (Q), (s), (v), (f), (I), (t), (d) and (X) are known, then its Sales Past must be:

$\$'= (\$-I-\{[Q+X]/I \cdot U\}/\{[1-t][1-d]\})/\{[1+s][v+f]\}$

Rule-21386:

If both (I), (U), $(\$)$, $(\$')$, (Q), (v), (f), (I), (t), (d) and (X) are known, then its Sales Growth Planned is:

$s= \$/\$'-1$

　　　　or it can also be found as

$s= (\$-I-\{[Q+X]/I \cdot U\}/\{[1-t][1-d]\})/\{\$'[v+f]\}-1$

Rule-21387:

If both (I), (U), $(\$)$, $(\$')$, (s), (Q), (f), (I), (t), (d) and (X) are known, then its Variable Portion Planned is:

$v= (\$-I-\{[Q+X]/I \cdot U\}/\{[1-t][1-d]\})/\{\$'[1+s]\}-f$

Rule-21388:

If both (I), (U), $(\$)$, $(\$')$, (s), (v), (Q), (I), (t), (d) and (X) are known, then its Fixed Portion Planned is:

$f= (\$-I-\{[Q+X]/I \cdot U\}/\{[1-t][1-d]\})/\{\$'[1+s]\}-v$

Rule-21389:

If both (I), (U), $(\$)$, $(\$')$, (s), (v), (f), (Q), (t), (d) and (X) are known, then its Interest Expense Planned is:

$I= \$-\$'[1+s][v+f]-\{[Q+X]/I \cdot U\}/\{[1-t][1-d]\}$

Rule-21390:

If both (I), (U), $(\$)$, $(\$')$, (s), (v), (f), (I), (Q), (d) and (X) are known, then its Tax Rate Planned is:

$t= 1-\{[Q+X]/I \cdot U\}/(\{\$-\$'[1+s][v+f]-I\}[1-d])$

Steve Asikin ISBN 14: 978-1515158691, ISBN 10: 1515158691

Rule-21391:

If both (Λ), (**U**), (**$**), (**$'**), (**s**), (**v**), (**f**), (**I**), (**t**), (**Q**) and (**X**) are known, then its Dividend Portion Planned is:

$$d= 1-\{[\mathbf{Q}+\mathbf{X}]/\mathbf{\Lambda U}\}/(\{\mathbf{\$}-\mathbf{\$'}[1+\mathbf{s}][\mathbf{v}+\mathbf{f}]-\mathbf{I}\}[1-\mathbf{t}])$$

Rule-21392:

If both (Λ), (**U**), (**$**), (**$'**), (**s**), (**v**), (**f**), (**I**), (**t**), (**d**) and (**Q**) are known, then its Xpress or Current Debt Planned is:

$$\mathbf{X}= \Lambda\mathbf{U}+\{\mathbf{\$}-\mathbf{\$'}[1+\mathbf{s}][\mathbf{v}+\mathbf{f}]-\mathbf{I}\}[1-\mathbf{t}][1-\mathbf{d}])-\mathbf{Q}$$

Rule-21393:

If both (Λ), (**U**), (**$**), (**$'**), (**s**), (**v**), (**f**), (**I**), (**t**), (**d**), (**P**), (**c**) and (**q**) are known, then its Quoted Longterm Debt Planned is:

$$\mathbf{Q}= \Lambda\mathbf{U}+\{\mathbf{\$}-\mathbf{\$'}[1+\mathbf{s}][\mathbf{v}+\mathbf{f}]-\mathbf{I}\}[1-\mathbf{t}][1-\mathbf{d}])-\mathbf{P}/[\mathbf{c}-\mathbf{q}]$$

Rule-21394:

If both (**Q**), (**U**), (**$**), (**$'**), (**s**), (**v**), (**f**), (**I**), (**t**), (**d**), (**P**), (**c**) and (**q**) are known, then its Leverage or Gearing Ratio Planned is:

$$\mathbf{\Lambda}= \{\mathbf{Q}+\mathbf{P}/[\mathbf{c}-\mathbf{q}]\}/(\mathbf{U}+\{\mathbf{\$}-\mathbf{\$'}[1+\mathbf{s}][\mathbf{v}+\mathbf{f}]-\mathbf{I}\}[1-\mathbf{t}][1-\mathbf{d}])$$

Rule-21395:

If both (Λ), (**Q**), (**$**), (**$'**), (**s**), (**v**), (**f**), (**I**), (**t**), (**d**), (**P**), (**c**) and (**q**) are known, then its Utilized or Starting Capital must be:

$$\mathbf{U}= \{\mathbf{Q}+\mathbf{P}/[\mathbf{c}-\mathbf{q}]\}/\mathbf{\Lambda}\{\mathbf{\$}-\mathbf{\$'}[1+\mathbf{s}][\mathbf{v}+\mathbf{f}]-\mathbf{I}\}[1-\mathbf{t}][1-\mathbf{d}]$$

Steve Asikin ISBN 14: 978-1515158691, ISBN 10: 1515158691

Rule-21396:

If both (f), (U), (Q), (S'), (s), (v), (f), (I), (t), (d), (P), (c) and (q) are known, then its Sales or Revenue Planned is:

$$S= S'[1+s][v+f]+I+(\{Q+P/[c-q]\}/f\cdot U)/\{[1-t][1-d]\}$$

Rule-21397:

If both (f), (U), (S), (Q), (s), (v), (f), (I), (t), (d), (P), (c) and (q) are known, then its Sales Past must be:

$$S'= [S-I-(\{Q+P/[c-q]\}/f\cdot U)/\{[1-t][1-d]\}]$$
$$/\{[1+s][v+f]\}$$

Rule-21398:

If both (f), (U), (S), (S'), (Q), (v), (f), (I), (t), (d), (P), (c) and (q) are known, then its sales Growth Planned is:

$$s= S/S'-1$$

or it can also be found as

$$s= [S-I-(\{Q+P/[c-q]\}/f\cdot U)/\{[1-t][1-d]\}]$$
$$/\{S'[v+f]\}-1$$

Rule-21399:

If both (f), (U), (S), (S'), (s), (Q), (f), (I), (t), (d), (P), (c) and (q) are known, then its Variable Portion Planned is:

$$v= [S-I-(\{Q+P/[c-q]\}/f\cdot U)/\{[1-t][1-d]\}]$$
$$/\{S'[1+s]\}-f$$

Steve Asikin ISBN 14: 978-1515158691, ISBN 10: 1515158691

Rule-21400:

If both (*l*), (**U**), (**$**), (**$'**), (**s**), (**v**), (**Q**), (**l**), (**t**), (**d**), (**P**), (**c**) and (**q**) are known, then its Fixed Portion Planned is:

$$f= [\$-I-(\{Q+P/[c-q]\}/\textit{l}U)$$
$$/\{[1-t][1-d]\}]/\{\$'[1+s]\}-v$$

Rule-21401:

If both (*l*), (**U**), (**$**), (**$'**), (**s**), (**v**), (**f**), (**Q**), (**t**), (**d**), (**P**), (**c**) and (**q**) are known, then its Interest Expense Planned is:

$$I= \$-\$'[1+s][v+f]-(\{Q+P/[c-q]\}/\textit{l}U)/\{[1-t][1-d]\}$$

Rule-21402:

If both (*l*), (**U**), (**$**), (**$'**), (**s**), (**v**), (**f**), (**l**), (**Q**), (**d**), (**P**), (**c**) and (**q**) are known, then its Tax Rate Planned is:

$$t= 1-(\{Q+P/[c-q]\}/\textit{l}U)/(\{\$-\$'[1+s][v+f]-I\}[1-d])$$

Rule-21403:

If both (*l*), (**U**), (**$**), (**$'**), (**s**), (**v**), (**f**), (**l**), (**t**), (**Q**), (**P**), (**c**) and (**q**) are known, then its Dividend Payout Planned is:

$$d= 1-(\{Q+P/[c-q]\}/\textit{l}U)/(\{\$-\$'[1+s][v+f]-I\}[1-t])$$

Rule-21404:

If both (*l*), (**U**), (**$**), (**$'**), (**s**), (**v**), (**f**), (**l**), (**t**), (**d**), (**Q**), (**c**) and (**q**) are known, then its procured Inventory Planned is:

$$P= [c-q][\textit{l}U+\{\$-\$'[1+s][v+f]-I\}[1-t][1-d])-Q]$$

Steve Asikin ISBN 14: 978-1515158691, ISBN 10: 1515158691

Rule-21405:

If both $(\textit{I})$, $(\textbf{U})$, $(\textbf{\$})$, $(\textbf{\$'})$, $(\textbf{s})$, $(\textbf{v})$, $(\textbf{f})$, $(\textbf{I})$, $(\textbf{t})$, $(\textbf{d})$, $(\textbf{P})$, $(\textbf{Q})$ and $(\textbf{q})$ are known, then its Current Ratio Planned is:

$$c = q + P/[\textit{I}(U + \{\$-\$'[1+s][v+f]-I\}[1-t][1-d]) - Q]$$

Rule-21406:

If both $(\textit{I})$, $(\textbf{U})$, $(\textbf{\$})$, $(\textbf{\$'})$, $(\textbf{s})$, $(\textbf{v})$, $(\textbf{f})$, $(\textbf{I})$, $(\textbf{t})$, $(\textbf{d})$, $(\textbf{P})$, $(\textbf{c})$ and $(\textbf{Q})$ are known, then itsQuick or Acid test Ratio Planned is:

$$q = c - P/[\textit{I}(U + \{\$-\$'[1+s][v+f]-I\}[1-t][1-d]) - Q]$$

Rule-21407:

If both $(\textit{I})$, $(\textbf{U})$, $(\textbf{\$})$, $(\textbf{\$'})$, $(\textbf{s})$, $(\textbf{v})$, $(\textbf{f})$, $(\textbf{I})$, $(\textbf{t})$, $(\textbf{d})$, $(\textbf{V})$, $(\textbf{p})$, $(\textbf{c})$ and $(\textbf{q})$ are known, then its Quoted Longterm Debt Planned is:

$$Q = \textit{I}(U + \{\$-\$'[1+s][v+f]-I\}[1-t][1-d])$$
$$-Vp/\{360[c-q]\}$$

Rule-21408:

If both $(\textbf{Q})$, $(\textbf{U})$, $(\textbf{\$})$, $(\textbf{\$'})$, $(\textbf{s})$, $(\textbf{v})$, $(\textbf{f})$, $(\textbf{I})$, $(\textbf{t})$, $(\textbf{d})$, $(\textbf{V})$, $(\textbf{p})$, $(\textbf{c})$ and $(\textbf{q})$ are known, then its Leverage or Gearing Ratio Planned is:

$$\textit{I} = (Q + Vp/\{360[c-q]\})$$
$$/(U + \{\$-\$'[1+s][v+f]-I\}[1-t][1-d])$$

Steve Asikin ISBN 14: 978-1515158691, ISBN 10: 1515158691

Rule-21409:

If both (I), (Q), (S), (S'), (s), (v), (f), (I), (t), (d), (V), (p), (c) and (q) are known, then its Utilized or Starting Capital must be:

$$U = (Q + Vp / \{360[c-q]\}) / I$$
$$- \{S - S'[1+s][v+f] - I\}[1-t][1-d]$$

Rule-21410:

If both (I), (U), (Q), (S'), (s), (v), (f), (I), (t), (d), (V), (p), (c) and (q) are known, then its Sales or Revenue Planned is:

$$S = S'[1+s][v+f] + I + [(Q+Vp/\{360[c-q]\})/I U]$$
$$/\{[1-t][1-d]\}$$

Rule-21411:

If both (I), (U), (S), (Q), (s), (v), (f), (I), (t), (d), (V), (p), (c) and (q) are known, then its Sales Past must be:

$$S' = \{S - I - [(Q+Vp/\{360[c-q]\})/I U]/\{[1-t][1-d]\}\}$$
$$/\{[1+s][v+f]\}$$

Rule-21412:

If both (I), (U), (S), (S'), (Q), (v), (f), (I), (t), (d), (V), (p), (c) and (q) are known, then its Sales Growth Planned is:

$$s = S/S' - 1$$

or it can also be found as

$$s = \{S - I - [(Q+Vp/\{360[c-q]\})/I U]/\{[1-t][1-d]\}\}$$
$$/\{S'[v+f]\} - 1$$

Steve Asikin ISBN 14: 978-1515158691, ISBN 10: 1515158691

Rule-21413:

If both (I), (U), (S), (S'), (s), (Q), (f), (I), (t), (d), (V), (p), (c) and (q) are known, then its Variable Portion Planned is:

$$v= \{S\text{-}I\text{-}[(Q+Vp/\{360[c\text{-}q]\})/I\text{-}U]/\{[1\text{-}t][1\text{-}d]\}\} /\{S'[1+s]\}\text{-}f$$

Rule-21414:

If both (I), (U), (S), (S'), (s), (v), (Q), (I), (t), (d), (V), (p), (c) and (q) are known, then its Fixed Portion Planned is:

$$f= \{S\text{-}I\text{-}[(Q+Vp/\{360[c\text{-}q]\})/I\text{-}U] /\{[1\text{-}t][1\text{-}d]\}\}/\{S'[1+s]\}\text{-}v$$

Rule-21415:

If both (I), (U), (S), (S'), (s), (v), (f), (Q), (t), (d), (V), (p), (c) and (q) are known, then its Interest Expense Planned is:

$$I= S\text{-}S'[1+s][v+f]\text{-}[(Q+Vp/\{360[c\text{-}q]\})/I\text{-}U] /\{[1\text{-}t][1\text{-}d]\}$$

Rule-21416:

If both (I), (U), (S), (S'), (s), (v), (f), (I), (Q), (d), (V), (p), (c) and (q) are known, then its Tax Rate Planned is:

$$t= 1\text{-}[(Q+Vp/\{360[c\text{-}q]\})/I\text{-}U] /(\{S\text{-}S'[1+s][v+f]\text{-}I\}[1\text{-}d])$$

Steve Asikin ISBN 14: 978-1515158691, ISBN 10: 1515158691

Rule-21417:

If both (l), $(\mathbf{U})$, $(\mathbf{S})$, $(\mathbf{S'})$, $(\mathbf{s})$, $(\mathbf{v})$, $(\mathbf{f})$, $(\mathbf{I})$, $(\mathbf{t})$, $(\mathbf{Q})$, $(\mathbf{V})$, $(\mathbf{p})$, $(\mathbf{c})$ and $(\mathbf{q})$ are known, then its Dividend Payout Planned is:

$$d = 1 - [(\mathbf{Q} + \mathbf{Vp}/\{360[\mathbf{c} - \mathbf{q}]\}) / \mathit{l}\mathbf{U}]$$
$$/(\{\mathbf{S} - \mathbf{S'}[1+\mathbf{s}][\mathbf{v}+\mathbf{f}] - \mathbf{I}\}[1-\mathbf{t}])$$

Rule-21418:

If both (l), $(\mathbf{U})$, $(\mathbf{S})$, $(\mathbf{S'})$, $(\mathbf{s})$, $(\mathbf{v})$, $(\mathbf{f})$, $(\mathbf{I})$, $(\mathbf{t})$, $(\mathbf{d})$, $(\mathbf{Q})$, $(\mathbf{p})$, $(\mathbf{c})$ and $(\mathbf{q})$ are known, then its Variable Cost Planned is:

$$V = 360[\mathbf{c} - \mathbf{q}][\mathit{l}\mathbf{U} + \{\mathbf{S} - \mathbf{S'}[1+\mathbf{s}][\mathbf{v}+\mathbf{f}] - \mathbf{I}\}$$
$$[1-\mathbf{t}][1-\mathbf{d}]) - \mathbf{Q}]/\mathbf{p}$$

Rule-21419:

If both (l), $(\mathbf{U})$, $(\mathbf{S})$, $(\mathbf{S'})$, $(\mathbf{s})$, $(\mathbf{v})$, $(\mathbf{f})$, $(\mathbf{I})$, $(\mathbf{t})$, $(\mathbf{d})$, $(\mathbf{V})$, $(\mathbf{Q})$, $(\mathbf{c})$ and $(\mathbf{q})$ are known, then its Procured Inventory Days Planned:

$$p = 360[\mathbf{c} - \mathbf{q}][\mathit{l}\mathbf{U} + \{\mathbf{S} - \mathbf{S'}[1+\mathbf{s}][\mathbf{v}+\mathbf{f}] - \mathbf{I}\}$$
$$[1-\mathbf{t}][1-\mathbf{d}]) - \mathbf{Q}]/\mathbf{V}$$

Rule-21420:

If both (l), $(\mathbf{U})$, $(\mathbf{S})$, $(\mathbf{S'})$, $(\mathbf{s})$, $(\mathbf{v})$, $(\mathbf{f})$, $(\mathbf{I})$, $(\mathbf{t})$, $(\mathbf{d})$, $(\mathbf{V})$, $(\mathbf{p})$, $(\mathbf{Q})$ and $(\mathbf{q})$ are known, then its Current Ratio Planned is:

$$c = \mathbf{q} + \mathbf{Vp}/\{360[\mathit{l}\mathbf{U} + \{\mathbf{S} - \mathbf{S'}[1+\mathbf{s}][\mathbf{v}+\mathbf{f}] - \mathbf{I}\}$$
$$[1-\mathbf{t}][1-\mathbf{d}]) - \mathbf{Q}]\}$$

Steve Asikin ISBN 14: 978-1515158691, ISBN 10: 1515158691

Rule-21421:

If both (I), (U), (S), (S'), (s), (v), (f), (I), (t), (d), (V), (p), (c) and (Q) are known, then its Quick or Acid Test Ratio Planned is:

$$q = c - Vp / \{360[I(U + \{S - S'[1+s][v+f] - I\} \\ [1-t][1-d]) - Q]\}$$

Rule-21422:

If both (I), (U), (S), (S'), (s), (v), (f), (I), (t), (d), (p), (c) and (q) are known, then its Quoted Longterm Debt Planned is:

$$Q = I(U + \{S - S'[1+s][v+f] - I\}[1-t][1-d]) \\ - Svp / \{360[c-q]\}$$

Rule-21423:

If both (Q), (U), (S), (S'), (s), (v), (f), (I), (t), (d), (p), (c) and (q) are known, then its Leverage or Gearing Ratio Planned is:

$$I = (Q + Svp / \{360[c-q]\}) \\ / (U + \{S - S'[1+s][v+f] - I\}[1-t][1-d])$$

Rule-21424:

If both (I), (Q), (S), (S'), (s), (v), (f), (I), (t), (d), (p), (c) and (q) are known, then its Utilized or Starting Capital must be:

$$U = (Q + Svp / \{360[c-q]\}) / I \\ - \{S - S'[1+s][v+f] - I\}[1-t][1-d]$$

Steve Asikin ISBN 14: 978-1515158691, ISBN 10: 1515158691

Rule-21425:

If both $(/)$, (U), (Q), (S'), (s), (v), (f), (I), (t), (d), (p), (c) and (q) are known, then its Sales or Revenue Planned is:

$$S= [/(U-\{S'[1+s][v+f]+I\}[1-t][1-d])-Q]$$
$$/(vp/\{360[c-q]\}-/[1-t][1-d])$$

Rule-21426:

If both $(/)$, (U), (S), (Q), (s), (v), (f), (I), (t), (d), (p), (c) and (q) are known, then its Sales Past must be:

$$S'= \{S-I-[(Q+Svp/\{360[c-q]\})//U]/\{[1-t][1-d]\}\}$$
$$/\{[1+s][v+f]\}$$

Rule-21427:

If both $(/)$, (U), (S), (S'), (Q), (v), (f), (I), (t), (d), (p), (c) and (q) are known, then its Sales Growth Planned is:

$$s= S/S'-1$$

or it can also be found as

$$s= \{S-I-[(Q+Svp/\{360[c-q]\})//U]/\{[1-t][1-d]\}\}$$
$$/\{S'[v+f]\}-1$$

Rule-21428:

If both $(/)$, (U), (S), (S'), (s), (Q), (f), (I), (t), (d), (p), (c) and (q) are known, then its Variable Portion Planned is:

$$v= [/(U+\{S-S'f[1+s]-I\}[1-t][1-d])-Q]$$
$$/(Sp/\{360[c-q]\}+S'f/[1+s] [1-t][1-d])$$

Steve Asikin ISBN 14: 978-1515158691, ISBN 10: 1515158691

Rule-21429:

If both (I), (**U**), (**$**), (**$'**), (**s**), (**v**), (**Q**), (**I**), (**t**), (**d**), (**p**), (**c**) and (**q**) are known, then its Fixed Portion Planned is:

$$f = \{\$-I-[(Q+\$vp/\{360[c-q]\})/I\cdot U]/\{[1-t][1-d]\}\}$$
$$/\{\$'[1+s]\}-v$$

Rule-21430:

If both (I), (**U**), (**$**), (**$'**), (**s**), (**v**), (**f**), (**Q**), (**t**), (**d**), (**p**), (**c**) and (**q**) are known, then its Interest Expense Planned is:

$$I = \$-\$'[1+s][v+f]-[(Q+\$vp/\{360[c-q]\})/I\cdot U]$$
$$/\{[1-t][1-d]\}$$

Rule-21431:

If both (I), (**U**), (**$**), (**$'**), (**s**), (**v**), (**f**), (**I**), (**Q**), (**d**), (**p**), (**c**) and (**q**) are known, then its Tax Rate Planned is:

$$t = 1-[(Q+\$vp/\{360[c-q]\})/I\cdot U]$$
$$/(\{\$-\$'[1+s][v+f]-I\}[1-d])$$

Rule-21432:

If both (I), (**U**), (**$**), (**$'**), (**s**), (**v**), (**f**), (**I**), (**t**), (**Q**), (**p**), (**c**) and (**q**) are known, then its Dividend Payout Planned is:

$$d = 1-[(Q+\$vp/\{360[c-q]\})/I\cdot U]$$
$$/(\{\$-\$'[1+s][v+f]-I\}[1-t])$$

Steve Asikin ISBN 14: 978-1515158691, ISBN 10: 1515158691

Rule-21433:

If both ($\mathbf{I}$), (**U**), (**$**), (**$'**), (**s**), (**v**), (**f**), (**I**), (**t**), (**d**), (**Q**), (**c**) and (**q**) are known, then its Procured Inventory Days Planned:

$$p= 360[c\text{-}q][(\mathbf{I}U+\{\$\text{-}\$'[1+s][v+f]\text{-}I\}[1\text{-}t][1\text{-}d])\text{-}Q]$$
$$/[\$v]$$

Rule-21434:

If both ($\mathbf{I}$), (**U**), (**$**), (**$'**), (**s**), (**v**), (**f**), (**I**), (**t**), (**d**), (**p**), (**Q**) and (**q**) are known, then its Current Ratio Planned is:

$$c= q+\$vp/\{360[(\mathbf{I}U+\{\$\text{-}\$'[1+s][v+f]\text{-}I\}$$
$$[1\text{-}t][1\text{-}d])\text{-}Q]\}$$

Rule-21435:

If both ($\mathbf{I}$), (**U**), (**$**), (**$'**), (**s**), (**v**), (**f**), (**I**), (**t**), (**d**), (**p**), (**c**) and (**Q**) are known, then its Quick or Acid test Ratio Planned is:

$$q= c\text{-}\$vp/\{360[(\mathbf{I}U+\{\$\text{-}\$'[1+s][v+f]\text{-}I\}$$
$$[1\text{-}t][1\text{-}d])\text{-}Q]\}$$

Rule-21436:

If both ($\mathbf{I}$), (**U**), (**$**), (**$'**), (**s**), (**v**), (**f**), (**I**), (**t**), (**d**), (**p**), (**c**) and (**q**) are known, then its Quoted Longterm Debt Planned is:

$$Q= (\mathbf{I}U+\{\$\text{-}\$'[1+s][v+f]\text{-}I\}$$
$$[1\text{-}t][1\text{-}d])\text{-}\$'vp[1+s]/\{360[c\text{-}q]\}$$

Steve Asikin ISBN 14: 978-1515158691, ISBN 10: 1515158691

Rule-21437:

If both $(\mathbf{Q})$, $(\mathbf{U})$, $(\mathbf{S})$, $(\mathbf{S'})$, $(\mathbf{s})$, $(\mathbf{v})$, $(\mathbf{f})$, $(\mathbf{I})$, $(\mathbf{t})$, $(\mathbf{d})$, $(\mathbf{p})$, $(\mathbf{c})$ and $(\mathbf{q})$ are known, then its Leverage or Gearing Ratio Planned is:

$$\mathbf{\mathit{I}} = (\mathbf{Q} + \mathbf{S'vp}[1+\mathbf{s}]/\{360[\mathbf{c} - \mathbf{q}]\})$$
$$/(\mathbf{U} + \{\mathbf{S} - \mathbf{S'}[1+\mathbf{s}][\mathbf{v} + \mathbf{f}] - \mathbf{I}\}[1-\mathbf{t}][1-\mathbf{d}])$$

Rule-21438:

If both $(\mathbf{\mathit{I}})$, $(\mathbf{Q})$, $(\mathbf{S})$, $(\mathbf{S'})$, $(\mathbf{s})$, $(\mathbf{v})$, $(\mathbf{f})$, $(\mathbf{I})$, $(\mathbf{t})$, $(\mathbf{d})$, $(\mathbf{p})$, $(\mathbf{c})$ and $(\mathbf{q})$ are known, then its Utilized or Starting Capital must be:

$$\mathbf{U} = (\mathbf{Q} + \mathbf{S'vp}[1+\mathbf{s}]/\{360[\mathbf{c} - \mathbf{q}]\})/\mathbf{\mathit{I}}$$
$$- \{\mathbf{S} - \mathbf{S'}[1+\mathbf{s}][\mathbf{v} + \mathbf{f}] - \mathbf{I}\}[1-\mathbf{t}][1-\mathbf{d}]$$

Rule-21439:

If both $(\mathbf{\mathit{I}})$, $(\mathbf{U})$, $(\mathbf{Q})$, $(\mathbf{S'})$, $(\mathbf{s})$, $(\mathbf{v})$, $(\mathbf{f})$, $(\mathbf{I})$, $(\mathbf{t})$, $(\mathbf{d})$, $(\mathbf{p})$, $(\mathbf{c})$ and $(\mathbf{q})$ are known, then its Sales or Revenue Planned is:

$$\mathbf{S} = [\mathbf{Q} - \mathbf{\mathit{I}}(\mathbf{U} - \{\mathbf{S'}[1+\mathbf{s}][\mathbf{v} + \mathbf{f}] + \mathbf{I}\}[1-\mathbf{t}][1-\mathbf{d}])$$
$$+ \mathbf{S'vp}[1+\mathbf{s}]/\{360[\mathbf{c} - \mathbf{q}]\}]/\{\mathbf{\mathit{I}}[1-\mathbf{t}][1-\mathbf{d}]\}$$

Rule-21440:

If both $(\mathbf{\mathit{I}})$, $(\mathbf{U})$, $(\mathbf{S})$, $(\mathbf{Q})$, $(\mathbf{s})$, $(\mathbf{v})$, $(\mathbf{f})$, $(\mathbf{I})$, $(\mathbf{t})$, $(\mathbf{d})$, $(\mathbf{p})$, $(\mathbf{c})$ and $(\mathbf{q})$ are known, then its Sales Past must be:

$$\mathbf{S'} = [\mathbf{\mathit{I}}(\mathbf{U} + \{\mathbf{S} - \mathbf{S'f}[1+\mathbf{s}] - \mathbf{I}\}[1-\mathbf{t}][1-\mathbf{d}]) - \mathbf{Q}]$$
$$/[\mathbf{v}[1+\mathbf{s}](\mathbf{p}/\{360[\mathbf{c} - \mathbf{q}]\} + \mathbf{f}\mathbf{\mathit{I}}/[1-\mathbf{t}][1-\mathbf{d}])]$$

Steve Asikin ISBN 14: 978-1515158691, ISBN 10: 1515158691

Rule-21441:

If both (**/**), (**U**), (**S**), (**S'**), (**Q**), (**v**), (**f**), (**I**), (**t**), (**d**), (**p**), (**c**) and (**q**) are known, then its Sales Growth Planned is:

$$s = S/S' - 1$$

or it can also be found

$$s = [\mathit{/}(U + \{S - S'f[1+s] - I\}[1-t][1-d]) - Q]$$
$$/[S'v(p/\{360[c-q]\} + f\mathit{/}[1-t][1-d])] - 1$$

Rule-21442:

If both (**/**), (**U**), (**S**), (**S'**), (**s**), (**Q**), (**f**), (**I**), (**t**), (**d**), (**p**), (**c**) and (**q**) are known, then its Variable Portion Planned is:

$$v = [\mathit{/}(U + \{S - S'f[1+s] - I\}[1-t][1-d]) - Q]$$
$$/[S'[1+s](p/\{360[c-q]\} + f\mathit{/}[1-t][1-d])]$$

Rule-21443:

If both (**/**), (**U**), (**S**), (**S'**), (**s**), (**v**), (**Q**), (**I**), (**t**), (**d**), (**p**), (**c**) and (**q**) are known, then its Fixed Portion Planned is:

$$f = \{S - I - [(Q + S'vp[1+s]/\{360[c-q]\})/\mathit{/}U]$$
$$/\{[1-t][1-d]\}\}/\{S'[1+s]\} - v$$

Rule-21444:

If both (**/**), (**U**), (**S**), (**S'**), (**s**), (**v**), (**f**), (**Q**), (**t**), (**d**), (**p**), (**c**) and (**q**) are known, then its Interest Expense Planned is:

$$I = S - S'[1+s][v+f] - [(Q + S'vp[1+s]/\{360[c-q]\})/\mathit{/}U]$$
$$/\{[1-t][1-d]\}$$

Steve Asikin ISBN 14: 978-1515158691, ISBN 10: 1515158691

Rule-21445:

If both (I), (U), (S), (S'), (s), (v), (f), (I), (Q), (d), (p), (c) and (q) are known, then its Tax Rate Planned is:

$$t = 1-[(Q+S'vp[1+s]/\{360[c-q]\})/I\cdot U]$$
$$/(\{S-S'[1+s][v+f]-I\}[1-d])$$

Rule-21446:

If both (I), (U), (S), (S'), (s), (v), (f), (I), (t), (Q), (p), (c) and (q) are known, then its Dividend Payout Planned is:

$$d = 1-[(Q+S'vp[1+s]/\{360[c-q]\})/I\cdot U]$$
$$/(\{S-S'[1+s][v+f]-I\}[1-t])$$

Rule-21447:

If both (I), (U), (S), (S'), (s), (v), (f), (I), (t), (d), (Q), (c) and (q) are known, then its Procured Inventory Days Planned:

$$p = 360[c-q][I\cdot U+\{S-S'[1+s][v+f]-I\}$$
$$[1-t][1-d])-Q]/\{S'v[1+s]\}$$

Rule-21448:

If both (I), (U), (S), (S'), (s), (v), (f), (I), (t), (d), (p), (Q) and (q) are known, then its Current Ratio Planned is:

$$c = q+S'vp[1+s]/\{360[I\cdot U+\{S-S'[1+s][v+f]-I\}$$
$$[1-t][1-d])-Q]\}$$

Steve Asikin ISBN 14: 978-1515158691, ISBN 10: 1515158691

Rule-21449:

If both (I), (U), (S), (S'), (s), (v), (f), (I), (t), (d), (p), (c) and (q) are known, then its Quoted Longterm Debt Planned is:

$$q= c\text{-}S'vp[1+s]/\{360[I(U+\{S\text{-}S'[1+s][v+f]\text{-}I\}$$
$$[1\text{-}t][1\text{-}d])\text{-}Q]\}$$

Rule-21450:

If both (I), (U), (S), (i), (S'), (s), (v), (f), (T), (D) and (X) are known, then its Quoted Longterm Debt Planned is:

$$Q= I\{U+S[1\text{-}i]\text{-}S'[1+s][v+f]\text{-}T\text{-}D\}\text{-}X$$

Rule-21451:

If both (Q), (U), (S), (i), (S'), (s), (v), (f), (T), (D) and (X) are known, then its Leverage or Gearing Ratio Planned is:

$$I= [Q+X]/\{U+S[1\text{-}i]\text{-}S'[1+s][v+f]\text{-}T\text{-}D\}$$

Rule-21452:

If both (I), (Q), (S), (i), (S'), (s), (v), (f), (T), (D) and (X) are known, then its Utilized or Starting Capital must be:

$$U= S'[1+s][v+f]+T+D+[Q+X]/I\text{-}S[1\text{-}i]$$

Rule-21453:

If both (I), (U), (Q), (i), (S'), (s), (v), (f), (T), (D) and (X) are known, then its Sales or Revenue Planned is:

$$S= \{S'[1+s][v+f]+T+D+[Q+X]/I\text{-}U\}/[1\text{-}i]$$

Steve Asikin ISBN 14: 978-1515158691, ISBN 10: 1515158691

Rule-21454:
   If both (I), (**U**), (**S**), (**Q**), (**S'**), (**s**), (**v**), (**f**), (**T**), (**D**) and
   (**X**) are known, then its Interest Portion Planned is:
   $$i = 1 - \{S'[1+s][v+f] + T + D + [Q+X]/I\cdot U\}/S$$

Rule-21455:
   If both (I), (**U**), (**S**), (**i**), (**Q**), (**s**), (**v**), (**f**), (**T**), (**D**) and
   (**X**) are known, then its Sales Past must be:
   $$S' = \{U + S[1-i] - T - D - [Q+X]/I\}/\{[1+s][v+f]\}$$

Rule-21456:
   If both (I), (**U**), (**S**), (**i**), (**S'**), (**Q**), (**v**), (**f**), (**T**), (**D**) and
   (**X**) are known, then its Sales Growth Planned is:
   $$s = S/S' - 1$$

   or it can also be found as
   $$s = \{U + S[1-i] - T - D - [Q+X]/I\}/\{S'[v+f]\} - 1$$

Rule-21457:
   If both (I), (**U**), (**S**), (**i**), (**S'**), (**s**), (**Q**), (**f**), (**T**), (**D**) and
   (**X**) are known, then its Variable Portion Planned is:
   $$v = \{U + S[1-i] - T - D - [Q+X]/I\}/\{S'[1+s]\} - f$$

Rule-21458:
   If both (I), (**U**), (**S**), (**i**), (**S'**), (**s**), (**v**), (**Q**), (**T**), (**D**) and
   (**X**) are known, then its Fixed Portion Planned is:
   $$f = \{U + S[1-i] - T - D - [Q+X]/I\}/\{S'[1+s]\} - v$$

Rule-21459:
   If both (I), (**U**), (**S**), (**i**), (**S'**), (**s**), (**v**), (**f**), (**Q**), (**D**) and
   (**X**) are known, then its Tax Planned is:
   $$T = U + S[1-i] - S'[1+s][v+f] - D - [Q+X]/I$$

Steve Asikin ISBN 14: 978-1515158691, ISBN 10: 1515158691

Rule-21460:

If both (*I*), (**U**), (**$**), (**i**), (**$'**), (**s**), (**v**), (**f**), (**T**), (**Q**) and (**X**) are known, then its Dividend Planned is:

$$D= U+\$[1-i]-\$'[1+s][v+f]-T-[Q+X]/I$$

Rule-21461:

If both (*I*), (**U**), (**$**), (**i**), (**$'**), (**s**), (**v**), (**f**), (**T**), (**D**) and (**Q**) are known, then its Xpress or Current Debt Planned is:

$$X= I\{U+\$[1-i]-\$'[1+s][v+f]-T-D\}-Q$$

Rule-21462:

If both (*I*), (**U**), (**$**), (**i**), (**$'**), (**s**), (**v**), (**f**), (**T**), (**D**), (**P**), (**c**) and (**q**) are known, then its Quoted Longterm Debt Planned is:

$$Q= I\{U+\$[1-i]-\$'[1+s][v+f]-T-D\}-P/[c-q]$$

Rule-21463:

If both (**Q**), (**U**), (**$**), (**i**), (**$'**), (**s**), (**v**), (**f**), (**T**), (**D**), (**P**), (**c**) and (**q**) are known, then its Leverage or Gearing Ratio Planned is:

$$I= \{Q+P/[c-q]\}/\{U+\$[1-i]-\$'[1+s][v+f]-T-D\}$$

Rule-21464:

If both (*I*), (**Q**), (**$**), (**i**), (**$'**), (**s**), (**v**), (**f**), (**T**), (**D**), (**P**), (**c**) and (**q**) are known, then its Utilized or Starting Capital must be:

$$U= \$'[1+s][v+f]+T+D+\{Q+P/[c-q]\}/I-\$[1-i]$$

Steve Asikin ISBN 14: 978-1515158691, ISBN 10: 1515158691

Rule-21465:

If both (I), (U), (Q), (i), (S'), (s), (v), (f), (T), (D), (P), (c) and (q) are known, then its Sales or Revenue Planned is:

$$S = (S'[1+s][v+f]+T+D+\{Q+P/[c-q]\}/I \cdot U)/[1-i]$$

Rule-21466:

If both (I), (U), (S), (Q), (S'), (s), (v), (f), (T), (D), (P), (c) and (q) are known, then its Interest Portion Planned is:

$$i = 1-(S'[1+s][v+f]+T+D+\{Q+P/[c-q]\}/I \cdot U)/S$$

Rule-21467:

If both (I), (U), (S), (i), (Q), (s), (v), (f), (T), (D), (P), (c) and (q) are known, then its Sales Past must be:

$$S' = (U+S[1-i]-T-D-\{Q+P/[c-q]\}/I)/\{[1+s][v+f]\}$$

Rule-21468:

If both (I), (U), (S), (i), (S'), (Q), (v), (f), (T), (D), (P), (c) and (q) are known, then its Sales Growth Planned is:

$$s = S/S'-1$$

or it can also be found as

$$s = (U+S[1-i]-T-D-\{Q+P/[c-q]\}/I)/\{S'[v+f]\}-1$$

Rule-21469:

If both (I), (U), (S), (i), (S'), (s), (Q), (f), (T), (D), (P), (c) and (q) are known, then its Variable Portion Planned is:

$$v = (U+S[1-i]-T-D-\{Q+P/[c-q]\}/I)/\{S'[1+s]\}-f$$

Steve Asikin ISBN 14: 978-1515158691, ISBN 10: 1515158691

Rule-21470:

If both (**/**), (**U**), (**\$**), (**i**), (**\$'**), (**s**), (**v**), (**Q**), (**T**), (**D**), (**P**), (**c**) and (**q**) are known, then its Fixed Portion Planned is:

$$f= (U+\$[1-i]-T-D-\{Q+P/[c-q]\}//)/\{\$'[1+s]\}-v$$

Rule-21471:

If both (**/**), (**U**), (**\$**), (**i**), (**\$'**), (**s**), (**v**), (**f**), (**Q**), (**D**), (**P**), (**c**) and (**q**) are known, then its Tax Planned is:

$$T= U+\$[1-i]-\$'[1+s][v+f]-D-\{Q+P/[c-q]\}//$$

Rule-21472:

If both (**/**), (**U**), (**\$**), (**i**), (**\$'**), (**s**), (**v**), (**f**), (**T**), (**Q**), (**P**), (**c**) and (**q**) are known, then its Dividend Planned is:

$$D= U+\$[1-i]-\$'[1+s][v+f]-T-\{Q+P/[c-q]\}//$$

Rule-21473:

If both (**/**), (**U**), (**\$**), (**i**), (**\$'**), (**s**), (**v**), (**f**), (**T**), (**D**), (**Q**), (**c**) and (**q**) are known, then its Procured Inventory Planned is:

$$P= [c-q](/\{U+\$[1-i]-\$'[1+s][v+f]-T-D\}-Q)$$

Rule-21474:

If both (**/**), (**U**), (**\$**), (**i**), (**\$'**), (**s**), (**v**), (**f**), (**T**), (**D**), (**P**), (**Q**) and (**q**) are known, then its Current Ratio Planned is:

$$c= q+P/(/\{U+\$[1-i]-\$'[1+s][v+f]-T-D\}-Q)$$

Steve Asikin ISBN 14: 978-1515158691, ISBN 10: 1515158691

Rule-21475:

If both (I), (U), (S), (i), (S'), (s), (v), (f), (T), (D), (P), (c) and (Q) are known, then its Quick or Acid test Ratio Planned is:

$$q= c\text{-}P/(I\{U+S[1\text{-}i]\text{-}S'[1+s][v+f]\text{-}T\text{-}D\}\text{-}Q)$$

Rule-21476:

If both (I), (U), (S), (i), (S'), (s), (v), (f), (T), (D), (V), (p), (c) and (q) are known, then its Quoted Longterm Debt Planned is:

$$Q= I\{U+S[1\text{-}i]\text{-}S'[1+s][v+f]\text{-}T\text{-}D\}\text{-}Vp/\{360[c\text{-}q]\}$$

Rule-21477:

If both (Q), (U), (S), (i), (S'), (s), (v), (f), (T), (D), (V), (p), (c) and (q) are known, then its Leverage or Gearing Ratio Planned is:

$$I= (Q+Vp/\{360[c\text{-}q]\})$$
$$/\{U+S[1\text{-}i]\text{-}S'[1+s][v+f]\text{-}T\text{-}D\}$$

Rule-21478:

If both (I), (Q), (S), (i), (S'), (s), (v), (f), (T), (D), (V), (p), (c) and (q) are known, then its Utilized or Starting Capital must be:

$$U= S'[1+s][v+f]+T+D+(Q+Vp/\{360[c\text{-}q]\})/I\text{-}S[1\text{-}i]$$

Rule-21479:

If both (I), (U), (Q), (i), (S'), (s), (v), (f), (T), (D), (V), (p), (c) and (q) are known, then its Sales or Revenue Planned is:

$$S= [S'[1+s][v+f]+T+D+(Q+Vp/\{360[c\text{-}q]\})/I\text{-}U]$$
$$/[1\text{-}i]$$

Steve Asikin ISBN 14: 978-1515158691, ISBN 10: 1515158691

Rule-21480:

If both $(\textbf{\textit{l}})$, $(\textbf{U})$, $(\textbf{\$})$, $(\textbf{Q})$, $(\textbf{\$'})$, $(\textbf{s})$, $(\textbf{v})$, $(\textbf{f})$, $(\textbf{T})$, $(\textbf{D})$, $(\textbf{V})$, $(\textbf{p})$, $(\textbf{c})$ and $(\textbf{q})$ are known, then its Interest Portion Planned is:

$$i= 1-[\$'[1+s][v+f]+T+D$$
$$+(Q+Vp/\{360[c-q]\})/\textit{l}U]/\$$$

Rule-21481:

If both $(\textbf{\textit{l}})$, $(\textbf{U})$, $(\textbf{\$})$, $(\textbf{i})$, $(\textbf{Q})$, $(\textbf{s})$, $(\textbf{v})$, $(\textbf{f})$, $(\textbf{T})$, $(\textbf{D})$, $(\textbf{V})$, $(\textbf{p})$, $(\textbf{c})$ and $(\textbf{q})$ are known, then its Sales Past must be:

$$\$'= [U+\$[1-i]-T-D-(Q+Vp/\{360[c-q]\})/\textit{l}]$$
$$/\{[1+s][v+f]\}$$

Rule-21482:

If both $(\textbf{\textit{l}})$, $(\textbf{U})$, $(\textbf{\$})$, $(\textbf{i})$, $(\textbf{\$'})$, $(\textbf{Q})$, $(\textbf{v})$, $(\textbf{f})$, $(\textbf{T})$, $(\textbf{D})$, $(\textbf{V})$, $(\textbf{p})$, $(\textbf{c})$ and $(\textbf{q})$ are known, then its Sales Growth Planned is:

$$s= \$/\$'-1$$

or it can also be found as

$$s= [U+\$[1-i]-T-D-(Q+Vp/\{360[c-q]\})/\textit{l}]$$
$$/\{\$'[v+f]\}-1$$

Rule-21483:

If both $(\textbf{\textit{l}})$, $(\textbf{U})$, $(\textbf{\$})$, $(\textbf{i})$, $(\textbf{\$'})$, $(\textbf{s})$, $(\textbf{Q})$, $(\textbf{f})$, $(\textbf{T})$, $(\textbf{D})$, $(\textbf{V})$, $(\textbf{p})$, $(\textbf{c})$ and $(\textbf{q})$ are known, then its Variable Portion Planned is:

$$v= V/\$$$

or it can also be found as

$$v= [U+\$[1-i]-T-D-(Q+Vp/\{360[c-q]\})/\textit{l}]$$
$$/\{\$'[1+s]\}-f$$

Steve Asikin ISBN 14: 978-1515158691, ISBN 10: 1515158691

Rule-21484:

If both (**ʃ**), (**U**), (**$**), (**i**), (**$'**), (**s**), (**v**), (**Q**), (**T**), (**D**), (**V**), (**p**), (**c**) and (**q**) are known, then its Fixed Portion Planned is:

$$f= [U+\$[1-i]-T-D-(Q+Vp/\{360[c-q]\})/ʃ]$$
$$/\{\$'[1+s]\}-v$$

Rule-21485:

If both (**ʃ**), (**U**), (**$**), (**i**), (**$'**), (**s**), (**v**), (**f**), (**Q**), (**D**), (**V**), (**p**), (**c**) and (**q**) are known, then its Tax Planned is:

$$T= U+\$[1-i]-\$'[1+s][v+f]-D-(Q+Vp/\{360[c-q]\})/ʃ$$

Rule-21486:

If both (**ʃ**), (**U**), (**$**), (**i**), (**$'**), (**s**), (**v**), (**f**), (**T**), (**Q**), (**V**), (**p**), (**c**) and (**q**) are known, then its Dividend Planned is:

$$D= U+\$[1-i]-\$'[1+s][v+f]-T-(Q+Vp/\{360[c-q]\})/ʃ$$

Rule-21487:

If both (**ʃ**), (**U**), (**$**), (**i**), (**$'**), (**s**), (**v**), (**f**), (**T**), (**D**), (**Q**), (**p**), (**c**) and (**q**) are known, then its Variable Cost Planned is:

$$V= 360[c-q](ʃ\{U+\$[1-i]-\$'[1+s][v+f]-T-D\}-Q)/p$$

Rule-21488:

If both (**ʃ**), (**U**), (**$**), (**i**), (**$'**), (**s**), (**v**), (**f**), (**T**), (**D**), (**V**), (**Q**), (**c**) and (**q**) are known, then its Procured Inventory Days Planned:

$$p= 360[c-q](ʃ\{U+\$[1-i]-\$'[1+s][v+f]-T-D\}-Q)/V$$

Steve Asikin ISBN 14: 978-1515158691, ISBN 10: 1515158691

Rule-21489:
  If both (**/**), (**U**), (**$**), (**i**), (**$'**), (**s**), (**v**), (**f**), (**T**), (**D**), (**V**),
  (**p**), (**Q**) and (**q**) are known, then its Current Ratio
  Planned is:
  $$c= q+Vp/[360(/\{U+\$[1-i]-\$'[1+s][v+f]-T-D\}-Q)]$$

Rule-21490:
  If both (**/**), (**U**), (**$**), (**i**), (**$'**), (**s**), (**v**), (**f**), (**T**), (**D**), (**V**),
  (**p**), (**c**) and (**Q**) are known, then its Quick or Acid test
  Ratio Planned is:
  $$q= c-Vp/[360(/\{U+\$[1-i]-\$'[1+s][v+f]-T-D\}-Q)]$$

Rule-21491:
  If both (**/**), (**U**), (**$**), (**i**), (**$'**), (**s**), (**v**), (**f**), (**T**), (**D**), (**p**),
  (**c**) and (**q**) are known, then its Quoted Longterm Debt
  Planned is:
  $$Q= /\{U+\$[1-i]-\$'[1+s][v+f]-T-D\}-\$vp/\{360[c-q]\}$$

Rule-21492:
  If both (**Q**), (**U**), (**$**), (**i**), (**$'**), (**s**), (**v**), (**f**), (**T**), (**D**), (**p**),
  (**c**) and (**q**) are known, then its Leverage or Gearing
  Ratio Planned is:
  $$/= (Q+\$vp/\{360[c-q]\})$$
  $$/\{U+\$[1-i]-\$'[1+s][v+f]-T-D\}$$

Rule-21493:
  If both (**/**), (**Q**), (**$**), (**i**), (**$'**), (**s**), (**v**), (**f**), (**T**), (**D**), (**p**),
  (**c**) and (**q**) are known, then its Utilized or Starting
  Capital must be:
  $$U= \$'[1+s][v+f]+T+D$$
  $$+(Q+\$vp/\{360[c-q]\})//\$[1-i]$$

Steve Asikin ISBN 14: 978-1515158691, ISBN 10: 1515158691

Rule-21494:

If both (*I*), (**U**), (**Q**), (**i**), (**S'**), (**s**), (**v**), (**f**), (**T**), (**D**), (**p**), (**c**) and (**q**) are known, then its Sales or Revenue Planned is:

$$S= (I\{U-S'[1+s][v+f]-T-D\}-Q)$$
$$/(vp/\{360[c-q]\}-I[1-i])$$

Rule-21495:

If both (*I*), (**U**), (**S**), (**Q**), (**S'**), (**s**), (**v**), (**f**), (**T**), (**D**), (**p**), (**c**) and (**q**) are known, then its Interest Portion Planned is:

$$i= 1-[S'[1+s][v+f]+T+D$$
$$+(Q+Svp/\{360[c-q]\})/I\text{-}U]/S$$

Rule-21496:

If both (*I*), (**U**), (**S**), (**i**), (**Q**), (**s**), (**v**), (**f**), (**T**), (**D**), (**p**), (**c**) and (**q**) are known, then its Sales Past must be:

$$S'= [U+S[1-i]-T-D-(Q+Svp/\{360[c-q]\})/I]$$
$$/\{[1+s][v+f]\}$$

Rule-21497:

If both (*I*), (**U**), (**S**), (**i**), (**S'**), (**Q**), (**v**), (**f**), (**T**), (**D**), (**p**), (**c**) and (**q**) are known, then its Sales Growth Planned is:

$$s= S/S'-1$$

or it can also be found as

$$s= [U+S[1-i]-T-D-(Q+Svp/\{360[c-q]\})/I]$$
$$/\{S'[v+f]\}-1$$

Steve Asikin ISBN 14: 978-1515158691, ISBN 10: 1515158691

Rule-21498:

If both (*l*), (**U**), (**$**), (**i**), (**$'**), (**s**), (**Q**), (**f**), (**T**), (**D**), (**p**), (**c**) and (**q**) are known, then its Variable Portion Planned is:

$$v = (l\{U+\$[1-i]-\$'f[1+s]-T-D\}-Q) / (\$p/\{360[c-q]\}+\$'l[1+s])$$

Rule-21499:

If both (*l*), (**U**), (**$**), (**i**), (**$'**), (**s**), (**v**), (**Q**), (**T**), (**D**), (**p**), (**c**) and (**q**) are known, then its Fixed Portion Planned is:

$$f = [U+\$[1-i]-T-D-(Q+\$vp/\{360[c-q]\})/l] / \{\$'[1+s]\}-v$$

Rule-21500:

If both (*l*), (**U**), (**$**), (**i**), (**$'**), (**s**), (**v**), (**f**), (**Q**), (**D**), (**p**), (**c**) and (**q**) are known, then its Tax Planned is:

$$T = U+\$[1-i]-\$'[1+s][v+f]-D-(Q+\$vp/\{360[c-q]\})/l$$

Rule-21501:

If both (*l*), (**U**), (**$**), (**i**), (**$'**), (**s**), (**v**), (**f**), (**T**), (**Q**), (**p**), (**c**) and (**q**) are known, then its Dividend Planned is:

$$D = U+\$[1-i]-\$'[1+s][v+f]-T-(Q+\$vp/\{360[c-q]\})/l$$

Rule-21502:

If both (*l*), (**U**), (**$**), (**i**), (**$'**), (**s**), (**v**), (**f**), (**T**), (**D**), (**Q**), (**c**) and (**q**) are known, then its Procured Inventory Days Planned:

$$p = 360[c-q](l\{U+\$[1-i]-\$'[1+s][v+f]-T-D\}-Q)/[\$v]$$

Steve Asikin ISBN 14: 978-1515158691, ISBN 10: 1515158691

Rule-21503:

If both (I), (U), (S), (i), (S'), (s), (v), (f), (T), (D), (p), (Q) and (q) are known, then its Current Ratio Planned is:

$$c = q + Svp/[360(I\{U+S[1-i]-S'[1+s][v+f]-T-D\}-Q)]$$

Rule-21504:

If both (I), (U), (S), (i), (S'), (s), (v), (f), (T), (D), (p), (c) and (Q) are known, then its Quick or Acid Test Ratio Planned is:

$$q = c - Svp/[360(I\{U+S[1-i]-S'[1+s][v+f]-T-D\}-Q)]$$

Rule-21505:

If both (I), (U), (S), (i), (S'), (s), (v), (f), (T), (D), (p), (c) and (q) are known, then its Quoted Longterm Debt Planned is:

$$Q = I\{U+S[1-i]-S'[1+s][v+f]-T-D\}$$
$$-S'vp[1+s]/\{360[c-q]\}$$

Rule-21506:

If both (Q), (U), (S), (i), (S'), (s), (v), (f), (T), (D), (p), (c) and (q) are known, then its Leverage or Gearing Ratio Planned is:

$$I = (Q+S'vp[1+s]/\{360[c-q]\})$$
$$/\{U+S[1-i]-S'[1+s][v+f]-T-D\}$$

Steve Asikin ISBN 14: 978-1515158691, ISBN 10: 1515158691

Rule-21507:

If both (**/**), (**Q**), (**$**), (**i**), (**$'**), (**s**), (**v**), (**f**), (**T**), (**D**), (**p**), (**c**) and (**q**) are known, then its Utilized or Starting Capital must be:

$$U= \$'[1+s][v+f]+T+D$$
$$+(Q+\$'vp[1+s]/\{360[c-q]\})//\$[1-i]$$

Rule-21508:

If both (**/**), (**U**), (**Q**), (**i**), (**$'**), (**s**), (**v**), (**f**), (**T**), (**D**), (**p**), (**c**) and (**q**) are known, then its Sales or Revenue Planned is:

$$\$= (Q-/\{U-\$'[1+s][v+f]-T-D\}$$
$$+\$'vp[1+s]/\{360[c-q]\})/\{/[1-i]\}$$

Rule-21509:

If both (**/**), (**U**), (**$**), (**Q**), (**$'**), (**s**), (**v**), (**f**), (**T**), (**D**), (**p**), (**c**) and (**q**) are known, then its Interest Portion Planned is:

$$i= 1-[\$'[1+s][v+f]+T+D$$
$$+(Q+\$'vp[1+s]/\{360[c-q]\})//U]/\$$$

Rule-21510:

If both (**/**), (**U**), (**$**), (**i**), (**Q**), (**s**), (**v**), (**f**), (**T**), (**D**), (**p**), (**c**) and (**q**) are known, then its Sales Past must be:

$$\$'= (/\{U+\$[1-i]-\$'f[1+s]-T-D\}-Q)$$
$$/[v[1+s](p/\{360[c-q]\}+/)]$$

Steve Asikin ISBN 14: 978-1515158691, ISBN 10: 1515158691

Rule-21511:

If both (I), (U), (S), (i), (S'), (Q), (v), (f), (T), (D), (p), (c) and (q) are known, then its Sales Growth Planned is:

$$s = S/S' - 1$$

or it can also be found as

$$s = (I\{U+S[1-i]-S'f[1+s]-T-D\}-Q)$$
$$/[S'v(p/\{360[c-q]\}+I)]-1$$

Rule-21512:

If both (I), (U), (S), (i), (S'), (s), (Q), (f), (T), (D), (p), (c) and (q) are known, then its Variable Portion Planned is:

$$v = (I\{U+S[1-i]-S'f[1+s]-T-D\}-Q)$$
$$/[S'v[1+s](p/\{360[c-q]\}+I)]$$

Rule-21513:

If both (I), (U), (S), (i), (S'), (s), (v), (Q), (T), (D), (p), (c) and (q) are known, then its Fixed Portion Planned is:

$$f = [U+S[1-i]-T-D-(Q+S'vp[1+s]/\{360[c-q]\})/I]$$
$$/\{S'[1+s]\}-v$$

Rule-21514:

If both (I), (U), (S), (i), (S'), (s), (v), (f), (Q), (D), (p), (c) and (q) are known, then its Tax Planned is:

$$T = U+S[1-i]-S'[1+s][v+f]-D$$
$$-(Q+S'vp[1+s]/\{360[c-q]\})/I$$

Steve Asikin ISBN 14: 978-1515158691, ISBN 10: 1515158691

Rule-21515:

If both (**/**), (**U**), (**$**), (**i**), (**$'**), (**s**), (**v**), (**f**), (**T**), (**Q**), (**p**), (**c**) and (**q**) are known, then its Dividend Planned is:

$$D= U+\$[1-i]-\$'[1+s][v+f]-T$$
$$-(Q+\$'vp[1+s]/\{360[c-q]\})/I$$

Rule-21516:

If both (**/**), (**U**), (**$**), (**i**), (**$'**), (**s**), (**v**), (**f**), (**T**), (**D**), (**Q**), (**c**) and (**q**) are known, then its Procured Inventory Days Planned:

$$p= 360[c-q](I\{U+\$[1-i]-\$'[1+s][v+f]-T-D\}-Q)$$
$$/\{\$'v[1+s]\}$$

Rule-21517:

If both (**/**), (**U**), (**$**), (**i**), (**$'**), (**s**), (**v**), (**f**), (**T**), (**D**), (**p**), (**Q**) and (**q**) are known, then its Current Ratio Planned is:

$$c= q+\$'vp[1+s]$$
$$/[360(I\{U+\$[1-i]-\$'[1+s][v+f]-T-D\}-Q)]$$

Rule-21518:

If both (**/**), (**U**), (**$**), (**i**), (**$'**), (**s**), (**v**), (**f**), (**T**), (**D**), (**p**), (**c**) and (**q**) are known, then its Quoted Longterm Debt Planned is:

$$q= c-\$'vp[1+s]$$
$$/[360(I\{U+\$[1-i]-\$'[1+s][v+f]-T-D\}-Q)]$$

Steve Asikin ISBN 14: 978-1515158691, ISBN 10: 1515158691

Rule-21519:

If both (*I*), (**U**), (**$**), (**i**), (**$'**), (**s**), (**v**), (**f**), (**T**), (**d**) and (**X**) are known, then its Quoted Longterm Debt Planned is:

$$Q= I U+\{\$[1-i]-\$'[1+s][v+f]-T\}[1-d])-X$$

Rule-21520:

If both (**Q**), (**U**), (**$**), (**i**), (**$'**), (**s**), (**v**), (**f**), (**T**), (**d**) and (**X**) are known, then its Leverage or Gearing Ratio Planned is:

$$I= [Q+X]/(U+\{\$[1-i]-\$'[1+s][v+f]-T\}[1-d])$$

Rule-21521:

If both (*I*), (**Q**), (**$**), (**i**), (**$'**), (**s**), (**v**), (**f**), (**T**), (**d**) and (**X**) are known, then its Utilized or Starting Capital must be:

$$U= [Q+X]/I\{\$[1-i]-\$'[1+s][v+f]-T\}[1-d]$$

Rule-21522:

If both (*I*), (**U**), (**Q**), (**i**), (**$'**), (**s**), (**v**), (**f**), (**T**), (**d**) and (**X**) are known, then its Sales or Revenue Planned is:

$$\$= (\$'[1+s][v+f]+T+\{[Q+X]/I U\}/[1-d])/[1-i]$$

Rule-21523:

If both (*I*), (**U**), (**$**), (**Q**), (**$'**), (**s**), (**v**), (**f**), (**T**), (**d**) and (**X**) are known, then its Interest Portion Planned is:

$$i= 1-(\$'[1+s][v+f]+T+\{[Q+X]/I U\}/[1-d])/\$$$

Steve Asikin ISBN 14: 978-1515158691, ISBN 10: 1515158691

Rule-21524:

If both (**/**), (**U**), (**$**), (**i**), (**Q**), (**s**), (**v**), (**f**), (**T**), (**d**) and (**X**) are known, then its Sales Past must be:

$S' = (S[1-i]-T-\{[Q+X]/\text{/-}U\}/[1-d])/\{[1+s][v+f]\}$

Rule-21525:

If both (**/**), (**U**), (**$**), (**i**), (**$'**), (**Q**), (**v**), (**f**), (**T**), (**d**) and (**X**) are known, then its Sales Growth Planned is:

$s = S/S'-1$

or it can also be found as

$s = (S[1-i]-T-\{[Q+X]/\text{/-}U\}/[1-d])/\{S'[v+f]\}-1$

Rule-21526:

If both (**/**), (**U**), (**$**), (**i**), (**$'**), (**s**), (**Q**), (**f**), (**T**), (**d**) and (**X**) are known, then its Variable Portion Planned is:

$v = (S[1-i]-T-\{[Q+X]/\text{/-}U\}/[1-d])/\{S'[1+s]\}-f$

Rule-21527:

If both (**/**), (**U**), (**$**), (**i**), (**$'**), (**s**), (**v**), (**Q**), (**T**), (**d**) and (**X**) are known, then its Fixed Portion Planned is:

$f = (S[1-i]-T-\{[Q+X]/\text{/-}U\}/[1-d])/\{S'[1+s]\}-v$

Rule-21528:

If both (**/**), (**U**), (**$**), (**i**), (**$'**), (**s**), (**v**), (**f**), (**Q**), (**d**) and (**X**) are known, then its Tax Planned is:

$T = S[1-i]-S'[1+s][v+f]-\{[Q+X]/\text{/-}U\}/[1-d]$

Rule-21529:

If both (**/**), (**U**), (**$**), (**i**), (**$'**), (**s**), (**v**), (**f**), (**T**), (**Q**) and (**X**) are known, then its Dividend Payout Planned is:

$d = 1-\{[Q+X]/\text{/-}U\}/\{S[1-i]-S'[1+s][v+f]-T\}$

Steve Asikin ISBN 14: 978-1515158691, ISBN 10: 1515158691

Rule-21530:

If both (I), (U), (S), (i), (S'), (s), (v), (f), (T), (d) and (Q) are known, then its Xpress or Gearing Ratio Planned is:

$$X= I(U+\{S[1-i]-S'[1+s][v+f]-T\}[1-d])-Q$$

Rule-21531:

If both (I), (U), (S), (i), (S'), (s), (v), (f), (T), (d), (P), (c) and (q) are known, then its Quoted Longterm Debt Planned is:

$$Q= I(U+\{S[1-i]-S'[1+s][v+f]-T\}[1-d])-P/[c-q]$$

Rule-21532:

If both (Q), (U), (S), (i), (S'), (s), (v), (f), (T), (d), (P), (c) and (q) are known, then its Leverage or Gearing Ratio Planned is:

$$I= \{Q+P/[c-q]\}/(U+\{S[1-i]-S'[1+s][v+f]-T\}[1-d])$$

Rule-21533:

If both (I), (Q), (S), (i), (S'), (s), (v), (f), (T), (d), (P), (c) and (q) are known, then its Utilized or Starting Capital must be:

$$U= \{Q+P/[c-q]\}/I-\{S[1-i]-S'[1+s][v+f]-T\}[1-d]$$

Rule-21534:

If both (I), (U), (Q), (i), (S'), (s), (v), (f), (T), (d), (P), (c) and (q) are known, then its Sales or Revenue Planned is:

$$S= [S'[1+s][v+f]+T+(\{Q+P/[c-q]\}/I-U)/[1-d]]/[1-i]$$

Steve Asikin ISBN 14: 978-1515158691, ISBN 10: 1515158691

Rule-21535:

If both ($\textit{\textbf{l}}$), (**U**), (**$**), (**Q**), (**$'**), (**s**), (**v**), (**f**), (**T**), (**d**), (**P**), (**c**) and (**q**) are known, then its Interest Portion Planned is:

$$i= 1-[\$'[1+s][v+f]+T+(\{Q+P/[c-q]\}/\textit{l}U)/[1-d]]/\$$$

Rule-21536:

If both ($\textit{\textbf{l}}$), (**U**), (**$**), (**i**), (**Q**), (**s**), (**v**), (**f**), (**T**), (**d**), (**P**), (**c**) and (**q**) are known, then its Sales Past must be:

$$\$'= [\$[1-i]-T-(\{Q+P/[c-q]\}/\textit{l}U)/[1-d]]$$
$$/\{[1+s][v+f]\}$$

Rule-21537:

If both ($\textit{\textbf{l}}$), (**U**), (**$**), (**i**), (**$'**), (**Q**), (**v**), (**f**), (**T**), (**d**), (**P**), (**c**) and (**q**) are known, then its Sales Growth Planned is:

$$s= \$/\$'-1$$

or it can also be found as

$$s= [\$[1-i]-T-(\{Q+P/[c-q]\}/\textit{l}U)/[1-d]]/\{\$'[v+f]\}-1$$

Rule-21538:

If both ($\textit{\textbf{l}}$), (**U**), (**$**), (**i**), (**$'**), (**s**), (**Q**), (**f**), (**T**), (**d**), (**P**), (**c**) and (**q**) are known, then its Variable Portion Planned is:

$$v= [\$[1-i]-T-(\{Q+P/[c-q]\}/\textit{l}U)/[1-d]]/\{\$'[1+s]\}-f$$

Rule-21539:

If both ($\textit{\textbf{l}}$), (**U**), (**$**), (**i**), (**$'**), (**s**), (**v**), (**Q**), (**T**), (**d**), (**P**), (**c**) and (**q**) are known, then its Fixed Portion Planned is:

$$f= [\$[1-i]-T-(\{Q+P/[c-q]\}/\textit{l}U)/[1-d]]/\{\$'[1+s]\}-v$$

Steve Asikin ISBN 14: 978-1515158691, ISBN 10: 1515158691

Rule-21540:

If both (I), (**U**), (**$**), (**i**), (**$'**), (**s**), (**v**), (**f**), (**Q**), (**d**), (**P**), (**c**) and (**q**) are known, then its Tax Planned is:

$$T= \$[1\text{-}i]\text{-}\$'[1\text{+}s][v\text{+}f]\text{-}(\{Q\text{+}P/[c\text{-}q]\}//U)/[1\text{-}d]$$

Rule-21541:

If both (I), (**U**), (**$**), (**i**), (**$'**), (**s**), (**v**), (**f**), (**T**), (**Q**), (**P**), (**c**) and (**q**) are known, then its Dividend Payout Planned is:

$$d= 1\text{-}(\{Q\text{+}P/[c\text{-}q]\}//U)/\{\$[1\text{-}i]\text{-}\$'[1\text{+}s][v\text{+}f]\text{-}T\}$$

Rule-21542:

If both (I), (**U**), (**$**), (**i**), (**$'**), (**s**), (**v**), (**f**), (**T**), (**d**), (**Q**), (**c**) and (**q**) are known, then its Procured Inventory Planned is:

$$P= [c\text{-}q][/U\text{+}\{\$[1\text{-}i]\text{-}\$'[1\text{+}s][v\text{+}f]\text{-}T\}[1\text{-}d])\text{-}Q]$$

Rule-21543:

If both (I), (**U**), (**$**), (**i**), (**$'**), (**s**), (**v**), (**f**), (**T**), (**d**), (**P**), (**Q**) and (**q**) are known, then its Current Ratio Planned is:

$$c= q\text{+}P/[/U\text{+}\{\$[1\text{-}i]\text{-}\$'[1\text{+}s][v\text{+}f]\text{-}T\}[1\text{-}d])\text{-}Q]$$

Rule-21544:

If both (I), (**U**), (**$**), (**i**), (**$'**), (**s**), (**v**), (**f**), (**T**), (**d**), (**P**), (**c**) and (**Q**) are known, then its Quick or Acid Test Ratio Planned is:

$$q= c\text{-}P/[/U\text{+}\{\$[1\text{-}i]\text{-}\$'[1\text{+}s][v\text{+}f]\text{-}T\}[1\text{-}d])\text{-}Q]$$

Steve Asikin ISBN 14: 978-1515158691, ISBN 10: 1515158691

Rule-21545:

If both (I), (U), (S), (i), (S'), (s), (v), (f), (T), (d), (V), (p), (c) and (q) are known, then its Quoted Longterm Debt Planned is:

$$Q = I(U + \{S[1-i]-S'[1+s][v+f]-T\}[1-d])$$
$$-Vp/\{360[c-q]\}$$

Rule-21546:

If both (Q), (U), (S), (i), (S'), (s), (v), (f), (T), (d), (V), (p), (c) and (q) are known, then its Leverage or Gearing Ratio Planned is:

$$I = (Q + Vp/\{360[c-q]\})$$
$$/(U + \{S[1-i]-S'[1+s][v+f]-T\}[1-d])$$

Rule-21547:

If both (I), (Q), (S), (i), (S'), (s), (v), (f), (T), (d), (V), (p), (c) and (q) are known, then its Utilized or Starting Capital must be:

$$U = (Q + Vp/\{360[c-q]\})/I$$
$$-\{S[1-i]-S'[1+s][v+f]-T\}[1-d]$$

Rule-21548:

If both (I), (U), (Q), (i), (S'), (s), (v), (f), (T), (d), (V), (p), (c) and (q) are known, then its Sales or Revenue Planned is:

$$S = \{S'[1+s][v+f]+T+[(Q+Vp/\{360[c-q]\})/I-U]$$
$$/[1-d]\}/[1-i]$$

Steve Asikin ISBN 14: 978-1515158691, ISBN 10: 1515158691

Rule-21549:
> If both (I), (U), (S), (Q), (S'), (s), (v), (f), (T), (d), (V), (p), (c) and (q) are known, then its Interest Portion Planned is:
>
> $$i= 1-\{S'[1+s][v+f]+T+[(Q+Vp/\{360[c-q]\})/I\text{-}U]/[1-d]\}/S$$

Rule-21550:
> If both (I), (U), (S), (i), (Q), (s), (v), (f), (T), (d), (V), (p), (c) and (q) are known, then its Sales Past must be:
>
> $$S'= \{S[1-i]-T-[(Q+Vp/\{360[c-q]\})/I\text{-}U]/[1-d]\}/\{[1+s][v+f]\}$$

Rule-21551:
> If both (I), (U), (S), (i), (S'), (Q), (v), (f), (T), (d), (V), (p), (c) and (q) are known, then its Sales Growth Planned is:
>
> $$s= S/S'-1$$
>
> or it can also be found as
>
> $$s= \{S[1-i]-T-[(Q+Vp/\{360[c-q]\})/I\text{-}U]/[1-d]\}/\{S'[v+f]\}-1$$

Rule-21552:
> If both (I), (U), (S), (i), (S'), (s), (Q), (f), (T), (d), (V), (p), (c) and (q) are known, then its Variable Portion Planned is:
>
> $$v= V/S$$
>
> or it can also be found as
>
> $$v= \{S[1-i]-T-[(Q+Vp/\{360[c-q]\})/I\text{-}U]/[1-d]\}/\{S'[1+s]\}-f$$

Steve Asikin ISBN 14: 978-1515158691, ISBN 10: 1515158691

Rule-21553:

If both $(\textbf{\textit{I}})$, $(\textbf{U})$, $(\textbf{\$})$, $(\textbf{i})$, $(\textbf{\$'})$, $(\textbf{s})$, $(\textbf{v})$, $(\textbf{Q})$, $(\textbf{T})$, $(\textbf{d})$, $(\textbf{V})$, $(\textbf{p})$, $(\textbf{c})$ and $(\textbf{q})$ are known, then its Fixed Portion Planned is:

$$\textbf{f} = \{\$[1\text{-}i]\text{-}\textbf{T}\text{-}[(\textbf{Q}+\textbf{Vp}/\{360[\textbf{c-q}]\})/\textbf{\textit{I}U}]/[1\text{-}\textbf{d}]\}$$
$$/\{\ \$'[1+\textbf{s}]\}\text{-}\textbf{v}$$

Rule-21554:

If both $(\textbf{\textit{I}})$, $(\textbf{U})$, $(\textbf{\$})$, $(\textbf{i})$, $(\textbf{\$'})$, $(\textbf{s})$, $(\textbf{v})$, $(\textbf{f})$, $(\textbf{Q})$, $(\textbf{d})$, $(\textbf{V})$, $(\textbf{p})$, $(\textbf{c})$ and $(\textbf{q})$ are known, then its Tax Planned is:

$$\textbf{T} = \$[1\text{-}i]\text{-}\$'[1+\textbf{s}][\textbf{v}+\textbf{f}]\text{-}[(\textbf{Q}+\textbf{Vp}/\{360[\textbf{c-q}]\})/\textbf{\textit{I}U}]$$
$$/[1\text{-}\textbf{d}]$$

Rule-21555:

If both $(\textbf{\textit{I}})$, $(\textbf{U})$, $(\textbf{\$})$, $(\textbf{i})$, $(\textbf{\$'})$, $(\textbf{s})$, $(\textbf{v})$, $(\textbf{f})$, $(\textbf{T})$, $(\textbf{Q})$, $(\textbf{V})$, $(\textbf{p})$, $(\textbf{c})$ and $(\textbf{q})$ are known, then its Dividend Payout Planned is:

$$\textbf{d} = 1\text{-}[(\textbf{Q}+\textbf{Vp}/\{360[\textbf{c-q}]\})/\textbf{\textit{I}U}]$$
$$/\{\$[1\text{-}i]\text{-}\$'[1+\textbf{s}][\textbf{v}+\textbf{f}]\text{-}\textbf{T}\}$$

Rule-21556:

If both $(\textbf{\textit{I}})$, $(\textbf{U})$, $(\textbf{\$})$, $(\textbf{i})$, $(\textbf{\$'})$, $(\textbf{s})$, $(\textbf{v})$, $(\textbf{f})$, $(\textbf{T})$, $(\textbf{d})$, $(\textbf{Q})$, $(\textbf{p})$, $(\textbf{c})$ and $(\textbf{q})$ are known, then its Variable Cost Planned is:

$$\textbf{V} = 360[\textbf{c-q}]$$
$$[\textbf{\textit{I}U}+\{\$[1\text{-}i]\text{-}\$'[1+\textbf{s}][\textbf{v}+\textbf{f}]\text{-}\textbf{T}\}[1\text{-}\textbf{d}])\text{-}\textbf{Q}]/\textbf{p}$$

Steve Asikin ISBN 14: 978-1515158691, ISBN 10: 1515158691

Rule-21557:

If both (I), $(\mathbf{U})$, $(\$)$, $(\mathbf{i})$, $(\$')$, $(\mathbf{s})$, $(\mathbf{v})$, $(\mathbf{f})$, $(\mathbf{T})$, $(\mathbf{d})$, $(\mathbf{V})$, $(\mathbf{Q})$, $(\mathbf{c})$ and $(\mathbf{q})$ are known, then its Procured Inventory Days Planned:

$$p= 360[c\text{-}q]$$
$$[I U+\{\$[1\text{-}i]\text{-}\$'[1+s][v+f]\text{-}T\}[1\text{-}d])\text{-}Q]/V$$

Rule-21558:

If both (I), $(\mathbf{U})$, $(\$)$, $(\mathbf{i})$, $(\$')$, $(\mathbf{s})$, $(\mathbf{v})$, $(\mathbf{f})$, $(\mathbf{T})$, $(\mathbf{d})$, $(\mathbf{V})$, $(\mathbf{p})$, $(\mathbf{Q})$ and $(\mathbf{q})$ are known, then its Current Ratio Planned is:

$$c= q+Vp$$
$$/\{360[I U+\{\$[1\text{-}i]\text{-}\$'[1+s][v+f]\text{-}T\}[1\text{-}d])\text{-}Q]\}$$

Rule-21559:

If both (I), $(\mathbf{U})$, $(\$)$, $(\mathbf{i})$, $(\$')$, $(\mathbf{s})$, $(\mathbf{v})$, $(\mathbf{f})$, $(\mathbf{T})$, $(\mathbf{d})$, $(\mathbf{V})$, $(\mathbf{p})$, $(\mathbf{c})$ and $(\mathbf{Q})$ are known, then its Quick or Acid Test Ratio Planned is:

$$q= c\text{-}Vp$$
$$/\{360[I U+\{\$[1\text{-}i]\text{-}\$'[1+s][v+f]\text{-}T\}[1\text{-}d])\text{-}Q]\}$$

Rule-21560:

If both (I), $(\mathbf{U})$, $(\$)$, $(\mathbf{i})$, $(\$')$, $(\mathbf{s})$, $(\mathbf{v})$, $(\mathbf{f})$, $(\mathbf{T})$, $(\mathbf{d})$, $(\mathbf{p})$, $(\mathbf{c})$ and $(\mathbf{q})$ are known, then its Quoted Longterm Debt Planned is:

$$Q= I U+\{\$[1\text{-}i]\text{-}\$'[1+s][v+f]\text{-}T\}[1\text{-}d])$$
$$\text{-}\$vp/\{360[c\text{-}q]\}$$

Steve Asikin ISBN 14: 978-1515158691, ISBN 10: 1515158691

Rule-21561:

If both $(\mathbf{Q})$, $(\mathbf{U})$, $(\mathbf{\$})$, $(\mathbf{i})$, $(\mathbf{\$'})$, $(\mathbf{s})$, $(\mathbf{v})$, $(\mathbf{f})$, $(\mathbf{T})$, $(\mathbf{d})$, $(\mathbf{p})$, $(\mathbf{c})$ and $(\mathbf{q})$ are known, then its Leverage or Gearing Ratio Planned is:

$$\mathbf{\mathit{l}} = (\mathbf{Q} + \mathbf{\$vp}/\{360[\mathbf{c}\text{-}\mathbf{q}]\})$$
$$/(\mathbf{U} + \{\mathbf{\$}[1\text{-}\mathbf{i}]\text{-}\mathbf{\$'}[1\text{+}\mathbf{s}][\mathbf{v}\text{+}\mathbf{f}]\text{-}\mathbf{T}\}[1\text{-}\mathbf{d}])$$

Rule-21562:

If both $(\mathbf{\mathit{l}})$, $(\mathbf{Q})$, $(\mathbf{\$})$, $(\mathbf{i})$, $(\mathbf{\$'})$, $(\mathbf{s})$, $(\mathbf{v})$, $(\mathbf{f})$, $(\mathbf{T})$, $(\mathbf{d})$, $(\mathbf{p})$, $(\mathbf{c})$ and $(\mathbf{q})$ are known, then its Utilized or Starting Capital must be:

$$\mathbf{U} = (\mathbf{Q} + \mathbf{\$vp}/\{360[\mathbf{c}\text{-}\mathbf{q}]\})/\mathbf{\mathit{l}}$$
$$- \{\mathbf{\$}[1\text{-}\mathbf{i}]\text{-}\mathbf{\$'}[1\text{+}\mathbf{s}][\mathbf{v}\text{+}\mathbf{f}]\text{-}\mathbf{T}\}[1\text{-}\mathbf{d}]$$

Rule-21563:

If both $(\mathbf{\mathit{l}})$, $(\mathbf{U})$, $(\mathbf{Q})$, $(\mathbf{i})$, $(\mathbf{\$'})$, $(\mathbf{s})$, $(\mathbf{v})$, $(\mathbf{f})$, $(\mathbf{T})$, $(\mathbf{d})$, $(\mathbf{p})$, $(\mathbf{c})$ and $(\mathbf{q})$ are known, then its Sales or Revenue Planned is:

$$\mathbf{\$} = [(\mathbf{\mathit{l}}\mathbf{U}\text{-}\mathbf{\$'}[1\text{+}\mathbf{s}][\mathbf{v}\text{+}\mathbf{f}]\text{-}\mathbf{T}\}[1\text{-}\mathbf{d}])\text{-}\mathbf{Q}]$$
$$/(\mathbf{vp}/\{360[\mathbf{c}\text{-}\mathbf{q}]\}\text{-}\mathbf{\mathit{l}}[1\text{-}\mathbf{i}][1\text{-}\mathbf{d}])$$

Rule-21564:

If both $(\mathbf{\mathit{l}})$, $(\mathbf{U})$, $(\mathbf{\$})$, $(\mathbf{Q})$, $(\mathbf{\$'})$, $(\mathbf{s})$, $(\mathbf{v})$, $(\mathbf{f})$, $(\mathbf{T})$, $(\mathbf{d})$, $(\mathbf{p})$, $(\mathbf{c})$ and $(\mathbf{q})$ are known, then its Interest Portion Planned is:

$$\mathbf{i} = 1 - \{\mathbf{\$'}[1\text{+}\mathbf{s}][\mathbf{v}\text{+}\mathbf{f}] + \mathbf{T} + [(\mathbf{Q} + \mathbf{\$vp}/\{360[\mathbf{c}\text{-}\mathbf{q}]\})/\mathbf{\mathit{l}}\mathbf{U}]$$
$$/[1\text{-}\mathbf{d}]\}/\mathbf{\$}$$

Steve Asikin ISBN 14: 978-1515158691, ISBN 10: 1515158691

Rule-21565:

If both (**/**), (**U**), (**$**), (**i**), (**Q**), (**s**), (**v**), (**f**), (**T**), (**d**), (**p**), (**c**) and (**q**) are known, then its Sales Past must be:

$$\textbf{\$'} = \{\textbf{\$}[1\textbf{-i}]\textbf{-T-}[(\textbf{Q+\$vp}/\{360[\textbf{c-q}]\})/\textbf{/U}]/[1\textbf{-d}]\}$$
$$/\{[1\textbf{+s}][\textbf{v+f}]\}$$

Rule-21566:

If both (**/**), (**U**), (**$**), (**i**), (**$'**), (**s**), (**v**), (**f**), (**T**), (**d**), (**p**), (**c**) and (**q**) are known, then its Quoted Longterm Debt Planned is:

$$\textbf{s} = \textbf{\$}/\textbf{\$'}\textbf{-}1$$

or it can also be found as

$$\textbf{s} = \{\textbf{\$}[1\textbf{-i}]\textbf{-T-}[(\textbf{Q+\$vp}/\{360[\textbf{c-q}]\})/\textbf{/U}]/[1\textbf{-d}]\}$$
$$/\{\textbf{\$'}[\textbf{v+f}]\}\textbf{-}1$$

Rule-21567:

If both (**/**), (**U**), (**$**), (**i**), (**$'**), (**s**), (**Q**), (**f**), (**T**), (**d**), (**p**), (**c**) and (**q**) are known, then its Variable Portion Planned is:

$$\textbf{v} = [\textbf{/U+}\{\textbf{\$}[1\textbf{-i}]\textbf{-\$'f}[1\textbf{+s}]\textbf{-T}\}[1\textbf{-d}])\textbf{-Q}]$$
$$/(\textbf{\$p}/\{360[\textbf{c-q}]\}\textbf{+\$'/}[1\textbf{+s}]\,[1\textbf{-d}])$$

Rule-21568:

If both (**/**), (**U**), (**$**), (**i**), (**$'**), (**s**), (**v**), (**Q**), (**T**), (**d**), (**p**), (**c**) and (**q**) are known, then its Fixed Portion Planned is:

$$\textbf{f} = \{\textbf{\$}[1\textbf{-i}]\textbf{-T-}[(\textbf{Q+\$vp}/\{360[\textbf{c-q}]\})/\textbf{/U}]/[1\textbf{-d}]\}$$
$$/\{\textbf{\$'}[1\textbf{+s}]\}\textbf{-v}$$

Steve Asikin ISBN 14: 978-1515158691, ISBN 10: 1515158691

Rule-21569:

If both (**/**), (**U**), (**$**), (**i**), (**$'**), (**s**), (**v**), (**f**), (**Q**), (**d**), (**p**), (**c**) and (**q**) are known, then its Tax Planned is:

$$T= \$[1\text{-}i]\text{-}\$'[1+s][v+f]$$
$$-[(Q+\$vp/\{360[c\text{-}q]\})/\text{/}U]/[1\text{-}d]$$

Rule-21570:

If both (**/**), (**U**), (**$**), (**i**), (**$'**), (**s**), (**v**), (**f**), (**T**), (**Q**), (**p**), (**c**) and (**q**) are known, then its Dividend Payout Planned is:

$$d= 1\text{-}[(Q+\$vp/\{360[c\text{-}q]\})/\text{/}U]$$
$$/\{\$[1\text{-}i]\text{-}\$'[1+s][v+f]\text{-}T\}$$

Rule-21571:

If both (**/**), (**U**), (**$**), (**i**), (**$'**), (**s**), (**v**), (**f**), (**T**), (**d**), (**Q**), (**c**) and (**q**) are known, then its Procured Inventory Days Planned:

$$p= 360[c\text{-}q]$$
$$[\text{/}U+\{\$[1\text{-}i]\text{-}\$'[1+s][v+f]\text{-}T\}[1\text{-}d])\text{-}Q]/[\$v]$$

Rule-21572:

If both (**/**), (**U**), (**$**), (**i**), (**$'**), (**s**), (**v**), (**f**), (**T**), (**d**), (**p**), (**Q**) and (**q**) are known, then its Current Ratio Planned is:

$$c= q+\$vp$$
$$/\{360[\text{/}U+\{\$[1\text{-}i]\text{-}\$'[1+s][v+f]\text{-}T\}[1\text{-}d])\text{-}Q]\}$$

Steve Asikin ISBN 14: 978-1515158691, ISBN 10: 1515158691

Rule-21573:

If both (I), (**U**), (**$**), (**i**), (**$'**), (**s**), (**v**), (**f**), (**T**), (**d**), (**p**), (**c**) and (**Q**) are known, then its Quick or Acid Test Ratio Planned is:

$q= c\text{-}\$vp$

$$/\{360[I U+\{\$[1\text{-}i]\text{-}\$'[1+s][v+f]\text{-}T\}[1\text{-}d])\text{-}Q]\}$$

Rule-21574:

If both (I), (**U**), (**$**), (**i**), (**$'**), (**s**), (**v**), (**f**), (**T**), (**d**), (**p**), (**c**) and (**q**) are known, then its Quoted Longterm Debt Planned is:

$$Q= I U+\{\$[1\text{-}i]\text{-}\$'[1+s][v+f]\text{-}T\}[1\text{-}d])$$

$$\text{-}\$'vp[1+s]/\{360[c\text{-}q]\}$$

Rule-21575:

If both (**Q**), (**U**), (**$**), (**i**), (**$'**), (**s**), (**v**), (**f**), (**T**), (**d**), (**p**), (**c**) and (**q**) are known, then its Leverage or Gearing Ratio Planned is:

$$I= (Q+\$'vp[1+s]/\{360[c\text{-}q]\})$$

$$/(U+\{\$[1\text{-}i]\text{-}\$'[1+s][v+f]\text{-}T\}[1\text{-}d])$$

Rule-21576:

If both (I), (**Q**), (**$**), (**i**), (**$'**), (**s**), (**v**), (**f**), (**T**), (**d**), (**p**), (**c**) and (**q**) are known, then its Utilized or Starting Capital must be:

$$U= (Q+\$'vp[1+s]/\{360[c\text{-}q]\})/I$$

$$\text{-}\{\$[1\text{-}i]\text{-}\$'[1+s][v+f]\text{-}T\}[1\text{-}d]$$

Steve Asikin ISBN 14: 978-1515158691, ISBN 10: 1515158691

Rule-21577:

If both (**/**), (**U**), (**Q**), (**i**), (**S'**), (**s**), (**v**), (**f**), (**T**), (**d**), (**p**), (**c**) and (**q**) are known, then its Sales or Revenue Planned is:

$$S= [Q-/(U-\{S'[1+s][v+f]+T\}[1-d])$$
$$-S'vp[1+s]/\{360[c-q]\}]/\{/[1-i]\,[1-d]\}$$

Rule-21578:

If both (**/**), (**U**), (**S**), (**Q**), (**S'**), (**s**), (**v**), (**f**), (**T**), (**d**), (**p**), (**c**) and (**q**) are known, then its Interest Portion Planned is:

$$i= 1-\{S'[1+s][v+f]+T+[(Q+S'vp[1+s]$$
$$/\{360[c-q]\})/\text{/}U]/[1-d]\}/S$$

Rule-21579:

If both (**/**), (**U**), (**S**), (**i**), (**Q**), (**s**), (**v**), (**f**), (**T**), (**d**), (**p**), (**c**) and (**q**) are known, then its Sales Past must be:

$$S'= [/(U+\{S[1-i]-S'f[1+s]-T\}[1-d])-Q]$$
$$/[S'v[1+s](p/\{360[c-q]\}+/[1-d])]$$

Rule-21580:

If both (**/**), (**U**), (**S**), (**i**), (**S'**), (**Q**), (**v**), (**f**), (**T**), (**d**), (**p**), (**c**) and (**q**) are known, then its Sales Growth Planned is:

$$s= S/S'-1$$

or it can also be found as

$$s= [/(U+\{S[1-i]-S'f[1+s]-T\}[1-d])-Q]$$
$$/[S'v\,(p/\{360[c-q]\}+/[1-d])]-1$$

Steve Asikin ISBN 14: 978-1515158691, ISBN 10: 1515158691

Rule-21581:
   If both (I), (U), $(\$)$, (i), $(\$')$, (s), (Q), (f), (T), (d), (p), (c) and (q) are known, then its Variable Portion Planned is:

$$v = [IU + \{\$[1-i] - \$'f[1+s] - T\}[1-d]) - Q] \; / [\$'[1+s](p/\{360[c-q]\} + I[1-d])]$$

Rule-21582:
   If both (I), (U), $(\$)$, (i), $(\$')$, (s), (v), (Q), (T), (d), (p), (c) and (q) are known, then its Fixed Portion Planned is:

$$f = \{\$[1-i] - T - [(Q + \$'vp[1+s]/\{360[c-q]\})/IU] \; / [1-d]\}/\{\$'[1+s]\} - v$$

Rule-21583:
   If both (I), (U), $(\$)$, (i), $(\$')$, (s), (v), (f), (Q), (d), (p), (c) and (q) are known, then its Tax Planned is:

$$T = \$[1-i] - \$'[1+s][v+f] - [(Q + \$'vp[1+s] \; /\{360[c-q]\})/IU]/[1-d]$$

Rule-21584:
   If both (I), (U), $(\$)$, (i), $(\$')$, (s), (v), (f), (T), (Q), (p), (c) and (q) are known, then its Dividend Payout Planned is:

$$d = 1 - [(Q + \$'vp[1+s]/\{360[c-q]\})/IU] \; /\{\$[1-i] - \$'[1+s][v+f] - T\}$$

Steve Asikin ISBN 14: 978-1515158691, ISBN 10: 1515158691

Rule-21585:

If both (**/**), (**U**), (**$**), (**i**), (**$'**), (**s**), (**v**), (**f**), (**T**), (**d**), (**Q**), (**c**) and (**q**) are known, then its Procured Inventory Days Planned:

$$p = 360[c\text{-}q][\textit{/}U + \{\$[1\text{-}i]\text{-}\$'[1\text{+}s][v\text{+}f]\text{-}T\}[1\text{-}d])\text{-}Q]$$
$$/\{\$'v[1\text{+}s]\}$$

Rule-21586:

If both (**/**), (**U**), (**$**), (**i**), (**$'**), (**s**), (**v**), (**f**), (**T**), (**d**), (**p**), (**Q**) and (**q**) are known, then its Current Ratio Planned is:

$$c = q + \$'vp[1\text{+}s]/\{360[\textit{/}U + \{\$[1\text{-}i]\text{-}\$'[1\text{+}s]$$
$$[v\text{+}f]\text{-}T\}[1\text{-}d])\text{-}Q]\}$$

Rule-21587:

If both (**/**), (**U**), (**$**), (**i**), (**$'**), (**s**), (**v**), (**f**), (**T**), (**d**), (**p**), (**c**) and (**Q**) are known, then its Quick or Acid Test Ratio Planned is:

$$q = c - \$'vp[1\text{+}s]/\{360[\textit{/}U + \{\$[1\text{-}i]\text{-}\$'[1\text{+}s]$$
$$[v\text{+}f]\text{-}T\}[1\text{-}d])\text{-}Q]\}$$

Rule-21588:

If both (**/**), (**U**), (**$**), (**i**), (**$'**), (**s**), (**v**), (**f**), (**T**), (**A**), (**d**) and (**X**) are known, then its Quoted Longterm Debt Planned is:

$$Q = \textit{/}\{U + \$[1\text{-}i]\text{-}\$'[1\text{+}s][v\text{+}f]\text{-}T\text{-}Ad\} - X$$

Steve Asikin ISBN 14: 978-1515158691, ISBN 10: 1515158691

Rule-21589:

If both $(\mathbf{Q})$, $(\mathbf{U})$, $(\mathbf{S})$, $(\mathbf{i})$, $(\mathbf{S'})$, $(\mathbf{s})$, $(\mathbf{v})$, $(\mathbf{f})$, $(\mathbf{T})$, $(\mathbf{A})$, $(\mathbf{d})$ and $(\mathbf{X})$ are known, then its Leverage or Gearing Ratio Planned is:

$$\boldsymbol{\mathit{l}} = [\mathbf{Q}+\mathbf{X}]/\{\mathbf{U}+\mathbf{S}[1-\mathbf{i}]-\mathbf{S'}[1+\mathbf{s}][\mathbf{v}+\mathbf{f}]-\mathbf{T}-\mathbf{Ad}\}$$

Rule-21590:

If both $(\boldsymbol{\mathit{l}})$, $(\mathbf{Q})$, $(\mathbf{S})$, $(\mathbf{i})$, $(\mathbf{S'})$, $(\mathbf{s})$, $(\mathbf{v})$, $(\mathbf{f})$, $(\mathbf{T})$, $(\mathbf{A})$, $(\mathbf{d})$ and $(\mathbf{X})$ are Utilized or Starting Capital must be:

$$\mathbf{U} = \mathbf{S'}[1+\mathbf{s}][\mathbf{v}+\mathbf{f}]+\mathbf{T}+\mathbf{Ad}+[\mathbf{Q}+\mathbf{X}]/\boldsymbol{\mathit{l}}-\mathbf{S}[1-\mathbf{i}]$$

Rule-21591:

If both $(\boldsymbol{\mathit{l}})$, $(\mathbf{U})$, $(\mathbf{Q})$, $(\mathbf{i})$, $(\mathbf{S'})$, $(\mathbf{s})$, $(\mathbf{v})$, $(\mathbf{f})$, $(\mathbf{T})$, $(\mathbf{A})$, $(\mathbf{d})$ and $(\mathbf{X})$ are known, then its Sales or Revenue Planned is:

$$\mathbf{S} = \{\mathbf{S'}[1+\mathbf{s}][\mathbf{v}+\mathbf{f}]+\mathbf{T}+\mathbf{Ad}+[\mathbf{Q}+\mathbf{X}]/\boldsymbol{\mathit{l}}-\mathbf{U}\}/[1-\mathbf{i}]$$

Rule-21592:

If both $(\boldsymbol{\mathit{l}})$, $(\mathbf{U})$, $(\mathbf{S})$, $(\mathbf{Q})$, $(\mathbf{S'})$, $(\mathbf{s})$, $(\mathbf{v})$, $(\mathbf{f})$, $(\mathbf{T})$, $(\mathbf{A})$, $(\mathbf{d})$ and $(\mathbf{X})$ are known, then its Interest Portion Planned is:

$$\mathbf{i} = 1-\{\mathbf{S'}[1+\mathbf{s}][\mathbf{v}+\mathbf{f}]+\mathbf{T}+\mathbf{Ad}+[\mathbf{Q}+\mathbf{X}]/\boldsymbol{\mathit{l}}-\mathbf{U}\}/\mathbf{S}$$

Rule-21593:

If both $(\boldsymbol{\mathit{l}})$, $(\mathbf{U})$, $(\mathbf{S})$, $(\mathbf{i})$, $(\mathbf{Q})$, $(\mathbf{s})$, $(\mathbf{v})$, $(\mathbf{f})$, $(\mathbf{T})$, $(\mathbf{A})$, $(\mathbf{d})$ and $(\mathbf{X})$ are known, then its Sales Past must be:

$$\mathbf{S'} = \{\mathbf{U}+\mathbf{S}[1-\mathbf{i}]-\mathbf{T}-\mathbf{Ad}-[\mathbf{Q}+\mathbf{X}]/\boldsymbol{\mathit{l}}\}/\{[1+\mathbf{s}][\mathbf{v}+\mathbf{f}]\}$$

Steve Asikin ISBN 14: 978-1515158691, ISBN 10: 1515158691

Rule-21594:
> If both (I), (**U**), (**$**), (**i**), (**$'**), (**Q**), (**v**), (**f**), (**T**), (**A**), (**d**) and (**X**) are known, then its Sales Growth Planned is:
> $$s= \$/\$'-1$$
> or it can also be found as
> $$s= \{U+\$[1-i]-T-Ad-[Q+X]/I\}/\{\$'[v+f]\}-1$$

Rule-21595:
> If both (I), (**U**), (**$**), (**i**), (**$'**), (**s**), (**Q**), (**f**), (**T**), (**A**), (**d**) and (**X**) are known, then its Variable Portion Planned is:
> $$v= \{U+\$[1-i]-T-Ad-[Q+X]/I\}/\{\$'[1+s]\}-f$$

Rule-21596:
> If both (I), (**U**), (**$**), (**i**), (**$'**), (**s**), (**v**), (**Q**), (**T**), (**A**), (**d**) and (**X**) are known, then its Fixed Portion Planned is:
> $$f= \{U+\$[1-i]-T-Ad-[Q+X]/I\}/\{\$'[1+s]\}-v$$

Rule-21597:
> If both (I), (**U**), (**$**), (**i**), (**$'**), (**s**), (**v**), (**f**), (**Q**), (**A**), (**d**) and (**X**) are known, then its Tax Planned is:
> $$T= U+\$[1-i]-\$'[1+s][v+f]-Ad-[Q+X]/I$$

Rule-21598:
> If both (I), (**U**), (**$**), (**i**), (**$'**), (**s**), (**v**), (**f**), (**T**), (**Q**), (**d**) and (**X**) are known, then its After Tax Income Planned is:
> $$A= \{U+\$[1-i]-\$'[1+s][v+f]-T-[Q+X]/I\}/d$$

Steve Asikin ISBN 14: 978-1515158691, ISBN 10: 1515158691

Rule-21599:

If both (I), (U), $(\$)$, (i), $(\$')$, (s), (v), (f), (T), (A), (Q) and (X) are known, then its Dividend Payout Planned is:

$$d= \{U+\$[1-i]-\$'[1+s][v+f]-T-[Q+X]/I\}/A$$

Rule-21600:

If both (I), (U), $(\$)$, (i), $(\$')$, (s), (v), (f), (T), (A), (d) and (Q) are known, then its Xpress or Current Debt Planned is:

$$X= I\{U+\$[1-i]-\$'[1+s][v+f]-T-Ad\}-Q$$

Rule-21601:

If both (I), (U), $(\$)$, (i), $(\$')$, (s), (v), (f), (T), (A), (d), (P), (c) and (q) are known, then its Quoted Longterm Debt Planned is:

$$Q= I\{U+\$[1-i]-\$'[1+s][v+f]-T-Ad\}-P/[c-q]$$

Rule-21602:

If both (Q), (U), $(\$)$, (i), $(\$')$, (s), (v), (f), (T), (A), (d), (P), (c) and (q) are known, then its Leverage or Gearing Ratio Planned is:

$$I= \{Q+P/[c-q]\}/\{U+\$[1-i]-\$'[1+s][v+f]-T-Ad\}$$

Rule-21603:

If both (I), (Q), $(\$)$, (i), $(\$')$, (s), (v), (f), (T), (A), (d), (P), (c) and (q) are known, then its Utilized or Starting Capital must be:

$$U= \$'[1+s][v+f]+T+Ad+\{Q+P/[c-q]\}/I-\$[1-i]$$

Steve Asikin ISBN 14: 978-1515158691, ISBN 10: 1515158691

Rule-21604:

If both (I), (U), (Q), (i), (S'), (s), (v), (f), (T), (A), (d), (P), (c) and (q) are known, then its Sales or Revenue Planned is:

$$S = (S'[1+s][v+f]+T+Ad+\{Q+P/[c-q]\}/I \cdot U)/[1-i]$$

Rule-21605:

If both (I), (U), (S), (Q), (S'), (s), (v), (f), (T), (A), (d), (P), (c) and (q) are known, then its Interest Portion Planned is:

$$i = 1-(S'[1+s][v+f]+T+Ad+\{Q+P/[c-q]\}/I \cdot U)/S$$

Rule-21606:

If both (I), (U), (S), (i), (Q), (s), (v), (f), (T), (A), (d), (P), (c) and (q) are known, then its Sales Past must be:

$$S' = (U+S[1-i]-T-Ad-\{Q+P/[c-q]\}/I)/\{[1+s][v+f]\}$$

Rule-21607:

If both (I), (U), (S), (i), (S'), (Q), (v), (f), (T), (A), (d), (P), (c) and (q) are known, then its Sales Growth Planned is:

$$s = S/S'-1$$

or it can also be found as

$$s = (U+S[1-i]-T-Ad-\{Q+P/[c-q]\}/I)/\{S'[v+f]\}-1$$

Rule-21608:

If both (I), (U), (S), (i), (S'), (s), (Q), (f), (T), (A), (d), (P), (c) and (q) are known, then its Variable Portion Planned is:

$$v = (U+S[1-i]-T-Ad-\{Q+P/[c-q]\}/I)/\{S'[1+s]\}-f$$

Steve Asikin ISBN 14: 978-1515158691, ISBN 10: 1515158691

Rule-21609:

If both (*l*), (**U**), (**S**), (**i**), (**S'**), (**s**), (**v**), (**Q**), (**T**), (**A**), (**d**), (**P**), (**c**) and (**q**) are known, then its Fixed Portion Planned is:

$$f= (U+S[1-i]-T-Ad-\{Q+P/[c-q]\}/l)/\{S'[1+s]\}-v$$

Rule-21610:

If both (*l*), (**U**), (**S**), (**i**), (**S'**), (**s**), (**v**), (**f**), (**Q**), (**A**), (**d**), (**P**), (**c**) and (**q**) are known, then its Tax Planned is:

$$T= U+S[1-i]-S'[1+s][v+f]-Ad-\{Q+P/[c-q]\}/l$$

Rule-21611:

If both (*l*), (**U**), (**S**), (**i**), (**S'**), (**s**), (**v**), (**f**), (**T**), (**Q**), (**d**), (**P**), (**c**) and (**q**) are known, then its After Tax Income Planned is:

$$A= (U+S[1-i]-S'[1+s][v+f]-T-\{Q+P/[c-q]\}/l)/d$$

Rule-21612:

If both (*l*), (**U**), (**S**), (**i**), (**S'**), (**s**), (**v**), (**f**), (**T**), (**A**), (**Q**), (**P**), (**c**) and (**q**) are known, then its Dividend Payout Planned is:

$$d= (U+S[1-i]-S'[1+s][v+f]-T-\{Q+P/[c-q]\}/l)/A$$

Rule-21613:

If both (*l*), (**U**), (**S**), (**i**), (**S'**), (**s**), (**v**), (**f**), (**T**), (**A**), (**d**), (**Q**), (**c**) and (**q**) are known, then its Procured Inventory Planned is:

$$P= [c-q](l\{U+S[1-i]-S'[1+s][v+f]-T-Ad\}-Q)$$

Steve Asikin ISBN 14: 978-1515158691, ISBN 10: 1515158691

Rule-21614:

If both (**/**), (**U**), (**$**), (**i**), (**$'**), (**s**), (**v**), (**f**), (**T**), (**A**), (**d**), (**P**), (**Q**) and (**q**) are known, then its Current Ratio Planned is:

$$c= q+P/(/\{U+\$[1-i]-\$'[1+s][v+f]-T-Ad\}-Q)$$

Rule-21615:

If both (**/**), (**U**), (**$**), (**i**), (**$'**), (**s**), (**v**), (**f**), (**T**), (**A**), (**d**), (**P**), (**c**) and (**Q**) are known, then its Quick or Acid Test Ratio Planned is:

$$q= c-P/(/\{U+\$[1-i]-\$'[1+s][v+f]-T-Ad\}-Q)$$

Rule-21616:

If both (**/**), (**U**), (**$**), (**i**), (**$'**), (**s**), (**v**), (**f**), (**T**), (**A**), (**d**), (**V**), (**p**), (**c**) and (**q**) are known, then its Quoted Longterm Debt Planned is:

$$Q= /\{U+\$[1-i]-\$'[1+s][v+f]-T-Ad\}-Vp/\{360[c-q]\}$$

Rule-21617:

If both (**/**), (**U**), (**$**), (**i**), (**$'**), (**s**), (**v**), (**f**), (**T**), (**A**), (**d**), (**V**), (**p**), (**c**) and (**q**) are known, then its Quoted Longterm Debt Planned is:

$$/= (Q+Vp/\{360[c-q]\})/\{U+\$[1-i]-\$'[1+s][v+f]-T-Ad\}$$

Rule-21618:

If both (**/**), (**U**), (**$**), (**i**), (**$'**), (**s**), (**v**), (**f**), (**T**), (**A**), (**d**), (**V**), (**p**), (**c**) and (**q**) are known, then its Quoted Longterm Debt Planned is:

$$U= \$'[1+s][v+f]+T+Ad+(Q+Vp/\{360[c-q]\})//\$[1-i]$$

Steve Asikin ISBN 14: 978-1515158691, ISBN 10: 1515158691

Rule-21619:

If both $(\pmb{\Lambda})$, $(\pmb{U})$, $(\pmb{S})$, $(\pmb{i})$, $(\pmb{S'})$, $(\pmb{s})$, $(\pmb{v})$, $(\pmb{f})$, $(\pmb{T})$, $(\pmb{A})$, $(\pmb{d})$, $(\pmb{V})$, $(\pmb{p})$, $(\pmb{c})$ and $(\pmb{q})$ are known, then its Quoted Longterm Debt Planned is:

$$\pmb{S}= [\pmb{S'}[1+\pmb{s}][\pmb{v}+\pmb{f}]+\pmb{T}+\pmb{Ad}+(\pmb{Q}+\pmb{Vp}/\{360[\pmb{c}-\pmb{q}]\})/\pmb{\Lambda}-\pmb{U}]/[1-\pmb{i}]$$

Rule-21620:

If both $(\pmb{\Lambda})$, $(\pmb{U})$, $(\pmb{S})$, $(\pmb{i})$, $(\pmb{S'})$, $(\pmb{s})$, $(\pmb{v})$, $(\pmb{f})$, $(\pmb{T})$, $(\pmb{A})$, $(\pmb{d})$, $(\pmb{V})$, $(\pmb{p})$, $(\pmb{c})$ and $(\pmb{q})$ are known, then its Quoted Longterm Debt Planned is:

$$\pmb{i}= 1-[\pmb{S'}[1+\pmb{s}][\pmb{v}+\pmb{f}]+\pmb{T}+\pmb{Ad}+(\pmb{Q}+\pmb{Vp}/\{360[\pmb{c}-\pmb{q}]\})/\pmb{\Lambda}-\pmb{U}]/\pmb{S}$$

Rule-21621:

If both $(\pmb{\Lambda})$, $(\pmb{U})$, $(\pmb{S})$, $(\pmb{i})$, $(\pmb{S'})$, $(\pmb{s})$, $(\pmb{v})$, $(\pmb{f})$, $(\pmb{T})$, $(\pmb{A})$, $(\pmb{d})$, $(\pmb{V})$, $(\pmb{p})$, $(\pmb{c})$ and $(\pmb{q})$ are known, then its Quoted Longterm Debt Planned is:

$$\pmb{S'}= [\pmb{U}+\pmb{S}[1-\pmb{i}]-\pmb{T}-\pmb{Ad}-(\pmb{Q}+\pmb{Vp}/\{360[\pmb{c}-\pmb{q}]\})/\pmb{\Lambda}]/\{[1+\pmb{s}][\pmb{v}+\pmb{f}]\}$$

Rule-21622:

If both $(\pmb{\Lambda})$, $(\pmb{U})$, $(\pmb{S})$, $(\pmb{i})$, $(\pmb{S'})$, $(\pmb{s})$, $(\pmb{v})$, $(\pmb{f})$, $(\pmb{T})$, $(\pmb{A})$, $(\pmb{d})$, $(\pmb{V})$, $(\pmb{p})$, $(\pmb{c})$ and $(\pmb{q})$ are known, then its Quoted Longterm Debt Planned is:

$$\pmb{s}= \pmb{S}/\pmb{S'}-1$$

or it can also be found as

$$\pmb{s}= [\pmb{U}+\pmb{S}[1-\pmb{i}]-\pmb{T}-\pmb{Ad}-(\pmb{Q}+\pmb{Vp}/\{360[\pmb{c}-\pmb{q}]\})/\pmb{\Lambda}]/\{\pmb{S'}[\pmb{v}+\pmb{f}]\}-1$$

Steve Asikin ISBN 14: 978-1515158691, ISBN 10: 1515158691

Rule-21623:

If both $(\textit{\textbf{I}})$, $(\textbf{U})$, $(\textbf{\$})$, $(\textbf{i})$, $(\textbf{\$'})$, $(\textbf{s})$, $(\textbf{v})$, $(\textbf{f})$, $(\textbf{T})$, $(\textbf{A})$, $(\textbf{d})$, $(\textbf{V})$, $(\textbf{p})$, $(\textbf{c})$ and $(\textbf{q})$ are known, then its Quoted Longterm Debt Planned is:

$$v = V/\$$$

or it can also be found as

$$v = [U + \$[1-i] - T - Ad - (Q + Vp/\{360[c-q]\})/I/\{\$'[1+s]\} - f$$

Rule-21624:

If both $(\textit{\textbf{I}})$, $(\textbf{U})$, $(\textbf{\$})$, $(\textbf{i})$, $(\textbf{\$'})$, $(\textbf{s})$, $(\textbf{v})$, $(\textbf{f})$, $(\textbf{T})$, $(\textbf{A})$, $(\textbf{d})$, $(\textbf{V})$, $(\textbf{p})$, $(\textbf{c})$ and $(\textbf{q})$ are known, then its Quoted Longterm Debt Planned is:

$$f = [U + \$[1-i] - T - Ad - (Q + Vp/\{360[c-q]\})/I/\{\$'[1+s]\} - v$$

Rule-21625:

If both $(\textit{\textbf{I}})$, $(\textbf{U})$, $(\textbf{\$})$, $(\textbf{i})$, $(\textbf{\$'})$, $(\textbf{s})$, $(\textbf{v})$, $(\textbf{f})$, $(\textbf{T})$, $(\textbf{A})$, $(\textbf{d})$, $(\textbf{V})$, $(\textbf{p})$, $(\textbf{c})$ and $(\textbf{q})$ are known, then its Quoted Longterm Debt Planned is:

$$T = U + \$[1-i] - \$'[1+s][v+f] - Ad - (Q + Vp/\{360[c-q]\})/I$$

Rule-21626:

If both $(\textit{\textbf{I}})$, $(\textbf{U})$, $(\textbf{\$})$, $(\textbf{i})$, $(\textbf{\$'})$, $(\textbf{s})$, $(\textbf{v})$, $(\textbf{f})$, $(\textbf{T})$, $(\textbf{A})$, $(\textbf{d})$, $(\textbf{V})$, $(\textbf{p})$, $(\textbf{c})$ and $(\textbf{q})$ are known, then its Quoted Longterm Debt Planned is:

$$A = [U + \$[1-i] - \$'[1+s][v+f] - T - (Q + Vp/\{360[c-q]\})/I/d$$

Steve Asikin ISBN 14: 978-1515158691, ISBN 10: 1515158691

Rule-21627:

> If both $(/)$, (U), $(\$)$, (i), $(\$')$, (s), (v), (f), (T), (A), (d), (V), (p), (c) and (q) are known, then its Quoted Longterm Debt Planned is:
>
> $$d= [U+\$[1-i]-\$'[1+s][v+f]-T-(Q+Vp/\{360[c-q]\})/\text{/}/A$$

Rule-21628:

> If both $(/)$, (U), $(\$)$, (i), $(\$')$, (s), (v), (f), (T), (A), (d), (V), (p), (c) and (q) are known, then its Quoted Longterm Debt Planned is:
>
> $$V= 360[c-q](/\{U+\$[1-i]-\$'[1+s][v+f]-T-Ad\}-Q)/p$$

Rule-21629:

> If both $(/)$, (U), $(\$)$, (i), $(\$')$, (s), (v), (f), (T), (A), (d), (V), (p), (c) and (q) are known, then its Quoted Longterm Debt Planned is:
>
> $$p= 360[c-q](/\{U+\$[1-i]-\$'[1+s][v+f]-T-Ad\}-Q)/V$$

Rule-21630:

> If both $(/)$, (U), $(\$)$, (i), $(\$')$, (s), (v), (f), (T), (A), (d), (V), (p), (c) and (q) are known, then its Quoted Longterm Debt Planned is:
>
> $$c= q+Vp/[360(/\{U+\$[1-i]-\$'[1+s][v+f]-T-Ad\}-Q)]$$

Rule-21631:

> If both $(/)$, (U), $(\$)$, (i), $(\$')$, (s), (v), (f), (T), (A), (d), (V), (p), (c) and (q) are known, then its Quoted Longterm Debt Planned is:
>
> $$q= c-Vp/[360(/\{U+\$[1-i]-\$'[1+s][v+f]-T-Ad\}-Q)]$$

Steve Asikin ISBN 14: 978-1515158691, ISBN 10: 1515158691

Rule-21632:

If both (I), (**U**), (**S**), (**i**), (**S'**), (**s**), (**v**), (**f**), (**T**), (**A**), (**d**), (**p**), (**c**) and (**q**) are known, then its Quoted Longterm Debt Planned is:

$$Q= I\{U+S[1-i]-S'[1+s][v+f]-T-Ad\}-Svp/\{360[c-q]\}$$

Rule-21633:

If both (**Q**), (**U**), (**S**), (**i**), (**S'**), (**s**), (**v**), (**f**), (**T**), (**A**), (**d**), (**p**), (**c**) and (**q**) are known, then its Leverage or Gearing Ratio Planned is:

$$I= (Q+Svp/\{360[c-q]\})$$
$$/\{U+S[1-i]-S'[1+s][v+f]-T-Ad\}$$

Rule-21634:

If both (I), (**Q**), (**S**), (**i**), (**S'**), (**s**), (**v**), (**f**), (**T**), (**A**), (**d**), (**p**), (**c**) and (**q**) are known, then its Utilized or Starting Capital must be:

$$U= S'[1+s][v+f]+T+Ad$$
$$+(Q+Svp/\{360[c-q]\})/IS[1-i]$$

Rule-21635:

If both (I), (**U**), (**Q**), (**i**), (**S'**), (**s**), (**v**), (**f**), (**T**), (**A**), (**d**), (**p**), (**c**) and (**q**) are known, then its Sales or Revenue Planned is:

$$S= (I\{U-S'[1+s][v+f]-T-Ad\}-Q)$$
$$/(vp/\{360[c-q]\}-I[1-i])$$

Steve Asikin ISBN 14: 978-1515158691, ISBN 10: 1515158691

Rule-21636:

If both (I), (U), (S), (Q), (S'), (s), (v), (f), (T), (A), (d), (p), (c) and (q) are known, then its Interest Portion Planned is:

$$i = 1 - [S'[1+s][v+f] + T + Ad$$
$$+ (Q + Svp/\{360[c-q]\})/I\text{-}U]/S$$

Rule-21637:

If both (I), (U), (S), (i), (Q), (s), (v), (f), (T), (A), (d), (p), (c) and (q) are known, then its Sales Past must be:

$$S' = [U + S[1-i] - T - Ad - (Q + Svp/\{360[c-q]\})/I]$$
$$/\{[1+s][v+f]\}$$

Rule-21638:

If both (I), (U), (S), (i), (S'), (Q), (v), (f), (T), (A), (d), (p), (c) and (q) are known, then its Sales Growth Planned is:

$$s = S/S' - 1$$

or it can also be found as

$$s = [U + S[1-i] - T - Ad - (Q + Svp/\{360[c-q]\})/I]$$
$$/\{S'[v+f]\} - 1$$

Rule-21639:

If both (I), (U), (S), (i), (S'), (s), (Q), (f), (T), (A), (d), (p), (c) and (q) are known, then its Variable Portion Planned is:

$$v = (I\{U + S[1-i] - S'f[1+s] - T - Ad\} - Q)$$
$$/(Sp/\{360[c-q]\} - S'I[1+s])$$

Steve Asikin ISBN 14: 978-1515158691, ISBN 10: 1515158691

Rule-21640:

If both (**∫**), (**U**), (**$**), (**i**), (**$'**), (**s**), (**v**), (**Q**), (**T**), (**A**), (**d**), (**p**), (**c**) and (**q**) are known, then its Fixed Portion Planned is:

$$f= [U+\$[1-i]-T-Ad-(Q+\$vp/\{360[c-q]\})/\!/]$$
$$/\{\$'[1+s]\}-v$$

Rule-21641:

If both (**∫**), (**U**), (**$**), (**i**), (**$'**), (**s**), (**v**), (**f**), (**Q**), (**A**), (**d**), (**p**), (**c**) and (**q**) are known, then its Tax Planned is:

$$T= U+\$[1-i]-\$'[1+s][v+f]-Ad$$
$$-(Q+\$vp/\{360[c-q]\})/\!/$$

Rule-21642:

If both (**∫**), (**U**), (**$**), (**i**), (**$'**), (**s**), (**v**), (**f**), (**T**), (**Q**), (**d**), (**p**), (**c**) and (**q**) are known, then its After Tax Income Planned is:

$$A= [U+\$[1-i]-\$'[1+s][v+f]-T$$
$$-(Q+\$vp/\{360[c-q]\})/\!/]/d$$

Rule-21643:

If both (**∫**), (**U**), (**$**), (**i**), (**$'**), (**s**), (**v**), (**f**), (**T**), (**A**), (**Q**), (**p**), (**c**) and (**q**) are known, then its Dividend Payout Planned is:

$$d= [U+\$[1-i]-\$'[1+s][v+f]-T$$
$$-(Q+\$vp/\{360[c-q]\})/\!/]/A$$

Steve Asikin ISBN 14: 978-1515158691, ISBN 10: 1515158691

Rule-21644:

If both (f), (U), $(\$)$, (i), $(\$')$, (s), (v), (f), (T), (A), (d), (Q), (c) and (q) are known, then its Procured Inventory Days Planned:

$$p= 360[c-q](f\{U+\$[1-i]-\$'[1+s][v+f]-T-Ad\}-Q)/[\$v]$$

Rule-21645:

If both (f), (U), $(\$)$, (i), $(\$')$, (s), (v), (f), (T), (A), (d), (p), (Q) and (q) are known, then its Current Ratio Planned is:

$$c= q+\$vp/[360(f\{U+\$[1-i]-\$'[1+s][v+f]-T-Ad\}-Q)]$$

Rule-21646:

If both (f), (U), $(\$)$, (i), $(\$')$, (s), (v), (f), (T), (A), (d), (p), (c) and (Q) are known, then its Quick or Acid Test Ratio Planned is:

$$q= c-\$vp/[360(f\{U+\$[1-i]-\$'[1+s][v+f]-T-Ad\}-Q)]$$

Rule-21647:

If both (f), (U), $(\$)$, (i), $(\$')$, (s), (v), (f), (T), (A), (d), (p), (c) and (q) are known, then its Quoted Longterm Debt Planned is:

$$Q= f\{U+\$[1-i]-\$'[1+s][v+f]-T-Ad\}-\$'vp[1+s]/\{360[c-q]\}$$

Steve Asikin ISBN 14: 978-1515158691, ISBN 10: 1515158691

Rule-21648:

If both $(\mathbf{Q})$, $(\mathbf{U})$, $(\mathbf{\$})$, $(\mathbf{i})$, $(\mathbf{\$'})$, $(\mathbf{s})$, $(\mathbf{v})$, $(\mathbf{f})$, $(\mathbf{T})$, $(\mathbf{A})$, $(\mathbf{d})$, $(\mathbf{p})$, $(\mathbf{c})$ and $(\mathbf{q})$ are known, then its Leverage or Gearing Ratio Planned is:

$$\mathit{l} = (\mathbf{Q} + \mathbf{\$'vp}[1+\mathbf{s}]/\{360[\mathbf{c\text{-}q}]\})/\{\mathbf{U} + \mathbf{\$}[1\text{-}\mathbf{i}]$$
$$-\mathbf{\$'}[1+\mathbf{s}][\mathbf{v+f}]\text{-}\mathbf{T}\text{-}\mathbf{Ad}\}$$

Rule-21649:

If both (l), $(\mathbf{Q})$, $(\mathbf{\$})$, $(\mathbf{i})$, $(\mathbf{\$'})$, $(\mathbf{s})$, $(\mathbf{v})$, $(\mathbf{f})$, $(\mathbf{T})$, $(\mathbf{A})$, $(\mathbf{d})$, $(\mathbf{p})$, $(\mathbf{c})$ and $(\mathbf{q})$ are known, then its Utilized or Starting Capital must be:

$$\mathbf{U} = \mathbf{\$'}[1+\mathbf{s}][\mathbf{v+f}] + \mathbf{T} + \mathbf{Ad}$$
$$+(\mathbf{Q} + \mathbf{\$'vp}[1+\mathbf{s}]/\{360[\mathbf{c\text{-}q}]\})/\mathit{l}\text{-}\mathbf{\$}[1\text{-}\mathbf{i}]$$

Rule-21650:

If both (l), $(\mathbf{U})$, $(\mathbf{Q})$, $(\mathbf{i})$, $(\mathbf{\$'})$, $(\mathbf{s})$, $(\mathbf{v})$, $(\mathbf{f})$, $(\mathbf{T})$, $(\mathbf{A})$, $(\mathbf{d})$, $(\mathbf{p})$, $(\mathbf{c})$ and $(\mathbf{q})$ are known, then its Sales or Revenue Planned is:

$$\mathbf{\$} = (\mathbf{Q} \text{-} \mathit{l}\{\mathbf{U} \text{-} \mathbf{\$'}[1+\mathbf{s}][\mathbf{v+f}]\text{-}\mathbf{T}\text{-}\mathbf{Ad}\}$$
$$+\mathbf{\$'vp}[1+\mathbf{s}]/\{360[\mathbf{c\text{-}q}]\})/\{\mathit{l}[1\text{-}\mathbf{i}]\}$$

Rule-21651:

If both (l), $(\mathbf{U})$, $(\mathbf{\$})$, $(\mathbf{Q})$, $(\mathbf{\$'})$, $(\mathbf{s})$, $(\mathbf{v})$, $(\mathbf{f})$, $(\mathbf{T})$, $(\mathbf{A})$, $(\mathbf{d})$, $(\mathbf{p})$, $(\mathbf{c})$ and $(\mathbf{q})$ are known, then its Interest Portion Planned is:

$$\mathbf{i} = 1 \text{-} [\mathbf{\$'}[1+\mathbf{s}][\mathbf{v+f}] + \mathbf{T} + \mathbf{Ad}$$
$$+(\mathbf{Q} + \mathbf{\$'vp}[1+\mathbf{s}]/\{360[\mathbf{c\text{-}q}]\})/\mathit{l}\text{-}\mathbf{U}]/\mathbf{\$}$$

Steve Asikin ISBN 14: 978-1515158691, ISBN 10: 1515158691

Rule-21652:

If both $(\textbf{\textit{l}})$, $(\textbf{U})$, $(\textbf{S})$, $(\textbf{i})$, $(\textbf{Q})$, $(\textbf{s})$, $(\textbf{v})$, $(\textbf{f})$, $(\textbf{T})$, $(\textbf{A})$, $(\textbf{d})$, $(\textbf{p})$, $(\textbf{c})$ and $(\textbf{q})$ are known, then its Sales Past must be:

$$S' = (l\{U+S[1-i]-S'f[1+s]-T-Ad\}-Q)$$
$$/[v[1+s](p/\{360[c-q]\}-l)]$$

Rule-21653:

If both $(\textbf{\textit{l}})$, $(\textbf{U})$, $(\textbf{S})$, $(\textbf{i})$, $(\textbf{S'})$, $(\textbf{Q})$, $(\textbf{v})$, $(\textbf{f})$, $(\textbf{T})$, $(\textbf{A})$, $(\textbf{d})$, $(\textbf{p})$, $(\textbf{c})$ and $(\textbf{q})$ are known, then its Sales Growth Planned is:

$$s = S/S'-1$$

or it can also be found as

$$s = (l\{U+S[1-i]-S'f[1+s]-T-Ad\}-Q)$$
$$/[S'v(p/\{360[c-q]\}-l)]-1$$

Rule-21654:

If both $(\textbf{\textit{l}})$, $(\textbf{U})$, $(\textbf{S})$, $(\textbf{i})$, $(\textbf{S'})$, $(\textbf{s})$, $(\textbf{Q})$, $(\textbf{f})$, $(\textbf{T})$, $(\textbf{A})$, $(\textbf{d})$, $(\textbf{p})$, $(\textbf{c})$ and $(\textbf{q})$ are known, then its Variable Portion Planned is:

$$v = (l\{U+S[1-i]-S'f[1+s]-T-Ad\}-Q)$$
$$/[S'[1+s](p/\{360[c-q]\}-l)]$$

Rule-21655:

If both $(\textbf{\textit{l}})$, $(\textbf{U})$, $(\textbf{S})$, $(\textbf{i})$, $(\textbf{S'})$, $(\textbf{s})$, $(\textbf{v})$, $(\textbf{Q})$, $(\textbf{T})$, $(\textbf{A})$, $(\textbf{d})$, $(\textbf{p})$, $(\textbf{c})$ and $(\textbf{q})$ are known, then its Fixed Portion Planned is:

$$f = [U+S[1-i]-T-Ad-(Q+S'vp[1+s]/\{360[c-q]\})/l]$$
$$/\{S'[1+s]\}-v$$

Steve Asikin ISBN 14: 978-1515158691, ISBN 10: 1515158691

Rule-21656:

If both (I), (U), $(\$)$, (i), $(\$')$, (s), (v), (f), (Q), (A), (d), (p), (c) and (q) are known, then its Tax Planned is:

$$T = U + \$[1-i] - \$'[1+s][v+f] - Ad$$
$$- (Q + \$'vp[1+s]/\{360[c-q]\})/I$$

Rule-21657:

If both (I), (U), $(\$)$, (i), $(\$')$, (s), (v), (f), (T), (Q), (d), (p), (c) and (q) are known, then its After Tax Income Planned is:

$$A = [U + \$[1-i] - \$'[1+s][v+f] - T$$
$$- (Q + \$'vp[1+s]/\{360[c-q]\})/I]/d$$

Rule-21658:

If both (I), (U), $(\$)$, (i), $(\$')$, (s), (v), (f), (T), (A), (Q), (p), (c) and (q) are known, then its Dividend Payout Planned is:

$$d = [U + \$[1-i] - \$'[1+s][v+f] - T$$
$$- (Q + \$'vp[1+s]/\{360[c-q]\})/I]/A$$

Rule-21659:

If both (I), (U), $(\$)$, (i), $(\$')$, (s), (v), (f), (T), (A), (d), (Q), (c) and (q) are known, then its Procured Inventory Days Planned:

$$p = 360[c-q](I\{U + \$[1-i] - \$'[1+s][v+f] - T - Ad\} - Q)$$
$$/\{\$'v[1+s]\}$$

Steve Asikin ISBN 14: 978-1515158691, ISBN 10: 1515158691

Rule-21660:

If both (**/**), (**U**), (**S**), (**i**), (**S'**), (**s**), (**v**), (**f**), (**T**), (**A**), (**d**), (**p**), (**Q**) and (**q**) are known, then its Current Ratio Planned is:

$$c = q + S'vp[1+s]$$
$$/[360(\textit{/}\{U + S[1-i] - S'[1+s][v+f] - T - Ad\} - Q)]$$

Rule-21661:

If both (**/**), (**U**), (**S**), (**i**), (**S'**), (**s**), (**v**), (**f**), (**T**), (**A**), (**d**), (**p**), (**c**) and (**Q**) are known, then its Quick or Acid Test Ratio Planned is:

$$q = c - S'vp[1+s]$$
$$/[360(\textit{/}\{U + S[1-i] - S'[1+s][v+f] - T - Ad\} - Q)]$$

Rule-21662:

If both (**/**), (**U**), (**S**), (**i**), (**S'**), (**s**), (**v**), (**f**), (**t**), (**D**) and (**X**) are known, then its Quoted Longterm Debt Planned is:

$$Q = \textit{/}U + \{S[1-i] - S'[1+s][v+f]\}[1-t] - D) - X$$

Rule-21663:

If both (**Q**), (**U**), (**S**), (**i**), (**S'**), (**s**), (**v**), (**f**), (**t**), (**D**) and (**X**) are known, then its Leverage or Gearing Ratio Planned is:

$$\textit{/} = [Q+X]/(U + \{S[1-i] - S'[1+s][v+f]\}[1-t] - D)$$

Rule-21664:

If both (**/**), (**Q**), (**S**), (**i**), (**S'**), (**s**), (**v**), (**f**), (**t**), (**D**) and (**X**) are known, then its Utilized or Starting Capital must be:

$$U = D + [Q+X]/\textit{/}\{S[1-i] - S'[1+s][v+f]\}[1-t]$$

Steve Asikin ISBN 14: 978-1515158691, ISBN 10: 1515158691

<u>Rule-21665</u>:

If both (f), (U), (Q), (i), (S'), (s), (v), (f), (t), (D) and (X) are known, then its Sales or Revenue Planned is:

$S= (S'[1+s][v+f]+\{D+[Q+X]/f\cdot U\}/[1-t])/[1-i]$

<u>Rule-21666</u>:

If both (f), (U), (S), (Q), (S'), (s), (v), (f), (t), (D) and (X) are known, then its Interest Portion Planned is:

$i= 1-(S'[1+s][v+f]+\{D+[Q+X]/f\cdot U\}/[1-t])/S$

<u>Rule-21667</u>:

If both (f), (U), (S), (i), (Q), (s), (v), (f), (t), (D) and (X) are known, then its Sales Past must be:

$S'= (S[1-i]-\{D+[Q+X]/f\cdot U\}/[1-t])/\{[1+s][v+f]\}$

<u>Rule-21668</u>:

If both (f), (U), (S), (i), (S'), (Q), (v), (f), (t), (D) and (X) are known, then its Sales Growth Planned is:

$s= S/S'-1$

or it can also be found as

$s= (S[1-i]-\{D+[Q+X]/f\cdot U\}/[1-t])/\{S'[v+f]\}-1$

<u>Rule-21669</u>:

If both (f), (U), (S), (i), (S'), (s), (Q), (f), (t), (D) and (X) are known, then its Variable Portion Planned is:

$v= (S[1-i]-\{D+[Q+X]/f\cdot U\}/[1-t])/\{S'[1+s]\}-f$

Steve Asikin ISBN 14: 978-1515158691, ISBN 10: 1515158691

Rule-21670:

If both (**/**), (**U**), (**$**), (**i**), (**$'**), (**s**), (**v**), (**Q**), (**t**), (**D**) and (**X**) are known, then its Fixed Portion Planned is:

$f = (\$[1\text{-}i]\text{-}\{D+[Q+X]/\text{/}U\}/[1\text{-}t])/\{\$'[1+s]\}\text{-}v$

Rule-21671:

If both (**/**), (**U**), (**$**), (**i**), (**$'**), (**s**), (**v**), (**f**), (**Q**), (**D**) and (**X**) are known, then its Tax Rate Planned is:

$t= 1\text{-}\{D+[Q+X]/\text{/}U\}/\{\$[1\text{-}i]\text{-}\$'[1+s][v+f]\}$

Rule-21672:

If both (**/**), (**U**), (**$**), (**i**), (**$'**), (**s**), (**v**), (**f**), (**t**), (**Q**) and (**X**) are known, then its Dividend Planned is:

$D= U+\{\$[1\text{-}i]\text{-}\$'[1+s][v+f]\}[1\text{-}t]\text{-}[Q+X]/\text{/}$

Rule-21673:

If both (**/**), (**U**), (**$**), (**i**), (**$'**), (**s**), (**v**), (**f**), (**t**), (**D**) and (**Q**) are known, then its Xpress or Current Debt Planned is:

$X= \text{/}(U+\{\$[1\text{-}i]\text{-}\$'[1+s][v+f]\}[1\text{-}t]\text{-}D)\text{-}Q$

Rule-21674:

If both (**/**), (**U**), (**$**), (**i**), (**$'**), (**s**), (**v**), (**f**), (**t**), (**D**), (**P**), (**c**) and (**q**) are known, then its Quoted Longterm Debt Planned is:

$Q= \text{/}(U+\{\$[1\text{-}i]\text{-}\$'[1+s][v+f]\}[1\text{-}t]\text{-}D)\text{-}P/[c\text{-}q]$

Steve Asikin ISBN 14: 978-1515158691, ISBN 10: 1515158691

Rule-21675:

If both (**Q**), (**U**), (**$**), (**i**), (**$'**), (**s**), (**v**), (**f**), (**t**), (**D**), (**P**), (**c**) and (**q**) are known, then its Leverage or Gearing Ratio Planned is:

$$\textit{l} = \{Q+P/[c\text{-}q]\}/(U+\{\$[1\text{-}i]\text{-}\$'[1+s][v+f]\}[1\text{-}t]\text{-}D)$$

Rule-21676:

If both (**l**), (**Q**), (**$**), (**i**), (**$'**), (**s**), (**v**), (**f**), (**t**), (**D**), (**P**), (**c**) and (**q**) are known, then its Utilized or Starting Capital must be:

$$U = D+\{Q+P/[c\text{-}q]\}/\textit{l}\{\$[1\text{-}i]\text{-}\$'[1+s][v+f]\}[1\text{-}t]$$

Rule-21677:

If both (**l**), (**U**), (**Q**), (**i**), (**$'**), (**s**), (**v**), (**f**), (**t**), (**D**), (**P**), (**c**) and (**q**) are known, then its Sales or Revenue Planned is:

$$\$ = [\$'[1+s][v+f]+(D+\{Q+P/[c\text{-}q]\}/\textit{l}U)/[1\text{-}t]]/[1\text{-}i]$$

Rule-21678:

If both (**l**), (**U**), (**$**), (**Q**), (**$'**), (**s**), (**v**), (**f**), (**t**), (**D**), (**P**), (**c**) and (**q**) are known, then its Interest Portion Planned is:

$$i = 1\text{-}[\$'[1+s][v+f]+(D+\{Q+P/[c\text{-}q]\}/\textit{l}U)/[1\text{-}t]]/\$$$

Rule-21679:

If both (**l**), (**U**), (**$**), (**i**), (**Q**), (**s**), (**v**), (**f**), (**t**), (**D**), (**P**), (**c**) and (**q**) are known, then its Sales Past must be:

$$\$' = [\$[1\text{-}i]\text{-}(D+\{Q+P/[c\text{-}q]\}/\textit{l}U)/[1\text{-}t]]$$
$$/\{[1+s][v+f]\}$$

Steve Asikin ISBN 14: 978-1515158691, ISBN 10: 1515158691

Rule-21680:

If both (**/**), (**U**), (**$**), (**i**), (**$'**), (**Q**), (**v**), (**f**), (**t**), (**D**), (**P**), (**c**) and (**q**) are known, then its Sales Growth Planned is:

$s = $/$'-1$

or it can also be found as

$s = [$[1-i]-(D+\{Q+P/[c-q]\}/IU)/[1-t]]/\{$'[v+f]\}-1$

Rule-21681:

If both (**/**), (**U**), (**$**), (**i**), (**$'**), (**s**), (**Q**), (**f**), (**t**), (**D**), (**P**), (**c**) and (**q**) are known, then its Variable Portion Planned is:

$v = [$[1-i]-(D+\{Q+P/[c-q]\}/IU)/[1-t]]/\{$'[1+s]\}-f$

Rule-21682:

If both (**/**), (**U**), (**$**), (**i**), (**$'**), (**s**), (**v**), (**Q**), (**t**), (**D**), (**P**), (**c**) and (**q**) are known, then its Fixed Portion Planned is:

$f = [$[1-i]-(D+\{Q+P/[c-q]\}/IU)/[1-t]]/\{$'[1+s]\}-v$

Rule-21683:

If both (**/**), (**U**), (**$**), (**i**), (**$'**), (**s**), (**v**), (**f**), (**Q**), (**D**), (**P**), (**c**) and (**q**) are known, then its Tax Rate Planned is:

$t = 1-(D+\{Q+P/[c-q]\}/IU)/\{$[1-i]-$'[1+s][v+f]\}$

Rule-21684:

If both (**/**), (**U**), (**$**), (**i**), (**$'**), (**s**), (**v**), (**f**), (**t**), (**Q**), (**P**), (**c**) and (**q**) are known, then its Dividend Planned is:

$D = U+\{$[1-i]-$'[1+s][v+f]\}[1-t]-\{Q+P/[c-q]\}/I$

Steve Asikin ISBN 14: 978-1515158691, ISBN 10: 1515158691

Rule-21685:

If both (I), (U), $(\$)$, (i), $(\$')$, (s), (v), (f), (t), (D), (Q), (c) and (q) are known, then its Procured Inventory Planned is:

$$P = [c\text{-}q][I(U + \{\$[1\text{-}i]\text{-}\$'[1+s][v+f]\}[1\text{-}t]\text{-}D)\text{-}Q]$$

Rule-21686:

If both (I), (U), $(\$)$, (i), $(\$')$, (s), (v), (f), (t), (D), (P), (Q) and (q) are known, then its Current Ratio Planned is:

$$c = q + P/[I(U + \{\$[1\text{-}i]\text{-}\$'[1+s][v+f]\}[1\text{-}t]\text{-}D)\text{-}Q]$$

Rule-21687:

If both (I), (U), $(\$)$, (i), $(\$')$, (s), (v), (f), (t), (D), (P), (c) and (Q) are known, then its Quick or Acid Test Ratio Planned is:

$$q = c - P/[I(U + \{\$[1\text{-}i]\text{-}\$'[1+s][v+f]\}[1\text{-}t]\text{-}D)\text{-}Q]$$

Rule-21688:

If both (I), (U), $(\$)$, (i), $(\$')$, (s), (v), (f), (t), (D), (V), (p), (c) and (q) are known, then its Quoted Longterm Debt Planned is:

$$Q = I(U + \{\$[1\text{-}i]\text{-}\$'[1+s][v+f]\}[1\text{-}t]\text{-}D)$$
$$-Vp/\{360[c\text{-}q]\}$$

Rule-21689:

If both (Q), (U), $(\$)$, (i), $(\$')$, (s), (v), (f), (t), (D), (V), (p), (c) and (q) are known, then its Leverage or Gearing Ratio Planned is:

$$I = (Q + Vp/\{360[c\text{-}q]\})$$
$$/(U + \{\$[1\text{-}i]\text{-}\$'[1+s][v+f]\}[1\text{-}t]\text{-}D)$$

Steve Asikin ISBN 14: 978-1515158691, ISBN 10: 1515158691

Rule-21690:

If both (I), (Q), (S), (i), (S'), (s), (v), (f), (t), (D), (V), (p), (c) and (q) are known, then its Utilized or Starting Capital must be:

$$U = D + (Q + Vp/\{360[c-q]\})/I$$
$$-\{S[1-i]-S'[1+s][v+f]\}[1-t]$$

Rule-21691:

If both (I), (U), (Q), (i), (S'), (s), (v), (f), (t), (D), (V), (p), (c) and (q) are known, then its Sales or Revenue Planned is:

$$S = \{S'[1+s][v+f]+[D+(Q+Vp/\{360[c-q]\})/I-U]$$
$$/[1-t]\}/[1-i]$$

Rule-21692:

If both (I), (U), (S), (Q), (S'), (s), (v), (f), (t), (D), (V), (p), (c) and (q) are known, then its Interest Portion Planned is:

$$i = 1-\{S'[1+s][v+f]$$
$$+[D+(Q+Vp/\{360[c-q]\})/I-U]/[1-t]\}/S$$

Rule-21693:

If both (I), (U), (S), (i), (Q), (s), (v), (f), (t), (D), (V), (p), (c) and (q) are known, then its Sales Past must be:

$$S' = \{S[1-i]-[D+(Q+Vp/\{360[c-q]\})/I-U]/[1-t]\}$$
$$/\{[1+s][v+f]\}$$

Steve Asikin ISBN 14: 978-1515158691, ISBN 10: 1515158691

Rule-21694:

If both $(/)$, $(\mathbf{U})$, $(\$)$, $(\mathbf{i})$, $(\$')$, $(\mathbf{Q})$, $(\mathbf{v})$, $(\mathbf{f})$, $(\mathbf{t})$, $(\mathbf{D})$, $(\mathbf{V})$, $(\mathbf{p})$, $(\mathbf{c})$ and $(\mathbf{q})$ are known, then its Sales Growth Planned is:

$$s = \$/\$' - 1$$

or it can also be found as

$$s = \{\$[1\text{-}i]\text{-}[D+(Q+Vp/\{360[c\text{-}q]\})/\textit{f}U]/[1\text{-}t]\} / \{\$'[v+f]\} - 1$$

Rule-21695:

If both $(/)$, $(\mathbf{U})$, $(\$)$, $(\mathbf{i})$, $(\$')$, (s), $(\mathbf{Q})$, $(\mathbf{f})$, $(\mathbf{t})$, $(\mathbf{D})$, $(\mathbf{V})$, $(\mathbf{p})$, $(\mathbf{c})$ and $(\mathbf{q})$ are known, then its Variable Portion Planned is:

$$v = V/\$$$

or it can also be found as

$$v = \{\$[1\text{-}i]\text{-}[D+(Q+Vp/\{360[c\text{-}q]\})/\textit{f}U]/[1\text{-}t]\} / \{\$'[1+s]\} - f$$

Rule-21696:

If both $(/)$, $(\mathbf{U})$, $(\$)$, $(\mathbf{i})$, $(\$')$, (s), $(\mathbf{v})$, $(\mathbf{Q})$, $(\mathbf{t})$, $(\mathbf{D})$, $(\mathbf{V})$, $(\mathbf{p})$, $(\mathbf{c})$ and $(\mathbf{q})$ are known, then its Fixed Portion Planned is:

$$f = \{\$[1\text{-}i]\text{-}[D+(Q+Vp/\{360[c\text{-}q]\})/\textit{f}U]/[1\text{-}t]\} / \{\$'[1+s]\} - v$$

Rule-21697:

If both $(/)$, $(\mathbf{U})$, $(\$)$, $(\mathbf{i})$, $(\$')$, (s), $(\mathbf{v})$, $(\mathbf{f})$, $(\mathbf{Q})$, $(\mathbf{D})$, $(\mathbf{V})$, $(\mathbf{p})$, $(\mathbf{c})$ and $(\mathbf{q})$ are known, then its Tax Rate Planned is:

$$t = 1 - [D+(Q+Vp/\{360[c\text{-}q]\})/\textit{f}U] / \{\$[1\text{-}i]\text{-}\$'[1+s][v+f]\}$$

Steve Asikin ISBN 14: 978-1515158691, ISBN 10: 1515158691

Rule-21698:

If both (I), (**U**), (**$**), (**i**), (**$'**), (**s**), (**v**), (**f**), (**t**), (**Q**), (**V**), (**p**), (**c**) and (**q**) are known, then its Dividend Planned is:

$$D= U+\{\$[1\text{-}i]\text{-}\$'[1+s][v+f]\}[1\text{-}t]$$
$$-(Q+Vp/\{360[c\text{-}q]\})/I$$

Rule-21699:

If both (I), (**U**), (**$**), (**i**), (**$'**), (**s**), (**v**), (**f**), (**t**), (**D**), (**Q**), (**p**), (**c**) and (**q**) are known, then its Variable Cost Planned is:

$$V= 360[c\text{-}q]$$
$$[(IU+\{\$[1\text{-}i]\text{-}\$'[1+s][v+f]\}[1\text{-}t]\text{-}D)\text{-}Q]/p$$

Rule-21700:

If both (I), (**U**), (**$**), (**i**), (**$'**), (**s**), (**v**), (**f**), (**t**), (**D**), (**V**), (**Q**), (**c**) and (**q**) are known, then its Procured Inventory Days Planned:

$$p= 360[c\text{-}q]$$
$$[(IU+\{\$[1\text{-}i]\text{-}\$'[1+s][v+f]\}[1\text{-}t]\text{-}D)\text{-}Q]/V$$

Rule-21701:

If both (I), (**U**), (**$**), (**i**), (**$'**), (**s**), (**v**), (**f**), (**t**), (**D**), (**V**), (**p**), (**Q**) and (**q**) are known, then its Current Ratio Planned is:

$$c= q+Vp$$
$$/\{360[(IU+\{\$[1\text{-}i]\text{-}\$'[1+s][v+f]\}[1\text{-}t]\text{-}D)\text{-}Q]\}$$

Steve Asikin ISBN 14: 978-1515158691, ISBN 10: 1515158691

Rule-21702:
  If both (*I*), (**U**), (**$**), (**i**), (**$'**), (**s**), (**v**), (**f**), (**t**), (**D**), (**V**),
  (**p**), (**c**) and (**Q**) are known, then its Quoted Longterm
  Debt Planned is:
  $$q = c\text{-}Vp/\{360[\textit{I}U + \{\$[1\text{-}i]\text{-}\$'[1+s][v+f]\} \\ [1\text{-}t]\text{-}D)\text{-}Q]\}$$

Rule-21703:
  If both (*I*), (**U**), (**$**), (**i**), (**$'**), (**s**), (**v**), (**f**), (**t**), (**D**), (**p**),
  (**c**) and (**q**) are known, then its Quoted Longterm Debt
  Planned is:
  $$Q = \textit{I}U + \{\$[1\text{-}i]\text{-}\$'[1+s][v+f]\}[1\text{-}t]\text{-}D) \\ \text{-}\$vp/\{360[c\text{-}q]\}$$

Rule-21704:
  If both (**Q**), (**U**), (**$**), (**i**), (**$'**), (**s**), (**v**), (**f**), (**t**), (**D**), (**p**),
  (**c**) and (**q**) are known, then its Leverage or Gearing
  Ratio Planned is:
  $$\textit{I} = (Q + \$vp/\{360[c\text{-}q]\}) \\ /(U + \{\$[1\text{-}i]\text{-}\$'[1+s][v+f]\}[1\text{-}t]\text{-}D)$$

Rule-21705:
  If both (*I*), (**Q**), (**$**), (**i**), (**$'**), (**s**), (**v**), (**f**), (**t**), (**D**), (**p**),
  (**c**) and (**q**) are known, then its Utilized or Starting
  Capital must be:
  $$U = D + (Q + \$vp/\{360[c\text{-}q]\})/\textit{I} \\ \text{-}\{\$[1\text{-}i]\text{-}\$'[1+s][v+f]\}[1\text{-}t]$$

Steve Asikin ISBN 14: 978-1515158691, ISBN 10: 1515158691

Rule-21706:

If both (I), (U), (Q), (i), (S'), (s), (v), (f), (t), (D), (p), (c) and (q) are known, then its Sales or Revenue Planned is:

$$S= (I\{U-S'[1+s][v+f][1-t]-D\}-Q)$$
$$/(vp/\{360[c-q]\}-I[1-i][1-t])$$

Rule-21707:

If both (I), (U), (S), (Q), (S'), (s), (v), (f), (t), (D), (p), (c) and (q) are known, then its Interest Portion Planned is:

$$i= 1-\{S'[1+s][v+f]$$
$$+[D+(Q+Svp/\{360[c-q]\})/IU]/[1-t]\}/S$$

Rule-21708:

If both (I), (U), (S), (i), (Q), (s), (v), (f), (t), (D), (p), (c) and (q) are known, then its Sales Past must be:

$$S'= \{S[1-i]-[D+(Q+Svp/\{360[c-q]\})/IU]/[1-t]\}$$
$$/\{[1+s][v+f]\}$$

Rule-21709:

If both (I), (U), (S), (i), (S'), (Q), (v), (f), (t), (D), (p), (c) and (q) are known, then its Sales Growth Planned is:

$$s= S/S'-1$$

or it can also be found as

$$s= \{S[1-i]-[D+(Q+Svp/\{360[c-q]\})/IU]/[1-t]\}$$
$$/\{S'[v+f]\}-1$$

Steve Asikin ISBN 14: 978-1515158691, ISBN 10: 1515158691

Rule-21710:

If both (l), (**U**), (**$**), (**i**), (**$'**), (**s**), (**Q**), (**f**), (**t**), (**D**), (**p**), (**c**) and (**q**) are known, then its Variable Portion Planned is:

$$v= [l(U+\{\$[1-i]-\$'f[1+s]\}[1-t]-D)-Q] /(\$p/\{360[c-q]\}+\$'l[1+s][1-t])$$

Rule-21711:

If both (l), (**U**), (**$**), (**i**), (**$'**), (**s**), (**v**), (**Q**), (**t**), (**D**), (**p**), (**c**) and (**q**) are known, then its Fixed Portion Planned is:

$$f = \{\$[1-i]-[D+(Q+\$vp/\{360[c-q]\})/lU]/[1-t]\} /\{\$'[1+s]\}-v$$

Rule-21712:

If both (l), (**U**), (**$**), (**i**), (**$'**), (**s**), (**v**), (**f**), (**Q**), (**D**), (**p**), (**c**) and (**q**) are known, then its Tax Rate Planned is:

$$t= 1-[D+(Q+\$vp/\{360[c-q]\})/lU] /\{\$[1-i]-\$'[1+s][v+f]\}$$

Rule-21713:

If both (l), (**U**), (**$**), (**i**), (**$'**), (**s**), (**v**), (**f**), (**t**), (**Q**), (**p**), (**c**) and (**q**) are known, then its Dividend Planned is:

$$D= U+\{\$[1-i]-\$'[1+s][v+f]\}[1-t] -(Q+\$vp/\{360[c-q]\})/l$$

Rule-21714:

If both (l), (**U**), (**$**), (**i**), (**$'**), (**s**), (**v**), (**f**), (**t**), (**D**), (**Q**), (**c**) and (**q**) are known, then its Procured Inventory Days Planned:

$$p= 360[c-q] [l(U+\{\$[1-i]-\$'[1+s][v+f]\}[1-t]-D)-Q]/[\$v]$$

Steve Asikin ISBN 14: 978-1515158691, ISBN 10: 1515158691

Rule-21715:

If both (I), (U), (S), (i), (S'), (s), (v), (f), (t), (D), (p), (Q) and (q) are known, then its Current Ratio Planned is:

$$c= q+Svp$$
$$/\{360[I U+\{S[1-i]-S'[1+s][v+f]\}[1-t]-D)-Q]\}$$

Rule-21716:

If both (I), (U), (S), (i), (S'), (s), (v), (f), (t), (D), (p), (c) and (Q) are known, then its Quick or Acid Test Ratio Planned is:

$$q= c-Svp$$
$$/\{360[I U+\{S[1-i]-S'[1+s][v+f]\}[1-t]-D)-Q]\}$$

Rule-21717:

If both (I), (U), (S), (i), (S'), (s), (v), (f), (t), (D), (p), (c) and (q) are known, then its Quoted Longterm Debt Planned is:

$$Q= I U+\{S[1-i]-S'[1+s][v+f]\}[1-t]-D)$$
$$-S'vp[1+s]/\{360[c-q]\}$$

Rule-21718:

If both (Q), (U), (S), (i), (S'), (s), (v), (f), (t), (D), (p), (c) and (q) are known, then its Leverage or Gearing Ratio Planned is:

$$I= (Q+S'vp[1+s]/\{360[c-q]\})$$
$$/(U+\{S[1-i]-S'[1+s][v+f]\}[1-t]-D)$$

Steve Asikin ISBN 14: 978-1515158691, ISBN 10: 1515158691

Rule-21719:

If both (**/**), (**Q**), (**$**), (**i**), (**$'**), (**s**), (**v**), (**f**), (**t**), (**D**), (**p**), (**c**) and (**q**) are known, then its Utilized or Starting Capital must be:

$$U = D + (Q + S'vp[1+s]/\{360[c-q]\})//$$
$$-\{S[1-i]-S'[1+s][v+f]\}[1-t]$$

Rule-21720:

If both (**/**), (**U**), (**Q**), (**i**), (**$'**), (**s**), (**v**), (**f**), (**t**), (**D**), (**p**), (**c**) and (**q**) are known, then its Sales or Revenue Planned is:

$$S = [Q-/(U+\{ -S'[1+s][v+f]\}[1-t]-D)$$
$$+S'vp[1+s]/\{360[c-q]\}]/\{/[1-i][1-t]\}$$

Rule-21721:

If both (**/**), (**U**), (**$**), (**Q**), (**$'**), (**s**), (**v**), (**f**), (**t**), (**D**), (**p**), (**c**) and (**q**) are known, then its Interest Portion Planned is:

$$i = 1-\{S'[1+s][v+f]+[D+(Q+S'vp[1+s]$$
$$/\{360[c-q]\})//U]/[1-t]\}/S$$

Rule-21722:

If both (**/**), (**U**), (**$**), (**i**), (**Q**), (**s**), (**v**), (**f**), (**t**), (**D**), (**p**), (**c**) and (**q**) are known, then its Sales Past must be:

$$S' = [/U+\{S[1-i]-S'f[1+s]\}[1-t]-D)-Q]$$
$$/[v[1+s](p/\{360[c-q]\}+/[1-t])]$$

Steve Asikin ISBN 14: 978-1515158691, ISBN 10: 1515158691

Rule-21723:

If both (*ƒ*), (**U**), (**$**), (**i**), (**$'**), (**Q**), (**v**), (**f**), (**t**), (**D**), (**p**), (**c**) and (**q**) are known, then its Sales Growth Planned is:

$s = \$/\$' - 1$

or it can also be found as

$$s = [\textit{ƒ}U + \{\$[1-i] - \$'f[1+s]\}[1-t] - D) - Q]$$
$$/[\$'v(p/\{360[c-q]\} + \textit{ƒ}[1-t])] - 1$$

Rule-21724:

If both (*ƒ*), (**U**), (**$**), (**i**), (**$'**), (**s**), (**Q**), (**f**), (**t**), (**D**), (**p**), (**c**) and (**q**) are known, then its Variable Portion Planned is:

$$v = [\textit{ƒ}U + \{\$[1-i] - \$'f[1+s]\}[1-t] - D) - Q]$$
$$/[\$'[1+s](p/\{360[c-q]\} + \textit{ƒ}[1-t])]$$

Rule-21725:

If both (*ƒ*), (**U**), (**$**), (**i**), (**$'**), (**s**), (**v**), (**Q**), (**t**), (**D**), (**p**), (**c**) and (**q**) are known, then its Fixed Portion Planned is:

$$f = \{\$[1-i] - [D + (Q + \$'vp[1+s]/\{360[c-q]\})/\textit{ƒ}U]$$
$$/[1-t]\}/\{\$'[1+s]\} - v$$

Rule-21726:

If both (*ƒ*), (**U**), (**$**), (**i**), (**$'**), (**s**), (**v**), (**f**), (**Q**), (**D**), (**p**), (**c**) and (**q**) are known, then its Tax Rate Planned is:

$$t = 1 - [D + (Q + \$'vp[1+s]/\{360[c-q]\})/\textit{ƒ}U]$$
$$/\{\$[1-i] - \$'[1+s][v+f]\}$$

Steve Asikin ISBN 14: 978-1515158691, ISBN 10: 1515158691

Rule-21727:

If both (I), (U), $(\$)$, (i), $(\$')$, (s), (v), (f), (t), (Q), (p), (c) and (q) are known, then its Dividend Planned is:

$$D = U + \{\$[1-i] - \$'[1+s][v+f]\}[1-t]$$
$$-(Q + \$'vp[1+s]/\{360[c-q]\})/I$$

Rule-21728:

If both (I), (U), $(\$)$, (i), $(\$')$, (s), (v), (f), (t), (D), (Q), (c) and (q) are known, then its Procured Inventory Days Planned:

$$p = 360[c-q][I(U + \{\$[1-i] - \$'[1+s][v+f]\}[1-t] - D) - Q]$$
$$/\{\$'v[1+s]\}$$

Rule-21729:

If both (I), (U), $(\$)$, (i), $(\$')$, (s), (v), (f), (t), (D), (p), (Q) and (q) are known, then its Current Ratio Planned is:

$$c = q + \$'vp[1+s]$$
$$/\{360[I(U + \{\$[1-i] - \$'[1+s][v+f]\}[1-t] - D) - Q]\}$$

Rule-21730:

If both (I), (U), $(\$)$, (i), $(\$')$, (s), (v), (f), (t), (D), (p), (c) and (q) are known, then its Quick or Acid Test Ratio Planned is:

$$q = c - \$'vp[1+s]$$
$$/\{360[I(U + \{\$[1-i] - \$'[1+s][v+f]\}[1-t] - D) - Q]\}$$

Steve Asikin ISBN 14: 978-1515158691, ISBN 10: 1515158691

Rule-21731:

If both (f), $(\mathbf{U})$, $(\$)$, (i), $(\$')$, (s), (v), (f), (t), $(\mathbf{A})$, (d) and $(\mathbf{X})$ are known, then its Quoted Longterm Debt Planned is:

$\mathbf{Q} = f(\mathbf{U} + \{\$[1-i] - \$'[1+s][v+f]\}[1-t] - \mathbf{Ad}) - \mathbf{X}$

Rule-21732:

If both $(\mathbf{Q})$, $(\mathbf{U})$, $(\$)$, (i), $(\$')$, (s), (v), (f), (t), $(\mathbf{A})$, (d) and $(\mathbf{X})$ are known, then its Leverage or Gearing Ratio Planned is:

$f = [\mathbf{Q} + \mathbf{X}]/(\mathbf{U} + \{\$[1-i] - \$'[1+s][v+f]\}[1-t] - \mathbf{Ad})$

Rule-21733:

If both (f), $(\mathbf{Q})$, $(\$)$, (i), $(\$')$, (s), (v), (f), (t), $(\mathbf{A})$, (d) and $(\mathbf{X})$ are known, then its Utilized or Starting Capital must be:

$\mathbf{U} = \$'[1+s][v+f]\}[1-t] + \mathbf{Ad} + [\mathbf{Q} + \mathbf{X}]/f\$[1-i]$

Rule-21734:

If both (f), $(\mathbf{U})$, $(\mathbf{Q})$, (i), $(\$')$, (s), (v), (f), (t), $(\mathbf{A})$, (d) and $(\mathbf{X})$ are known, then its Sales or Revenue Planned is:

$\$ = \{\$'[1+s][v+f][1-t] + \mathbf{Ad} + [\mathbf{Q} + \mathbf{X}]/f\mathbf{U}]/[1-i]$

Rule-21735:

If both (f), $(\mathbf{U})$, $(\$)$, $(\mathbf{Q})$, $(\$')$, (s), (v), (f), (t), $(\mathbf{A})$, (d) and $(\mathbf{X})$ are known, then its Interest Portion Planned is:

$i = 1 - \{\$'[1+s][v+f][1-t] + \mathbf{Ad} + [\mathbf{Q} + \mathbf{X}]/f\mathbf{U}]/\$$

Steve Asikin ISBN 14: 978-1515158691, ISBN 10: 1515158691

Rule-21736:

If both (l), $(\mathbf{U})$, $(\mathbf{S})$, $(\mathbf{i})$, $(\mathbf{Q})$, $(\mathbf{s})$, $(\mathbf{v})$, $(\mathbf{f})$, $(\mathbf{t})$, $(\mathbf{A})$, $(\mathbf{d})$ and $(\mathbf{X})$ are known, then its Sales Past must be:

$$S'= \{U+S[1\text{-}i]\text{-}Ad\text{-}[Q+X]/\mathit{l}\}/\{[1+s][v+f][1\text{-}t]\}$$

Rule-21737:

If both (l), $(\mathbf{U})$, $(\mathbf{S})$, $(\mathbf{i})$, $(\mathbf{S'})$, $(\mathbf{Q})$, $(\mathbf{v})$, $(\mathbf{f})$, $(\mathbf{t})$, $(\mathbf{A})$, $(\mathbf{d})$ and $(\mathbf{X})$ are known, then its Sales Growth Planned is:

$$s= S/S'\text{-}1$$

or it can also be found as

$$s= \{U+S[1\text{-}i]\text{-}Ad\text{-}[Q+X]/\mathit{l}\}/\{S'[v+f][1\text{-}t]\}\text{-}1$$

Rule-21738:

If both (l), $(\mathbf{U})$, $(\mathbf{S})$, $(\mathbf{i})$, $(\mathbf{S'})$, $(\mathbf{s})$, $(\mathbf{Q})$, $(\mathbf{f})$, $(\mathbf{t})$, $(\mathbf{A})$, $(\mathbf{d})$ and $(\mathbf{X})$ are known, then its Variable Portion Planned is:

$$v= \{U+S[1\text{-}i]\text{-}Ad\text{-}[Q+X]/\mathit{l}\}/\{S'[1+s][1\text{-}t]\}\text{-}f$$

Rule-21739:

If both (l), $(\mathbf{U})$, $(\mathbf{S})$, $(\mathbf{i})$, $(\mathbf{S'})$, $(\mathbf{s})$, $(\mathbf{v})$, $(\mathbf{Q})$, $(\mathbf{t})$, $(\mathbf{A})$, $(\mathbf{d})$ and $(\mathbf{X})$ are known, then its Fixed Portion Planned is:

$$f= \{U+S[1\text{-}i]\text{-}Ad\text{-}[Q+X]/\mathit{l}\}/\{S'[1+s][1\text{-}t]\}\text{-}v$$

Rule-21740:

If both (l), $(\mathbf{U})$, $(\mathbf{S})$, $(\mathbf{i})$, $(\mathbf{S'})$, $(\mathbf{s})$, $(\mathbf{v})$, $(\mathbf{f})$, $(\mathbf{Q})$, $(\mathbf{A})$, $(\mathbf{d})$ and $(\mathbf{X})$ are known, then its Tax Rate Planned is:

$$t= 1\text{-}\{U+S[1\text{-}i]\text{-}Ad\text{-}[Q+X]/\mathit{l}\}/\{S'[1+s][v+f]\}$$

Steve Asikin ISBN 14: 978-1515158691, ISBN 10: 1515158691

Rule-21741:

If both (I), (U), (S), (i), (S'), (s), (v), (f), (t), (Q), (d) and (X) are known, then its After Tax Income Planned is:

$$A= (U+\{S[1-i]-S'[1+s][v+f]\}[1-t]-[Q+X]/I\}/d$$

Rule-21742:

If both (I), (U), (S), (i), (S'), (s), (v), (f), (t), (A), (Q) and (X) are known, then its Dividend Payout Planned is:

$$d= (U+\{S[1-i]-S'[1+s][v+f]\}[1-t]-[Q+X]/I\}/A$$

Rule-21743:

If both (I), (U), (S), (i), (S'), (s), (v), (f), (t), (A), (d) and (Q) are known, then its Xpress or Current Debt Planned is:

$$X= I(U+\{S[1-i]-S'[1+s][v+f]\}[1-t]-Ad)-Q$$

Rule-21744:

If both (I), (U), (S), (i), (S'), (s), (v), (f), (t), (A), (d), (P), (c) and (q) are known, then its Quoted Longterm Debt Planned is:

$$Q= I(U+\{S[1-i]-S'[1+s][v+f]\}[1-t]-Ad)-P/[c-q]$$

Rule-21745:

If both (Q), (U), (S), (i), (S'), (s), (v), (f), (t), (A), (d), (P), (c) and (q) are known, then its Leverage or Gearing Ratio Planned is:

$$I= \{Q+P/[c-q]\}/(U+\{S[1-i]-S'[1+s][v+f]\}[1-t]-Ad)$$

Steve Asikin ISBN 14: 978-1515158691, ISBN 10: 1515158691

Rule-21746:

If both (**/**), (**Q**), (**$**), (**i**), (**$'**), (**s**), (**v**), (**f**), (**t**), (**A**), (**d**), (**P**), (**c**) and (**q**) are known, then its Utilized or Starting Capital must be:

$$U = \$'[1+s][v+f]\}[1-t]+Ad+\{Q+P/[c-q]\}//\$[1-i]$$

Rule-21747:

If both (**/**), (**U**), (**Q**), (**i**), (**$'**), (**s**), (**v**), (**f**), (**t**), (**A**), (**d**), (**P**), (**c**) and (**q**) are known, then its Sales Growth Planned is:

$$\$ = (\$'[1+s][v+f][1-t]+Ad+\{Q+P/[c-q]\}//U)/[1-i]$$

Rule-21748:

If both (**/**), (**U**), (**$**), (**Q**), (**$'**), (**s**), (**v**), (**f**), (**t**), (**A**), (**d**), (**P**), (**c**) and (**q**) are known, then its Interest Portion Debt Planned is:

$$i = 1-(\$'[1+s][v+f][1-t]+Ad+\{Q+P/[c-q]\}//U)/\$$$

Rule-21749:

If both (**/**), (**U**), (**$**), (**i**), (**Q**), (**s**), (**v**), (**f**), (**t**), (**A**), (**d**), (**P**), (**c**) and (**q**) are known, then its Sales Past must be:

$$\$' = (U+\$[1-i]-Ad-\{Q+P/[c-q]\}//)$$
$$/\{[1+s][v+f][1-t]\}$$

Rule-21750:

If both (**/**), (**U**), (**$**), (**i**), (**$'**), (**s**), (**v**), (**f**), (**t**), (**A**), (**d**), (**P**), (**c**) and (**q**) are known, then its Sales Growth Planned is:

$$s = \$/\$'-1$$

or it can also be found as

$$s = (U+\$[1-i]-Ad-\{Q+P/[c-q]\}//)/\{\$'[v+f][1-t]\}-1$$

Steve Asikin ISBN 14: 978-1515158691, ISBN 10: 1515158691

Rule-21751:

If both (*ʃ*, (**U**), (**$**), (**i**), (**$'**), (**s**), (**Q**), (**f**), (**t**), (**A**), (**d**), (**P**), (**c**) and (**q**) are known, then its Variable Portion Planned is:

$$v = (U + \$[1-i] - Ad - \{Q + P/[c-q]\}/\textit{ʃ}/\{\$'[1+s][1-t]\} - f$$

Rule-21752:

If both (*ʃ*, (**U**), (**$**), (**i**), (**$'**), (**s**), (**v**), (**Q**), (**t**), (**A**), (**d**), (**P**), (**c**) and (**q**) are known, then its Fixed Portion Planned is:

$$f = (U + \$[1-i] - Ad - \{Q + P/[c-q]\}/\textit{ʃ}/\{\$'[1+s][1-t]\} - v$$

Rule-21753:

If both (*ʃ*, (**U**), (**$**), (**i**), (**$'**), (**s**), (**v**), (**f**), (**Q**), (**A**), (**d**), (**P**), (**c**) and (**q**) are known, then its Tax Rate Planned is:

$$t = 1 - (U + \$[1-i] - Ad - \{Q + P/[c-q]\}/\textit{ʃ}/\{\$'[1+s][v+f]\}$$

Rule-21754:

If both (*ʃ*, (**U**), (**$**), (**i**), (**$'**), (**s**), (**v**), (**f**), (**t**), (**Q**), (**d**), (**P**), (**c**) and (**q**) are known, then its After Tax Income Planned is:

$$A = (U + \{\$[1-i] - \$'[1+s][v+f]\}[1-t] - \{Q + P/[c-q]\}/\textit{ʃ}/d$$

Rule-21755:

If both (*ʃ*, (**U**), (**$**), (**i**), (**$'**), (**s**), (**v**), (**f**), (**t**), (**A**), (**Q**), (**P**), (**c**) and (**q**) are known, then its Dividend Payout Planned is:

$$d = (U + \{\$[1-i] - \$'[1+s][v+f]\}[1-t] - \{Q + P/[c-q]\}/\textit{ʃ}/A$$

Steve Asikin ISBN 14: 978-1515158691, ISBN 10: 1515158691

Rule-21756:

   If both $(\textit{I})$, $(\mathbf{U})$, $(\mathbf{\$})$, $(\mathbf{i})$, $(\mathbf{\$'})$, $(\mathbf{s})$, $(\mathbf{v})$, $(\mathbf{f})$, $(\mathbf{t})$, $(\mathbf{A})$, $(\mathbf{d})$, $(\mathbf{Q})$, $(\mathbf{c})$ and $(\mathbf{q})$ are known, then its Procured Inventory Planned is:

$$P= [c\text{-}q][\textit{I}U+\{\$[1\text{-}i]\text{-}\$'[1+s][v+f]\}[1\text{-}t]\text{-}Ad)\text{-}Q]$$

Rule-21757:

   If both $(\textit{I})$, $(\mathbf{U})$, $(\mathbf{\$})$, $(\mathbf{i})$, $(\mathbf{\$'})$, $(\mathbf{s})$, $(\mathbf{v})$, $(\mathbf{f})$, $(\mathbf{t})$, $(\mathbf{A})$, $(\mathbf{d})$, $(\mathbf{P})$, $(\mathbf{Q})$ and $(\mathbf{q})$ are known, then its Current Ratio Planned is:

$$c= q+P/[\textit{I}U+\{\$[1\text{-}i]\text{-}\$'[1+s][v+f]\}[1\text{-}t]\text{-}Ad)\text{-}Q]$$

Rule-21758:

   If both $(\textit{I})$, $(\mathbf{U})$, $(\mathbf{\$})$, $(\mathbf{i})$, $(\mathbf{\$'})$, $(\mathbf{s})$, $(\mathbf{v})$, $(\mathbf{f})$, $(\mathbf{t})$, $(\mathbf{A})$, $(\mathbf{d})$, $(\mathbf{P})$, $(\mathbf{c})$ and $(\mathbf{Q})$ are known, then its Quick or Acid Test Ratio Planned is:

$$q= c\text{-}P/[\textit{I}U+\{\$[1\text{-}i]\text{-}\$'[1+s][v+f]\}[1\text{-}t]\text{-}Ad)\text{-}Q]$$

Rule-21759:

   If both $(\textit{I})$, $(\mathbf{U})$, $(\mathbf{\$})$, $(\mathbf{i})$, $(\mathbf{\$'})$, $(\mathbf{s})$, $(\mathbf{v})$, $(\mathbf{f})$, $(\mathbf{t})$, $(\mathbf{A})$, $(\mathbf{d})$, $(\mathbf{V})$, $(\mathbf{p})$, $(\mathbf{c})$ and $(\mathbf{q})$ are known, then its Quoted Longterm Debt Planned is:

$$Q= \textit{I}U+\{\$[1\text{-}i]\text{-}\$'[1+s][v+f]\}[1\text{-}t]\text{-}Ad)$$
$$-Vp/\{360[c\text{-}q]\}$$

Rule-21760:

   If both $(\mathbf{Q})$, $(\mathbf{U})$, $(\mathbf{\$})$, $(\mathbf{i})$, $(\mathbf{\$'})$, $(\mathbf{s})$, $(\mathbf{v})$, $(\mathbf{f})$, $(\mathbf{t})$, $(\mathbf{A})$, $(\mathbf{d})$, $(\mathbf{V})$, $(\mathbf{p})$, $(\mathbf{c})$ and $(\mathbf{q})$ are known, then its Leverage or Gearing Ratio Planned is:

$$\textit{I}= (Q+Vp/\{360[c\text{-}q]\})$$
$$/(U+\{\$[1\text{-}i]\text{-}\$'[1+s][v+f]\}[1\text{-}t]\text{-}Ad)$$

Steve Asikin ISBN 14: 978-1515158691, ISBN 10: 1515158691

Rule-21761:

If both (I), (Q), (S), (i), (S'), (s), (v), (f), (t), (A), (d), (V), (p), (c) and (q) are known, then its Utilized or Starting Capital must be:

$$U = S'[1+s][v+f]\}[1-t]+Ad$$
$$+(Q+Vp/\{360[c-q]\})/I\text{-}S[1-i]$$

Rule-21762:

If both (I), (U), (Q), (i), (S'), (s), (v), (f), (t), (A), (d), (V), (p), (c) and (q) are known, then its Sales or Revenue Planned is:

$$S = [S'[1+s][v+f][1-t]+Ad$$
$$+(Q+Vp/\{360[c-q]\})/I\text{-}U]/[1-i]$$

Rule-21763:

If both (I), (U), (S), (Q), (S'), (s), (v), (f), (t), (A), (d), (V), (p), (c) and (q) are known, then its Interest Portion Planned is:

$$i = 1-[S'[1+s][v+f][1-t]+Ad$$
$$+(Q+Vp/\{360[c-q]\})/I\text{-}U]/S$$

Rule-21764:

If both (I), (U), (S), (i), (Q), (s), (v), (f), (t), (A), (d), (V), (p), (c) and (q) are known, then its Sales Past must be:

$$S' = [U+S[1-i]-Ad-(Q+Vp/\{360[c-q]\})/I]$$
$$/\{[1+s][v+f][1-t]\}$$

Steve Asikin ISBN 14: 978-1515158691, ISBN 10: 1515158691

Rule-21765:

If both (I), (**U**), (**$**), (**i**), (**$'**), (**Q**), (**v**), (**f**), (**t**), (**A**), (**d**), (**V**), (**p**), (**c**) and (**q**) are known, then its Sales Growth Planned is:

$s = \$/\$'-1$

> or it can also be found as

$s = [U+\$[1-i]-Ad-(Q+Vp/\{360[c-q]\})/I]$
$/\{\$'[v+f][1-t]\}-1$

Rule-21766:

If both (I), (**U**), (**$**), (**i**), (**$'**), (**s**), (**Q**), (**f**), (**t**), (**A**), (**d**), (**V**), (**p**), (**c**) and (**q**) are known, then its Variable Portion Planned is:

$v = V/\$$

> or it can also be found as

$v = [U+\$[1-i]-Ad-(Q+Vp/\{360[c-q]\})/I]$
$/\{\$'[1+s][1-t]\}-f$

Rule-21767:

If both (I), (**U**), (**$**), (**i**), (**$'**), (**s**), (**v**), (**Q**), (**t**), (**A**), (**d**), (**V**), (**p**), (**c**) and (**q**) are known, then its Fixed Portion Planned is:

$f = [U+\$[1-i]-Ad-(Q+Vp/\{360[c-q]\})/I]$
$/\{\$'[1+s][1-t]\}-v$

Rule-21768:

If both (I), (**U**), (**$**), (**i**), (**$'**), (**s**), (**v**), (**f**), (**Q**), (**A**), (**d**), (**V**), (**p**), (**c**) and (**q**) are known, then its Tax Rate Planned is:

$t = 1-[U+\$[1-i]-Ad-(Q+Vp/\{360[c-q]\})/I]$
$/\{\$'[1+s][v+f]\}$

Steve Asikin ISBN 14: 978-1515158691, ISBN 10: 1515158691

Rule-21769:

If both $(\textbf{\textit{l}})$, $(\textbf{U})$, $(\textbf{\$})$, $(\textbf{i})$, $(\textbf{\$'})$, $(\textbf{s})$, $(\textbf{v})$, $(\textbf{f})$, $(\textbf{t})$, $(\textbf{Q})$, $(\textbf{d})$, $(\textbf{V})$, $(\textbf{p})$, $(\textbf{c})$ and $(\textbf{q})$ are known, then its After Tax Income Planned is:

$$\textbf{A}= [\textbf{U}+\{\textbf{\$}[1\text{-}\textbf{i}]\text{-}\textbf{\$'}[1+\textbf{s}][\textbf{v}+\textbf{f}]\}[1\text{-}\textbf{t}]$$
$$-(\textbf{Q}+\textbf{Vp}/\{360[\textbf{c-q}]\})/\textbf{\textit{l}}/\textbf{d}$$

Rule-21770:

If both $(\textbf{\textit{l}})$, $(\textbf{U})$, $(\textbf{\$})$, $(\textbf{i})$, $(\textbf{\$'})$, $(\textbf{s})$, $(\textbf{v})$, $(\textbf{f})$, $(\textbf{t})$, $(\textbf{A})$, $(\textbf{Q})$, $(\textbf{V})$, $(\textbf{p})$, $(\textbf{c})$ and $(\textbf{q})$ are known, then its Dividend Payout Planned is:

$$\textbf{d}= [\textbf{U}+\{\textbf{\$}[1\text{-}\textbf{i}]\text{-}\textbf{\$'}[1+\textbf{s}][\textbf{v}+\textbf{f}]\}[1\text{-}\textbf{t}]$$
$$-(\textbf{Q}+\textbf{Vp}/\{360[\textbf{c-q}]\})/\textbf{\textit{l}}/\textbf{A}$$

Rule-21771:

If both $(\textbf{\textit{l}})$, $(\textbf{U})$, $(\textbf{\$})$, $(\textbf{i})$, $(\textbf{\$'})$, $(\textbf{s})$, $(\textbf{v})$, $(\textbf{f})$, $(\textbf{t})$, $(\textbf{A})$, $(\textbf{d})$, $(\textbf{Q})$, $(\textbf{p})$, $(\textbf{c})$ and $(\textbf{q})$ are known, then its Variable Cost Planned is:

$$\textbf{V}= 360[\textbf{c-q}]$$
$$[\textbf{\textit{l}}(\textbf{U}+\{\textbf{\$}[1\text{-}\textbf{i}]\text{-}\textbf{\$'}[1+\textbf{s}][\textbf{v}+\textbf{f}]\}[1\text{-}\textbf{t}]\text{-}\textbf{Ad})\text{-}\textbf{Q}]/\textbf{p}$$

Rule-21772:

If both $(\textbf{\textit{l}})$, $(\textbf{U})$, $(\textbf{\$})$, $(\textbf{i})$, $(\textbf{\$'})$, $(\textbf{s})$, $(\textbf{v})$, $(\textbf{f})$, $(\textbf{t})$, $(\textbf{A})$, $(\textbf{d})$, $(\textbf{V})$, $(\textbf{Q})$, $(\textbf{c})$ and $(\textbf{q})$ are known, then its Procured Inventory Days Planned:

$$\textbf{p}= 360[\textbf{c-q}]$$
$$[\textbf{\textit{l}}(\textbf{U}+\{\textbf{\$}[1\text{-}\textbf{i}]\text{-}\textbf{\$'}[1+\textbf{s}][\textbf{v}+\textbf{f}]\}[1\text{-}\textbf{t}]\text{-}\textbf{Ad})\text{-}\textbf{Q}]/\textbf{V}$$

Steve Asikin ISBN 14: 978-1515158691, ISBN 10: 1515158691

Rule-21773:

If both (I), (U), (S), (i), (S'), (s), (v), (f), (t), (A), (d), (V), (p), (Q) and (q) are known, then its Current Ratio Planned is:

$$c = q + Vp/\{360[I\{U + \{S[1-i]-S'[1+s][v+f]\}[1-t]-Ad)-Q]\}$$

Rule-21774:

If both (I), (U), (S), (i), (S'), (s), (v), (f), (t), (A), (d), (V), (p), (c) and (Q) are known, then its Quick or Acid Test Ratio Planned is:

$$q = c - Vp/\{360[I\{U + \{S[1-i]-S'[1+s][v+f]\}[1-t]-Ad)-Q]\}$$

Rule-21775:

If both (I), (U), (S), (i), (S'), (s), (v), (f), (t), (A), (d), (p), (c) and (q) are known, then its Quoted Longterm Debt Planned is:

$$Q = I\{U + \{S[1-i]-S'[1+s][v+f]\}[1-t]-Ad) -Svp/\{360[c-q]\}$$

Rule-21776:

If both (Q), (U), (S), (i), (S'), (s), (v), (f), (t), (A), (d), (p), (c) and (q) are known, then its Leverage or Gearing Ratio Planned is:

$$I = (Q + Svp/\{360[c-q]\})/(U + \{S[1-i] -S'[1+s][v+f]\}[1-t]-Ad)$$

Steve Asikin ISBN 14: 978-1515158691, ISBN 10: 1515158691

Rule-21777:

If both $(\textbf{\textit{I}})$, $(\textbf{Q})$, $(\textbf{S})$, $(\textbf{i})$, $(\textbf{S'})$, $(\textbf{s})$, $(\textbf{v})$, $(\textbf{f})$, $(\textbf{t})$, $(\textbf{A})$, $(\textbf{d})$, $(\textbf{p})$, $(\textbf{c})$ and $(\textbf{q})$ are known, then its Utilized or Starting Capital must be:

$$U = S'[1+s][v+f]\}[1-t]+Ad$$
$$+(Q+Svp/\{360[c-q]\})/\textit{I}S[1-i]$$

Rule-21778:

If both $(\textbf{\textit{I}})$, $(\textbf{U})$, $(\textbf{Q})$, $(\textbf{i})$, $(\textbf{S'})$, $(\textbf{s})$, $(\textbf{v})$, $(\textbf{f})$, $(\textbf{t})$, $(\textbf{A})$, $(\textbf{d})$, $(\textbf{p})$, $(\textbf{c})$ and $(\textbf{q})$ are known, then its Sales or Revenue Planned is:

$$S = (\textit{I}\{U-S'[1+s][v+f][1-t]-Ad\}-S)$$
$$/(vp/\{360[c-q]\}-\textit{I}[1-i]\,[1-t])$$

Rule-21779:

If both $(\textbf{\textit{I}})$, $(\textbf{U})$, $(\textbf{S})$, $(\textbf{Q})$, $(\textbf{S'})$, $(\textbf{s})$, $(\textbf{v})$, $(\textbf{f})$, $(\textbf{t})$, $(\textbf{A})$, $(\textbf{d})$, $(\textbf{p})$, $(\textbf{c})$ and $(\textbf{q})$ are known, then its Interest Portion Planned is:

$$i = 1-[S'[1+s][v+f][1-t]+Ad$$
$$+(Q+Svp/\{360[c-q]\})/\textit{I}U]/S$$

Rule-21780:

If both $(\textbf{\textit{I}})$, $(\textbf{U})$, $(\textbf{S})$, $(\textbf{i})$, $(\textbf{Q})$, $(\textbf{s})$, $(\textbf{v})$, $(\textbf{f})$, $(\textbf{t})$, $(\textbf{A})$, $(\textbf{d})$, $(\textbf{p})$, $(\textbf{c})$ and $(\textbf{q})$ are known, then its Sales Past must be:

$$S' = [U+S[1-i]-Ad-(Q+Vp/\{360[c-q]\})/\textit{I}]$$
$$/\{[1+s][v+f][1-t]\}$$

Steve Asikin ISBN 14: 978-1515158691, ISBN 10: 1515158691

Rule-21781:

If both (**/**), (**U**), (**$**), (**i**), (**$'**), (**Q**), (**v**), (**f**), (**t**), (**A**), (**d**), (**p**), (**c**) and (**q**) are known, then its Sales Growth Planned is:

$s = \$/\$' - 1$

or it can also be found as

$$s = [U+\$[1-i]-Ad-(Q+\$vp/\{360[c-q]\})/\text{/}]$$
$$/\{ \$'[v+f][1-t]\} - 1$$

Rule-21782:

If both (**/**), (**U**), (**$**), (**i**), (**$'**), (**s**), (**Q**), (**f**), (**t**), (**A**), (**d**), (**p**), (**c**) and (**q**) are known, then its Variable Portion Planned is:

$$v = [\text{/}U+\{\$[1-i]-\$'f[1+s]\}[1-t]-Ad)-Q]$$
$$/(\$p/\{360[c-q]\}+\$'\text{/}[1+s][1-t])$$

Rule-21783:

If both (**/**), (**U**), (**$**), (**i**), (**$'**), (**s**), (**v**), (**Q**), (**t**), (**A**), (**d**), (**p**), (**c**) and (**q**) are known, then its Fixed Portion Planned is:

$$f = [U+\$[1-i]-Ad-(Q+\$vp/\{360[c-q]\})/\text{/}]$$
$$/\{ \$'[1+s][1-t]\} - v$$

Rule-21784:

If both (**/**), (**U**), (**$**), (**i**), (**$'**), (**s**), (**v**), (**f**), (**Q**), (**A**), (**d**), (**p**), (**c**) and (**q**) are known, then its Tax Rate Planned is:

$$t = 1 - [U+\$[1-i]-Ad-(Q+\$vp/\{360[c-q]\})/\text{/}]$$
$$/\{ \$'[1+s][v+f]\}$$

Steve Asikin ISBN 14: 978-1515158691, ISBN 10: 1515158691

Rule-21785:

If both $(\pmb{\mathit{I}})$, $(\pmb{U})$, $(\pmb{\$})$, $(\pmb{i})$, $(\pmb{\$'})$, $(\pmb{s})$, $(\pmb{v})$, $(\pmb{f})$, $(\pmb{t})$, $(\pmb{Q})$, $(\pmb{d})$, $(\pmb{p})$, $(\pmb{c})$ and $(\pmb{q})$ are known, then its After Tax Income Planned is:

$$A = [U + \{\$[1\text{-}i]\text{-}\$'[1+s][v+f]\}[1\text{-}t]$$
$$-(Q + \$vp / \{360[c\text{-}q]\}) / \pmb{\mathit{I}}]/d$$

Rule-21786:

If both $(\pmb{\mathit{I}})$, $(\pmb{U})$, $(\pmb{\$})$, $(\pmb{i})$, $(\pmb{\$'})$, $(\pmb{s})$, $(\pmb{v})$, $(\pmb{f})$, $(\pmb{t})$, $(\pmb{A})$, $(\pmb{Q})$, $(\pmb{p})$, $(\pmb{c})$ and $(\pmb{q})$ are known, then its Dividend Payout Planned is:

$$d = [U + \{\$[1\text{-}i]\text{-}\$'[1+s][v+f]\}[1\text{-}t]$$
$$-(Q + \$vp / \{360[c\text{-}q]\}) / \pmb{\mathit{I}}]/A$$

Rule-21787:

If both $(\pmb{\mathit{I}})$, $(\pmb{U})$, $(\pmb{\$})$, $(\pmb{i})$, $(\pmb{\$'})$, $(\pmb{s})$, $(\pmb{v})$, $(\pmb{f})$, $(\pmb{t})$, $(\pmb{A})$, $(\pmb{d})$, $(\pmb{Q})$, $(\pmb{c})$ and $(\pmb{q})$ are known, then its Procured Inventory Days Planned:

$$p = 360[c\text{-}q]$$
$$[(\pmb{\mathit{I}}U + \{\$[1\text{-}i]\text{-}\$'[1+s][v+f]\}[1\text{-}t]\text{-}Ad)\text{-}Q] / [\$v]$$

Rule-21788:

If both $(\pmb{\mathit{I}})$, $(\pmb{U})$, $(\pmb{\$})$, $(\pmb{i})$, $(\pmb{\$'})$, $(\pmb{s})$, $(\pmb{v})$, $(\pmb{f})$, $(\pmb{t})$, $(\pmb{A})$, $(\pmb{d})$, $(\pmb{p})$, $(\pmb{Q})$ and $(\pmb{q})$ are known, then its Current Ratio Planned is:

$$c = q + \$vp / \{360[(\pmb{\mathit{I}}U + \{\$[1\text{-}i]\text{-}\$'[1+s][v+f]\}$$
$$[1\text{-}t]\text{-}Ad)\text{-}Q]\}$$

Steve Asikin ISBN 14: 978-1515158691, ISBN 10: 1515158691

Rule-21789:

If both (**/**), (**U**), (**$**), (**i**), (**$'**), (**s**), (**v**), (**f**), (**t**), (**A**), (**d**), (**p**), (**c**) and (**Q**) are known, then its Quick or Acid Test Ratio Planned is:

q= **c-$vp**/{360[**/U**+{**$**[1-**i**]-**$'**[1+**s**][**v**+**f**]}
[1-**t**]-**Ad**)-**Q**]}

Rule-21790:

If both (**/**), (**U**), (**$**), (**i**), (**$'**), (**s**), (**v**), (**f**), (**t**), (**A**), (**d**), (**p**), (**c**) and (**q**) are known, then its Quoted Longterm Debt Planned is:

Q= **/U**+{**$**[1-**i**]-**$'**[1+**s**][**v**+**f**]}[1-**t**]-**Ad**)
-**$'vp**[1+**s**]/{360[**c-q**]}

Rule-21791:

If both (**Q**), (**U**), (**$**), (**i**), (**$'**), (**s**), (**v**), (**f**), (**t**), (**A**), (**d**), (**p**), (**c**) and (**q**) are known, then its Leverage or Gearing Ratio Planned is:

/= (**Q**+**$'vp**[1+**s**]/{360[**c-q**]})
/(**U**+{**$**[1-**i**]-**$'**[1+**s**][**v**+**f**]}[1-**t**]-**Ad**)

Rule-21792:

If both (**/**), (**Q**), (**$**), (**i**), (**$'**), (**s**), (**v**), (**f**), (**t**), (**A**), (**d**), (**p**), (**c**) and (**q**) are known, then its Utilized or Starting Capital must be:

U= **$'**[1+**s**][**v**+**f**]}[1-**t**]+**Ad**
+(**Q**+**$'vp**[1+**s**]/{360[**c-q**]})/**/$**[1-**i**]

Steve Asikin ISBN 14: 978-1515158691, ISBN 10: 1515158691

Rule-21793:

If both (*I*), (**U**), (**Q**), (**i**), (**S'**), (**s**), (**v**), (**f**), (**t**), (**A**), (**d**), (**p**), (**c**) and (**q**) are known, then its Sales or Revenue Planned is:

$$S= (Q-I\{U-S'[1+s][v+f][1-t]-Ad\}$$
$$+S'vp[1+s]/\{360[c-q]\})/\{I[1-i][1-t]\}$$

Rule-21794:

If both (*I*), (**U**), (**S**), (**Q**), (**S'**), (**s**), (**v**), (**f**), (**t**), (**A**), (**d**), (**p**), (**c**) and (**q**) are known, then its Interest Portion Planned is:

$$i= 1-[S'[1+s][v+f][1-t]+Ad$$
$$+(Q+S'vp[1+s]/\{360[c-q]\})/I\text{-}U]/S$$

Rule-21795:

If both (*I*), (**U**), (**S**), (**i**), (**Q**), (**s**), (**v**), (**f**), (**t**), (**A**), (**d**), (**p**), (**c**) and (**q**) are known, then its Sales Past must be:

$$S'= [I\!(U+\{S[1-i]-S'f[1+s]\}[1-t]-Ad)-Q]$$
$$/[v[1+s](p/\{360[c-q]\}+I[1-t])]$$

Rule-21796:

If both (*I*), (**U**), (**S**), (**i**), (**S'**), (**Q**), (**v**), (**f**), (**t**), (**A**), (**d**), (**p**), (**c**) and (**q**) are known, then its Sales Growth Planned is:

$$s= S/S'-1$$

or it can also be found as

$$s= [I\!(U+\{S[1-i]-S'f[1+s]\}[1-t]-Ad)-Q]$$
$$/[S'v(p/\{360[c-q]\}+I[1-t])]-1$$

Steve Asikin ISBN 14: 978-1515158691, ISBN 10: 1515158691

Rule-21797:

If both $(\textit{f})$, $(\textbf{U})$, $(\textbf{\$})$, $(\textbf{i})$, $(\textbf{\$'})$, $(\textbf{s})$, $(\textbf{Q})$, $(\textbf{f})$, $(\textbf{t})$, $(\textbf{A})$, $(\textbf{d})$, $(\textbf{p})$, $(\textbf{c})$ and $(\textbf{q})$ are known, then its Variable Portion Planned is:

$$\textbf{v} = [\textit{f}\textbf{U} + \{\textbf{\$}[1-\textbf{i}] - \textbf{\$'f}[1+\textbf{s}]\}[1-\textbf{t}] - \textbf{Ad}) - \textbf{Q}]$$
$$/[\textbf{\$'}[1+\textbf{s}](\textbf{p}/\{360[\textbf{c}-\textbf{q}]\} + \textit{f}[1-\textbf{t}])]$$

Rule-21798:

If both $(\textit{f})$, $(\textbf{U})$, $(\textbf{\$})$, $(\textbf{i})$, $(\textbf{\$'})$, $(\textbf{s})$, $(\textbf{v})$, $(\textbf{Q})$, $(\textbf{t})$, $(\textbf{A})$, $(\textbf{d})$, $(\textbf{p})$, $(\textbf{c})$ and $(\textbf{q})$ are known, then its Fixed Portion Planned is:

$$\textbf{f} = [\textbf{U} + \textbf{\$}[1-\textbf{i}] - \textbf{Ad} - (\textbf{Q} + \textbf{\$'vp}[1+\textbf{s}]/\{360[\textbf{c}-\textbf{q}]\})/\textit{f}]$$
$$/\{\textbf{\$'}[1+\textbf{s}][1-\textbf{t}]\} - \textbf{v}$$

Rule-21799:

If both $(\textit{f})$, $(\textbf{U})$, $(\textbf{\$})$, $(\textbf{i})$, $(\textbf{\$'})$, $(\textbf{s})$, $(\textbf{v})$, $(\textbf{f})$, $(\textbf{Q})$, $(\textbf{A})$, $(\textbf{d})$, $(\textbf{p})$, $(\textbf{c})$ and $(\textbf{q})$ are known, then its Tax Rate Planned is:

$$\textbf{t} = 1 - [\textbf{U} + \textbf{\$}[1-\textbf{i}] - \textbf{Ad} - (\textbf{Q} + \textbf{\$'vp}[1+\textbf{s}]/\{360[\textbf{c}-\textbf{q}]\})/\textit{f}]$$
$$/\{\textbf{\$'}[1+\textbf{s}][\textbf{v}+\textbf{f}]\}$$

Rule-21800:

If both $(\textit{f})$, $(\textbf{U})$, $(\textbf{\$})$, $(\textbf{i})$, $(\textbf{\$'})$, $(\textbf{s})$, $(\textbf{v})$, $(\textbf{f})$, $(\textbf{t})$, $(\textbf{Q})$, $(\textbf{d})$, $(\textbf{p})$, $(\textbf{c})$ and $(\textbf{q})$ are known, then its After Tax Income Planned is:

$$\textbf{A} = [\textbf{U} + \{\textbf{\$}[1-\textbf{i}] - \textbf{\$'}[1+\textbf{s}][\textbf{v}+\textbf{f}]\}[1-\textbf{t}]$$
$$-(\textbf{Q} + \textbf{\$'vp}[1+\textbf{s}]\textbf{p}/\{360[\textbf{c}-\textbf{q}]\})/\textit{f}]/\textbf{d}$$

Steve Asikin ISBN 14: 978-1515158691, ISBN 10: 1515158691

Rule-21801:

If both (J), (**U**), (**$**), (**i**), (**$'**), (**s**), (**v**), (**f**), (**t**), (**A**), (**Q**), (**p**), (**c**) and (**q**) are known, then its Dividend Payout Planned is:

$$d= [U+\{\$[1-i]-\$'[1+s][v+f]\}[1-t]$$
$$-(Q+\$'vp[1+s]/\{360[c-q]\})/J/A$$

Rule-21802:

If both (J), (**U**), (**$**), (**i**), (**$'**), (**s**), (**v**), (**f**), (**t**), (**A**), (**d**), (**Q**), (**c**) and (**q**) are known, then its Procured Inventory Days Planned:

$$p= 360[c-q][JU+\{\$[1-i]-\$'[1+s][v+f]\}$$
$$[1-t]-Ad)-Q]/\{\$'v[1+s]\}$$

Rule-21803:

If both (J), (**U**), (**$**), (**i**), (**$'**), (**s**), (**v**), (**f**), (**t**), (**A**), (**d**), (**p**), (**Q**) and (**q**) are known, then its Current Ratio Planned is:

$$c= q+\$'vp[1+s]/\{360[JU+\{\$[1-i]-\$'[1+s][v+f]\}$$
$$[1-t]-Ad)-Q]\}$$

Rule-21804:

If both (J), (**U**), (**$**), (**i**), (**$'**), (**s**), (**v**), (**f**), (**t**), (**A**), (**d**), (**p**), (**c**) and (**Q**) are known, then its Quick or Acid Test Ratio Planned is:

$$q= c-\$'vp[1+s]/\{360[JU+\{\$[1-i]-\$'[1+s][v+f]\}$$
$$[1-t]-Ad)-Q]\}$$

Steve Asikin ISBN 14: 978-1515158691, ISBN 10: 1515158691

Rule-21805:
    If both (f), (U), $(\$)$, (i), $(\$')$, (s), (v), (f), (t), (d) and (X) are known, then its Quoted Longterm Debt Planned is:
$$Q= f(U+\{\$[1-i]-\$'[1+s][v+f]\}[1-t][1-d])-X$$

Rule-21806:
    If both (Q), (U), $(\$)$, (i), $(\$')$, (s), (v), (f), (t), (d) and (X) are known, then its Leverage or Gearing Ratio Planned is:
$$f= [Q+X]/(U+\{\$[1-i]-\$'[1+s][v+f]\}[1-t][1-d])$$

Rule-21807:
    If both (f), (Q), $(\$)$, (i), $(\$')$, (s), (v), (f), (t), (d) and (X) are known, then its Utilized or Starting Capital must be:
$$U= \$'[1+s][v+f]\}[1-t][1-d]+[Q+X]/f-\$[1-i]$$

Rule-21808:
    If both (f), (U), (Q), (i), $(\$')$, (s), (v), (f), (t), (d) and (X) are known, then its Sales or Revenue Planned is:
$$\$= \{\$'[1+s][v+f][1-t][1-d]+[Q+X]/f-U\}/[1-i]$$

Rule-21809:
    If both (f), (U), $(\$)$, (Q), $(\$')$, (s), (v), (f), (t), (d) and (X) are known, then its Interest Portion Planned is:
$$i= 1-\{\$'[1+s][v+f][1-t][1-d]+[Q+X]/f-U\}/\$$$

Rule-21810:
    If both (f), (U), $(\$)$, (i), (Q), (s), (v), (f), (t), (d) and (X) are known, then its Sales Past must be:
$$\$'= \{U+\$[1-i]-[Q+X]/f\}/\{[1+s][v+f][1-t][1-d]\}$$

Steve Asikin ISBN 14: 978-1515158691, ISBN 10: 1515158691

Rule-21811:

If both (**/**), (**U**), (**$**), (**i**), (**$'**), (**Q**), (**v**), (**f**), (**t**), (**d**) and (**X**) are known, then its Sales Growth Planned is:

$s= \$/\$'-1$

or it can also be found as

$s= \{U+\$[1-i]-[Q+X]/\!\!/\}/\{\$'[v+f][1-t][1-d]\}-1$

Rule-21812:

If both (**/**), (**U**), (**$**), (**i**), (**$'**), (**s**), (**Q**), (**f**), (**t**), (**d**) and (**X**) are known, then its Variable Portion Planned is:

$v= \{U+\$[1-i]-[Q+X]/\!\!/\}/\{ \$'[1+s][1-t][1-d]\}-f$

Rule-21813:

If both (**/**), (**U**), (**$**), (**i**), (**$'**), (**s**), (**v**), (**Q**), (**t**), (**d**) and (**X**) are known, then its Fixed Portion Planned is:

$f= \{U+\$[1-i]-[Q+X]/\!\!/\}/\{ \$'[1+s][1-t][1-d]\}-v$

Rule-21814:

If both (**/**), (**U**), (**$**), (**i**), (**$'**), (**s**), (**v**), (**f**), (**Q**), (**d**) and (**X**) are known, then its Tax Rate Planned is:

$t= 1-\{U+\$[1-i]-[Q+X]/\!\!/\}/\{ \$'[1+s][v+f][1-d]\}$

Rule-21815:

If both (**/**), (**U**), (**$**), (**i**), (**$'**), (**s**), (**v**), (**f**), (**t**), (**Q**) and (**X**) are known, then its Dividend Payout Planned is:

$d= 1-\{U+\$[1-i]-[Q+X]/\!\!/\}/\{ \$'[1+s][v+f][1-t]\}$

Rule-21816:

If both (**/**), (**U**), (**$**), (**i**), (**$'**), (**s**), (**v**), (**f**), (**t**), (**d**) and (**Q**) are known, then its Xpress or Current Debt Planned is:

$X= /\!\!/(U+\{\$[1-i]-\$'[1+s][v+f]\}[1-t][1-d])-Q$

Steve Asikin ISBN 14: 978-1515158691, ISBN 10: 1515158691

Rule-21817:

> If both (I), $(\mathbf{U})$, $(\mathbf{S})$, $(\mathbf{i})$, $(\mathbf{S'})$, $(\mathbf{s})$, $(\mathbf{v})$, $(\mathbf{f})$, $(\mathbf{t})$, $(\mathbf{d})$, $(\mathbf{P})$, $(\mathbf{c})$ and $(\mathbf{q})$ are known, then its Quoted Longterm Debt Planned is:
>
> $$\mathbf{Q}= \mathit{I}(\mathbf{U}+\{\mathbf{S}[1\text{-}\mathbf{i}]\text{-}\mathbf{S'}[1\text{+}\mathbf{s}][\mathbf{v}\text{+}\mathbf{f}]\}[1\text{-}\mathbf{t}][1\text{-}\mathbf{d}])\text{-}\mathbf{P}/[\mathbf{c}\text{-}\mathbf{q}]$$

Rule-21818:

> If both $(\mathbf{Q})$, $(\mathbf{U})$, $(\mathbf{S})$, $(\mathbf{i})$, $(\mathbf{S'})$, $(\mathbf{s})$, $(\mathbf{v})$, $(\mathbf{f})$, $(\mathbf{t})$, $(\mathbf{d})$, $(\mathbf{P})$, $(\mathbf{c})$ and $(\mathbf{q})$ are known, then its Leverage or Gearing Ratio Planned is:
>
> $$\mathit{I}= \{\mathbf{Q}+\mathbf{P}/[\mathbf{c}\text{-}\mathbf{q}]\}/(\mathbf{U}+\{\mathbf{S}[1\text{-}\mathbf{i}]$$
> $$-\mathbf{S'}[1\text{+}\mathbf{s}][\mathbf{v}\text{+}\mathbf{f}]\}[1\text{-}\mathbf{t}][1\text{-}\mathbf{d}])$$

Rule-21819:

> If both (I), $(\mathbf{Q})$, $(\mathbf{S})$, $(\mathbf{i})$, $(\mathbf{S'})$, $(\mathbf{s})$, $(\mathbf{v})$, $(\mathbf{f})$, $(\mathbf{t})$, $(\mathbf{d})$, $(\mathbf{P})$, $(\mathbf{c})$ and $(\mathbf{q})$ are known, then its Utilized or Starting Capital must be:
>
> $$\mathbf{U}= \mathbf{S'}[1\text{+}\mathbf{s}][\mathbf{v}\text{+}\mathbf{f}]\}[1\text{-}\mathbf{t}][1\text{-}\mathbf{d}]+\{\mathbf{Q}+\mathbf{P}/[\mathbf{c}\text{-}\mathbf{q}]\}/\mathit{I}\text{+}\mathbf{S}[1\text{-}\mathbf{i}]$$

Rule-21820:

> If both (I), $(\mathbf{U})$, $(\mathbf{Q})$, $(\mathbf{i})$, $(\mathbf{S'})$, $(\mathbf{s})$, $(\mathbf{v})$, $(\mathbf{f})$, $(\mathbf{t})$, $(\mathbf{d})$, $(\mathbf{P})$, $(\mathbf{c})$ and $(\mathbf{q})$ are known, then its Sales or Revenue Planned is:
>
> $$\mathbf{S}= (\mathbf{S'}[1\text{+}\mathbf{s}][\mathbf{v}\text{+}\mathbf{f}][1\text{-}\mathbf{t}][1\text{-}\mathbf{d}]+\{\mathbf{Q}+\mathbf{P}/[\mathbf{c}\text{-}\mathbf{q}]\}/\mathit{I}\text{-}\mathbf{U})/[1\text{-}\mathbf{i}]$$

Rule-21821:

> If both (I), $(\mathbf{U})$, $(\mathbf{S})$, $(\mathbf{Q})$, $(\mathbf{S'})$, $(\mathbf{s})$, $(\mathbf{v})$, $(\mathbf{f})$, $(\mathbf{t})$, $(\mathbf{d})$, $(\mathbf{P})$, $(\mathbf{c})$ and $(\mathbf{q})$ are known, then its Interest Portion Planned is:
>
> $$\mathbf{i}= 1\text{-}(\mathbf{S'}[1\text{+}\mathbf{s}][\mathbf{v}\text{+}\mathbf{f}][1\text{-}\mathbf{t}][1\text{-}\mathbf{d}]+\{\mathbf{Q}+\mathbf{P}/[\mathbf{c}\text{-}\mathbf{q}]\}/\mathit{I}\text{-}\mathbf{U})/\mathbf{S}$$

Steve Asikin ISBN 14: 978-1515158691, ISBN 10: 1515158691

Rule-21822:

If both (I), (U), (S), (i), (Q), (s), (v), (f), (t), (d), (P), (c) and (q) are known, then its Sales Past must be:

$$S' = (U + S[1-i] - \{Q + P/[c-q]\}/I)$$
$$/\{[1+s][v+f][1-t][1-d]\}$$

Rule-21823:

If both (I), (U), (S), (i), (S'), (Q), (v), (f), (t), (d), (P), (c) and (q) are known, then its Sales Growth Planned is:

$$s = S/S' - 1$$

or it can also be found as

$$s = (U + S[1-i] - \{Q + P/[c-q]\}/I)/\{S'[v+f][1-t][1-d]\} - 1$$

Rule-21824:

If both (I), (U), (S), (i), (S'), (s), (Q), (f), (t), (d), (P), (c) and (q) are known, then its Variable Portion Planned is:

$$v = (U + S[1-i] - \{Q + P/[c-q]\}/I)/\{S'[1+s][1-t][1-d]\} - f$$

Rule-21825:

If both (I), (U), (S), (i), (S'), (s), (v), (Q), (t), (d), (P), (c) and (q) are known, then its Fixed Portion Planned is:

$$f = (U + S[1-i] - \{Q + P/[c-q]\}/I)/\{S'[1+s][1-t][1-d]\} - v$$

Steve Asikin ISBN 14: 978-1515158691, ISBN 10: 1515158691

Rule-21826:

If both (I), (**U**), (**\$**), (**i**), (**\$'**), (**s**), (**v**), (**f**), (**Q**), (**d**), (**P**), (**c**) and (**q**) are known, then its Tax Rate Planned is:
$$t= 1-(U+\$[1-i]-\{Q+P/[c-q]\}/I)/\{\$'[1+s][v+f][1-d]\}$$

Rule-21827:

If both (I), (**U**), (**\$**), (**i**), (**\$'**), (**s**), (**v**), (**f**), (**t**), (**Q**), (**P**), (**c**) and (**q**) are known, then its Dividend Payout Planned is:
$$d= 1-(U+\$[1-i]-\{Q+P/[c-q]\}/I)/\{\$'[1+s][v+f][1-t]\}$$

Rule-21828:

If both (I), (**U**), (**\$**), (**i**), (**\$'**), (**s**), (**v**), (**f**), (**t**), (**d**), (**Q**), (**c**) and (**q**) are known, then its Procured Inventory Planned is:
$$P= [c-q][I(U+\{\$[1-i]-\$'[1+s][v+f]\}[1-t][1-d])-Q]$$

Rule-21829:

If both (I), (**U**), (**\$**), (**i**), (**\$'**), (**s**), (**v**), (**f**), (**t**), (**d**), (**P**), (**Q**) and (**q**) are known, then its Current Ratio Planned is:
$$c= q+P/[I(U+\{\$[1-i]-\$'[1+s][v+f]\}[1-t][1-d])-Q]$$

Rule-21830:

If both (I), (**U**), (**\$**), (**i**), (**\$'**), (**s**), (**v**), (**f**), (**t**), (**d**), (**P**), (**c**) and (**Q**) are known, then its Quick or Acid Test Ratio Planned is:
$$q= c-P/[I(U+\{\$[1-i]-\$'[1+s][v+f]\}[1-t][1-d])-Q]$$

Steve Asikin ISBN 14: 978-1515158691, ISBN 10: 1515158691

Rule-21831:
 If both (**/**), (**U**), (**$**), (**i**), (**$'**), (**s**), (**v**), (**f**), (**t**), (**d**), (**V**), (**p**), (**c**) and (**q**) are known, then its Quoted Longterm Debt Planned is:
 $$Q= \textit{/}(U+\{\$[1-i]-\$'[1+s][v+f]\}[1-t][1-d])$$
 $$-Vp/\{360[c-q]\}$$

Rule-21832:
 If both (**Q**), (**U**), (**$**), (**i**), (**$'**), (**s**), (**v**), (**f**), (**t**), (**d**), (**V**), (**p**), (**c**) and (**q**) are known, then its Leverage or Gearing Ratio Planned is:
 $$\textit{/}= (Q+Vp/\{360[c-q]\})$$
 $$/(U+\{\$[1-i]-\$'[1+s][v+f]\}[1-t][1-d])$$

Rule-21833:
 If both (**/**), (**Q**), (**$**), (**i**), (**$'**), (**s**), (**v**), (**f**), (**t**), (**d**), (**V**), (**p**), (**c**) and (**q**) are known, then its Utilized or Starting Capital must be:
 $$U= \$'[1+s][v+f]\}[1-t][1-d]$$
 $$+(Q+Vp/\{360[c-q]\})/\textit{/}+\$[1-i]$$

Rule-21834:
 If both (**/**), (**U**), (**Q**), (**i**), (**$'**), (**s**), (**v**), (**f**), (**t**), (**d**), (**V**), (**p**), (**c**) and (**q**) are known, then its Sales or Revenue Debt Planned is:
 $$\$= [\$'[1+s][v+f][1-t][1-d]$$
 $$+(Q+Vp/\{360[c-q]\})/\textit{/}-U]/[1-i]$$

Steve Asikin ISBN 14: 978-1515158691, ISBN 10: 1515158691

Rule-21835:

If both (f), (U), $(\$)$, (Q), $(\$')$, (s), (v), (f), (t), (d), (V), (p), (c) and (q) are known, then its Interest Portion Planned is:

$$i = 1 - [\$'[1+s][v+f][1-t][1-d] \\ +(Q+Vp/\{360[c-q]\})/f-U]/\$$$

Rule-21836:

If both (f), (U), $(\$)$, (i), (Q), (s), (v), (f), (t), (d), (V), (p), (c) and (q) are known, then its Sales Past must be:

$$\$' = [U+\$[1-i]-(Q+Vp/\{360[c-q]\})/f] \\ /\{[1+s][v+f][1-t][1-d]\}$$

Rule-21837:

If both (f), (U), $(\$)$, (i), $(\$')$, (Q), (v), (f), (t), (d), (V), (p), (c) and (q) are known, then its Sales Growth Planned is:

$$s = \$/\$' - 1$$

or it can also be found as

$$s = [U+\$[1-i]-(Q+Vp/\{360[c-q]\})/f] \\ /\{\$'[v+f][1-t][1-d]\} - 1$$

Rule-21838:

If both (f), (U), $(\$)$, (i), $(\$')$, (s), (Q), (f), (t), (d), (V), (p), (c) and (q) are known, then its Variable Portion Planned is:

$$v = V/\$$$

or it can also be found as

$$v = [U+\$[1-i]-(Q+Vp/\{360[c-q]\})/f] \\ /\{\$'[1+s][1-t][1-d]\} - f$$

Steve Asikin ISBN 14: 978-1515158691, ISBN 10: 1515158691

Rule-21839:

If both $(\textit{f})$, $(\mathbf{U})$, $(\$)$, $(\mathbf{i})$, $(\$')$, $(\mathbf{s})$, $(\mathbf{v})$, $(\mathbf{Q})$, $(\mathbf{t})$, $(\mathbf{d})$, $(\mathbf{V})$, $(\mathbf{p})$, $(\mathbf{c})$ and $(\mathbf{q})$ are known, then its Fixed Portion Planned is:

$$f= [U+\$[1\text{-}i]\text{-}(Q+Vp/\{360[c\text{-}q]\})/\textit{f}]$$
$$/\{\$'[1+s][1\text{-}t][1\text{-}d]\}\text{-}v$$

Rule-21840:

If both $(\textit{f})$, $(\mathbf{U})$, $(\$)$, $(\mathbf{i})$, $(\$')$, $(\mathbf{s})$, $(\mathbf{v})$, $(\mathbf{f})$, $(\mathbf{Q})$, $(\mathbf{d})$, $(\mathbf{V})$, $(\mathbf{p})$, $(\mathbf{c})$ and $(\mathbf{q})$ are known, then its Tax Rate Planned is:

$$t= 1\text{-}[U+\$[1\text{-}i]\text{-}(Q+Vp/\{360[c\text{-}q]\})/\textit{f}]$$
$$/\{\$'[1+s][v+f][1\text{-}d]\}$$

Rule-21841:

If both $(\textit{f})$, $(\mathbf{U})$, $(\$)$, $(\mathbf{i})$, $(\$')$, $(\mathbf{s})$, $(\mathbf{v})$, $(\mathbf{f})$, $(\mathbf{t})$, $(\mathbf{Q})$, $(\mathbf{V})$, $(\mathbf{p})$, $(\mathbf{c})$ and $(\mathbf{q})$ are known, then its Dividend Payout Planned is:

$$d= 1\text{-}[U+\$[1\text{-}i]\text{-}(Q+Vp/\{360[c\text{-}q]\})/\textit{f}]$$
$$/\{\$'[1+s][v+f][1\text{-}t]\}$$

Rule-21842:

If both $(\textit{f})$, $(\mathbf{U})$, $(\$)$, $(\mathbf{i})$, $(\$')$, $(\mathbf{s})$, $(\mathbf{v})$, $(\mathbf{f})$, $(\mathbf{t})$, $(\mathbf{d})$, $(\mathbf{Q})$, $(\mathbf{p})$, $(\mathbf{c})$ and $(\mathbf{q})$ are known, then its Variable Cost Planned is:

$$V= 360[c\text{-}q]$$
$$[(\textit{f}U+\{\$[1\text{-}i]\text{-}\$'[1+s][v+f]\}[1\text{-}t][1\text{-}d])\text{-}Q]/p$$

Steve Asikin ISBN 14: 978-1515158691, ISBN 10: 1515158691

Rule-21843:

If both (**/**), (**U**), (**$**), (**i**), (**$'**), (**s**), (**v**), (**f**), (**t**), (**d**), (**V**), (**Q**), (**c**) and (**q**) are known, then its Procured Inventory Days Planned:

$$\mathbf{p}= 360[\mathbf{c\text{-}q}]$$
$$[\mathbf{/U}+\{\mathbf{\$}[1\text{-}\mathbf{i}]\text{-}\mathbf{\$'}[1+\mathbf{s}][\mathbf{v}+\mathbf{f}]\}[1\text{-}\mathbf{t}][1\text{-}\mathbf{d}])\text{-}\mathbf{Q}]/\mathbf{V}$$

Rule-21844:

If both (**/**), (**U**), (**$**), (**i**), (**$'**), (**s**), (**v**), (**f**), (**t**), (**d**), (**V**), (**p**), (**Q**) and (**q**) are known, then its Current Ratio Planned is:

$$\mathbf{c}= \mathbf{q}+\mathbf{Vp}/\{360[\mathbf{/U}+\{\mathbf{\$}[1\text{-}\mathbf{i}]\text{-}\mathbf{\$'}[1+\mathbf{s}][\mathbf{v}+\mathbf{f}]\}$$
$$[1\text{-}\mathbf{t}][1\text{-}\mathbf{d}])\text{-}\mathbf{Q}]\}$$

Rule-21845:

If both (**/**), (**U**), (**$**), (**i**), (**$'**), (**s**), (**v**), (**f**), (**t**), (**d**), (**V**), (**p**), (**c**) and (**Q**) are known, then its Quick or Acid Test Ratio Planned is:

$$\mathbf{q}= \mathbf{c}\text{-}\mathbf{Vp}/\{360[\mathbf{/U}+\{\mathbf{\$}[1\text{-}\mathbf{i}]\text{-}\mathbf{\$'}[1+\mathbf{s}][\mathbf{v}+\mathbf{f}]\}$$
$$[1\text{-}\mathbf{t}][1\text{-}\mathbf{d}])\text{-}\mathbf{Q}]\}$$

Rule-21846:

If both (**/**), (**U**), (**$**), (**i**), (**$'**), (**s**), (**v**), (**f**), (**t**), (**d**), (**p**), (**c**) and (**q**) are known, then its Quoted Longterm Debt Planned is:

$$\mathbf{Q}= \mathbf{/U}+\{\mathbf{\$}[1\text{-}\mathbf{i}]\text{-}\mathbf{\$'}[1+\mathbf{s}][\mathbf{v}+\mathbf{f}]\}[1\text{-}\mathbf{t}][1\text{-}\mathbf{d}])$$
$$\text{-}\mathbf{\$vp}/\{360[\mathbf{c}\text{-}\mathbf{q}]\}$$

Steve Asikin ISBN 14: 978-1515158691, ISBN 10: 1515158691

Rule-21847:

If both $(\mathbf{Q})$, $(\mathbf{U})$, $(\mathbf{S})$, $(\mathbf{i})$, $(\mathbf{S'})$, $(\mathbf{s})$, $(\mathbf{v})$, $(\mathbf{f})$, $(\mathbf{t})$, $(\mathbf{d})$, $(\mathbf{p})$, $(\mathbf{c})$ and $(\mathbf{q})$ are known, then its Leverage or Gearing Ratio Planned is:

$\mathbf{\mathit{l}} = (\mathbf{Q} + \mathbf{Svp}/\{360[\mathbf{c} - \mathbf{q}]\})$
$/(\mathbf{U} + \{\mathbf{S}[1 - \mathbf{i}] - \mathbf{S'}[1 + \mathbf{s}][\mathbf{v} + \mathbf{f}]\}[1 - \mathbf{t}][1 - \mathbf{d}])$

Rule-21848:

If both $(\mathbf{\mathit{l}})$, $(\mathbf{Q})$, $(\mathbf{S})$, $(\mathbf{i})$, $(\mathbf{S'})$, $(\mathbf{s})$, $(\mathbf{v})$, $(\mathbf{f})$, $(\mathbf{t})$, $(\mathbf{d})$, $(\mathbf{p})$, $(\mathbf{c})$ and $(\mathbf{q})$ are known, then its Utilized or Starting Capital must be:

$\mathbf{U} = \mathbf{S'}[1 + \mathbf{s}][\mathbf{v} + \mathbf{f}]\}[1 - \mathbf{t}][1 - \mathbf{d}]$
$+(\mathbf{Q} + \mathbf{Svp}/\{360[\mathbf{c} - \mathbf{q}]\})/\mathbf{\mathit{l}} + \mathbf{S}[1 - \mathbf{i}]$

Rule-21849:

If both $(\mathbf{\mathit{l}})$, $(\mathbf{U})$, $(\mathbf{Q})$, $(\mathbf{i})$, $(\mathbf{S'})$, $(\mathbf{s})$, $(\mathbf{v})$, $(\mathbf{f})$, $(\mathbf{t})$, $(\mathbf{d})$, $(\mathbf{p})$, $(\mathbf{c})$ and $(\mathbf{q})$ are known, then its Sales or Revenue Planned is:

$\mathbf{S} = (\mathbf{\mathit{l}}\{\mathbf{U} - \mathbf{S'}[1 + \mathbf{s}][\mathbf{v} + \mathbf{f}][1 - \mathbf{t}][1 - \mathbf{d}]\} - \mathbf{Q})$
$/(\mathbf{vp}/\{360[\mathbf{c} - \mathbf{q}]\} - \mathbf{\mathit{l}}[1 - \mathbf{i}] [1 - \mathbf{t}][1 - \mathbf{d}])$

Rule-21850:

If both $(\mathbf{\mathit{l}})$, $(\mathbf{U})$, $(\mathbf{S})$, $(\mathbf{Q})$, $(\mathbf{S'})$, $(\mathbf{s})$, $(\mathbf{v})$, $(\mathbf{f})$, $(\mathbf{t})$, $(\mathbf{d})$, $(\mathbf{p})$, $(\mathbf{c})$ and $(\mathbf{q})$ are known, then its Interest Portion Planned is:

$\mathbf{i} = 1 - [\mathbf{S'}[1 + \mathbf{s}][\mathbf{v} + \mathbf{f}][1 - \mathbf{t}][1 - \mathbf{d}]$
$+(\mathbf{Q} + \mathbf{Svp}/\{360[\mathbf{c} - \mathbf{q}]\})/\mathbf{\mathit{l}} - \mathbf{U}]/\mathbf{S}$

Steve Asikin ISBN 14: 978-1515158691, ISBN 10: 1515158691

Rule-21851:

If both (I), (**U**), (**$**), (**i**), (**Q**), (**s**), (**v**), (**f**), (**t**), (**d**), (**p**), (**c**) and (**q**) are known, then its Sales Past must be:

$$\$' = [U+\$[1\text{-}i]\text{-}(Q+\$vp/\{360[c\text{-}q]\})/I]$$
$$/\{[1+s][v+f][1\text{-}t][1\text{-}d]\}$$

Rule-21852:

If both (I), (**U**), (**$**), (**i**), (**$'**), (**Q**), (**v**), (**f**), (**t**), (**d**), (**p**), (**c**) and (**q**) are known, then its Sales Past Planned is:

$$s = \$/\$'\text{-}1$$

or it can also be found as

$$s = [U+\$[1\text{-}i]\text{-}(Q+\$vp/\{360[c\text{-}q]\})/I]$$
$$/\{\$'[v+f][1\text{-}t][1\text{-}d]\}\text{-}1$$

Rule-21853:

If both (I), (**U**), (**$**), (**i**), (**$'**), (**s**), (**Q**), (**f**), (**t**), (**d**), (**p**), (**c**) and (**q**) are known, then its Variable Portion Planned is:

$$v = [I(U+\{\$[1\text{-}i]\text{-}\$'f[1+s]\}[1\text{-}t][1\text{-}d])\text{-}Q]$$
$$/(\$p/\{360[c\text{-}q]\}\ \text{-}\$'I[1+s][1\text{-}t][1\text{-}d])$$

Rule-21854:

If both (I), (**U**), (**$**), (**i**), (**$'**), (**s**), (**v**), (**Q**), (**t**), (**d**), (**p**), (**c**) and (**q**) are known, then its Fixed Portion Planned is:

$$f = [U+\$[1\text{-}i]\text{-}(Q+\$vp/\{360[c\text{-}q]\})/I]$$
$$/\{\$'[1+s][1\text{-}t][1\text{-}d]\}\text{-}v$$

Steve Asikin ISBN 14: 978-1515158691, ISBN 10: 1515158691

Rule-21855:

If both (**/**), (**U**), (**$**), (**i**), (**$'**), (**s**), (**v**), (**f**), (**Q**), (**d**), (**p**), (**c**) and (**q**) are known, then its Tax Rate Planned is:

$$t = 1-[U+\$[1-i]-(Q+\$vp/\{360[c-q]\})//]$$
$$/\{\$'[1+s][v+f][1-d]\}$$

Rule-21856:

If both (**/**), (**U**), (**$**), (**i**), (**$'**), (**s**), (**v**), (**f**), (**t**), (**Q**), (**p**), (**c**) and (**q**) are known, then its Dividend Payout Planned is:

$$d = 1-[U+\$[1-i]-(Q+\$vp/\{360[c-q]\})//]$$
$$/\{\$'[1+s][v+f][1-t]\}$$

Rule-21857:

If both (**/**), (**U**), (**$**), (**i**), (**$'**), (**s**), (**v**), (**f**), (**t**), (**d**), (**Q**), (**c**) and (**q**) are known, then its Procured Inventory Days Planned:

$$p = 360[c-q][/U+\{\$[1-i]-\$'[1+s][v+f]\}$$
$$[1-t][1-d])-Q]/[\$v]$$

Rule-21858:

If both (**/**), (**U**), (**$**), (**i**), (**$'**), (**s**), (**v**), (**f**), (**t**), (**d**), (**p**), (**Q**) and (**q**) are known, then its Current Ratio Planned is:

$$c = q+\$vp/\{360[/U+\{\$[1-i]-\$'[1+s][v+f]\}$$
$$[1-t][1-d])-Q]\}$$

Steve Asikin ISBN 14: 978-1515158691, ISBN 10: 1515158691

Rule-21859:

If both ($\int$), (**U**), (**$**), (**i**), (**$'**), (**s**), (**v**), (**f**), (**t**), (**d**), (**p**), (**c**) and (**Q**) are known, then its Quick or Acid Test Ratio Planned is:

$$q= c\text{-}\$vp/\{360[\int U+\{\$[1\text{-}i]\text{-}\$'[1+s][v+f]\} \\ [1\text{-}t][1\text{-}d])\text{-}Q]\}$$

Rule-21860:

If both ($\int$), (**U**), (**$**), (**i**), (**$'**), (**s**), (**v**), (**f**), (**t**), (**d**), (**p**), (**c**) and (**q**) are known, then its Quoted Longterm Debt Planned is:

$$Q= \int U+\{\$[1\text{-}i]\text{-}\$'[1+s][v+f]\}[1\text{-}t][1\text{-}d]) \\ -\$'vp[1+s]/\{360[c\text{-}q]\}$$

Rule-21861:

If both (**Q**), (**U**), (**$**), (**i**), (**$'**), (**s**), (**v**), (**f**), (**t**), (**d**), (**p**), (**c**) and (**q**) are known, then its Leverage or Gearing Ratio Planned is:

$$\int= (Q+\$'vp[1+s]/\{360[c\text{-}q]\}) \\ /(U+\{\$[1\text{-}i]\text{-}\$'[1+s][v+f]\}[1\text{-}t][1\text{-}d])$$

Rule-21862:

If both ($\int$), (**Q**), (**$**), (**i**), (**$'**), (**s**), (**v**), (**f**), (**t**), (**d**), (**p**), (**c**) and (**q**) are known, then its Utilized or Starting Capital must be:

$$U= \$'[1+s][v+f]\}[1\text{-}t][1\text{-}d] \\ +(Q+\$'vp[1+s]/\{360[c\text{-}q]\})/\int\text{-}\$[1\text{-}i]$$

Steve Asikin ISBN 14: 978-1515158691, ISBN 10: 1515158691

Rule-21863:

If both (I), (U), (Q), (i), (S'), (s), (v), (f), (t), (d), (p), (c) and (q) are known, then its Sales or Revenue Planned is:

$$S= [Q-I(U+\{ -S'[1+s][v+f]\}[1-t][1-d])$$
$$+S'vp[1+s]/\{360[c-q]\}]/\{I[1-i][1-t][1-d]\}$$

Rule-21864:

If both (I), (U), (S), (Q), (S'), (s), (v), (f), (t), (d), (p), (c) and (q) are known, then its Interest Portion Planned is:

$$i= 1-[S'[1+s][v+f][1-t][1-d]$$
$$+(Q+S'vp[1+s]/\{360[c-q]\})/I-U]/S$$

Rule-21865:

If both (I), (U), (S), (i), (Q), (s), (v), (f), (t), (d), (p), (c) and (q) are known, then its Sales Past must be:

$$S'= [I(U+\{S[1-i]-S'f[1+s]\}[1-t][1-d])-Q]$$
$$/[v[1+s](p/\{360[c-q]\}-I[1-t][1-d])]$$

Rule-21866:

If both (I), (U), (S), (i), (S'), (Q), (v), (f), (t), (d), (p), (c) and (q) are known, then its Sales Growth Planned is:

$$s= S/S'-1$$

or it can also be found as

$$s= [I(U+\{S[1-i]-S'f[1+s]\}[1-t][1-d])-Q]$$
$$/[S'v(p/\{360[c-q]\}-I[1-t][1-d])]-1$$

Steve Asikin ISBN 14: 978-1515158691, ISBN 10: 1515158691

Rule-21867:

If both (**/**), (**U**), (**$**), (**i**), (**$'**), (**s**), (**Q**), (**f**), (**t**), (**d**), (**p**), (**c**) and (**q**) are known, then its Variable Portion Planned is:

$$v= [\textbf{/}(\textbf{U}+\{\textbf{\$}[1\text{-}\textbf{i}]\text{-}\textbf{\$'f}[1+\textbf{s}]\}[1\text{-}\textbf{t}][1\text{-}\textbf{d}])\text{-}\textbf{Q}]$$
$$/[\textbf{\$'}[1+\textbf{s}](\textbf{p}/\{360[\textbf{c-q}]\}\text{-}\textbf{/}[1\text{-}\textbf{t}][1\text{-}\textbf{d}])]$$

Rule-21868:

If both (**/**), (**U**), (**$**), (**i**), (**$'**), (**s**), (**v**), (**Q**), (**t**), (**d**), (**p**), (**c**) and (**q**) are known, then its Fixed Portion Planned is:

$$f= [\textbf{U}+\textbf{\$}[1\text{-}\textbf{i}]\text{-}(\textbf{Q}+\textbf{\$'vp}[1+\textbf{s}]/\{360[\textbf{c-q}]\})/\textbf{/}]$$
$$/\{\textbf{\$'}[1+\textbf{s}][1\text{-}\textbf{t}][1\text{-}\textbf{d}]\}\text{-}\textbf{v}$$

Rule-21869:

If both (**/**), (**U**), (**$**), (**i**), (**$'**), (**s**), (**v**), (**f**), (**Q**), (**d**), (**p**), (**c**) and (**q**) are known, then its Tax Rate Planned is:

$$t= 1\text{-}[\textbf{U}+\textbf{\$}[1\text{-}\textbf{i}]\text{-}(\textbf{Q}+\textbf{\$'vp}[1+\textbf{s}]/\{360[\textbf{c-q}]\})/\textbf{/}]$$
$$/\{\textbf{\$'}[1+\textbf{s}][\textbf{v+f}][1\text{-}\textbf{d}]\}$$

Rule-21870:

If both (**/**), (**U**), (**$**), (**i**), (**$'**), (**s**), (**v**), (**f**), (**t**), (**Q**), (**p**), (**c**) and (**q**) are known, then its Dividend Payout Planned is:

$$d= 1\text{-}[\textbf{U}+\textbf{\$}[1\text{-}\textbf{i}]\text{-}(\textbf{Q}+\textbf{\$'vp}[1+\textbf{s}]/\{360[\textbf{c-q}]\})/\textbf{/}]$$
$$/\{\textbf{\$'}[1+\textbf{s}][\textbf{v+f}][1\text{-}\textbf{t}]\}$$

Steve Asikin ISBN 14: 978-1515158691, ISBN 10: 1515158691

Rule-21871:

If both (**/**), (**U**), (**$**), (**i**), (**$'**), (**s**), (**v**), (**f**), (**t**), (**d**), (**Q**), (**c**) and (**q**) are known, then its Procured Inventory Days Planned:

$$\mathbf{p}= 360[\mathbf{c}\text{-}\mathbf{q}][\mathbf{/U}+\{\mathbf{\$}[1\text{-}\mathbf{i}]\text{-}\mathbf{\$'}[1+\mathbf{s}][\mathbf{v}+\mathbf{f}]\}$$
$$[1\text{-}\mathbf{t}][1\text{-}\mathbf{d}])\text{-}\mathbf{Q}]/\{\mathbf{\$'v}[1+\mathbf{s}]\}$$

Rule-21872:

If both (**/**), (**U**), (**$**), (**i**), (**$'**), (**s**), (**v**), (**f**), (**t**), (**d**), (**p**), (**Q**) and (**q**) are known, then its Current Ratio Planned is:

$$\mathbf{c}= \mathbf{q}+\mathbf{\$'vp}[1+\mathbf{s}]/\{360[\mathbf{/U}+\{\mathbf{\$}[1\text{-}\mathbf{i}]\text{-}\mathbf{\$'}[1+\mathbf{s}][\mathbf{v}+\mathbf{f}]\}$$
$$[1\text{-}\mathbf{t}][1\text{-}\mathbf{d}])\text{-}\mathbf{Q}]\}$$

Rule-21873:

If both (**/**), (**U**), (**$**), (**i**), (**$'**), (**s**), (**v**), (**f**), (**t**), (**d**), (**p**), (**c**) and (**Q**) are known, then its Quick or Acid Test Ratio Planned is:

$$\mathbf{q}= \mathbf{c}\text{-}\mathbf{\$'vp}[1+\mathbf{s}]/\{360[\mathbf{/U}+\{\mathbf{\$}[1\text{-}\mathbf{i}]\text{-}\mathbf{\$'}[1+\mathbf{s}][\mathbf{v}+\mathbf{f}]\}$$
$$[1\text{-}\mathbf{t}][1\text{-}\mathbf{d}])\text{-}\mathbf{Q}]\}$$

Rule-21874:

If both (**/**), (**U**), (**$**), (**$'**), (**s**), (**v**), (**f**), (**i**), (**T**), (**D**) and (**X**) are known, then its Quoted Longterm Debt Planned is:

$$\mathbf{Q}= \mathbf{/}\{\mathbf{U}+\mathbf{\$}\text{-}\mathbf{\$'}[1+\mathbf{s}][\mathbf{v}+\mathbf{f}+\mathbf{i}]\text{-}\mathbf{T}\text{-}\mathbf{D}\}\text{-}\mathbf{X}$$

Rule-21875:

If both (**Q**), (**U**), (**$**), (**$'**), (**s**), (**v**), (**f**), (**i**), (**T**), (**D**) and (**X**) are known, then its Leverage or Gearing Rtaio Planned is:

$$\mathbf{/}= [\mathbf{Q}+\mathbf{X}]/\{\mathbf{U}+\mathbf{\$}\text{-}\mathbf{\$'}[1+\mathbf{s}][\mathbf{v}+\mathbf{f}+\mathbf{i}]\text{-}\mathbf{T}\text{-}\mathbf{D}\}$$

Steve Asikin ISBN 14: 978-1515158691, ISBN 10: 1515158691

Rule-21876:

If both (*I*), (**Q**), (**$**), (**$'**), (**s**), (**v**), (**f**), (**i**), (**T**), (**D**) and (**X**) are known, then its Utilized or Starting Capital must be:

U = **$'**[1+**s**][**v**+**f**+**i**]+**T**+**D**+[**Q**+**X**]/*I*-**$**

Rule-21877:

If both (*I*), (**U**), (**Q**), (**$'**), (**s**), (**v**), (**f**), (**i**), (**T**), (**D**) and (**X**) are known, then its Sales or Revenue Planned is:

$**= **$'[1+**s**][**v**+**f**+**i**]+**T**+**D**+[**Q**+**X**]/*I*-**U**

Rule-21878:

If both (*I*), (**U**), (**$**), (**Q**), (**s**), (**v**), (**f**), (**i**), (**T**), (**D**) and (**X**) are known, then its Sales Past must be Planned is:

$'**= {**U**+**$-**T**-**D**-[**Q**+**X**]/*I*}/{[1+**s**][**v**+**f**+**i**]}

Rule-21879:

If both (*I*), (**U**), (**$**), (**$'**), (**Q**), (**v**), (**f**), (**i**), (**T**), (**D**) and (**X**) are known, then its Sales Growth Planned is:

s= **$**/**$'**-1

or it can also be found as

s= {**U**+**$**-**T**-**D**-[**Q**+**X**]/*I*}/{**$'**[**v**+**f**+**i**]}-1

Rule-21880:

If both (*I*), (**U**), (**$**), (**$'**), (**s**), (**Q**), (**f**), (**i**), (**T**), (**D**) and (**X**) are known, then its Variable Portion Planned is:

v= {**U**+**$**-**T**-**D**-[**Q**+**X**]/*I*}/{**$'**[1+**s**]}-**f**-**i**

Steve Asikin ISBN 14: 978-1515158691, ISBN 10: 1515158691

<u>Rule-21881</u>:

If both (I, (**U**), (**$**), (**$'**), (**s**), (**v**), (**Q**), (**i**), (**T**), (**D**) and (**X**) are known, then its Fixed Portion Planned is:

$$f= \{U+\$-T-D-[Q+X]/I\}/\{\$'[1+s]\}-v-i$$

<u>Rule-21882</u>:

If both (I, (**U**), (**$**), (**$'**), (**s**), (**v**), (**f**), (**Q**), (**T**), (**D**) and (**X**) are known, then its Interest Portion Planned is:

$$i= \{U+\$-T-D-[Q+X]/I\}/\{\$'[1+s]\}-v-f$$

<u>Rule-21883</u>:

If both (I, (**U**), (**$**), (**$'**), (**s**), (**v**), (**f**), (**i**), (**Q**), (**D**) and (**X**) are known, then its Tax Planned is:

$$T= U+\$-\$'[1+s][v+f+i]\}-D-[Q+X]/I$$

<u>Rule-21884</u>:

If both (I, (**U**), (**$**), (**$'**), (**s**), (**v**), (**f**), (**i**), (**T**), (**Q**) and (**X**) are known, then its Dividend Planned is:

$$D= U+\$-\$'[1+s][v+f+i]\}-T-[Q+X]/I$$

<u>Rule-21885</u>:

If both (I, (**U**), (**$**), (**$'**), (**s**), (**v**), (**f**), (**i**), (**T**), (**D**) and (**Q**) are known, then its Xpress or Current Debt Planned is:

$$X= I\{U+\$-\$'[1+s][v+f+i]\}-T-D\}-Q$$

Steve Asikin ISBN 14: 978-1515158691, ISBN 10: 1515158691

Rule-21886:

If both (**l**), (**U**), (**$**), (**$'**), (**s**), (**v**), (**f**), (**i**), (**T**), (**D**), (**P**), (**c**) and (**q**) are known, then its Quoted Longterm Debt Planned is:

$$Q= l\{U+\$-\$'[1+s][v+f+i]-T-D\}-P/[c-q]$$

Rule-21887:

If both (**Q**), (**U**), (**$**), (**$'**), (**s**), (**v**), (**f**), (**i**), (**T**), (**D**), (**P**), (**c**) and (**q**) are known, then its Leverage or Gearing Ratio Planned is:

$$l= \{Q+P/[c-q]\}/\{U+\$-\$'[1+s][v+f+i]-T-D\}$$

Rule-21888:

If both (**l**), (**Q**), (**$**), (**$'**), (**s**), (**v**), (**f**), (**i**), (**T**), (**D**), (**P**), (**c**) and (**q**) are known, then its Utilized or Starting Capital must be:

$$U = \$'[1+s][v+f+i]+T+D+\{Q+P/[c-q]\}/l-\$$$

Rule-21889:

If both (**l**), (**U**), (**Q**), (**$'**), (**s**), (**v**), (**f**), (**i**), (**T**), (**D**), (**P**), (**c**) and (**q**) are known, then its Sales or Revenue Planned is:

$$\$= \$'[1+s][v+f+i]+T+D+\{Q+P/[c-q]\}/l-U$$

Rule-21890:

If both (**l**), (**U**), (**$**), (**Q**), (**s**), (**v**), (**f**), (**i**), (**T**), (**D**), (**P**), (**c**) and (**q**) are known, then its Sales Past must be:

$$\$'= (U+\$-T-D-\{Q+P/[c-q]\}/l)/\{[1+s][v+f+i]\}$$

Steve Asikin ISBN 14: 978-1515158691, ISBN 10: 1515158691

Rule-21891:

If both ($\int$), (**U**), (**S**), (**S'**), (**Q**), (**v**), (**f**), (**i**), (**T**), (**D**), (**P**), (**c**) and (**q**) are known, then its Sales Growth Planned is:

$s = S/S'-1$

or it can also be found as

$s = (U+S-T-D-\{Q+P/[c-q]\}//)/\{S'[v+f+i]\}-1$

Rule-21892:

If both ($\int$), (**U**), (**S**), (**S'**), (**s**), (**Q**), (**f**), (**i**), (**T**), (**D**), (**P**), (**c**) and (**q**) are known, then its Variable Portion Planned is:

$v = (U+S-T-D-\{Q+P/[c-q]\}//)/\{S'[1+s]\}-f-i$

Rule-21893:

If both ($\int$), (**U**), (**S**), (**S'**), (**s**), (**v**), (**Q**), (**i**), (**T**), (**D**), (**P**), (**c**) and (**q**) are known, then its Fixed Portion Planned is:

$f = (U+S-T-D-\{Q+P/[c-q]\}//)/\{S'[1+s]\}-v-i$

Rule-21894:

If both ($\int$), (**U**), (**S**), (**S'**), (**s**), (**v**), (**f**), (**Q**), (**T**), (**D**), (**P**), (**c**) and (**q**) are known, then its Interest Portion Planned is:

$i = (U+S-T-D-\{Q+P/[c-q]\}//)/\{S'[1+s]\}-v-f$

Rule-21895:

If both ($\int$), (**U**), (**S**), (**S'**), (**s**), (**v**), (**f**), (**i**), (**Q**), (**D**), (**P**), (**c**) and (**q**) are known, then its Tax Planned is:

$T = U+S-S'[1+s][v+f+i]\}-D-\{Q+P/[c-q]\}//$

Steve Asikin ISBN 14: 978-1515158691, ISBN 10: 1515158691

Rule-21896:

If both (**/**), (**U**), (**$**), (**$'**), (**s**), (**v**), (**f**), (**i**), (**T**), (**Q**), (**P**), (**c**) and (**q**) are known, then its Dividend Planned is:

$$D= U+\$-\$'[1+s][v+f+i]\}-T-\{Q+P/[c-q]\}//$$

Rule-21897:

If both (**/**), (**U**), (**$**), (**$'**), (**s**), (**v**), (**f**), (**i**), (**T**), (**D**), (**Q**), (**c**) and (**q**) are known, then its Procured Inventory Planned is:

$$P= [c-q][/\{U+\$-\$'[1+s][v+f+i]\}-T-D\}-Q]$$

Rule-21898:

If both (**/**), (**U**), (**$**), (**$'**), (**s**), (**v**), (**f**), (**i**), (**T**), (**D**), (**P**), (**Q**) and (**q**) are known, then its Current Ratio Planned is:

$$c= q+P/[/\{U+\$-\$'[1+s][v+f+i]\}-T-D\}-Q]$$

Rule-21899:

If both (**/**), (**U**), (**$**), (**$'**), (**s**), (**v**), (**f**), (**i**), (**T**), (**D**), (**P**), (**c**) and (**Q**) are known, then its Quick or Acid Test Ratio Planned is:

$$q= c-P/[/\{U+\$-\$'[1+s][v+f+i]\}-T-D\}-Q]$$

Rule-21900:

If both (**/**), (**U**), (**$**), (**$'**), (**s**), (**v**), (**f**), (**i**), (**T**), (**D**), (**V**), (**p**), (**c**) and (**q**) are known, then its Quoted Longterm Debt Planned is:

$$Q= /\{U+\$-\$'[1+s][v+f+i]-T-D\}-Vp/\{360[c-q]\}$$

Steve Asikin ISBN 14: 978-1515158691, ISBN 10: 1515158691

Rule-21901:

If both **(Q)**, **(U)**, **($)**, **($')**, **(s)**, **(v)**, **(f)**, **(i)**, **(T)**, **(D)**, **(V)**, **(p)**, **(c)** and **(q)** are known, then its Leverage or Gearing Ratio Planned is:

$$l = (Q+Vp/\{360[c-q]\})/\{U+\$-\$'[1+s][v+f+i]-T-D\}$$

Rule-21902:

If both **(l)**, **(Q)**, **($)**, **($')**, **(s)**, **(v)**, **(f)**, **(i)**, **(T)**, **(D)**, **(V)**, **(p)**, **(c)** and **(q)** are known, then its Utilized or Starting Capital must be:

$$U = \$'[1+s][v+f+i]+T+D+(Q+Vp/\{360[c-q]\})/l-\$$$

Rule-21903:

If both **(l)**, **(U)**, **(Q)**, **($')**, **(s)**, **(v)**, **(f)**, **(i)**, **(T)**, **(D)**, **(V)**, **(p)**, **(c)** and **(q)** are known, then its Sales or Revenue Planned is:

$$\$= \$'[1+s][v+f+i]+T+D+(Q+Vp/\{360[c-q]\})/l-U$$

Rule-21904:

If both **(l)**, **(U)**, **($)**, **(Q)**, **(s)**, **(v)**, **(f)**, **(i)**, **(T)**, **(D)**, **(V)**, **(p)**, **(c)** and **(q)** are known, then its Sales Past must be:

$$\$'= [U+\$-T-D-(Q+Vp/\{360[c-q]\})/l]$$
$$/\{[1+s][v+f+i]\}$$

Rule-21905:

If both **(l)**, **(U)**, **($)**, **($')**, **(Q)**, **(v)**, **(f)**, **(i)**, **(T)**, **(D)**, **(V)**, **(p)**, **(c)** and **(q)** are known, then its Sales Growth Planned is:

$$s= \$/\$'-1$$

or it can also be found as

$$s= [U+\$-T-D-(Q+Vp/\{360[c-q]\})/l]/\{\$'[v+f+i]\}-1$$

Steve Asikin ISBN 14: 978-1515158691, ISBN 10: 1515158691

Rule-21906:

    If both (**/**), (**U**), (**$**), (**$'**), (**s**), (**Q**), (**f**), (**i**), (**T**), (**D**), (**V**), (**p**), (**c**) and (**q**) are known, then its Variable Portion Planned is:

    $v = V/\$$

        or it can also be found as

    $v = [U+\$-T-D-(Q+Vp/\{360[c-q]\})/\textit{/}]/\{\$'[1+s]\}-f-i$

Rule-21907:

    If both (**/**), (**U**), (**$**), (**$'**), (**s**), (**v**), (**Q**), (**i**), (**T**), (**D**), (**V**), (**p**), (**c**) and (**q**) are known, then its Fixed Portion Planned is:

    $f = [U+\$-T-D-(Q+Vp/\{360[c-q]\})/\textit{/}]/\{\$'[1+s]\}-v-i$

Rule-21908:

    If both (**/**), (**U**), (**$**), (**$'**), (**s**), (**v**), (**f**), (**Q**), (**T**), (**D**), (**V**), (**p**), (**c**) and (**q**) are known, then its Interest Portion Planned is:

    $i = [U+\$-T-D-(Q+Vp/\{360[c-q]\})/\textit{/}]/\{\$'[1+s]\}-v-f$

Rule-21909:

    If both (**/**), (**U**), (**$**), (**$'**), (**s**), (**v**), (**f**), (**i**), (**Q**), (**D**), (**V**), (**p**), (**c**) and (**q**) are known, then its Tax Planned is:

    $T = U+\$-\$'[1+s][v+f+i]\}-D-(Q+Vp/\{360[c-q]\})/\textit{/}$

Rule-21910:

    If both (**/**), (**U**), (**$**), (**$'**), (**s**), (**v**), (**f**), (**i**), (**T**), (**Q**), (**V**), (**p**), (**c**) and (**q**) are known, then its Dividend Planned is:

    $D = U+\$-\$'[1+s][v+f+i]\}-T-(Q+Vp/\{360[c-q]\})/\textit{/}$

Steve Asikin ISBN 14: 978-1515158691, ISBN 10: 1515158691

Rule-21911:

If both (*I*), (**U**), (**S**), (**S'**), (**s**), (**v**), (**f**), (**i**), (**T**), (**D**), (**Q**), (**p**), (**c**) and (**q**) are known, then its Variable Cost Planned is:

$$V = 360[c-q][I\{U+S-S'[1+s][v+f+i]\}-T-D\}-Q]/p$$

Rule-21912:

If both (*I*), (**U**), (**S**), (**S'**), (**s**), (**v**), (**f**), (**i**), (**T**), (**D**), (**V**), (**Q**), (**c**) and (**q**) are known, then its Procured Inventory Days Planned:

$$p = 360[c-q][I\{U+S-S'[1+s][v+f+i]\}-T-D\}-Q]/V$$

Rule-21913:

If both (*I*), (**U**), (**S**), (**S'**), (**s**), (**v**), (**f**), (**i**), (**T**), (**D**), (**V**), (**p**), (**Q**) and (**q**) are known, then its Current Ratio Planned is:

$$c = q+Vp/\{360[I\{U+S-S'[1+s][v+f+i]\}-T-D\}-Q]\}$$

Rule-21914:

If both (*I*), (**U**), (**S**), (**S'**), (**s**), (**v**), (**f**), (**i**), (**T**), (**D**), (**V**), (**p**), (**c**) and (**Q**) are known, then its Quick or Acid Test Ratio Planned is:

$$q = c-Vp/\{360[I\{U+S-S'[1+s][v+f+i]\}-T-D\}-Q]\}$$

Rule-21915:

If both (*I*), (**U**), (**S**), (**S'**), (**s**), (**v**), (**f**), (**i**), (**T**), (**D**), (**p**), (**c**) and (**q**) are known, then its Quoted Longterm Debt Planned is:

$$Q = I\{U+S-S'[1+s][v+f+i]-T-D\}-Svp/\{360[c-q]\}$$

Steve Asikin ISBN 14: 978-1515158691, ISBN 10: 1515158691

Rule-21916:

If both (**Q**), (**U**), (**$**), (**$'**), (**s**), (**v**), (**f**), (**i**), (**T**), (**D**), (**p**), (**c**) and (**q**) are known, then its Leverage or Gearing Ratio Planned is:

$$l = (Q+\$vp/\{360[c\text{-}q]\})/\{U+\$\text{-}\$'[1+s][v+f+i]\text{-}T\text{-}D\}$$

Rule-21917:

If both (**l**), (**Q**), (**$**), (**$'**), (**s**), (**v**), (**f**), (**i**), (**T**), (**D**), (**p**), (**c**) and (**q**) are known, then its Utilized or Starting Capital must be:

$$U = \$'[1+s][v+f+i]+T+D+(Q+\$vp/\{360[c\text{-}q]\})/l\text{-}\$$$

Rule-21918:

If both (**l**), (**U**), (**Q**), (**$'**), (**s**), (**v**), (**f**), (**i**), (**T**), (**D**), (**p**), (**c**) and (**q**) are known, then its Sales Growth Planned is:

$$\$= (l\{U\text{-}\$'[1+s][v+f+i]\text{-}T\text{-}D\}\text{-}Q)/(vp/\{360[c\text{-}q]\}\text{-}l)$$

Rule-21919:

If both (**l**), (**U**), (**$**), (**Q**), (**s**), (**v**), (**f**), (**i**), (**T**), (**D**), (**p**), (**c**) and (**q**) are known, then its Sales Past must be:

$$\$'= [U+\$\text{-}T\text{-}D\text{-}(Q+\$vp/\{360[c\text{-}q]\})/l]$$
$$/\{[1+s][v+f+i]\}$$

Steve Asikin ISBN 14: 978-1515158691, ISBN 10: 1515158691

Rule-21920:

If both (f), (U), (S), (S'), (Q), (v), (f), (i), (T), (D), (p), (c) and (q) are known, then its Sales Growth Planned is:

$s = S/S' - 1$

or it can also be found as

$s = [U+S-T-D-(Q+Svp/\{360[c-q]\})/f]/\{S'[v+f+i]\} - 1$

Rule-21921:

If both (f), (U), (S), (S'), (s), (Q), (f), (i), (T), (D), (p), (c) and (q) are known, then its Variable Portion Planned is:

$v = (f\{U+S-S'[1+s][f+i]-T-D\}-Q)$
$\quad /(Sp/\{360[c-q]\}+S'f[1+s])$

Rule-21922:

If both (f), (U), (S), (S'), (s), (v), (Q), (i), (T), (D), (p), (c) and (q) are known, then its Fixed Portion Planned is:

$f = [U+S-T-D-(Q+Svp/\{360[c-q]\})/f]/\{S'[1+s]\}-v-i$

Rule-21923:

If both (f), (U), (S), (S'), (s), (v), (f), (Q), (T), (D), (p), (c) and (q) are known, then its Interest Portion Planned is:

$i = [U+S-T-D-(Q+Svp/\{360[c-q]\})/f]/\{S'[1+s]\}-v-f$

Steve Asikin ISBN 14: 978-1515158691, ISBN 10: 1515158691

Rule-21924:

If both (I, (**U**), (**$**), (**$'**), (**s**), (**v**), (**f**), (**i**), (**Q**), (**D**), (**p**), (**c**) and (**q**) are known, then its Tax Planned is:

$$T= U+\$-\$'[1+s][v+f+i]\}-D-(Q+\$vp/\{360[c-q]\})/I$$

Rule-21925:

If both (I, (**U**), (**$**), (**$'**), (**s**), (**v**), (**f**), (**i**), (**T**), (**Q**), (**p**), (**c**) and (**q**) are known, then its Dividend Planned is:

$$D= U+\$-\$'[1+s][v+f+i]\}-T-(Q+\$vp/\{360[c-q]\})/I$$

Rule-21926:

If both (I, (**U**), (**$**), (**$'**), (**s**), (**v**), (**f**), (**i**), (**T**), (**D**), (**Q**), (**c**) and (**q**) are known, then its Procured Inventory Days Planned:

$$p= 360[c-q][I\{U+\$-\$'[1+s][v+f+i]\}-T-D\}-Q]/[\$v]$$

Rule-21927:

If both (I, (**U**), (**$**), (**$'**), (**s**), (**v**), (**f**), (**i**), (**T**), (**D**), (**p**), (**Q**) and (**q**) are known, then its Current Ratio Planned is:

$$c= q+\$vp/\{360[I\{U+\$-\$'[1+s][v+f+i]\}-T-D\}-Q]\}$$

Rule-21928:

If both (I, (**U**), (**$**), (**$'**), (**s**), (**v**), (**f**), (**i**), (**T**), (**D**), (**p**), (**c**) and (**Q**) are known, then its Quick or Acid Test Ratio Planned is:

$$q= c-\$vp/\{360[I\{U+\$-\$'[1+s][v+f+i]\}-T-D\}-Q]\}$$

Steve Asikin ISBN 14: 978-1515158691, ISBN 10: 1515158691

Rule-21929:

If both (**l**), (**U**), (**S**), (**S'**), (**s**), (**v**), (**f**), (**i**), (**T**), (**D**), (**p**), (**c**) and (**q**) are known, then its Quoted Longterm Debt Planned is:

$$Q= l\{U+S\text{-}S'[1+s][v+f+i]\text{-}T\text{-}D\}$$
$$\text{-}S'vp[1+s]/\{360[c\text{-}q]\}$$

Rule-21930:

If both (**Q**), (**U**), (**S**), (**S'**), (**s**), (**v**), (**f**), (**i**), (**T**), (**D**), (**p**), (**c**) and (**q**) are known, then its Leverage or Gearing Ratio Planned is:

$$l= (Q+S'vp[1+s]/\{360[c\text{-}q]\})$$
$$/\{U+S\text{-}S'[1+s][v+f+i]\text{-}T\text{-}D\}$$

Rule-21931:

If both (**l**), (**Q**), (**S**), (**S'**), (**s**), (**v**), (**f**), (**i**), (**T**), (**D**), (**p**), (**c**) and (**q**) are known, then its Utilized or Starting Capital must be:

$$U = S'[1+s][v+f+i]+T+D$$
$$+(Q+S'vp[1+s]/\{360[c\text{-}q]\})/l\text{-}S$$

Rule-21932:

If both (**l**), (**U**), (**Q**), (**S'**), (**s**), (**v**), (**f**), (**i**), (**T**), (**D**), (**p**), (**c**) and (**q**) are known, then its Sales or Revenue Planned is:

$$S= (l\{U\text{-}S'[1+s][v+f+i]\text{-}T\text{-}D\}\text{-}Q)/(vp/\{360[c\text{-}q]\}\text{-}l)$$

Steve Asikin ISBN 14: 978-1515158691, ISBN 10: 1515158691

Rule-21933:

If both (*l*), (**U**), (**$**), (**Q**), (**s**), (**v**), (**f**), (**i**), (**T**), (**D**), (**p**), (**c**) and (**q**) are known, then its Sales Past must be:

$$\$' = (l\{U+\$-\$'[1+s][f+i]-T-D\}-Q)$$
$$/[v[1+s](p/\{360[c-q]\}+l)]$$

Rule-21934:

If both (*l*), (**U**), (**$**), (**$'**), (**Q**), (**v**), (**f**), (**i**), (**T**), (**D**), (**p**), (**c**) and (**q**) are known, then its Sales Growth Planned is:

$$s = \$/\$'-1$$

or it can also be found as

$$s = (l\{U+\$-\$'[1+s][f+i]-T-D\}-Q)$$
$$/[\$'v(p/\{360[c-q]\}+l)]-1$$

Rule-21935:

If both (*l*), (**U**), (**$**), (**$'**), (**s**), (**Q**), (**f**), (**i**), (**T**), (**D**), (**p**), (**c**) and (**q**) are known, then its Variable Portion Planned is:

$$v = (l\{U+\$-\$'[1+s][f+i]-T-D\}-Q)$$
$$/[\$'[1+s](p/\{360[c-q]\}+l)]$$

Rule-21936:

If both (*l*), (**U**), (**$**), (**$'**), (**s**), (**v**), (**Q**), (**i**), (**T**), (**D**), (**p**), (**c**) and (**q**) are known, then its Fixed Portion Planned is:

$$f = [U+\$-T-D-(Q+\$'vp[1+s]/\{360[c-q]\})/l]$$
$$/\{\$'[1+s]\}-v-i$$

Steve Asikin ISBN 14: 978-1515158691, ISBN 10: 1515158691

Rule-21937:

If both (**/**), (**U**), (**$**), (**$'**), (**s**), (**v**), (**f**), (**Q**), (**T**), (**D**), (**p**), (**c**) and (**q**) are known, then its Interest Portion Planned is:

$$i= [U+\$-T-D-(Q+\$'vp[1+s]/\{360[c-q]\})/\textbf{/}]$$
$$/\{\$'[1+s]\}-v-f$$

Rule-21938:

If both (**/**), (**U**), (**$**), (**$'**), (**s**), (**v**), (**f**), (**i**), (**Q**), (**D**), (**p**), (**c**) and (**q**) are known, then its Tax Planned is:

$$T= U+\$-\$'[1+s][v+f+i]\}-D$$
$$-(Q+\$'vp[1+s]/\{360[c-q]\})/\textbf{/}$$

Rule-21939:

If both (**/**), (**U**), (**$**), (**$'**), (**s**), (**v**), (**f**), (**i**), (**T**), (**Q**), (**p**), (**c**) and (**q**) are known, then its Dividend Planned is:

$$D= U+\$-\$'[1+s][v+f+i]\}-T$$
$$-(Q+\$'vp[1+s]/\{360[c-q]\})/\textbf{/}$$

Rule-21940:

If both (**/**), (**U**), (**$**), (**$'**), (**s**), (**v**), (**f**), (**i**), (**T**), (**D**), (**Q**), (**c**) and (**q**) are known, then its Procured Inventory Days Planned:

$$p= 360[c-q][/\{U+\$-\$'[1+s][v+f+i]\}-T-D\}-Q]$$
$$/\{\$'v[1+s]\}$$

Steve Asikin ISBN 14: 978-1515158691, ISBN 10: 1515158691

Rule-21941:

If both (**/**), (**U**), (**$**), (**$'**), (**s**), (**v**), (**f**), (**i**), (**T**), (**D**), (**p**), (**Q**) and (**q**) are known, then its Current Ratio Planned is:

$$c= q+\$'vp[1+s]$$
$$/\{360[/\{U+\$-\$'[1+s][v+f+i]\}-T-D\}-Q]\}$$

Rule-21942:

If both (**/**), (**U**), (**$**), (**$'**), (**s**), (**v**), (**f**), (**i**), (**T**), (**D**), (**p**), (**c**) and (**Q**) are known, then its Quick or Acid Test Ratio Planned is:

$$q= c-\$'vp[1+s]$$
$$/\{360[/\{U+\$-\$'[1+s][v+f+i]\}-T-D\}-Q]\}$$

Rule-21943:

If both (**/**), (**U**), (**$**), (**$'**), (**s**), (**v**), (**f**), (**i**), (**T**), (**d**) and (**X**) are known, then its Quoted Longterm Debt Planned is:

$$Q= /(U+\{\$-\$'[1+s][v+f+i]-T\}[1-d])-X$$

Rule-21944:

If both (**Q**), (**U**), (**$**), (**$'**), (**s**), (**v**), (**f**), (**i**), (**T**), (**d**) and (**X**) are known, then its Leverage or Gearing Ratio Planned is:

$$/= [Q+X]/(U+\{\$-\$'[1+s][v+f+i]-T\}[1-d])$$

Rule-21945:

If both (**/**), (**Q**), (**$**), (**$'**), (**s**), (**v**), (**f**), (**i**), (**T**), (**d**) and (**X**) are known, then its Utilized or Starting Capital must be:

$$U= [Q+X]//\{\$-\$'[1+s][v+f+i]-T\}[1-d]$$

Steve Asikin ISBN 14: 978-1515158691, ISBN 10: 1515158691

Rule-21946:

If both $(\textit{f})$, $(\textbf{U})$, $(\textbf{Q})$, $(\textbf{S'})$, $(\textbf{s})$, $(\textbf{v})$, $(\textbf{f})$, $(\textbf{i})$, $(\textbf{T})$, $(\textbf{d})$ and $(\textbf{X})$ are known, then its Sales or Revenue Planned is:

$$S= S'[1+s][v+f+i]+T+\{[Q+X]/\textit{f}U\}/[1-d]$$

Rule-21947:

If both $(\textit{f})$, $(\textbf{U})$, $(\textbf{S})$, $(\textbf{Q})$, $(\textbf{s})$, $(\textbf{v})$, $(\textbf{f})$, $(\textbf{i})$, $(\textbf{T})$, $(\textbf{d})$ and $(\textbf{X})$ are known, then its Sales Past must be:

$$S'= (S-T-\{[Q+X]/\textit{f}U\}/[1-d])/\{[1+s][v+f+i]\}$$

Rule-21948:

If both $(\textit{f})$, $(\textbf{U})$, $(\textbf{S})$, $(\textbf{S'})$, $(\textbf{Q})$, $(\textbf{v})$, $(\textbf{f})$, $(\textbf{i})$, $(\textbf{T})$, $(\textbf{d})$ and $(\textbf{X})$ are known, then its Sales Growth Planned is:

$$s= S/S'-1$$

or it can also be found as

$$s= (S-T-\{[Q+X]/\textit{f}U\}/[1-d])/\{S'[v+f+i]\}-1$$

Rule-21949:

If both $(\textit{f})$, $(\textbf{U})$, $(\textbf{S})$, $(\textbf{S'})$, $(\textbf{s})$, $(\textbf{Q})$, $(\textbf{f})$, $(\textbf{i})$, $(\textbf{T})$, $(\textbf{d})$ and $(\textbf{X})$ are known, then its Variable Portion Planned is:

$$v= (S-T-\{[Q+X]/\textit{f}U\}/[1-d])/\{S'[1+s]\}-f-i$$

Rule-21950:

If both $(\textit{f})$, $(\textbf{U})$, $(\textbf{S})$, $(\textbf{S'})$, $(\textbf{s})$, $(\textbf{v})$, $(\textbf{Q})$, $(\textbf{i})$, $(\textbf{T})$, $(\textbf{d})$ and $(\textbf{X})$ are known, then its Fixed Portion Planned is:

$$f= (S-T-\{[Q+X]/\textit{f}U\}/[1-d])/\{S'[1+s]\}-v-i$$

Steve Asikin ISBN 14: 978-1515158691, ISBN 10: 1515158691

Rule-21951:

If both (**/**), (**U**), (**$**), (**$'**), (**s**), (**v**), (**f**), (**Q**), (**T**), (**d**) and (**X**) are known, then its Interest Portion Planned is:

$$i= (\$-T-\{[Q+X]/\!/U\}/[1-d])/\{\$'[1+s]\}-v-f$$

Rule-21952:

If both (**/**), (**U**), (**$**), (**$'**), (**s**), (**v**), (**f**), (**i**), (**Q**), (**d**) and (**X**) are known, then its Tax Planned is:

$$T= \$-\$'[1+s][v+f+i]-\{[Q+X]/\!/U\}/[1-d]$$

Rule-21953:

If both (**/**), (**U**), (**$**), (**$'**), (**s**), (**v**), (**f**), (**i**), (**T**), (**Q**) and (**X**) are known, then its Dividend Payout Planned is:

$$d= 1-\{[Q+X]/\!/U\}/\{\$-\$'[1+s][v+f+i]-T\}$$

Rule-21954:

If both (**/**), (**U**), (**$**), (**$'**), (**s**), (**v**), (**f**), (**i**), (**T**), (**d**) and (**Q**) are known, then its Xpress or Current Debt Planned is:

$$X= /\!U+\{\$-\$'[1+s][v+f+i]-T\}[1-d])-Q$$

Rule-21955:

If both (**/**), (**U**), (**$**), (**$'**), (**s**), (**v**), (**f**), (**i**), (**T**), (**d**), (**P**), (**c**) and (**q**) are known, then its Quoted Longterm Debt Planned is:

$$Q= /\!U+\{\$-\$'[1+s][v+f+i]-T\}[1-d])-P/[c-q]$$

Steve Asikin ISBN 14: 978-1515158691, ISBN 10: 1515158691

Rule-21956:

If both $(\mathbf{Q})$, $(\mathbf{U})$, $(\mathbf{S})$, $(\mathbf{S'})$, $(\mathbf{s})$, $(\mathbf{v})$, $(\mathbf{f})$, $(\mathbf{i})$, $(\mathbf{T})$, $(\mathbf{d})$, $(\mathbf{P})$, $(\mathbf{c})$ and $(\mathbf{q})$ are known, then its Leverage or Gearing Ratio Planned is:

$$l = \{Q+P/[c-q]\}/(U+\{S-S'[1+s][v+f+i]-T\}[1-d])$$

Rule-21957:

If both $(\mathbf{l})$, $(\mathbf{Q})$, $(\mathbf{S})$, $(\mathbf{S'})$, $(\mathbf{s})$, $(\mathbf{v})$, $(\mathbf{f})$, $(\mathbf{i})$, $(\mathbf{T})$, $(\mathbf{d})$, $(\mathbf{P})$, $(\mathbf{c})$ and $(\mathbf{q})$ are known, then its Utilized or Starting Capital must be:

$$U = \{Q+P/[c-q]\}/l-\{S-S'[1+s][v+f+i]-T\}[1-d]$$

Rule-21958:

If both $(\mathbf{l})$, $(\mathbf{U})$, $(\mathbf{Q})$, $(\mathbf{S'})$, $(\mathbf{s})$, $(\mathbf{v})$, $(\mathbf{f})$, $(\mathbf{i})$, $(\mathbf{T})$, $(\mathbf{d})$, $(\mathbf{P})$, $(\mathbf{c})$ and $(\mathbf{q})$ are known, then its Sales or Revenue Planned is:

$$S = S'[1+s][v+f+i]+T+(\{Q+P/[c-q]\}/l-U)/[1-d]$$

Rule-21959:

If both $(\mathbf{l})$, $(\mathbf{U})$, $(\mathbf{S})$, $(\mathbf{Q})$, $(\mathbf{s})$, $(\mathbf{v})$, $(\mathbf{f})$, $(\mathbf{i})$, $(\mathbf{T})$, $(\mathbf{d})$, $(\mathbf{P})$, $(\mathbf{c})$ and $(\mathbf{q})$ are known, then its Sales Past must be:

$$S' = [S-T-(\{Q+P/[c-q]\}/l-U)/[1-d]]/\{[1+s][v+f+i]\}$$

Rule-21960:

If both $(\mathbf{l})$, $(\mathbf{U})$, $(\mathbf{S})$, $(\mathbf{S'})$, $(\mathbf{Q})$, $(\mathbf{v})$, $(\mathbf{f})$, $(\mathbf{i})$, $(\mathbf{T})$, $(\mathbf{d})$, $(\mathbf{P})$, $(\mathbf{c})$ and $(\mathbf{q})$ are known, then its Sales Growth Planned is:

$$s = S/S'-1$$

or it can also be found as

$$s = [S-T-(\{Q+P/[c-q]\}/l-U)/[1-d]]/\{S'[v+f+i]\}-1$$

Steve Asikin ISBN 14: 978-1515158691, ISBN 10: 1515158691

Rule-21961:

If both (**/**), (**U**), (**\$**), (**\$'**), (**s**), (**Q**), (**f**), (**i**), (**T**), (**d**), (**P**), (**c**) and (**q**) are known, then its Variable Portion Planned is:

$$v= [\$-T-(\{Q+P/[c-q]\}/fU)/[1-d]]/\{\$'[1+s]\}-f-i$$

Rule-21962:

If both (**/**), (**U**), (**\$**), (**\$'**), (**s**), (**v**), (**Q**), (**i**), (**T**), (**d**), (**P**), (**c**) and (**q**) are known, then its Fixed Portion Planned is:

$$f= [\$-T-(\{Q+P/[c-q]\}/fU)/[1-d]]/\{\$'[1+s]\}-v-i$$

Rule-21963:

If both (**/**), (**U**), (**\$**), (**\$'**), (**s**), (**v**), (**f**), (**Q**), (**T**), (**d**), (**P**), (**c**) and (**q**) are known, then its Interest Portion Planned is:

$$i= [\$-T-(\{Q+P/[c-q]\}/fU)/[1-d]]/\{\$'[1+s]\}-v-f$$

Rule-21964:

If both (**/**), (**U**), (**\$**), (**\$'**), (**s**), (**v**), (**f**), (**i**), (**Q**), (**d**), (**P**), (**c**) and (**q**) are known, then its Tax Planned is:

$$T= \$-\$'[1+s][v+f+i]-(\{Q+P/[c-q]\}/fU)/[1-d]$$

Rule-21965:

If both (**/**), (**U**), (**\$**), (**\$'**), (**s**), (**v**), (**f**), (**i**), (**T**), (**Q**), (**P**), (**c**) and (**q**) are known, then its Dividend Payout Planned is:

$$d= 1-(\{Q+P/[c-q]\}/fU)/\{\$-\$'[1+s][v+f+i]-T\}$$

Steve Asikin ISBN 14: 978-1515158691, ISBN 10: 1515158691

Rule-21966:

If both (I), (**U**), (**$**), (**$'**), (**$**), (**v**), (**f**), (**i**), (**T**), (**d**), (**Q**), (**c**) and (**q**) are known, then its Procured Inventory Planned is:

$$P= [c\text{-}q][I(U+\{\$\text{-}\$'[1+s][v+f+i]\text{-}T\}[1\text{-}d])\text{-}Q]$$

Rule-21967:

If both (I), (**U**), (**$**), (**$'**), (**$**), (**v**), (**f**), (**i**), (**T**), (**d**), (**P**), (**Q**) and (**q**) are known, then its Current Ratio Planned is:

$$c= q+P/[I(U+\{\$\text{-}\$'[1+s][v+f+i]\text{-}T\}[1\text{-}d])\text{-}Q]$$

Rule-21968:

If both (I), (**U**), (**$**), (**$'**), (**$**), (**v**), (**f**), (**i**), (**T**), (**d**), (**P**), (**c**) and (**Q**) are known, then its Quick or Acid Test Ratio Planned is:

$$q= c\text{-}P/[I(U+\{\$\text{-}\$'[1+s][v+f+i]\text{-}T\}[1\text{-}d])\text{-}Q]$$

Rule-21969:

If both (I), (**U**), (**$**), (**$'**), (**$**), (**v**), (**f**), (**i**), (**T**), (**d**), (**V**), (**p**), (**c**) and (**q**) are known, then its Quoted Longterm Debt Planned is:

$$Q= I(U+\{\$\text{-}\$'[1+s][v+f+i]\text{-}T\}[1\text{-}d])\text{-}Vp/\{360[c\text{-}q]\}$$

Rule-21970:

If both (**Q**), (**U**), (**$**), (**$'**), (**$**), (**v**), (**f**), (**i**), (**T**), (**d**), (**V**), (**p**), (**p**), (**c**) and (**q**) are known, then its Leverage or Gearing Ratio Planned is:

$$I= (Q+Vp/\{360[c\text{-}q]\})$$
$$/(U+\{\$\text{-}\$'[1+s][v+f+i]\text{-}T\}[1\text{-}d])$$

Steve Asikin ISBN 14: 978-1515158691, ISBN 10: 1515158691

Rule-21971:

If both (I), (Q), (S), (S'), (s), (v), (f), (i), (T), (d), (V), (p), (c) and (q) are known, then its Utilized or Starting Capital must be:

$$U= (Q+Vp/\{360[c-q]\})/I$$
$$-\{S-S'[1+s][v+f+i]-T\}[1-d]$$

Rule-21972:

If both (I), (U), (Q), (S'), (s), (v), (f), (i), (T), (d), (V), (p), (c) and (q) are known, then its Sales or Revenue Planned is:

$$S= S'[1+s][v+f+i]+T+[(Q+Vp/\{360[c-q]\})/I-U]$$
$$/[1-d]$$

Rule-21973:

If both (I), (U), (S), (Q), (s), (v), (f), (i), (T), (d), (V), (p), (c) and (q) are known, then its Sales Past must be:

$$S'= \{S-T-[(Q+Vp/\{360[c-q]\})/I-U]/[1-d]\}$$
$$/\{[1+s][v+f+i]\}$$

Rule-21974:

If both (I), (U), (S), (S'), (Q), (v), (f), (i), (T), (d), (V), (p), (c) and (q) are known, then its Sales Growth Planned is:

$$s= S/S'-1$$

or it can also be found as

$$s= \{S-T-[(Q+Vp/\{360[c-q]\})/I-U]/[1-d]\}$$
$$/\{S'[v+f+i]\}-1$$

Steve Asikin ISBN 14: 978-1515158691, ISBN 10: 1515158691

Rule-21975:

If both (I), (U), (S), (S'), (s), (Q), (f), (i), (T), (d), (V), (p), (c) and (q) are known, then its Variable Portion Planned is:

$$v = \{S\text{-}T\text{-}[(Q+Vp/\{360[c\text{-}q]\})/I\text{-}U]/[1\text{-}d]\} / \{S'[1+s]\}\text{-}f\text{-}i$$

Rule-21976:

If both (I), (U), (S), (S'), (s), (v), (Q), (i), (T), (d), (V), (p), (c) and (q) are known, then its Fixed Portion Planned is:

$$f = \{S\text{-}T\text{-}[(Q+Vp/\{360[c\text{-}q]\})/I\text{-}U]/[1\text{-}d]\} / \{S'[1+s]\}\text{-}v\text{-}i$$

Rule-21977:

If both (I), (U), (S), (S'), (s), (v), (f), (Q), (T), (d), (V), (p), (c) and (q) are known, then its Interest Portion Planned is:

$$i = \{S\text{-}T\text{-}[(Q+Vp/\{360[c\text{-}q]\})/I\text{-}U]/[1\text{-}d]\} / \{S'[1+s]\}\text{-}v\text{-}f$$

Rule-21978:

If both (I), (U), (S), (S'), (s), (v), (f), (i), (Q), (d), (V), (p), (c) and (q) are known, then its Tax Planned is:

$$T = S\text{-}S'[1+s][v+f+i][(Q+Vp/\{360[c\text{-}q]\})/I\text{-}U] / [1\text{-}d]$$

Steve Asikin ISBN 14: 978-1515158691, ISBN 10: 1515158691

Rule-21979:

If both (I), (**U**), (**$**), (**$'**), (**s**), (**v**), (**f**), (**i**), (**T**), (**Q**), (**V**), (**p**), (**c**) and (**q**) are known, then its Dividend Payout Planned is:

$$d= 1-[(Q+Vp/\{360[c-q]\})/IU]$$
$$/\{\$-\$'[1+s][v+f+i]-T\}$$

Rule-21980:

If both (I), (**U**), (**$**), (**$'**), (**s**), (**v**), (**f**), (**i**), (**T**), (**d**), (**Q**), (**p**), (**c**) and (**q**) are known, then its Variable Cost Planned is:

$$V= 360[c-q][(IU+\{\$-\$'[1+s][v+f+i]-T\}[1-d])-Q]/p$$

Rule-21981:

If both (I), (**U**), (**$**), (**$'**), (**s**), (**v**), (**f**), (**i**), (**T**), (**d**), (**V**), (**Q**), (**c**) and (**q**) are known, then its Procured Inventory Days Planned:

$$p= 360[c-q][(IU+\{\$-\$'[1+s][v+f+i]-T\}[1-d])-Q]/V$$

Rule-21982:

If both (I), (**U**), (**$**), (**$'**), (**s**), (**v**), (**f**), (**i**), (**T**), (**d**), (**V**), (**p**), (**Q**) and (**q**) are known, then its Current Ratio Planned is:

$$c= q+Vp$$
$$/\{360[(IU+\{\$-\$'[1+s][v+f+i]-T\}[1-d])-Q]\}$$

Rule-21983:

If both (I), (**U**), (**$**), (**$'**), (**s**), (**v**), (**f**), (**i**), (**T**), (**d**), (**V**), (**p**), (**c**) and (**Q**) are known, then its Quick or Acid Test Ratio Planned is:

$$q= c-Vp/\{360[(IU+\{\$-\$'[1+s][v+f+i]-T\}[1-d])-Q]\}$$

Steve Asikin ISBN 14: 978-1515158691, ISBN 10: 1515158691

Rule-21984:

If both (I), (U), (S), (S'), (s), (v), (f), (i), (T), (d), (p), (c) and (q) are known, then its Quoted Longterm Debt Planned is:

$$Q= I(U+\{S-S'[1+s][v+f+i]-T\}[1-d])$$
$$-Svp/\{360[c-q]\}$$

Rule-21985:

If both (Q), (U), (S), (S'), (s), (v), (f), (i), (T), (d), (p), (c) and (q) are known, then its Leverage or Gearing Ratio Planned is:

$$I= (Q+Svp/\{360[c-q]\})$$
$$/(U+\{S-S'[1+s][v+f+i]-T\}[1-d])$$

Rule-21986:

If both (I), (Q), (S), (S'), (s), (v), (f), (i), (T), (d), (p), (c) and (q) are known, then its Utilized or Starting Capital must be:

$$U= (Q+Svp/\{360[c-q]\})/I$$
$$-\{S-S'[1+s][v+f+i]-T\}[1-d]$$

Rule-21987:

If both (I), (U), (Q), (S'), (s), (v), (f), (i), (T), (d), (p), (c) and (q) are known, then its Sales or Revenue Planned is:

$$S= [I(U+\{S-S'[1+s][v+f+i]-T\}[1-d])-Q]$$
$$/(vp/\{360[c-q]\}-I[1-d])$$

Steve Asikin ISBN 14: 978-1515158691, ISBN 10: 1515158691

Rule-21988:

If both (**/**), (**U**), (**$**), (**Q**), (**s**), (**v**), (**f**), (**i**), (**T**), (**d**), (**p**), (**c**) and (**q**) are known, then its Sales Past must be:

$$\$' = \{\$-T-[(Q+\$vp/\{360[c-q]\})/\textit{/}U]/[1-d]\}$$
$$/\{[1+s][v+f+i]\}$$

Rule-21989:

If both (**/**), (**U**), (**$**), (**$'**), (**Q**), (**v**), (**f**), (**i**), (**T**), (**d**), (**p**), (**c**) and (**q**) are known, then its Sales Growth Planned is:

$$s = \$/\$'-1$$

or it can also be found as

$$s = \{\$-T-[(Q+\$vp/\{360[c-q]\})/\textit{/}U]$$
$$/[1-d]\}/\{\$'[v+f+i]\}-1$$

Rule-21990:

If both (**/**), (**U**), (**$**), (**$'**), (**s**), (**Q**), (**f**), (**i**), (**T**), (**d**), (**p**), (**c**) and (**q**) are known, then its Variable Portion Planned is:

$$v = [\textit{/}U+\{\$-\$'[1+s][f+i]-T\}[1-d])-Q]$$
$$/(\$p/\{360[c-q]\}+\$'\textit{/}[1+s][1-d])$$

Rule-21991:

If both (**/**), (**U**), (**$**), (**$'**), (**s**), (**v**), (**Q**), (**i**), (**T**), (**d**), (**p**), (**c**) and (**q**) are known, then its Fixed Portion Planned is:

$$f = \{\$-T-[(Q+\$vp/\{360[c-q]\})/\textit{/}U]/[1-d]\}$$
$$/\{\$'[1+s]\}-v-i$$

Steve Asikin ISBN 14: 978-1515158691, ISBN 10: 1515158691

Rule-21992:

If both (f), (U), (S), (S'), (s), (v), (f), (Q), (T), (d), (p), (c) and (q) are known, then its Interest Portion Planned is:

$$i = \{S-T-[(Q+Svp/\{360[c-q]\})/fU]/[1-d]\}$$
$$/\{S'[1+s]\}-v-f$$

Rule-21993:

If both (f), (U), (S), (S'), (s), (v), (f), (i), (Q), (d), (p), (c) and (q) are known, then its Tax Planned is:

$$T = S-S'[1+s][v+f+i][(Q+Svp/\{360[c-q]\})/fU]$$
$$/[1-d]$$

Rule-21994:

If both (f), (U), (S), (S'), (s), (v), (f), (i), (T), (Q), (p), (c) and (q) are known, then its Dividend Payout Planned is:

$$d = 1-[(Q+Svp/\{360[c-q]\})/fU]$$
$$/\{S-S'[1+s][v+f+i]-T\}$$

Rule-21995:

If both (f), (U), (S), (S'), (s), (v), (f), (i), (T), (d), (Q), (c) and (q) are known, then its Procured Inventory Days Planned:

$$p = 360[c-q]$$
$$[(fU+\{S-S'[1+s][v+f+i]-T\}[1-d])-Q]/[Sv]$$

Steve Asikin ISBN 14: 978-1515158691, ISBN 10: 1515158691

Rule-21996:

If both (**/**), (**U**), (**$**), (**$'**), (**s**), (**v**), (**f**), (**i**), (**T**), (**d**), (**p**), (**Q**) and (**q**) are known, then its Current Ratio Planned is:

$$c = q + \$vp$$
$$/\{360[/(U+\{\$-\$'[1+s][v+f+i]-T\}[1-d])-Q]\}$$

Rule-21997:

If both (**/**), (**U**), (**$**), (**$'**), (**s**), (**v**), (**f**), (**i**), (**T**), (**d**), (**p**), (**c**) and (**Q**) are known, then its Quick or Acid Test Ratio Planned is:

$$q = c - \$vp$$
$$/\{360[/(U+\{\$-\$'[1+s][v+f+i]-T\}[1-d])-Q]\}$$

Rule-21998:

If both (**/**), (**U**), (**$**), (**$'**), (**s**), (**v**), (**f**), (**i**), (**T**), (**d**), (**p**), (**c**) and (**q**) are known, then its Quoted Longterm Debt Planned is:

$$Q = /(U+\{\$-\$'[1+s][v+f+i]-T\}[1-d])$$
$$-\$'vp[1+s]/\{360[c-q]\}$$

Rule-21999:

If both (**Q**), (**U**), (**$**), (**$'**), (**s**), (**v**), (**f**), (**i**), (**T**), (**d**), (**p**), (**c**) and (**q**) are known, then its Leverage or Gearing Ratio Planned is:

$$/ = (Q+\$'vp[1+s]/\{360[c-q]\})$$
$$/(U+\{\$-\$'[1+s][v+f+i]-T\}[1-d])$$

Steve Asikin ISBN 14: 978-1515158691, ISBN 10: 1515158691

Rule-22000:

If both (**/**), (**Q**), (**S**), (**S'**), (**s**), (**v**), (**f**), (**i**), (**T**), (**d**), (**p**), (**c**) and (**q**) are known, then its Utilized or Starting Capital must be:

$$U = (Q + S'vp[1+s]/\{360[c-q]\})/I$$
$$-\{S-S'[1+s][v+f+i]-T\}[1-d]$$

Rule-22001:

If both (**/**), (**U**), (**Q**), (**S'**), (**s**), (**v**), (**f**), (**i**), (**T**), (**d**), (**p**), (**c**) and (**q**) are known, then its Sales or Revenue Planned is:

$$S = [Q - I(U - \{S'[1+s][v+f+i]+T\}[1-d])$$
$$+ S'vp[1+s]/\{360[c-q]\}]/\{I[1-d]\}$$

Rule-22002:

If both (**/**), (**U**), (**S**), (**Q**), (**s**), (**v**), (**f**), (**i**), (**T**), (**d**), (**p**), (**c**) and (**q**) are known, then its Sales Past must be:

$$S' = (I\{U + [S-T][1-d]\} - Q)$$
$$/[[1+s](vp/\{360[c-q]\} + I[v+f+i][1-d]]$$

Rule-22003:

If both (**/**), (**U**), (**S**), (**S'**), (**Q**), (**v**), (**f**), (**i**), (**T**), (**d**), (**p**), (**c**) and (**q**) are known, then its Sales Growth Planned is:

$$s = S/S' - 1$$

or it can also be found as

$$s = = (I\{U + [S-T][1-d]\} - Q)$$
$$/[S'(vp/\{360[c-q]\} + I[v+f+i][1-d]] - 1$$

Steve Asikin ISBN 14: 978-1515158691, ISBN 10: 1515158691

Rule-22004:
If both (I), $(\mathbf{U})$, $(\$)$, $(\$')$, (s), $(\mathbf{Q})$, $(\mathbf{f})$, $(\mathbf{i})$, $(\mathbf{T})$, $(\mathbf{d})$, $(\mathbf{p})$, $(\mathbf{c})$ and $(\mathbf{q})$ are known, then its Variable Portion Planned is:

$$v= [I(U+\{\$-\$'[1+s][f+i]-T\}[1-d])-Q] / [\$'[1+s](p/\{360[c-q]\}+I[1-d])]$$

Rule-22005:
If both (I), $(\mathbf{U})$, $(\$)$, $(\$')$, (s), (v), $(\mathbf{Q})$, $(\mathbf{i})$, $(\mathbf{T})$, $(\mathbf{d})$, $(\mathbf{p})$, $(\mathbf{c})$ and $(\mathbf{q})$ are known, then its Fixed Portion Planned is:

$$f=\{\$-T-[(Q+\$'vp[1+s]/\{360[c-q]\})/IU]/[1-d]\} / \{\$'[1+s]\}-v-i$$

Rule-22006:
If both (I), $(\mathbf{U})$, $(\$)$, $(\$')$, (s), (v), $(\mathbf{f})$, $(\mathbf{Q})$, $(\mathbf{T})$, $(\mathbf{d})$, $(\mathbf{p})$, $(\mathbf{c})$ and $(\mathbf{q})$ are known, then its Interest Portion Planned is:

$$i= \{\$-T-[(Q+\$'vp[1+s]/\{360[c-q]\})/IU]/[1-d]\} / \{\$'[1+s]\}-v-f$$

Rule-22007:
If both (I), $(\mathbf{U})$, $(\$)$, $(\$')$, (s), (v), $(\mathbf{f})$, $(\mathbf{i})$, $(\mathbf{Q})$, $(\mathbf{d})$, $(\mathbf{p})$, $(\mathbf{c})$ and $(\mathbf{q})$ are known, then its Tax Planned is:

$$T= \$-\$'[1+s][v+f+i] [(Q+\$'vp[1+s]/\{360[c-q]\})/IU]/ [1-d]$$

Steve Asikin ISBN 14: 978-1515158691, ISBN 10: 1515158691

Rule-22008:

If both (I), (U), (S), (S'), (s), (v), (f), (i), (T), (Q), (p), (c) and (q) are known, then its Dividend Payout Planned is:

$$d= 1-[(Q+S'vp[1+s]/\{360[c-q]\})/I-U]$$
$$/\{S-S'[1+s][v+f+i]-T\}$$

Rule-22009:

If both (I), (U), (S), (S'), (s), (v), (f), (i), (T), (d), (Q), (c) and (q) are known, then its Procured Inventory Days Planned:

$$p= 360[c-q][I(U+\{S-S'[1+s][v+f+i]-T\}[1-d])-Q]$$
$$/\{S'v[1+s]\}$$

Rule-22010:

If both (I), (U), (S), (S'), (s), (v), (f), (i), (T), (d), (p), (Q) and (q) are known, then its Current Ratio Planned is:

$$c= q+S'vp[1+s]$$
$$/\{360[I(U+\{S-S'[1+s][v+f+i]-T\}[1-d])-Q]\}$$

Rule-22011:

If both (I), (U), (S), (S'), (s), (v), (f), (i), (T), (d), (p), (c) and (Q) are known, then its Quick or Acid Test Ratio Planned is:

$$q= c-S'vp[1+s]$$
$$/\{360[I(U+\{S-S'[1+s][v+f+i]-T\}[1-d])-Q]\}$$

Steve Asikin ISBN 14: 978-1515158691, ISBN 10: 1515158691

Rule-22012:

    If both (*I*), (**U**), (**S**), (**S'**), (**s**), (**v**), (**f**), (**i**), (**T**), (**A**), (**d**) and (**X**) are known, then its Quoted Longterm Debt Planned is:

$$Q = I\{U+S-S'[1+s][v+f+i]\}-T-Ad\}-X$$

Rule-22013:

    If both (**Q**), (**U**), (**S**), (**S'**), (**s**), (**v**), (**f**), (**i**), (**T**), (**A**), (**d**) and (**X**) are known, then its Leverage or Gearing Ratio Planned is:

$$I = [Q+X]/\{U+S-S'[1+s][v+f+i]\}-T-Ad\}$$

Rule-22014:

    If both (*I*), (**Q**), (**S**), (**S'**), (**s**), (**v**), (**f**), (**i**), (**T**), (**A**), (**d**) and (**X**) are known, then its Utilized or Starting Capital must be:

$$U = S'[1+s][v+f+i]+T+Ad+[Q+X]/I-S$$

Rule-22015:

    If both (*I*), (**U**), (**Q**), (**S'**), (**s**), (**v**), (**f**), (**i**), (**T**), (**A**), (**d**) and (**X**) are known, then its Sales or Revenue Planned is:

$$S = S'[1+s][v+f+i]\}+T+Ad+[Q+X]/I-U$$

Rule-22016:

    If both (*I*), (**U**), (**S**), (**Q**), (**s**), (**v**), (**f**), (**i**), (**T**), (**A**), (**d**) and (**X**) are known, then its Sales Past must be:

$$S' = \{U+S-T-Ad-[Q+X]/I\}/\{[1+s][v+f+i]\}$$

Steve Asikin ISBN 14: 978-1515158691, ISBN 10: 1515158691

Rule-22017:

If both (I), $(\mathbf{U})$, $(\mathbf{\$})$, $(\mathbf{\$'})$, $(\mathbf{Q})$, $(\mathbf{v})$, $(\mathbf{f})$, $(\mathbf{i})$, $(\mathbf{T})$, $(\mathbf{A})$, $(\mathbf{d})$ and $(\mathbf{X})$ are known, then its Sales Growth Planned is:

$s = \$/\$' - 1$

or it can also be found as

$s = \{U+\$-T-Ad-[Q+X]/I\}/\{\$'[v+f+i]\} - 1$

Rule-22018:

If both (I), $(\mathbf{U})$, $(\mathbf{\$})$, $(\mathbf{\$'})$, (s), $(\mathbf{Q})$, $(\mathbf{f})$, $(\mathbf{i})$, $(\mathbf{T})$, $(\mathbf{A})$, $(\mathbf{d})$ and $(\mathbf{X})$ are known, then its Variable Portion Planned is:

$v = \{U+\$-T-Ad-[Q+X]/I\}/\{\$'[1+s]\} - f - i$

Rule-22019:

If both (I), $(\mathbf{U})$, $(\mathbf{\$})$, $(\mathbf{\$'})$, (s), $(\mathbf{v})$, $(\mathbf{Q})$, $(\mathbf{i})$, $(\mathbf{T})$, $(\mathbf{A})$, $(\mathbf{d})$ and $(\mathbf{X})$ are known, then its Fixed Portion Planned is:

$f = \{U+\$-T-Ad-[Q+X]/I\}/\{\$'[1+s]\} - v - i$

Rule-22020:

If both (I), $(\mathbf{U})$, $(\mathbf{\$})$, $(\mathbf{\$'})$, (s), $(\mathbf{v})$, $(\mathbf{f})$, $(\mathbf{Q})$, $(\mathbf{T})$, $(\mathbf{A})$, $(\mathbf{d})$ and $(\mathbf{X})$ are known, then its Interest Portion Planned is:

$i = \{U+\$-T-Ad-[Q+X]/I\}/\{\$'[1+s]\} - v - f$

Rule-22021:

If both (I), $(\mathbf{U})$, $(\mathbf{\$})$, $(\mathbf{\$'})$, (s), $(\mathbf{v})$, $(\mathbf{f})$, $(\mathbf{i})$, $(\mathbf{Q})$, $(\mathbf{A})$, $(\mathbf{d})$ and $(\mathbf{X})$ are known, then its Tax Planned is:

$T = U+\$-\$'[1+s][v+f+i]\} - Ad - [Q+X]/I$

Steve Asikin ISBN 14: 978-1515158691, ISBN 10: 1515158691

Rule-22022:
 If both (**/**), (**U**), (**\$**), (**\$'**), (**s**), (**v**), (**f**), (**i**), (**T**), (**Q**), (**d**) and (**X**) are known, then its After Tax Income Planned is:
 $$A= \{U+\$-\$'[1+s][v+f+i]\}-T-[Q+X]/\!\!\!/ /d$$

Rule-22023:
 If both (**/**), (**U**), (**\$**), (**\$'**), (**s**), (**v**), (**f**), (**i**), (**T**), (**A**), (**Q**) and (**X**) are known, then its Dividend Payout Planned is:
 $$d= \{U+\$-\$'[1+s][v+f+i]\}-T-[Q+X]/\!\!\!/ /A$$

Rule-22024:
 If both (**/**), (**U**), (**\$**), (**\$'**), (**s**), (**v**), (**f**), (**i**), (**T**), (**A**), (**d**) and (**Q**) are known, then its Xpress or Current Debt Planned is:
 $$X= /\!\!\!/\{U+\$-\$'[1+s][v+f+i]\}-T-Ad\}-Q$$

Rule-22025:
 If both (**/**), (**U**), (**\$**), (**\$'**), (**s**), (**v**), (**f**), (**i**), (**T**), (**A**), (**d**), (**P**), (**c**) and (**q**) are known, then its Quoted Longterm Debt Planned is:
 $$Q= /\!\!\!/\{U+\$-\$'[1+s][v+f+i]\}-T-Ad\}-P/[c-q]$$

Rule-22026:
 If both (**Q**), (**U**), (**\$**), (**\$'**), (**s**), (**v**), (**f**), (**i**), (**T**), (**A**), (**d**), (**P**), (**c**) and (**q**) are known, then its Leverage or Gearing Ratio Planned is:
 $$/\!\!\!/= \{Q+P/[c-q]\}/\{U+\$-\$'[1+s][v+f+i]\}-T-Ad\}$$

Steve Asikin ISBN 14: 978-1515158691, ISBN 10: 1515158691

Rule-22027:

If both (**I**), (**Q**), (**S**), (**S'**), (**s**), (**v**), (**f**), (**i**), (**T**), (**A**), (**d**), (**P**), (**c**) and (**q**) are known, then its Utilized or Starting Capital must be:

$$U= S'[1+s][v+f+i]+T+Ad+\{Q+P/[c-q]\}/I\text{-}S$$

Rule-22028:

If both (**I**), (**U**), (**Q**), (**S'**), (**s**), (**v**), (**f**), (**i**), (**T**), (**A**), (**d**), (**P**), (**c**) and (**q**) are known, then its Sales or Revenue Planned is:

$$S= S'[1+s][v+f+i]\}+T+Ad+\{Q+P/[c-q]\}/I\text{-}U$$

Rule-22029:

If both (**I**), (**U**), (**S**), (**Q**), (**s**), (**v**), (**f**), (**i**), (**T**), (**A**), (**d**), (**P**), (**c**) and (**q**) are known, then its Sales Past must be:

$$S'= (U+S\text{-}T\text{-}Ad\text{-}\{Q+P/[c-q]\}/I)/\{[1+s][v+f+i]\}$$

Rule-22030:

If both (**I**), (**U**), (**S**), (**S'**), (**Q**), (**v**), (**f**), (**i**), (**T**), (**A**), (**d**), (**P**), (**c**) and (**q**) are known, then its Sales Growth Planned is:

$$s= S/S'\text{-}1$$

or it can also be found as

$$s= (U+S\text{-}T\text{-}Ad\text{-}\{Q+P/[c-q]\}/I)/\{S'[v+f+i]\}\text{-}1$$

Rule-22031:

If both (**I**), (**U**), (**S**), (**S'**), (**s**), (**Q**), (**f**), (**i**), (**T**), (**A**), (**d**), (**P**), (**c**) and (**q**) are known, then its Variable Portion Planned is:

$$v= (U+S\text{-}T\text{-}Ad\text{-}\{Q+P/[c-q]\}/I)/\{S'[1+s]\}\text{-}f\text{-}i$$

Steve Asikin ISBN 14: 978-1515158691, ISBN 10: 1515158691

Rule-22032:

If both $(\pmb{\mathit{I}})$, $(\pmb{U})$, $(\pmb{S})$, $(\pmb{S'})$, $(\pmb{s})$, $(\pmb{v})$, $(\pmb{Q})$, $(\pmb{i})$, $(\pmb{T})$, $(\pmb{A})$, $(\pmb{d})$, $(\pmb{P})$, $(\pmb{c})$ and $(\pmb{q})$ are known, then its Fixed Portion Planned is:

$$f = (U + S - T - Ad - \{Q + P/[c-q]\}/\pmb{\mathit{I}})/\{S'[1+s]\} - v - i$$

Rule-22033:

If both $(\pmb{\mathit{I}})$, $(\pmb{U})$, $(\pmb{S})$, $(\pmb{S'})$, $(\pmb{s})$, $(\pmb{v})$, $(\pmb{f})$, $(\pmb{Q})$, $(\pmb{T})$, $(\pmb{A})$, $(\pmb{d})$, $(\pmb{P})$, $(\pmb{c})$ and $(\pmb{q})$ are known, then its Interest Portion Planned is:

$$i = (U + S - T - Ad - \{Q + P/[c-q]\}/\pmb{\mathit{I}})/\{S'[1+s]\} - v - f$$

Rule-22034:

If both $(\pmb{\mathit{I}})$, $(\pmb{U})$, $(\pmb{S})$, $(\pmb{S'})$, $(\pmb{s})$, $(\pmb{v})$, $(\pmb{f})$, $(\pmb{i})$, $(\pmb{Q})$, $(\pmb{A})$, $(\pmb{d})$, $(\pmb{P})$, $(\pmb{c})$ and $(\pmb{q})$ are known, then its Tax Planned is:

$$T = U + S - S'[1+s][v+f+i]\} - Ad - \{Q + P/[c-q]\}/\pmb{\mathit{I}}$$

Rule-22035:

If both $(\pmb{\mathit{I}})$, $(\pmb{U})$, $(\pmb{S})$, $(\pmb{S'})$, $(\pmb{s})$, $(\pmb{v})$, $(\pmb{f})$, $(\pmb{i})$, $(\pmb{T})$, $(\pmb{Q})$, $(\pmb{d})$, $(\pmb{P})$, $(\pmb{c})$ and $(\pmb{q})$ are known, then its After Tax Income Planned is:

$$A = (U + S - S'[1+s][v+f+i]\} - T - \{Q + P/[c-q]\}/\pmb{\mathit{I}})/d$$

Rule-22036:

If both $(\pmb{\mathit{I}})$, $(\pmb{U})$, $(\pmb{S})$, $(\pmb{S'})$, $(\pmb{s})$, $(\pmb{v})$, $(\pmb{f})$, $(\pmb{i})$, $(\pmb{T})$, $(\pmb{A})$, $(\pmb{Q})$, $(\pmb{P})$, $(\pmb{c})$ and $(\pmb{q})$ are known, then its Dividend Payout Planned is:

$$d = (U + S - S'[1+s][v+f+i]\} - T - \{Q + P/[c-q]\}/\pmb{\mathit{I}})/A$$

Steve Asikin ISBN 14: 978-1515158691, ISBN 10: 1515158691

Rule-22037:

If both (I), (**U**), (**$**), (**$'**), (**s**), (**v**), (**f**), (**i**), (**T**), (**A**), (**d**), (**Q**), (**c**) and (**q**) are known, then its Procured Inventory Planned is:

$$P= [c\text{-}q](I\{U\text{+}\$\text{-}\$'[1\text{+}s][v\text{+}f\text{+}i]\}\text{-}T\text{-}Ad\}\text{-}Q)$$

Rule-22038:

If both (I), (**U**), (**$**), (**$'**), (**s**), (**v**), (**f**), (**i**), (**T**), (**A**), (**d**), (**P**), (**Q**) and (**q**) are known, then its Current Ratio Planned is:

$$c= q\text{+}P/(I\{U\text{+}\$\text{-}\$'[1\text{+}s][v\text{+}f\text{+}i]\}\text{-}T\text{-}Ad\}\text{-}Q)$$

Rule-22039:

If both (I), (**U**), (**$**), (**$'**), (**s**), (**v**), (**f**), (**i**), (**T**), (**A**), (**d**), (**P**), (**c**) and (**Q**) are known, then its Quick or Acid Test Ratio Planned is:

$$q= c\text{-}P/(I\{U\text{+}\$\text{-}\$'[1\text{+}s][v\text{+}f\text{+}i]\}\text{-}T\text{-}Ad\}\text{-}Q)$$

Rule-22040:

If both (I), (**U**), (**$**), (**$'**), (**s**), (**v**), (**f**), (**i**), (**T**), (**A**), (**d**), (**V**), (**p**), (**c**) and (**q**) are known, then its Quoted Longterm Debt Planned is:

$$Q= I\{U\text{+}\$\text{-}\$'[1\text{+}s][v\text{+}f\text{+}i]\}\text{-}T\text{-}Ad\}\text{-}Vp/\{360[c\text{-}q]\}$$

Rule-22041:

If both (**Q**), (**U**), (**$**), (**$'**), (**s**), (**v**), (**f**), (**i**), (**T**), (**A**), (**d**), (**V**), (**p**), (**c**) and (**q**) are known, then its Leverage or Gearing Ratio Planned is:

$$I= (Q\text{+}Vp/\{360[c\text{-}q]\})$$
$$/\{U\text{+}\$\text{-}\$'[1\text{+}s][v\text{+}f\text{+}i]\}\text{-}T\text{-}Ad\}$$

Steve Asikin ISBN 14: 978-1515158691, ISBN 10: 1515158691

Rule-22042:

If both (I), $(\mathbf{Q})$, $(\mathbf{\$})$, $(\mathbf{\$'})$, $(\mathbf{s})$, $(\mathbf{v})$, $(\mathbf{f})$, $(\mathbf{i})$, $(\mathbf{T})$, $(\mathbf{A})$, $(\mathbf{d})$, $(\mathbf{V})$, $(\mathbf{p})$, $(\mathbf{c})$ and $(\mathbf{q})$ are known, then its Utilized or Starting Capital must be:

$$U= \$'[1+s][v+f+i]+T+Ad+(Q+Vp/\{360[c-q]\})/I\text{-}\$$$

Rule-22043:

If both (I), $(\mathbf{U})$, $(\mathbf{Q})$, $(\mathbf{\$'})$, $(\mathbf{s})$, $(\mathbf{v})$, $(\mathbf{f})$, $(\mathbf{i})$, $(\mathbf{T})$, $(\mathbf{A})$, $(\mathbf{d})$, $(\mathbf{V})$, $(\mathbf{p})$, $(\mathbf{c})$ and $(\mathbf{q})$ are known, then its Sales or Revenue Planned is:

$$\$= \$'[1+s][v+f+i]\}+T+Ad+(Q+Vp/\{360[c-q]\})/I\text{-}U$$

Rule-22044:

If both (I), $(\mathbf{U})$, $(\mathbf{\$})$, $(\mathbf{Q})$, $(\mathbf{s})$, $(\mathbf{v})$, $(\mathbf{f})$, $(\mathbf{i})$, $(\mathbf{T})$, $(\mathbf{A})$, $(\mathbf{d})$, $(\mathbf{V})$, $(\mathbf{p})$, $(\mathbf{c})$ and $(\mathbf{q})$ are known, then its Sales Past must be:

$$\$'= [U+\$\text{-}T\text{-}Ad\text{-}(Q+Vp/\{360[c-q]\})/I] \\ /\{[1+s][v+f+i]\}$$

Rule-22045:

If both (I), $(\mathbf{U})$, $(\mathbf{\$})$, $(\mathbf{\$'})$, $(\mathbf{Q})$, $(\mathbf{v})$, $(\mathbf{f})$, $(\mathbf{i})$, $(\mathbf{T})$, $(\mathbf{A})$, $(\mathbf{d})$, $(\mathbf{V})$, $(\mathbf{p})$, $(\mathbf{c})$ and $(\mathbf{q})$ are known, then its Sales Growth Planned is:

$$s= \$/\$'\text{-}1$$

or it can also be found as

$$s= [U+\$\text{-}T\text{-}Ad\text{-}(Q+Vp/\{360[c-q]\})/I] \\ /\{\$'[v+f+i]\}\text{-}1$$

Steve Asikin ISBN 14: 978-1515158691, ISBN 10: 1515158691

Rule-22046:

If both (**/**), (**U**), (**$**), (**$'**), (**s**), (**Q**), (**f**), (**i**), (**T**), (**A**), (**d**), (**V**), (**p**), (**c**) and (**q**) are known, then its Variable Portion Planned is:

$$v= V/\$$$

or it can also be found as

$$v= [U+\$-T-Ad-(Q+Vp/\{360[c-q]\})/I/\{\$'[1+s]\}-f-i$$

Rule-22047:

If both (**/**), (**U**), (**$**), (**$'**), (**s**), (**v**), (**Q**), (**i**), (**T**), (**A**), (**d**), (**V**), (**p**), (**c**) and (**q**) are known, then its Fixed Portion Planned is:

$$f= [U+\$-T-Ad-(Q+Vp/\{360[c-q]\})/I/\{\$'[1+s]\}-v-i$$

Rule-22048:

If both (**/**), (**U**), (**$**), (**$'**), (**s**), (**v**), (**f**), (**Q**), (**T**), (**A**), (**d**), (**V**), (**p**), (**c**) and (**q**) are known, then its Interest Portion Planned is:

$$i= [U+\$-T-Ad-(Q+Vp/\{360[c-q]\})/I/\{\$'[1+s]\}-v-f$$

Rule-22049:

If both (**/**), (**U**), (**$**), (**$'**), (**s**), (**v**), (**f**), (**i**), (**Q**), (**A**), (**d**), (**V**), (**p**), (**c**) and (**q**) are known, then its Tax Planned is:

$$T= U+\$-\$'[1+s][v+f+i]\}-Ad-(Q+Vp/\{360[c-q]\})/I$$

Rule-22050:

If both (**/**), (**U**), (**$**), (**$'**), (**s**), (**v**), (**f**), (**i**), (**T**), (**Q**), (**d**), (**V**), (**p**), (**c**) and (**q**) are known, then its After Tax Income Planned is:

$$A= [U+\$-\$'[1+s][v+f+i]\}-T$$
$$-(Q+Vp/\{360[c-q]\})/I/d$$

Steve Asikin ISBN 14: 978-1515158691, ISBN 10: 1515158691

Rule-22051:

If both (I), (U), $(\$)$, $(\$')$, (s), (v), (f), (i), (T), (A), (Q), (V), (p), (c) and (q) are known, then its Dividend Payout Planned is:

$$d= [U+\$-\$'[1+s][v+f+i]\}-T$$
$$-(Q+Vp/\{360[c-q]\})/I]/A$$

Rule-22052:

If both (I), (U), $(\$)$, $(\$')$, (s), (v), (f), (i), (T), (A), (d), (Q), (p), (c) and (q) are known, then its Variable Cost Planned is:

$$V= 360[c-q](I\{U+\$-\$'[1+s][v+f+i]\}-T-Ad\}-Q)/p$$

Rule-22053:

If both (I), (U), $(\$)$, $(\$')$, (s), (v), (f), (i), (T), (A), (d), (V), (Q), (c) and (q) are known, then its Procured Inventory Days Planned:

$$p= 360[c-q](I\{U+\$-\$'[1+s][v+f+i]\}-T-Ad\}-Q)/V$$

Rule-22054:

If both (I), (U), $(\$)$, $(\$')$, (s), (v), (f), (i), (T), (A), (d), (V), (p), (Q) and (q) are known, then its Current Ratio Planned is:

$$c= q+Vp/[360(I\{U+\$-\$'[1+s][v+f+i]\}-T-Ad\}-Q)]$$

Rule-22055:

If both (I), (U), $(\$)$, $(\$')$, (s), (v), (f), (i), (T), (A), (d), (V), (p), (c) and (Q) are known, then its Quick or Acid Test Ratio Planned is:

$$q= c-Vp/[360(I\{U+\$-\$'[1+s][v+f+i]\}-T-Ad\}-Q)]$$

Steve Asikin ISBN 14: 978-1515158691, ISBN 10: 1515158691

Rule-22056:

If both (*I*), (**U**), (**S**), (**S'**), (**s**), (**v**), (**f**), (**i**), (**T**), (**A**), (**d**), (**p**), (**c**) and (**q**) are known, then its Quoted Longterm Debt Planned is:

$$Q= I\{U+S-S'[1+s][v+f+i]\}-T-Ad\}-Svp/\{360[c-q]\}$$

Rule-22057:

If both (**Q**), (**U**), (**S**), (**S'**), (**s**), (**v**), (**f**), (**i**), (**T**), (**A**), (**d**), (**p**), (**c**) and (**q**) are known, then its Leverage or Gearing Ratio Planned is:

$$I= (Q+Svp/\{360[c-q]\})$$
$$/\{U+S-S'[1+s][v+f+i]\}-T-Ad\}$$

Rule-22058:

If both (*I*), (**Q**), (**S**), (**S'**), (**s**), (**v**), (**f**), (**i**), (**T**), (**A**), (**d**), (**p**), (**c**) and (**q**) are known, then its Utilized or Starting Capital must be:

$$U= S'[1+s][v+f+i]+T+Ad+(Q+Svp/\{360[c-q]\})/I-S$$

Rule-22059:

If both (*I*), (**U**), (**Q**), (**S'**), (**s**), (**v**), (**f**), (**i**), (**T**), (**A**), (**d**), (**p**), (**c**) and (**q**) are known, then its Sales or Revenue Planned is:

$$S= (I\{U+S-S'[1+s][v+f+i]\}-T-Ad\}-Q)$$
$$/(vp/\{360[c-q]\}-I)$$

Rule-22060:

If both (*I*), (**U**), (**S**), (**Q**), (**s**), (**v**), (**f**), (**i**), (**T**), (**A**), (**d**), (**p**), (**c**) and (**q**) are known, then its Sales Past must be:

$$S'= [U+S-T-Ad-(Q+Svp/\{360[c-q]\})/I]$$
$$/\{[1+s][v+f+i]\}$$

Steve Asikin ISBN 14: 978-1515158691, ISBN 10: 1515158691

Rule-22061:

If both (**/**), (**U**), (**$**), (**$'**), (**Q**), (**v**), (**f**), (**i**), (**T**), (**A**), (**d**), (**p**), (**c**) and (**q**) are known, then its Sales Growth Planned is:

$s = \$/\$' - 1$

or it can also be found as

$s = [U+\$-T-Ad-(Q+\$vp/\{360[c-q]\})/\textit{l}]$
$/\{\$'[v+f+i]\} - 1$

Rule-22062:

If both (**/**), (**U**), (**$**), (**$'**), (**s**), (**Q**), (**f**), (**i**), (**T**), (**A**), (**d**), (**p**), (**c**) and (**q**) are known, then its Variable Portion Planned is:

$v = (\textit{l}\{U+\$-\$'[1+s][f+i]\}-T-Ad\}-Q)$
$/(\$p/\{360[c-q]\}+\$'\textit{l}[1+s])$

Rule-22063:

If both (**/**), (**U**), (**$**), (**$'**), (**s**), (**v**), (**Q**), (**i**), (**T**), (**A**), (**d**), (**p**), (**c**) and (**q**) are known, then its Fixed Portion Planned is:

$f = [U+\$-T-Ad-(Q+\$vp/\{360[c-q]\})/\textit{l}]$
$/\{\$'[1+s]\} - v - i$

Rule-22064:

If both (**/**), (**U**), (**$**), (**$'**), (**s**), (**v**), (**f**), (**Q**), (**T**), (**A**), (**d**), (**p**), (**c**) and (**q**) are known, then its Interest Portion Planned is:

$i = [U+\$-T-Ad-(Q+\$vp/\{360[c-q]\})/\textit{l}]$
$/\{\$'[1+s]\} - v - f$

Steve Asikin ISBN 14: 978-1515158691, ISBN 10: 1515158691

Rule-22065:

If both (I), (U), (S), (S'), (s), (v), (f), (i), (Q), (A), (d), (p), (c) and (q) are known, then its Tax Planned is:

$$T = U+S-S'[1+s][v+f+i]\}-Ad-(Q+Svp/\{360[c-q]\})/I$$

Rule-22066:

If both (I), (U), (S), (S'), (s), (v), (f), (i), (T), (Q), (d), (p), (c) and (q) are known, then its After Tax Income Planned is:

$$A = [U+S-S'[1+s][v+f+i]\}-T$$
$$-(Q+Svp/\{360[c-q]\})/I]/d$$

Rule-22067:

If both (I), (U), (S), (S'), (s), (v), (f), (i), (T), (A), (Q), (p), (c) and (q) are known, then its Dividend Payout Planned is:

$$d = [U+S-S'[1+s][v+f+i]\}-T$$
$$-(Q+Svp/\{360[c-q]\})/I]/A$$

Rule-22068:

If both (I), (U), (S), (S'), (s), (v), (f), (i), (T), (A), (d), (Q), (c) and (q) are known, then its Procured Inventory Days Planned:

$$p = 360[c-q](I\{U+S-S'[1+s][v+f+i]\}-T-Ad\}-Q)/[Sv]$$

Rule-22069:

If both (I), (U), (S), (S'), (s), (v), (f), (i), (T), (A), (d), (p), (Q) and (q) are known, then its Current Ratio Planned is:

$$c = q+Svp/[360(I\{U+S-S'[1+s][v+f+i]\}-T-Ad\}-Q)]$$

Steve Asikin ISBN 14: 978-1515158691, ISBN 10: 1515158691

Rule-22070:

If both ($\textit{l}$), (**U**), (**$**), (**$'**), (**$**), (**v**), (**f**), (**i**), (**T**), (**A**), (**d**), (**p**), (**c**) and (**Q**) are known, then its Quick or Acid Test Ratio Planned is:

$$q= c\text{-}\$vp/[360(\textit{l}\{U+\$-\$'[1+s][v+f+i]\}\text{-}T\text{-}Ad\}\text{-}Q)]$$

Rule-22071:

If both ($\textit{l}$), (**U**), (**$**), (**$'**), (**$**), (**v**), (**f**), (**i**), (**T**), (**A**), (**d**), (**p**), (**c**) and (**q**) are known, then its Quoted Longterm Debt Planned is:

$$Q= \textit{l}\{U+\$-\$'[1+s][v+f+i]\}\text{-}T\text{-}Ad\}$$
$$-\$'vp[1+s]/\{360[c\text{-}q]\}$$

Rule-22072:

If both (**Q**), (**U**), (**$**), (**$'**), (**$**), (**v**), (**f**), (**i**), (**T**), (**A**), (**d**), (**p**), (**c**) and (**q**) are known, then its Leverage or Gearing Ratio Planned is:

$$\textit{l}= (Q+\$'vp[1+s]/\{360[c\text{-}q]\})$$
$$/\{U+\$-\$'[1+s][v+f+i]\}\text{-}T\text{-}Ad\}$$

Rule-22073:

If both ($\textit{l}$), (**Q**), (**$**), (**$'**), (**$**), (**v**), (**f**), (**i**), (**T**), (**A**), (**d**), (**p**), (**c**) and (**q**) are known, then its Utilized or Starting Capital must be:

$$U= \$'[1+s][v+f+i]+T+Ad$$
$$+(Q+\$'vp[1+s]/\{360[c\text{-}q]\})/\textit{l}\text{-}\$$$

Steve Asikin ISBN 14: 978-1515158691, ISBN 10: 1515158691

Rule-22074:

If both (**/**), (**U**), (**Q**), (**S'**), (**s**), (**v**), (**f**), (**i**), (**T**), (**A**), (**d**), (**p**), (**c**) and (**q**) are known, then its Sales or Revenue Planned is:

$$S= (Q\text{-}/\{U\text{-}S'[1+s][v+f+i]\}\text{-}T\text{-}Ad\}$$
$$+S'vp[1+s]/\{360[c\text{-}q]\})//$$

Rule-22075:

If both (**/**), (**U**), (**S**), (**Q**), (**s**), (**v**), (**f**), (**i**), (**T**), (**A**), (**d**), (**p**), (**c**) and (**q**) are known, then its Sales Past must be:

$$S'= (/\{U+S\text{-}S'[1+s][f+i]\}\text{-}T\text{-}Ad\}\text{-}Q)$$
$$/[v[1+s](p/\{360[c\text{-}q]\}+/)]$$

Rule-22076:

If both (**/**), (**U**), (**S**), (**S'**), (**Q**), (**v**), (**f**), (**i**), (**T**), (**A**), (**d**), (**p**), (**c**) and (**q**) are known, then its Sales Growth Planned is:

$$s= S/S'\text{-}1$$

    or it can also be found as

$$s= (/\{U+S\text{-}S'[1+s][f+i]\}\text{-}T\text{-}Ad\}\text{-}Q)$$
$$/[S'v(p/\{360[c\text{-}q]\}+/)]\text{-}1$$

Rule-22077:

If both (**/**), (**U**), (**S**), (**S'**), (**s**), (**Q**), (**f**), (**i**), (**T**), (**A**), (**d**), (**p**), (**c**) and (**q**) are known, then its Variable Portion Planned is:

$$v= (/\{U+S\text{-}S'[1+s][f+i]\}\text{-}T\text{-}Ad\}\text{-}Q)$$
$$/[S'v[1+s](p/\{360[c\text{-}q]\}+/)]$$

Steve Asikin ISBN 14: 978-1515158691, ISBN 10: 1515158691

Rule-22078:

If both (f), (U), $(\$)$, $(\$')$, (s), (v), (Q), (i), (T), (A), (d), (p), (c) and (q) are known, then its Fixed Portion Planned is:

$$f= [U+\$-T-Ad-(Q+\$'vp[1+s]/\{360[c-q]\})/f]$$
$$/\{\$'[1+s]\}-v-i$$

Rule-22079:

If both (f), (U), $(\$)$, $(\$')$, (s), (v), (f), (Q), (T), (A), (d), (p), (c) and (q) are known, then its Interest Portion Planned is:

$$i= [U+\$-T-Ad-(Q+\$'vp[1+s]/\{360[c-q]\})/f]$$
$$/\{\$'[1+s]\}-v-f$$

Rule-22080:

If both (f), (U), $(\$)$, $(\$')$, (s), (v), (f), (i), (Q), (A), (d), (p), (c) and (q) are known, then its Tax Planned is:

$$T= U+\$-\$'[1+s][v+f+i]\}-Ad$$
$$-(Q+\$'vp[1+s]/\{360[c-q]\})/f$$

Rule-22081:

If both (f), (U), $(\$)$, $(\$')$, (s), (v), (f), (i), (T), (Q), (d), (p), (c) and (q) are known, then its After Tax Income Planned is:

$$A= [U+\$-\$'[1+s][v+f+i]\}-T$$
$$-(Q+\$'vp[1+s]/\{360[c-q]\})/f]/d$$

Steve Asikin ISBN 14: 978-1515158691, ISBN 10: 1515158691

Rule-22082:

If both (I), (U), (S), (S'), (s), (v), (f), (i), (T), (A), (Q), (p), (c) and (q) are known, then its Dividend Payout Planned is:

$$d= [U+S-S'[1+s][v+f+i]\}-T$$
$$-(Q+S'vp[1+s]/\{360[c-q]\})/I]/A$$

Rule-22083:

If both (I), (U), (S), (S'), (s), (v), (f), (i), (T), (A), (d), (Q), (c) and (q) are known, then its Procured Inventory Days Planned:

$$p= 360[c-q](I\{U+S-S'[1+s][v+f+i]\}-T-Ad\}-Q)$$
$$/\{S'v[1+s]\}$$

Rule-22084:

If both (I), (U), (S), (S'), (s), (v), (f), (i), (T), (A), (d), (p), (Q) and (q) are known, then its Current Ratio Planned is:

$$c= q+S'vp[1+s]$$
$$/[360(I\{U+S-S'[1+s][v+f+i]\}-T-Ad\}-Q)]$$

Rule-22085:

If both (I), (U), (S), (S'), (s), (v), (f), (i), (T), (A), (d), (p), (c) and (Q) are known, then its Quick or Acid Test Ratio Planned is:

$$q= c-S'vp[1+s]$$
$$/[360(I\{U+S-S'[1+s][v+f+i]\}-T-Ad\}-Q)]$$

Steve Asikin ISBN 14: 978-1515158691, ISBN 10: 1515158691

Rule-22086:
  If both (*ƒ*), (**U**), (**$**), (**$'**), (**s**), (**v**), (**f**), (**i**), (**t**), (**D**) and (**X**) are known, then its Quoted Longterm Debt Planned is:
  $$Q= ƒ(U+\{$-$'[1+s][v+f+i]\}[1-t]-D)-X$$

Rule-22087:
  If both (**Q**), (**U**), (**$**), (**$'**), (**s**), (**v**), (**f**), (**i**), (**t**), (**D**) and (**X**) are known, then its Leverage or Gearing Ratio Planned is:
  $$ƒ= [Q+X]/(U+\{$-$'[1+s][v+f+i]\}[1-t]-D)$$

Rule-22088:
  If both (*ƒ*), (**Q**), (**$**), (**$'**), (**s**), (**v**), (**f**), (**i**), (**t**), (**D**) and (**X**) are known, then its Utilized or Starting Capital must be:
  $$U= D+[Q+X]/ƒ\{$-$'[1+s][v+f+i]\}[1-t]$$

Rule-22089:
  If both (*ƒ*), (**U**), (**Q**), (**$'**), (**s**), (**v**), (**f**), (**i**), (**t**), (**D**) and (**X**) are known, then its Sales or Revenue Planned is:
  $$$= $'[1+s][v+f+i]+\{D+[Q+X]/ƒU\}/[1-t]$$

Rule-22090:
  If both (*ƒ*), (**U**), (**$**), (**Q**), (**s**), (**v**), (**f**), (**i**), (**t**), (**D**) and (**X**) are known, then its Sales Past must be:
  $$$'= ($-\{D+[Q+X]/ƒU\}/[1-t])/\{[1+s][v+f+i]\}$$

Steve Asikin ISBN 14: 978-1515158691, ISBN 10: 1515158691

Rule-22091:

If both (**/**), (**U**), (**$**), (**$'**), (**Q**), (**v**), (**f**), (**i**), (**t**), (**D**) and (**X**) are known, then its Sales Growth Planned is:

$s = \$/\$' - 1$

or it can also be found as

$s = (\$ - \{D + [Q+X]/\text{/U}\}/[1-t])/\{\$'[v+f+i]\} - 1$

Rule-22092:

If both (**/**), (**U**), (**$**), (**$'**), (**s**), (**Q**), (**f**), (**i**), (**t**), (**D**) and (**X**) are known, then its Variable Portion Planned is:

$v = (\$ - \{D + [Q+X]/\text{/U}\}/[1-t])/\{\$'[1+s]\} - f - i$

Rule-22093:

If both (**/**), (**U**), (**$**), (**$'**), (**s**), (**v**), (**Q**), (**i**), (**t**), (**D**) and (**X**) are known, then its Fixed Portion Planned is:

$f = (\$ - \{D + [Q+X]/\text{/U}\}/[1-t])/\{\$'[1+s]\} - v - i$

Rule-22094:

If both (**/**), (**U**), (**$**), (**$'**), (**s**), (**v**), (**f**), (**Q**), (**t**), (**D**) and (**X**) are known, then its Interest Portion Planned is:

$i = (\$ - \{D + [Q+X]/\text{/U}\}/[1-t])/\{\$'[1+s]\} - v - f$

Rule-22095:

If both (**/**), (**U**), (**$**), (**$'**), (**s**), (**v**), (**f**), (**i**), (**Q**), (**D**) and (**X**) are known, then its Tax Rate Planned is:

$t = 1 - \{D + [Q+X]/\text{/U}\}/\{\$ - \$'[1+s][v+f+i]\}$

Rule-22096:

If both (**/**), (**U**), (**$**), (**$'**), (**s**), (**v**), (**f**), (**i**), (**t**), (**Q**) and (**X**) are known, then its Dividend Planned is:

$D = U + \{\$ - \$'[1+s][v+f+i]\}[1-t] - [Q+X]/\text{/}$

Steve Asikin ISBN 14: 978-1515158691, ISBN 10: 1515158691

Rule-22097:

If both (*l*), (**U**), (**S**), (**S'**), (**s**), (**v**), (**f**), (**i**), (**t**), (**D**) and
(**Q**) are known, then its Xpress or Current Debt
Planned is:

$$X= l(U+\{S-S'[1+s][v+f+i]\}[1-t]-D)-Q$$

Rule-22098:

If both (*l*), (**U**), (**S**), (**S'**), (**s**), (**v**), (**f**), (**i**), (**t**), (**D**), (**P**),
(**c**) and (**q**) are known, then its Quoted Longterm Debt
Planned is:

$$Q= l(U+\{S-S'[1+s][v+f+i]\}[1-t]-D)-P/[c-q]$$

Rule-22099:

If both (**Q**), (**U**), (**S**), (**S'**), (**s**), (**v**), (**f**), (**i**), (**t**), (**D**), (**P**),
(**c**) and (**q**) are known, then its Leverage or Gearing
Ratio Planned is:

$$l= \{Q+P/[c-q]\}/(U+\{S-S'[1+s][v+f+i]\}[1-t]-D)$$

Rule-22100:

If both (*l*), (**Q**), (**S**), (**S'**), (**s**), (**v**), (**f**), (**i**), (**t**), (**D**), (**P**),
(**c**) and (**q**) are known, then its Utilized or Starting
Capital must be:

$$U= D+\{Q+P/[c-q]\}/l-\{S-S'[1+s][v+f+i]\}[1-t]$$

Rule-22101:

If both (*l*), (**U**), (**Q**), (**S'**), (**s**), (**v**), (**f**), (**i**), (**t**), (**D**), (**P**),
(**c**) and (**q**) are known, then its Sales or Revenue
Planned is:

$$S= S'[1+s][v+f+i]+(D+\{Q+P/[c-q]\}/l-U)/[1-t]$$

Steve Asikin ISBN 14: 978-1515158691, ISBN 10: 1515158691

Rule-22102:
   If both (*ʃ*), (**U**), (**$**), (**Q**), (**s**), (**v**), (**f**), (**i**), (**t**), (**D**), (**P**), (**c**) and (**q**) are known, then its Sales Past must be:
   $$\$'= [\$-(D+\{Q+P/[c-q]\}/\!\!\!\!\!\;ʃU)/[1-t]]/\{[1+s][v+f+i]\}$$

Rule-22103:
   If both (*ʃ*), (**U**), (**$**), (**$'**), (**Q**), (**v**), (**f**), (**i**), (**t**), (**D**), (**P**), (**c**) and (**q**) are known, then its Sales Growth Planned is:
   $$s= \$/\$'-1$$
   or it can also be found as
   $$s= [\$-(D+\{Q+P/[c-q]\}/\!\!\!\!\!\;ʃU)/[1-t]]/\{\$'[v+f+i]\}-1$$

Rule-22104:
   If both (*ʃ*), (**U**), (**$**), (**$'**), (**s**), (**Q**), (**f**), (**i**), (**t**), (**D**), (**P**), (**c**) and (**q**) are known, then its Variable Portion Planned is:
   $$v= [\$-(D+\{Q+P/[c-q]\}/\!\!\!\!\!\;ʃU)/[1-t]]/\{\$'[1+s]\}-f-i$$

Rule-22105:
   If both (*ʃ*), (**U**), (**$**), (**$'**), (**s**), (**v**), (**Q**), (**i**), (**t**), (**D**), (**P**), (**c**) and (**q**) are known, then its Fixed Portion Planned is:
   $$f= [\$-(D+\{Q+P/[c-q]\}/\!\!\!\!\!\;ʃU)/[1-t]]/\{\$'[1+s]\}-v-i$$

Rule-22106:
   If both (*ʃ*), (**U**), (**$**), (**$'**), (**s**), (**v**), (**f**), (**Q**), (**t**), (**D**), (**P**), (**c**) and (**q**) are known, then its Interest Portion Planned is:
   $$i= [\$-(D+\{Q+P/[c-q]\}/\!\!\!\!\!\;ʃU)/[1-t]]/\{\$'[1+s]\}-v-f$$

Steve Asikin ISBN 14: 978-1515158691, ISBN 10: 1515158691

Rule-22107:

If both (I), (**U**), (**$**), (**$'**), (**s**), (**v**), (**f**), (**i**), (**Q**), (**D**), (**P**),
(**c**) and (**q**) are known, then its Tax Rate Planned is:

$$t = 1 - (D + \{Q + P/[c-q]\}/I \cdot U)/\{S-S'[1+s][v+f+i]\}$$

Rule-22108:

If both (I), (**U**), (**$**), (**$'**), (**s**), (**v**), (**f**), (**i**), (**t**), (**Q**), (**P**),
(**c**) and (**q**) are known, then its Dividend Planned is:

$$D = U + \{S-S'[1+s][v+f+i]\}[1-t] - \{Q+P/[c-q]\}/I$$

Rule-22109:

If both (I), (**U**), (**$**), (**$'**), (**s**), (**v**), (**f**), (**i**), (**t**), (**D**), (**Q**),
(**c**) and (**q**) are known, then its Procured Inventory
Planned is:

$$P = [c-q][I(U + \{S-S'[1+s][v+f+i]\}[1-t] - D) - Q]$$

Rule-22110:

If both (I), (**U**), (**$**), (**$'**), (**s**), (**v**), (**f**), (**i**), (**t**), (**D**), (**P**),
(**Q**) and (**q**) are known, then its Current Ratio Planned
is:

$$c = q + P/[I(U + \{S-S'[1+s][v+f+i]\}[1-t] - D) - Q]$$

Rule-22111:

If both (I), (**U**), (**$**), (**$'**), (**s**), (**v**), (**f**), (**i**), (**t**), (**D**), (**P**),
(**c**) and (**Q**) are known, then its Quick or Acid Test
Ratio Planned is:

$$q = c - P/[I(U + \{S-S'[1+s][v+f+i]\}[1-t] - D) - Q]$$

Steve Asikin ISBN 14: 978-1515158691, ISBN 10: 1515158691

Rule-22112:

If both (**/**), (**U**), (**$**), (**$'**), (**s**), (**v**), (**f**), (**i**), (**t**), (**D**), (**V**), (**p**), (**c**) and (**q**) are known, then its Quoted Longterm Debt Planned is:

$$Q= /\{U+\{\$-\$'[1+s][v+f+i]\}[1-t]-D)-Vp/\{360[c-q]\}$$

Rule-22113:

If both (**Q**), (**U**), (**$**), (**$'**), (**s**), (**v**), (**f**), (**i**), (**t**), (**D**), (**V**), (**p**), (**c**) and (**q**) are known, then its Leverage or Gearing Ratio Planned is:

$$/= (Q+Vp/\{360[c-q]\})$$
$$/ (U+\{\$-\$'[1+s][v+f+i]\}[1-t]-D)$$

Rule-22114:

If both (**/**), (**Q**), (**$**), (**$'**), (**s**), (**v**), (**f**), (**i**), (**t**), (**D**), (**V**), (**p**), (**c**) and (**q**) are known, then its Utilized or Starting Capital must be:

$$U= D+(Q+Vp/\{360[c-q]\})//$$
$$-\{\$-\$'[1+s][v+f+i]\}[1-t]$$

Rule-22115:

If both (**/**), (**U**), (**Q**), (**$'**), (**s**), (**v**), (**f**), (**i**), (**t**), (**D**), (**V**), (**p**), (**c**) and (**q**) are known, then its Sales Growth Planned is:

$$\$= \$'[1+s][v+f+i]+[D+(Q+Vp/\{360[c-q]\})//-U]$$
$$/[1-t]$$

Steve Asikin ISBN 14: 978-1515158691, ISBN 10: 1515158691

Rule-22116:

If both ($\textit{I}$), (**U**), (**\$**), (**Q**), (**s**), (**v**), (**f**), (**i**), (**t**), (**D**), (**V**), (**p**), (**c**) and (**q**) are known, then its Sales Past must be:

$\$'= \{\$-[D+(Q+Vp/\{360[c-q]\})/\textit{I}U]/[1-t]\}$
$/\{[1+s][v+f+i]\}$

Rule-22117:

If both ($\textit{I}$), (**U**), (**\$**), (**\$'**), (**Q**), (**v**), (**f**), (**i**), (**t**), (**D**), (**V**), (**p**), (**c**) and (**q**) are known, then its Sales Growth Planned is:

$s= \$/\$'-1$

or it can also be found as

$s= \{\$-[D+(Q+Vp/\{360[c-q]\})/\textit{I}U]/[1-t]\}$
$/\{\$'[v+f+i]\}-1$

Rule-22118:

If both ($\textit{I}$), (**U**), (**\$**), (**\$'**), (**Q**), (**v**), (**f**), (**i**), (**t**), (**D**), (**V**), (**p**), (**c**) and (**q**) are known, then its Sales Growth Planned is:

$v= V/\$$

or it can also be found as

$v= \{\$-[D+(Q+Vp/\{360[c-q]\})/\textit{I}U]/[1-t]\}$
$/\{\$'[1+s]\}-f-i$

Rule-22119:

If both ($\textit{I}$), (**U**), (**\$**), (**\$'**), (**s**), (**v**), (**Q**), (**i**), (**t**), (**D**), (**V**), (**p**), (**c**) and (**q**) are known, then its Fixed Portion Planned is:

$f= \{\$-[D+(Q+Vp/\{360[c-q]\})/\textit{I}U]/[1-t]\}$
$/\{\$'[1+s]\}-v-i$

Steve Asikin ISBN 14: 978-1515158691, ISBN 10: 1515158691

Rule-22120:

If both (I), (U), (S), (S'), (s), (v), (f), (Q), (t), (D), (V), (p), (c) and (q) are known, then its Interest Portion Planned is:

$$i = \{S - [D + (Q + Vp/\{360[c-q]\})/I\!U]/[1-t]\} / \{S'[1+s]\} - v - f$$

Rule-22121:

If both (I), (U), (S), (S'), (s), (v), (f), (i), (Q), (D), (V), (p), (c) and (q) are known, then its Tax Rate Planned is:

$$t = 1 - [D + (Q + Vp/\{360[c-q]\})/I\!U] / \{S - S'[1+s][v+f+i]\}$$

Rule-22122:

If both (I), (U), (S), (S'), (s), (v), (f), (i), (t), (Q), (V), (p), (c) and (q) are known, then its Diviudend Planned is:

$$D = U + \{S - S'[1+s][v+f+i]\}[1-t] - (Q + Vp/\{360[c-q]\})/I$$

Rule-22123:

If both (I), (U), (S), (S'), (s), (v), (f), (i), (t), (D), (Q), (p), (c) and (q) are known, then its Variable Cost Planned is:

$$V = 360[c-q][I\!U + \{S - S'[1+s][v+f+i]\}[1-t] - D) - Q]/p$$

Rule-22124:

If both (I), (U), (S), (S'), (s), (v), (f), (i), (t), (D), (V), (Q), (c) and (q) are known, then its Procured Inventory Days Planned:

$$p = 360[c-q][I\!U + \{S - S'[1+s][v+f+i]\}[1-t] - D) - Q]/V$$

Steve Asikin ISBN 14: 978-1515158691, ISBN 10: 1515158691

Rule-22125:

If both (*l*), (**U**), (**$**), (**$'**), (**s**), (**v**), (**f**), (**i**), (**t**), (**D**), (**V**), (**p**), (**Q**) and (**q**) are known, then its Current Ratio Planned is:

$$c= q+Vp$$
$$/\{360[\textit{l}U+\{\$-\$'[1+s][v+f+i]\}[1-t]-D)-Q]\}$$

Rule-22126:

If both (*l*), (**U**), (**$**), (**$'**), (**s**), (**v**), (**f**), (**i**), (**t**), (**D**), (**V**), (**p**), (**c**) and (**Q**) are known, then its Quick or Acid Test Ratio Planned is:

$$q= c-Vp/\{360[\textit{l}U+\{\$-\$'[1+s][v+f+i]\}[1-t]-D)-Q]\}$$

Rule-22127:

If both (*l*), (**U**), (**$**), (**$'**), (**s**), (**v**), (**f**), (**i**), (**t**), (**D**), (**p**), (**c**) and (**q**) are known, then its Quoted Longterm Debt Planned is:

$$Q= \textit{l}U+\{\$-\$'[1+s][v+f+i]\}[1-t]-D)$$
$$-\$vp/\{360[c-q]\}$$

Rule-22128:

If both (**Q**), (**U**), (**$**), (**$'**), (**s**), (**v**), (**f**), (**i**), (**t**), (**D**), (**p**), (**c**) and (**q**) are known, then its Leverage or Gearing Ratio Planned is:

$$\textit{l}= (Q+\$vp/\{360[c-q]\})$$
$$/(U+\{\$-\$'[1+s][v+f+i]\}[1-t]-D)$$

Steve Asikin ISBN 14: 978-1515158691, ISBN 10: 1515158691

Rule-22129:

If both (**/**), (**Q**), (**S**), (**S'**), (**s**), (**v**), (**f**), (**i**), (**t**), (**D**), (**p**), (**c**) and (**q**) are known, then its Utilized or Starting Capital must be:

$$U = D + (Q + Svp/\{360[c-q]\})/I$$
$$- \{S-S'[1+s][v+f+i]\}[1-t]$$

Rule-22130:

If both (**/**), (**U**), (**Q**), (**S'**), (**s**), (**v**), (**f**), (**i**), (**t**), (**D**), (**p**), (**c**) and (**q**) are known, then its Sales or Revenue Planned is:

$$S = (I\{U-S'[1+s][v+f+i][1-t]-D\}-Q)$$
$$/(vp/\{360[c-q]\}-I[1-t])$$

Rule-22131:

If both (**/**), (**U**), (**S**), (**Q**), (**s**), (**v**), (**f**), (**i**), (**t**), (**D**), (**p**), (**c**) and (**q**) are known, then its Sales Past must be:

$$S' = \{S-[D+(Q+Svp/\{360[c-q]\})/I-U]/[1-t]\}$$
$$/\{[1+s][v+f+i]\}$$

Rule-22132:

If both (**/**), (**U**), (**S**), (**S'**), (**Q**), (**v**), (**f**), (**i**), (**t**), (**D**), (**p**), (**c**) and (**q**) are known, then its Sales Growth Planned is:

$$s = S/S' - 1$$

or it can also be found as

$$s = \{S-[D+(Q+Svp/\{360[c-q]\})/I-U]/[1-t]\}$$
$$/\{S'[v+f+i]\} - 1$$

Steve Asikin ISBN 14: 978-1515158691, ISBN 10: 1515158691

Rule-22133:

If both (**/**), (**U**), (**$**), (**$'**), (**s**), (**Q**), (**f**), (**i**), (**t**), (**D**), (**p**), (**c**) and (**q**) are known, then its Variable Portion Planned is:

$$v= [\text{/U}+\{\$-\$'[1+s][f+i]\}[1-t]-D)-Q]$$
$$/(\$p/\{360[c-q]\}+\$'/[1+s])$$

Rule-22134:

If both (**/**), (**U**), (**$**), (**$'**), (**s**), (**v**), (**Q**), (**i**), (**t**), (**D**), (**p**), (**c**) and (**q**) are known, then its Fixed Portion Planned is:

$$f= \{\$-[D+(Q+\$vp/\{360[c-q]\})/\text{/U}]/[1-t]\}$$
$$/\{\$'[1+s]\}-v-i$$

Rule-22135:

If both (**/**), (**U**), (**$**), (**$'**), (**s**), (**v**), (**f**), (**Q**), (**t**), (**D**), (**p**), (**c**) and (**q**) are known, then its Interest Portion Planned is:

$$i=\{\$-[D+(Q+\$vp/\{360[c-q]\})/\text{/U}]/[1-t]\}$$
$$/\{\$'[1+s]\}-v-f$$

Rule-22136:

If both (**/**), (**U**), (**$**), (**$'**), (**s**), (**v**), (**f**), (**i**), (**Q**), (**D**), (**p**), (**c**) and (**q**) are known, then its Tax Rate Planned is:

$$t= 1-[D+(Q+\$vp/\{360[c-q]\})/\text{/U}]$$
$$/\{\$-\$'[1+s][v+f+i]\}$$

Steve Asikin ISBN 14: 978-1515158691, ISBN 10: 1515158691

Rule-22137:

If both (*I*), (**U**), (**$**), (**$'**), (**s**), (**v**), (**f**), (**i**), (**t**), (**Q**), (**p**), (**c**) and (**q**) are known, then its Dividend Planned is:

$$D= U+\{\$-\$'[1+s][v+f+i]\}[1-t]$$
$$-(Q+\$vp/\{360[c-q]\})/I$$

Rule-22138:

If both (*I*), (**U**), (**$**), (**$'**), (**s**), (**v**), (**f**), (**i**), (**t**), (**D**), (**Q**), (**c**) and (**q**) are known, then its Procured Inventory Days Planned:

$$p= 360[c-q][I U+\{\$-\$'[1+s][v+f+i]\}[1-t]-D)-Q]$$
$$/[\$v]$$

Rule-22139:

If both (*I*), (**U**), (**$**), (**$'**), (**s**), (**v**), (**f**), (**i**), (**t**), (**D**), (**p**), (**Q**) and (**q**) are known, then its Current Ratio Planned is:

$$c= q+\$vp$$
$$/\{360[I U+\{\$-\$'[1+s][v+f+i]\}[1-t]-D)-Q]\}$$

Rule-22140:

If both (*I*), (**U**), (**$**), (**$'**), (**s**), (**v**), (**f**), (**i**), (**t**), (**D**), (**p**), (**c**) and (**Q**) are known, then its Quick or Acid Test Ratio Planned is:

$$q= c-\$vp$$
$$/\{360[I U+\{\$-\$'[1+s][v+f+i]\}[1-t]-D)-Q]\}$$

Steve Asikin ISBN 14: 978-1515158691, ISBN 10: 1515158691

Rule-22141:
> If both (I), (U), (S), (S'), (s), (v), (f), (i), (t), (D), (p),
> (c) and (q) are known, then its Quoted Longterm Debt
> Planned is:
> $$Q = I(U + \{S-S'[1+s][v+f+i]\}[1-t]-D)$$
> $$-S'vp[1+s]/\{360[c-q]\}$$

Rule-22142:
> If both (Q), (U), (S), (S'), (s), (v), (f), (i), (t), (D), (p),
> (c) and (q) are known, then its Leverage or Gearing
> Ratio Planned is:
> $$I = (Q + S'vp[1+s]/\{360[c-q]\})$$
> $$/(U + \{S-S'[1+s][v+f+i]\}[1-t]-D)$$

Rule-22143:
> If both (I), (Q), (S), (S'), (s), (v), (f), (i), (t), (D), (p),
> (c) and (q) are known, then its Utilized or Starting
> Capital must be:
> $$U = D + (Q + S'vp[1+s]/\{360[c-q]\})/I$$
> $$-\{S-S'[1+s][v+f+i]\}[1-t]$$

Rule-22144:
> If both (I), (U), (Q), (S'), (s), (v), (f), (i), (t), (D), (p),
> (c) and (q) are known, then its Sales or Revenue
> Planned is:
> $$S = [Q - I(U + \{S-S'[1+s][v+f+i]\}[1-t]-D)$$
> $$+S'vp[1+s]/\{360[c-q]\}]/\{I[1-t]\}$$

Steve Asikin ISBN 14: 978-1515158691, ISBN 10: 1515158691

Rule-22145:

If both (**/**), (**U**), (**$**), (**Q**), (**s**), (**v**), (**f**), (**i**), (**t**), (**D**), (**p**), (**c**) and (**q**) are known, then its Sales Past must be:

$$\$'= [(\text{/U}+\{\$-\$'[1+s][f+i]\}[1-t]-D)-Q]$$
$$/[v[1+s](p/\{360[c-q]\}+\text{/})]$$

Rule-22146:

If both (**/**), (**U**), (**$**), (**$'**), (**Q**), (**v**), (**f**), (**i**), (**t**), (**D**), (**p**), (**c**) and (**q**) are known, then its Sales Growth Planned is:

$$s= \$/\$'-1$$

> or it can also be found as

$$s= [(\text{/U}+\{\$-\$'[1+s][f+i]\}[1-t]-D)-Q]$$
$$/[\$'v(p/\{360[c-q]\}+\text{/})]-1$$

Rule-22147:

If both (**/**), (**U**), (**$**), (**$'**), (**s**), (**Q**), (**f**), (**i**), (**t**), (**D**), (**p**), (**c**) and (**q**) are known, then its Variable Portion Planned is:

$$v= [(\text{/U}+\{\$-\$'[1+s][f+i]\}[1-t]-D)-Q]$$
$$/[\$'[1+s](p/\{360[c-q]\}+\text{/})]$$

Rule-22148:

If both (**/**), (**U**), (**$**), (**$'**), (**s**), (**v**), (**Q**), (**i**), (**t**), (**D**), (**p**), (**c**) and (**q**) are known, then its Fixed Portion Planned is:

$$f= \{\$-[D+(Q+\$'vp[1+s]/\{360[c-q]\})/\text{/U}]/[1-t]\}$$
$$/\{\$'[1+s]\}-v-i$$

Steve Asikin ISBN 14: 978-1515158691, ISBN 10: 1515158691

Rule-22149:

If both (I), (U), $(\$)$, $(\$')$, (s), (v), (f), (Q), (t), (D), (p), (c) and (q) are known, then its Interest Portion Planned is:

$$i=\{\$-[D+(Q+\$'vp[1+s]/\{360[c-q]\})/I \cdot U]/[1-t]\}/\{\$'[1+s]\}-v-f$$

Rule-22150:

If both (I), (U), $(\$)$, $(\$')$, (s), (v), (f), (i), (Q), (D), (p), (c) and (q) are known, then its Tax Rate Planned is:

$$t= 1-[D+(Q+\$'vp[1+s]/\{360[c-q]\})/I \cdot U]/\{\$-\$'[1+s][v+f+i]\}$$

Rule-22151:

If both (I), (U), $(\$)$, $(\$')$, (s), (v), (f), (i), (t), (Q), (p), (c) and (q) are known, then its Dividend Payout Planned is:

$$D= U+\{\$-\$'[1+s][v+f+i]\}[1-t] -(Q+\$'vp[1+s]/\{360[c-q]\})/I$$

Rule-22152:

If both (I), (U), $(\$)$, $(\$')$, (s), (v), (f), (i), (t), (D), (Q), (c) and (q) are known, then its Procured Inventory Days Planned:

$$p= 360[c-q][I(U+\{\$-\$'[1+s][v+f+i]\}[1-t]-D)-Q]/\{\$'v[1+s]\}$$

Steve Asikin ISBN 14: 978-1515158691, ISBN 10: 1515158691

Rule-22153:

If both (**ʃ**), (**U**), (**S**), (**S'**), (**s**), (**v**), (**f**), (**i**), (**t**), (**D**), (**p**), (**Q**) and (**q**) are known, then its Current Ratio Planned is:

$$c = q + S'vp[1+s]$$
$$/\{360[ʃ(U+\{S-S'[1+s][v+f+i]\}[1-t]-D)-Q]\}$$

Rule-22154:

If both (**ʃ**), (**U**), (**S**), (**S'**), (**s**), (**v**), (**f**), (**i**), (**t**), (**D**), (**p**), (**c**) and (**Q**) are known, then its Quick or Acid Test Ratio Planned is:

$$q = c - S'vp[1+s]$$
$$/\{360[ʃ(U+\{S-S'[1+s][v+f+i]\}[1-t]-D)-Q]\}$$

Rule-22155:

If both (**ʃ**), (**U**), (**S**), (**S'**), (**s**), (**v**), (**f**), (**i**), (**t**), (**A**), (**d**) and (**X**) are known, then its Quoted Longterm Debt Planned is:

$$Q = ʃ(U+\{S-S'[1+s][v+f+i]\}[1-t]-Ad)-X$$

Rule-22156:

If both (**Q**), (**U**), (**S**), (**S'**), (**s**), (**v**), (**f**), (**i**), (**t**), (**A**), (**d**) and (**X**) are known, then its Leverage or Gearing Ratio Planned is:

$$ʃ = [Q+X]/(U+\{S-S'[1+s][v+f+i]\}[1-t]-Ad)$$

Rule-22157:

If both (**ʃ**), (**Q**), (**S**), (**S'**), (**s**), (**v**), (**f**), (**i**), (**t**), (**A**), (**d**) and (**X**) are known, then its Utilized or Starting Capital must be:

$$U = Ad + [Q+X]/ʃ\{S-S'[1+s][v+f+i]\}[1-t]$$

Steve Asikin ISBN 14: 978-1515158691, ISBN 10: 1515158691

Rule-22158:

If both (I), (U), (Q), (S'), (s), (v), (f), (i), (t), (A), (d) and (X) are known, then its Sales or Revenue Planned is:

$$S= S'[1+s][v+f+i]\} + \{Ad+[Q+X]/I+U\}/[1-t]$$

Rule-22159:

If both (I), (U), (S), (Q), (s), (v), (f), (i), (t), (A), (d) and (X) are known, then its Sales Past must be:

$$S'= (S-\{Ad+[Q+X]/I+U\}/[1-t])/\{[1+s][v+f+i]\}$$

Rule-22160:

If both (I), (U), (S), (S'), (Q), (v), (f), (i), (t), (A), (d) and (X) are known, then its Sales Growth Planned is:

$$s= S/S'-1$$

or it can also be found as

$$s= (S-\{Ad+[Q+X]/I+U\}/[1-t])/\{S'[v+f+i]\}-1$$

Rule-22161:

If both (I), (U), (S), (S'), (s), (Q), (f), (i), (t), (A), (d) and (X) are known, then its Variable Portion Planned is:

$$v= (S-\{Ad+[Q+X]/I+U\}/[1-t])/\{S'[1+s]\}-f-i$$

Rule-22162:

If both (I), (U), (S), (S'), (s), (v), (Q), (i), (t), (A), (d) and (X) are known, then its Fixed Portion Planned is:

$$f= (S-\{Ad+[Q+X]/I+U\}/[1-t])/\{S'[1+s]\}-v-i$$

Steve Asikin ISBN 14: 978-1515158691, ISBN 10: 1515158691

Rule-22163:

If both $(\textit{f})$, $(\textbf{U})$, $(\textbf{\$})$, $(\textbf{\$'})$, $(\textbf{s})$, $(\textbf{v})$, $(\textbf{f})$, $(\textbf{Q})$, $(\textbf{t})$, $(\textbf{A})$, $(\textbf{d})$ and $(\textbf{X})$ are known, then its Interest Portion Planned is:

$$i= (\$-\{\textbf{Ad}+[\textbf{Q}+\textbf{X}]//+\textbf{U}\}/[1-\textbf{t}])/\{\$'[1+\textbf{s}]\}-\textbf{v}-\textbf{f}$$

Rule-22164:

If both $(\textit{f})$, $(\textbf{U})$, $(\textbf{\$})$, $(\textbf{\$'})$, $(\textbf{s})$, $(\textbf{v})$, $(\textbf{f})$, $(\textbf{i})$, $(\textbf{Q})$, $(\textbf{A})$, $(\textbf{d})$ and $(\textbf{X})$ are known, then its Tax Rate Planned is:

$$t= 1-\{\textbf{Ad}+[\textbf{Q}+\textbf{X}]//+\textbf{U}\}/\{\$-\$'[1+\textbf{s}][\textbf{v}+\textbf{f}+\textbf{i}]\}$$

Rule-22165:

If both $(\textit{f})$, $(\textbf{U})$, $(\textbf{\$})$, $(\textbf{\$'})$, $(\textbf{s})$, $(\textbf{v})$, $(\textbf{f})$, $(\textbf{i})$, $(\textbf{t})$, $(\textbf{Q})$, $(\textbf{d})$ and $(\textbf{X})$ are known, then its After Tax Income Planned is:

$$A= (\textbf{U}+\{\$-\$'[1+\textbf{s}][\textbf{v}+\textbf{f}+\textbf{i}]\}[1-\textbf{t}]-[\textbf{Q}+\textbf{X}]//)/\textbf{d}$$

Rule-22166:

If both $(\textit{f})$, $(\textbf{U})$, $(\textbf{\$})$, $(\textbf{\$'})$, $(\textbf{s})$, $(\textbf{v})$, $(\textbf{f})$, $(\textbf{i})$, $(\textbf{t})$, $(\textbf{A})$, $(\textbf{Q})$ and $(\textbf{X})$ are known, then its Dividend Payout Planned is:

$$d= (\textbf{U}+\{\$-\$'[1+\textbf{s}][\textbf{v}+\textbf{f}+\textbf{i}]\}[1-\textbf{t}]-[\textbf{Q}+\textbf{X}]//)/\textbf{A}$$

Rule-22167:

If both $(\textit{f})$, $(\textbf{U})$, $(\textbf{\$})$, $(\textbf{\$'})$, $(\textbf{s})$, $(\textbf{v})$, $(\textbf{f})$, $(\textbf{i})$, $(\textbf{t})$, $(\textbf{A})$, $(\textbf{d})$ and $(\textbf{Q})$ are known, then its Xpress or Current Debt Planned is:

$$X= f(\textbf{U}+\{\$-\$'[1+\textbf{s}][\textbf{v}+\textbf{f}+\textbf{i}]\}[1-\textbf{t}]-\textbf{Ad})-\textbf{Q}$$

Steve Asikin ISBN 14: 978-1515158691, ISBN 10: 1515158691

Rule-22168:

If both (I), (U), (S), (S'), (s), (v), (f), (i), (t), (A), (d), (P), (c) and (q) are known, then its Quoted Longterm Debt Planned is:

$Q = I(U + \{S - S'[1+s][v+f+i]\}[1-t] - Ad) - P/[c-q]$

Rule-22169:

If both (Q), (U), (S), (S'), (s), (v), (f), (i), (t), (A), (d), (P), (c) and (q) are known, then its Leverage or Gearing Ratio Planned is:

$I = \{Q + P/[c-q]\}/(U + \{S - S'[1+s][v+f+i]\}[1-t] - Ad)$

Rule-22170:

If both (I), (Q), (S), (S'), (s), (v), (f), (i), (t), (A), (d), (P), (c) and (q) are known, then its Utilized or Starting Capital must be:

$U = Ad + \{Q + P/[c-q]\}/I - \{S - S'[1+s][v+f+i]\}[1-t]$

Rule-22171:

If both (I), (U), (Q), (S'), (s), (v), (f), (i), (t), (A), (d), (P), (c) and (q) are known, then its Sales or Revenue Planned is:

$S = S'[1+s][v+f+i]\} + (Ad + \{Q + P/[c-q]\}/I + U)/[1-t]$

Rule-22172:

If both (I), (U), (S), (Q), (s), (v), (f), (i), (t), (A), (d), (P), (c) and (q) are known, then its Sales Past must be:

$S' = [S - (Ad + \{Q + P/[c-q]\}/I + U)/[1-t]]$
$\quad /\{[1+s][v+f+i]\}$

Steve Asikin ISBN 14: 978-1515158691, ISBN 10: 1515158691

Rule-22173:

If both (I), (U), (S), (S'), (Q), (v), (f), (i), (t), (A), (d), (P), (c) and (q) are known, then its Sales Growth Planned is:

$s = S/S' - 1$

or it can also be found as

$s = [S-(Ad+\{Q+P/[c-q]\}/I+U)/[1-t]]/\{S'[v+f+i]\} - 1$

Rule-22174:

If both (I), (U), (S), (S'), (s), (Q), (f), (i), (t), (A), (d), (P), (c) and (q) are known, then its Variable Portion Planned is:

$v = [S-(Ad+\{Q+P/[c-q]\}/I+U)/[1-t]]/\{S'[1+s]\} - f - i$

Rule-22175:

If both (I), (U), (S), (S'), (s), (v), (Q), (i), (t), (A), (d), (P), (c) and (q) are known, then its Fixed Portion Planned is:

$f = [S-(Ad+\{Q+P/[c-q]\}/I+U)/[1-t]]/\{S'[1+s]\} - v - i$

Rule-22176:

If both (I), (U), (S), (S'), (s), (v), (f), (Q), (t), (A), (d), (P), (c) and (q) are known, then its Interest Portion Planned is:

$i = [S-(Ad+\{Q+P/[c-q]\}/I+U)/[1-t]]/\{S'[1+s]\} - v - f$

Rule-22177:

If both (I), (U), (S), (S'), (s), (v), (f), (i), (Q), (A), (d), (P), (c) and (q) are known, then its Tax Rate Planned is:

$t = 1-(Ad+\{Q+P/[c-q]\}/I+U)/\{S-S'[1+s][v+f+i]\}$

Steve Asikin ISBN 14: 978-1515158691, ISBN 10: 1515158691

Rule-22178:

If both $(\textbf{\textit{I}})$, $(\textbf{U})$, $(\textbf{\$})$, $(\textbf{\$'})$, $(\textbf{s})$, $(\textbf{v})$, $(\textbf{f})$, $(\textbf{i})$, $(\textbf{t})$, $(\textbf{Q})$, $(\textbf{d})$, $(\textbf{P})$, $(\textbf{c})$ and $(\textbf{q})$ are known, then its After Tax Income Planned is:

$$A = (U + \{\$-\$'[1+s][v+f+i]\}[1-t] - \{Q+P/[c-q]\}/I)/d$$

Rule-22179:

If both $(\textbf{\textit{I}})$, $(\textbf{U})$, $(\textbf{\$})$, $(\textbf{\$'})$, $(\textbf{s})$, $(\textbf{v})$, $(\textbf{f})$, $(\textbf{i})$, $(\textbf{t})$, $(\textbf{A})$, $(\textbf{Q})$, $(\textbf{P})$, $(\textbf{c})$ and $(\textbf{q})$ are known, then its Dividend Payout Planned is:

$$d = (U + \{\$-\$'[1+s][v+f+i]\}[1-t] - \{Q+P/[c-q]\}/I)/A$$

Rule-22180:

If both $(\textbf{\textit{I}})$, $(\textbf{U})$, $(\textbf{\$})$, $(\textbf{\$'})$, $(\textbf{s})$, $(\textbf{v})$, $(\textbf{f})$, $(\textbf{i})$, $(\textbf{t})$, $(\textbf{A})$, $(\textbf{d})$, $(\textbf{Q})$, $(\textbf{c})$ and $(\textbf{q})$ are known, then its Procured Inventory Planned is:

$$P = [c-q][I(U + \{\$-\$'[1+s][v+f+i]\}[1-t] - Ad) - Q]$$

Rule-22181:

If both $(\textbf{\textit{I}})$, $(\textbf{U})$, $(\textbf{\$})$, $(\textbf{\$'})$, $(\textbf{s})$, $(\textbf{v})$, $(\textbf{f})$, $(\textbf{i})$, $(\textbf{t})$, $(\textbf{A})$, $(\textbf{d})$, $(\textbf{P})$, $(\textbf{Q})$ and $(\textbf{q})$ are known, then its Current Ratio Planned is:

$$c = q + P/[I(U + \{\$-\$'[1+s][v+f+i]\}[1-t] - Ad) - Q]$$

Rule-22182:

If both $(\textbf{\textit{I}})$, $(\textbf{U})$, $(\textbf{\$})$, $(\textbf{\$'})$, $(\textbf{s})$, $(\textbf{v})$, $(\textbf{f})$, $(\textbf{i})$, $(\textbf{t})$, $(\textbf{A})$, $(\textbf{d})$, $(\textbf{P})$, $(\textbf{c})$ and $(\textbf{Q})$ are known, then its Quick or Acid Test Ratio Planned is:

$$q = c - P/[I(U + \{\$-\$'[1+s][v+f+i]\}[1-t] - Ad) - Q]$$

Steve Asikin ISBN 14: 978-1515158691, ISBN 10: 1515158691

Rule-22183:

If both (**I**), (**U**), (**S**), (**S'**), (**s**), (**v**), (**f**), (**i**), (**t**), (**A**), (**d**), (**V**), (**p**), (**c**) and (**q**) are known, then its Quoted Longterm Debt Planned is:

$$Q = I(U + \{S-S'[1+s][v+f+i]\}[1-t]-Ad) - Vp/\{360[c-q]\}$$

Rule-22184:

If both (**Q**), (**U**), (**S**), (**S'**), (**s**), (**v**), (**f**), (**i**), (**t**), (**A**), (**d**), (**V**), (**p**), (**c**) and (**q**) are known, then its Leverage or Gearing Ratio Planned is:

$$I = (Q + Vp/\{360[c-q]\}) /(U + \{S-S'[1+s][v+f+i]\}[1-t]-Ad)$$

Rule-22185:

If both (**I**), (**Q**), (**S**), (**S'**), (**s**), (**v**), (**f**), (**i**), (**t**), (**A**), (**d**), (**V**), (**p**), (**c**) and (**q**) are known, then its Utilized or Starting Capital must be:

$$U = Ad + (Q + Vp/\{360[c-q]\})/I - \{S-S'[1+s][v+f+i]\}[1-t]$$

Rule-22186:

If both (**I**), (**U**), (**Q**), (**S'**), (**s**), (**v**), (**f**), (**i**), (**t**), (**A**), (**d**), (**V**), (**p**), (**c**) and (**q**) are known, then its Sales or Revenue Planned is:

$$S = S'[1+s][v+f+i] + [Ad + (Q + Vp/\{360[c-q]\})/I + U]/[1-t]$$

Steve Asikin ISBN 14: 978-1515158691, ISBN 10: 1515158691

Rule-22187:

If both (I), (U), (S), (Q), (s), (v), (f), (i), (t), (A), (d), (V), (p), (c) and (q) are known, then its Sales Past must be:

$$S' = \{S-[Ad+(Q+Vp/\{360[c-q]\})/I+U]/[1-t]\}/\{[1+s][v+f+i]\}$$

Rule-22188:

If both (I), (U), (S), (S'), (Q), (v), (f), (i), (t), (A), (d), (V), (p), (c) and (q) are known, then its Sales Growth Planned is:

$$s = S/S' - 1$$

or it can also be found as

$$s = \{S-[Ad+(Q+Vp/\{360[c-q]\})/I+U]/[1-t]\}/\{S'[v+f+i]\} - 1$$

Rule-22189:

If both (I), (U), (S), (S'), (s), (Q), (f), (i), (t), (A), (d), (V), (p), (c) and (q) are known, then its Variable Portion Planned is:

$$v = \{S-[Ad+(Q+Vp/\{360[c-q]\})/I+U]/[1-t]\}/\{S'[1+s]\} - f - i$$

Rule-22190:

If both (I), (U), (S), (S'), (s), (v), (Q), (i), (t), (A), (d), (V), (p), (c) and (q) are known, then its Fixed Portion Planned is:

$$f = \{S-[Ad+(Q+Vp/\{360[c-q]\})/I+U]/[1-t]\}/\{S'[1+s]\} - v - i$$

Steve Asikin ISBN 14: 978-1515158691, ISBN 10: 1515158691

Rule-22191:

  If both (**/**), (**U**), (**$**), (**$'**), (**s**), (**v**), (**f**), (**Q**), (**t**), (**A**), (**d**),
  (**V**), (**p**), (**c**) and (**q**) are known, then its Interest
  Portion Planned is:

  $$i = \{ \$-[\textbf{Ad}+(\textbf{Q}+\textbf{Vp}/\{360[\textbf{c-q}]\})/\textbf{/}+\textbf{U}]/[1\text{-}\textbf{t}]\}$$
  $$/\{\$'[1+\textbf{s}]\}\text{-}\textbf{v-f}$$

Rule-22192:

  If both (**/**), (**U**), (**$**), (**$'**), (**s**), (**v**), (**f**), (**i**), (**Q**), (**A**), (**d**),
  (**V**), (**p**), (**c**) and (**q**) are known, then its Tax Rate
  Planned is:

  $$t = 1\text{-}[\textbf{Ad}+(\textbf{Q}+\textbf{Vp}/\{360[\textbf{c-q}]\})/\textbf{/}+\textbf{U}]$$
  $$/\{\$\text{-}\$'[1+\textbf{s}][\textbf{v+f+i}]\}$$

Rule-22193:

  If both (**/**), (**U**), (**$**), (**$'**), (**s**), (**v**), (**f**), (**i**), (**t**), (**Q**), (**d**),
  (**V**), (**p**), (**c**) and (**q**) are known, then its After Tax
  Income Planned is:

  $$A = [\textbf{U}+\{\$\text{-}\$'[1+\textbf{s}][\textbf{v+f+i}]\}[1\text{-}\textbf{t}]$$
  $$-(\textbf{Q}+\textbf{Vp}/\{360[\textbf{c-q}]\})/\textbf{/}]/\textbf{d}$$

Rule-22194:

  If both (**/**), (**U**), (**$**), (**$'**), (**s**), (**v**), (**f**), (**i**), (**t**), (**A**), (**Q**),
  (**V**), (**p**), (**c**) and (**q**) are known, then its Dividend
  Payout Planned is:

  $$d = [\textbf{U}+\{\$\text{-}\$'[1+\textbf{s}][\textbf{v+f+i}]\}[1\text{-}\textbf{t}]$$
  $$-(\textbf{Q}+\textbf{Vp}/\{360[\textbf{c-q}]\})/\textbf{/}]/\textbf{A}$$

Steve Asikin ISBN 14: 978-1515158691, ISBN 10: 1515158691

Rule-22195:

If both (*l*), (**U**), (**\$**), (**\$'**), (**s**), (**v**), (**f**), (**i**), (**t**), (**A**), (**d**), (**Q**), (**p**), (**c**) and (**q**) are known, then its Variable Portion Planned is:

$$V= 360[c\text{-}q]$$
$$[\textit{l}U+\{\$\text{-}\$'[1+s][v+f+i]\}[1\text{-}t]\text{-}Ad)\text{-}Q]/p$$

Rule-22196:

If both (*l*), (**U**), (**\$**), (**\$'**), (**s**), (**v**), (**f**), (**i**), (**t**), (**A**), (**d**), (**V**), (**Q**), (**c**) and (**q**) are known, then its Procured Inventory Days Planned:

$$p= 360[c\text{-}q]$$
$$[\textit{l}U+\{\$\text{-}\$'[1+s][v+f+i]\}[1\text{-}t]\text{-}Ad)\text{-}Q]/V$$

Rule-22197:

If both (*l*), (**U**), (**\$**), (**\$'**), (**s**), (**v**), (**f**), (**i**), (**t**), (**A**), (**d**), (**V**), (**p**), (**Q**) and (**q**) are known, then its Current Ratio Planned is:

$$c= q+Vp$$
$$/\{360[\textit{l}U+\{\$\text{-}\$'[1+s][v+f+i]\}[1\text{-}t]\text{-}Ad)\text{-}Q]\}$$

Rule-22198:

If both (*l*), (**U**), (**\$**), (**\$'**), (**s**), (**v**), (**f**), (**i**), (**t**), (**A**), (**d**), (**V**), (**p**), (**c**) and (**Q**) are known, then its Quick or Acid Test Ratio Planned is:

$$q= c\text{-}Vp$$
$$/\{360[\textit{l}U+\{\$\text{-}\$'[1+s][v+f+i]\}[1\text{-}t]\text{-}Ad)\text{-}Q]\}$$

Steve Asikin ISBN 14: 978-1515158691, ISBN 10: 1515158691

Rule-22199:

 If both (I), (U), (S), (S'), (s), (v), (f), (i), (t), (A), (d), (p), (c) and (q) are known, then its Quoted Longterm Debt Planned is:

$$Q = I(U + \{S - S'[1+s][v+f+i]\}[1-t] - Ad)$$
$$- Svp/\{360[c-q]\}$$

Rule-22200:

 If both (Q), (U), (S), (S'), (s), (v), (f), (i), (t), (A), (d), (p), (c) and (q) are known, then its Leverage or Gearing Ratio Planned is:

$$I = (Q + Svp/\{360[c-q]\})$$
$$/(U + \{S - S'[1+s][v+f+i]\}[1-t] - Ad)$$

Rule-22201:

 If both (I), (Q), (S), (S'), (s), (v), (f), (i), (t), (A), (d), (p), (c) and (q) are known, then its Utilized or Starting Capital must be:

$$U = Ad + (Q + Svp/\{360[c-q]\})/I$$
$$- \{S - S'[1+s][v+f+i]\}[1-t]$$

Rule-22202:

 If both (I), (U), (Q), (S'), (s), (v), (f), (i), (t), (A), (d), (p), (c) and (q) are known, then its Sales or Revenue Planned is:

$$S = (I\{U - S'[1+s][v+f+i][1-t] - Ad\} - Q)$$
$$/(vp/\{360[c-q]\} - I[1-t])$$

Steve Asikin ISBN 14: 978-1515158691, ISBN 10: 1515158691

Rule-22203:

If both (**/**), (**U**), (**S**), (**Q**), (**s**), (**v**), (**f**), (**i**), (**t**), (**A**), (**d**), (**p**), (**c**) and (**q**) are known, then its Sales Past must be:

$$S' = \{S - [Ad + (Q + Svp/\{360[c-q]\})/I + U]/[1-t]\} / \{[1+s][v+f+i]\}$$

Rule-22204:

If both (**/**), (**U**), (**S**), (**S'**), (**Q**), (**v**), (**f**), (**i**), (**t**), (**A**), (**d**), (**p**), (**c**) and (**q**) are known, then its Sales Growth Planned is:

$$s = S/S' - 1$$

or it can also be found as

$$s = \{S - [Ad + (Q + Svp/\{360[c-q]\})/I + U]/[1-t]\} / \{S'[v+f+i]\} - 1$$

Rule-22205:

If both (**/**), (**U**), (**S**), (**S'**), (**s**), (**Q**), (**f**), (**i**), (**t**), (**A**), (**d**), (**p**), (**c**) and (**q**) are known, then its Variable Portion Planned is:

$$v = (I\{U - S'[1+s][v+f+i]\}[1-t] - Ad) - Q) / (Sp/\{360[c-q]\} + S'I[1+s][1-t])$$

Rule-22206:

If both (**/**), (**U**), (**S**), (**S'**), (**s**), (**v**), (**Q**), (**i**), (**t**), (**A**), (**d**), (**p**), (**c**) and (**q**) are known, then its Fixed Portion Planned is:

$$f = \{S - [Ad + (Q + Svp/\{360[c-q]\})/I + U]/[1-t]\} / \{S'[1+s]\} - v - i$$

Steve Asikin ISBN 14: 978-1515158691, ISBN 10: 1515158691

Rule-22207:

If both (I), (U), (S), (S'), (s), (v), (f), (Q), (t), (A), (d), (p), (c) and (q) are known, then its Interest Portion Planned is:

$$i= \{S-[Ad+(Q+Svp/\{360[c-q]\})/I+U]/[1-t]\}$$
$$/\{S'[1+s]\}-v-f$$

Rule-22208:

If both (I), (U), (S), (S'), (s), (v), (f), (i), (Q), (A), (d), (p), (c) and (q) are known, then its Tax Rate Planned is:

$$t= 1-[Ad+(Q+Svp/\{360[c-q]\})/I+U]$$
$$/\{S-S'[1+s][v+f+i]\}$$

Rule-22209:

If both (I), (U), (S), (S'), (s), (v), (f), (i), (t), (Q), (d), (p), (c) and (q) are known, then its After Tax Income Planned is:

$$A= [U+\{S-S'[1+s][v+f+i]\}[1-t]$$
$$-(Q+Svp/\{360[c-q]\})/I]/d$$

Rule-22210:

If both (I), (U), (S), (S'), (s), (v), (f), (i), (t), (A), (Q), (p), (c) and (q) are known, then its Dividend Payout Planned is:

$$d= [U+\{S-S'[1+s][v+f+i]\}[1-t]$$
$$-(Q+Svp/\{360[c-q]\})/I]/A$$

Steve Asikin ISBN 14: 978-1515158691, ISBN 10: 1515158691

Rule-22211:

If both (**f**), (**U**), (**$**), (**$'**), (**s**), (**v**), (**f**), (**i**), (**t**), (**A**), (**d**), (**Q**), (**c**) and (**q**) are known, then its Procured Inventory Days Planned:

$$p = 360[c\text{-}q]$$
$$[f U + \{\$\text{-}\$'[1+s][v+f+i]\}[1\text{-}t]\text{-}Ad)\text{-}Q]/[\$v]$$

Rule-22212:

If both (**f**), (**U**), (**$**), (**$'**), (**s**), (**v**), (**f**), (**i**), (**t**), (**A**), (**d**), (**p**), (**Q**) and (**q**) are known, then its Current Ratio Planned is:

$$c = q + \$vp$$
$$/\{360[f U + \{\$\text{-}\$'[1+s][v+f+i]\}[1\text{-}t]\text{-}Ad)\text{-}Q]\}$$

Rule-22213:

If both (**f**), (**U**), (**$**), (**$'**), (**s**), (**v**), (**f**), (**i**), (**t**), (**A**), (**d**), (**p**), (**c**) and (**Q**) are known, then its Quick or Acid Test Ratio Planned is:

$$q = c - \$vp$$
$$/\{360[f U + \{\$\text{-}\$'[1+s][v+f+i]\}[1\text{-}t]\text{-}Ad)\text{-}Q]\}$$

Rule-22214:

If both (**f**), (**U**), (**$**), (**$'**), (**s**), (**v**), (**f**), (**i**), (**t**), (**A**), (**d**), (**p**), (**c**) and (**q**) are known, then its Quoted Longterm Debt Planned is:

$$Q = f U + \{\$\text{-}\$'[1+s][v+f+i]\}[1\text{-}t]\text{-}Ad)$$
$$-\$'vp[1+s]/\{360[c\text{-}q]\}$$

Steve Asikin ISBN 14: 978-1515158691, ISBN 10: 1515158691

Rule-22215:

If both **(Q)**, **(U)**, **(S)**, **(S')**, **(s)**, **(v)**, **(f)**, **(i)**, **(t)**, **(A)**, **(d)**, **(p)**, **(c)** and **(q)** are known, then its Leverage or Gearing Ratio Planned is:

$$\textbf{I} = (\textbf{Q}+\textbf{S'vp}[1+\textbf{s}]/\{360[\textbf{c-q}]\})$$
$$/(\textbf{U}+\{\textbf{S-S'}[1+\textbf{s}][\textbf{v+f+i}]\}[1-\textbf{t}]-\textbf{Ad})$$

Rule-22216:

If both **(I)**, **(Q)**, **(S)**, **(S')**, **(s)**, **(v)**, **(f)**, **(i)**, **(t)**, **(A)**, **(d)**, **(p)**, **(c)** and **(q)** are known, then its Utilized or Starting Capital must be:

$$\textbf{U} = \textbf{Ad}+(\textbf{Q}+\textbf{S'vp}[1+\textbf{s}]/\{360[\textbf{c-q}]\})/\textbf{I}$$
$$-\{\textbf{S-S'}[1+\textbf{s}][\textbf{v+f+i}]\}[1-\textbf{t}]$$

Rule-22217:

If both **(I)**, **(U)**, **(Q)**, **(S')**, **(s)**, **(v)**, **(f)**, **(i)**, **(t)**, **(A)**, **(d)**, **(p)**, **(c)** and **(q)** are known, then its Sales or Revenue Planned is:

$$\textbf{S} = (\textbf{Q-I}\{\textbf{U-S'}[1+\textbf{s}][\textbf{v+f+i}]\}[1-\textbf{t}]-\textbf{Ad}\}$$
$$+\textbf{S'vp}[1+\textbf{s}]/\{360[\textbf{c-q}]\})/\{\textbf{I}[1-\textbf{t}]\}$$

Rule-22218:

If both **(I)**, **(U)**, **(S)**, **(Q)**, **(s)**, **(v)**, **(f)**, **(i)**, **(t)**, **(A)**, **(d)**, **(p)**, **(c)** and **(q)** are known, then its Sales Past must be:

$$\textbf{S'} = (\textbf{I}\{\textbf{U-S'}[1+\textbf{s}][\textbf{v+f+i}]\}[1-\textbf{t}]-\textbf{Ad})-\textbf{Q})$$
$$/[\textbf{v}[1+\textbf{s}](\textbf{p}/\{360[\textbf{c-q}]\}+\textbf{I}[1-\textbf{t}])]$$

Steve Asikin ISBN 14: 978-1515158691, ISBN 10: 1515158691

Rule-22219:

If both (I), (U), (S), (S'), (Q), (v), (f), (i), (t), (A), (d), (p), (c) and (q) are known, then its Sales Growth Planned is:

$s = S/S' - 1$

or it can also be found as

$$s = (I\{U - S'[1+s][v+f+i]\}[1-t] - Ad) - Q) / [S'v(p/\{360[c-q]\} + I[1-t])] - 1$$

Rule-22220:

If both (I), (U), (S), (S'), (s), (Q), (f), (i), (t), (A), (d), (p), (c) and (q) are known, then its Variable Portion Planned is:

$$v = (I\{U - S'[1+s][v+f+i]\}[1-t] - Ad) - Q) / [S'[1+s](p/\{360[c-q]\} + I[1-t])]$$

Rule-22221:

If both (I), (U), (S), (S'), (s), (v), (Q), (i), (t), (A), (d), (p), (c) and (q) are known, then its Fixed Portion Planned is:

$$f = \{S - [Ad + (Q + S'vp[1+s]/\{360[c-q]\})/I + U]/[1-t]\} / \{S'[1+s]\} - v - i$$

Rule-22222:

If both (I), (U), (S), (S'), (s), (v), (f), (Q), (t), (A), (d), (p), (c) and (q) are known, then its Interest Portion Planned is:

$$i = \{S - [Ad + (Q + S'vp[1+s]/\{360[c-q]\})/I + U]/[1-t]\} / \{S'[1+s]\} - v - f$$

Steve Asikin ISBN 14: 978-1515158691, ISBN 10: 1515158691

Rule-22223:

If both (*/*), (**U**), (**$**), (**$'**), (**s**), (**v**), (**f**), (**i**), (**Q**), (**A**), (**d**), (**p**), (**c**) and (**q**) are known, then its Tax Rate Planned is:

$$t= 1-[\mathbf{Ad}+(\mathbf{Q}+\mathbf{\$'vp}[1+s]/\{360[\mathbf{c\text{-}q}]\})/\mathbf{/}+\mathbf{U}]$$
$$/\{\mathbf{\$\text{-}\$'}[1+s][\mathbf{v+f+i}]\}$$

Rule-22224:

If both (*/*), (**U**), (**$**), (**$'**), (**s**), (**v**), (**f**), (**i**), (**t**), (**Q**), (**d**), (**p**), (**c**) and (**q**) are known, then its After Tax Income Planned is:

$$\mathbf{A}= [\mathbf{U}+\{\mathbf{\$\text{-}\$'}[1+s][\mathbf{v+f+i}]\}[1\text{-}\mathbf{t}]$$
$$-(\mathbf{Q}+\mathbf{\$'vp}[1+s]/\{360[\mathbf{c\text{-}q}]\})/\mathbf{/}]/\mathbf{d}$$

Rule-22225:

If both (*/*), (**U**), (**$**), (**$'**), (**s**), (**v**), (**f**), (**i**), (**t**), (**A**), (**Q**), (**p**), (**c**) and (**q**) are known, then its Dividend Payout Planned is:

$$\mathbf{d}= [\mathbf{U}+\{\mathbf{\$\text{-}\$'}[1+s][\mathbf{v+f+i}]\}[1\text{-}\mathbf{t}]$$
$$-(\mathbf{Q}+\mathbf{\$'vp}[1+s]/\{360[\mathbf{c\text{-}q}]\})/\mathbf{/}]/\mathbf{A}$$

Rule-22226:

If both (*/*), (**U**), (**$**), (**$'**), (**s**), (**v**), (**f**), (**i**), (**t**), (**A**), (**d**), (**Q**), (**c**) and (**q**) are known, then its Procured Inventory Days Planned:

$$\mathbf{p}= 360[\mathbf{c\text{-}q}][\mathbf{/U}+\{\mathbf{\$\text{-}\$'}[1+s][\mathbf{v+f+i}]\}[1\text{-}\mathbf{t}]\text{-}\mathbf{Ad})\text{-}\mathbf{Q}]$$
$$/\{\mathbf{\$'v}[1+s]\}$$

Steve Asikin ISBN 14: 978-1515158691, ISBN 10: 1515158691

Rule-22227:
> If both (I), (U), (S), (S'), (s), (v), (f), (i), (t), (A), (d), (p), (Q) and (q) are known, then its Current Ratio Planned is:
> $$c= q+S'vp[1+s]$$
> $$/\{360[I(U+\{S-S'[1+s][v+f+i]\}[1-t]-Ad)-Q]\}$$

Rule-22228:
> If both (I), (U), (S), (S'), (s), (v), (f), (i), (t), (A), (d), (p), (c) and (Q) are known, then its Quick or Acid Test Ratio Planned is:
> $$q= c-S'vp[1+s]$$
> $$/\{360[I(U+\{S-S'[1+s][v+f+i]\}[1-t]-Ad)-Q]\}$$

Rule-22229:
> If both (I), (U), (S), (S'), (s), (v), (f), (i), (t), (d) and (X) are known, then its Quoted Longterm Debt Planned is:
> $$Q= I(U+\{S-S'[1+s][v+f+i]\}[1-t][1-d])-X$$

Rule-22230:
> If both (Q), (U), (S), (S'), (s), (v), (f), (i), (t), (d) and (X) are known, then its Leverage or Gearing Ratio Planned is:
> $$I= [Q+X]/(U+\{S-S'[1+s][v+f+i]\}[1-t][1-d])$$

Rule-22231:
> If both (I), (Q), (S), (S'), (s), (v), (f), (i), (t), (d) and (X) are known, then its Utilized or Starting Capital must be:
> $$U= [Q+X]/I\{S-S'[1+s][v+f+i]\}[1-t][1-d]$$

Steve Asikin ISBN 14: 978-1515158691, ISBN 10: 1515158691

Rule-22232:

If both (*ʃ*, (**U**), (**Q**), (**S'**), (**s**), (**v**), (**f**), (**i**), (**t**), (**d**) and (**X**) are known, then its Sales or Revenue Planned is:

$$S= S'[1+s][v+f+i]+\{[Q+X]/I\!U\}/\{[1-t][1-d]\}$$

Rule-22233:

If both (*ʃ*, (**U**), (**S**), (**Q**), (**s**), (**v**), (**f**), (**i**), (**t**), (**d**) and (**X**) are known, then its Sales Past must be:

$$S'= (S-\{[Q+X]/I\!U\}/\{[1-t][1-d]\})/\{[1+s][v+f+i]\}\}$$

Rule-22234:

If both (*ʃ*, (**U**), (**S**), (**S'**), (**Q**), (**v**), (**f**), (**i**), (**t**), (**d**) and (**X**) are known, then its Sales Growth Planned is:

$$s= S/S'-1$$

or it can also be found as

$$s= (S-\{[Q+X]/I\!U\}/\{[1-t][1-d]\})/\{S'[v+f+i]\}\}-1$$

Rule-22235:

If both (*ʃ*, (**U**), (**S**), (**S'**), (**s**), (**Q**), (**f**), (**i**), (**t**), (**d**) and (**X**) are known, then its Variable Portion Planned is:

$$v= (S-\{[Q+X]/I\!U\}/\{[1-t][1-d]\})/\{S'[1+s]\}- f-i$$

Rule-22236:

If both (*ʃ*, (**U**), (**S**), (**S'**), (**s**), (**v**), (**Q**), (**i**), (**t**), (**d**) and (**X**) are known, then its Fixed Portion Planned is:

$$f= (S-\{[Q+X]/I\!U\}/\{[1-t][1-d]\})/\{S'[1+s]\}-v-i$$

Rule-22237:

If both (*ʃ*, (**U**), (**S**), (**S'**), (**s**), (**v**), (**f**), (**Q**), (**t**), (**d**) and (**X**) are known, then its Interest Portion Planned is:

$$i= (S-\{[Q+X]/I\!U\}/\{[1-t][1-d]\})/\{S'[1+s]\}-v-f$$

Steve Asikin ISBN 14: 978-1515158691, ISBN 10: 1515158691

Rule-22238:

    If both $(\textit{\textbf{l}})$, $(\textbf{U})$, $(\textbf{\$})$, $(\textbf{\$'})$, $(\textbf{s})$, $(\textbf{v})$, $(\textbf{f})$, $(\textbf{i})$, $(\textbf{Q})$, $(\textbf{d})$ and $(\textbf{X})$ are known, then its Tax Rate Planned is:

$$\textbf{t}= 1-\{[\textbf{Q}+\textbf{X}]/\textbf{\textit{l}U}\}/(\{\textbf{\$}-\textbf{\$'}[1+\textbf{s}][\textbf{v}+\textbf{f}+\textbf{i}]\}[1-\textbf{d}])$$

Rule-22239:

    If both $(\textit{\textbf{l}})$, $(\textbf{U})$, $(\textbf{\$})$, $(\textbf{\$'})$, $(\textbf{s})$, $(\textbf{v})$, $(\textbf{f})$, $(\textbf{i})$, $(\textbf{t})$, $(\textbf{Q})$ and $(\textbf{X})$ are known, then its Dividend Payout Planned is:

$$\textbf{d}= 1-\{[\textbf{Q}+\textbf{X}]/\textbf{\textit{l}U}\}/(\{\textbf{\$}-\textbf{\$'}[1+\textbf{s}][\textbf{v}+\textbf{f}+\textbf{i}]\}[1-\textbf{t}])$$

Rule-22240:

    If both $(\textit{\textbf{l}})$, $(\textbf{U})$, $(\textbf{\$})$, $(\textbf{\$'})$, $(\textbf{s})$, $(\textbf{v})$, $(\textbf{f})$, $(\textbf{i})$, $(\textbf{t})$, $(\textbf{d})$ and $(\textbf{Q})$ are known, then its Xpress or Current Debt Planned is:

$$\textbf{X}= \textbf{\textit{l}U}+\{\textbf{\$}-\textbf{\$'}[1+\textbf{s}][\textbf{v}+\textbf{f}+\textbf{i}]\}[1-\textbf{t}][1-\textbf{d}])-\textbf{Q}$$

Rule-22241:

    If both $(\textit{\textbf{l}})$, $(\textbf{U})$, $(\textbf{\$})$, $(\textbf{\$'})$, $(\textbf{s})$, $(\textbf{v})$, $(\textbf{f})$, $(\textbf{i})$, $(\textbf{t})$, $(\textbf{d})$, $(\textbf{P})$, $(\textbf{c})$ and $(\textbf{q})$ are known, then its Quoted Longterm Debt Planned is:

$$\textbf{Q}= \textbf{\textit{l}U}+\{\textbf{\$}-\textbf{\$'}[1+\textbf{s}][\textbf{v}+\textbf{f}+\textbf{i}]\}[1-\textbf{t}][1-\textbf{d}])-\textbf{P}/[\textbf{c}-\textbf{q}]$$

Rule-22242:

    If both $(\textbf{Q})$, $(\textbf{U})$, $(\textbf{\$})$, $(\textbf{\$'})$, $(\textbf{s})$, $(\textbf{v})$, $(\textbf{f})$, $(\textbf{i})$, $(\textbf{t})$, $(\textbf{d})$, $(\textbf{P})$, $(\textbf{c})$ and $(\textbf{q})$ are known, then its Leverage or Gearing Ratio Planned is:

$$\textbf{\textit{l}}= \{\textbf{Q}+\textbf{P}/[\textbf{c}-\textbf{q}]\}/(\textbf{U}+\{\textbf{\$}-\textbf{\$'}[1+\textbf{s}][\textbf{v}+\textbf{f}+\textbf{i}]\}[1-\textbf{t}][1-\textbf{d}])$$

Steve Asikin ISBN 14: 978-1515158691, ISBN 10: 1515158691

Rule-22243:

If both (I), (Q), (S), (S'), (s), (v), (f), (i), (t), (d), (P), (c) and (q) are known, then its Utilized or Startin g Capital must be:

$$U= \{Q+P/[c-q]\}/I \{S-S'[1+s][v+f+i]\}[1-t][1-d]$$

Rule-22244:

If both (I), (U), (Q), (S'), (s), (v), (f), (i), (t), (d), (P), (c) and (q) are known, then its Sales or Revenue Planned is:

$$S= S'[1+s][v+f+i]+(\{Q+P/[c-q]\}/I U)/\{[1-t][1-d]\}$$

Rule-22245:

If both (I), (U), (S), (Q), (s), (v), (f), (i), (t), (d), (P), (c) and (q) are known, then its Sales Past must be:

$$S'= [S-(\{Q+P/[c-q]\}/I U)/\{[1-t][1-d]\}]$$
$$/\{[1+s][v+f+i]\}\}$$

Rule-22246:

If both (I), (U), (S), (S'), (Q), (v), (f), (i), (t), (d), (P), (c) and (q) are known, then its Sales Growth Planned is:

$$s= S/S'-1$$

or it can also be found as

$$s= [S-(\{Q+P/[c-q]\}/I U)/\{[1-t][1-d]\}]$$
$$/\{S'[v+f+i]\}\}-1$$

Steve Asikin ISBN 14: 978-1515158691, ISBN 10: 1515158691

Rule-22247:

If both (**/**), (**U**), (**$**), (**$'**), (**s**), (**Q**), (**f**), (**i**), (**t**), (**d**), (**P**), (**c**) and (**q**) are known, then its Variable PortionPlanned is:

$$v= [\$-(\{Q+P/[c-q]\}/\text{/-}U)/\{[1-t][1-d]\}]$$
$$/\{\$'[1+s]\}- f\text{-}i$$

Rule-22248:

If both (**/**), (**U**), (**$**), (**$'**), (**s**), (**v**), (**Q**), (**i**), (**t**), (**d**), (**P**), (**c**) and (**q**) are known, then its Fixed Portion Planned is:

$$f= [\$-(\{Q+P/[c-q]\}/\text{/-}U)/\{[1-t][1-d]\}]$$
$$/\{\$'[1+s]\}\text{-}v\text{-}i$$

Rule-22249:

If both (**/**), (**U**), (**$**), (**$'**), (**s**), (**v**), (**f**), (**Q**), (**t**), (**d**), (**P**), (**c**) and (**q**) are known, then its Interest Portion Planned is:

$$i= [\$-(\{Q+P/[c-q]\}/\text{/-}U)/\{[1-t][1-d]\}]$$
$$/\{\$'[1+s]\}\text{-}v\text{-}f$$

Rule-22250:

If both (**/**), (**U**), (**$**), (**$'**), (**s**), (**v**), (**f**), (**i**), (**Q**), (**d**), (**P**), (**c**) and (**q**) are known, then its Tax Rate Planned is:

$$t= 1-(\{Q+P/[c-q]\}/\text{/-}U)/(\{\$-\$'[1+s][v+f+i]\}[1-d])$$

Rule-22251:

If both (**/**), (**U**), (**$**), (**$'**), (**s**), (**v**), (**f**), (**i**), (**t**), (**Q**), (**P**), (**c**) and (**q**) are known, then its Dividend Payout Planned is:

$$d= 1-(\{Q+P/[c-q]\}/\text{/-}U)/(\{\$-\$'[1+s][v+f+i]\}[1-t])$$

Steve Asikin ISBN 14: 978-1515158691, ISBN 10: 1515158691

Rule-22252:

If both (**/**), (**U**), (**$**), (**$'**), (**s**), (**v**), (**f**), (**i**), (**t**), (**d**), (**Q**), (**c**) and (**q**) are known, then its Procured Inventory Planned is:

$$\mathbf{P}= [\mathbf{c\text{-}q}][\mathbf{/(U}+\{\mathbf{\$\text{-}\$'}[1+\mathbf{s}][\mathbf{v+f+i}]\}[1\text{-}\mathbf{t}][1\text{-}\mathbf{d}])\text{-}\mathbf{Q}]$$

Rule-22253:

If both (**/**), (**U**), (**$**), (**$'**), (**s**), (**v**), (**f**), (**i**), (**t**), (**d**), (**P**), (**Q**) and (**q**) are known, then its Current Ratio Planned is:

$$\mathbf{c}= \mathbf{q+P}/[\mathbf{/(U}+\{\mathbf{\$\text{-}\$'}[1+\mathbf{s}][\mathbf{v+f+i}]\}[1\text{-}\mathbf{t}][1\text{-}\mathbf{d}])\text{-}\mathbf{Q}]$$

Rule-22254:

If both (**/**), (**U**), (**$**), (**$'**), (**s**), (**v**), (**f**), (**i**), (**t**), (**d**), (**P**), (**c**) and (**Q**) are known, then its Quick or Acid Test Ratio Planned is:

$$\mathbf{q}= \mathbf{c\text{-}P}/[\mathbf{/(U}+\{\mathbf{\$\text{-}\$'}[1+\mathbf{s}][\mathbf{v+f+i}]\}[1\text{-}\mathbf{t}][1\text{-}\mathbf{d}])\text{-}\mathbf{Q}]$$

Rule-22255:

If both (**/**), (**U**), (**$**), (**$'**), (**s**), (**v**), (**f**), (**i**), (**t**), (**d**), (**V**), (**p**), (**c**) and (**q**) are known, then its Quoted Longterm Debt Planned is:

$$\mathbf{Q}= \mathbf{/(U}+\{\mathbf{\$\text{-}\$'}[1+\mathbf{s}][\mathbf{v+f+i}]\}[1\text{-}\mathbf{t}][1\text{-}\mathbf{d}])$$
$$-\mathbf{Vp}/\{360[\mathbf{c\text{-}q}]\}$$

Rule-22256:

If both (**Q**), (**U**), (**$**), (**$'**), (**s**), (**v**), (**f**), (**i**), (**t**), (**d**), (**V**), (**p**), (**c**) and (**q**) are known, then its Leverage or Gearing Ratio Planned is:

$$\mathbf{/}= (\mathbf{Q+Vp}/\{360[\mathbf{c\text{-}q}]\})$$
$$/(\mathbf{U}+\{\mathbf{\$\text{-}\$'}[1+\mathbf{s}][\mathbf{v+f+i}]\}[1\text{-}\mathbf{t}][1\text{-}\mathbf{d}])$$

Steve Asikin ISBN 14: 978-1515158691, ISBN 10: 1515158691

Rule-22257:

If both (*l*), (**Q**), (**S**), (**S'**), (**s**), (**v**), (**f**), (**i**), (**t**), (**d**), (**V**), (**p**), (**c**) and (**q**) are known, then its Utilized or Starting Capital must be:

$$U= (Q+Vp/\{360[c-q]\})/l$$
$$-\{S-S'[1+s][v+f+i]\}[1-t][1-d]$$

Rule-22258:

If both (*l*), (**U**), (**Q**), (**S'**), (**s**), (**v**), (**f**), (**i**), (**t**), (**d**), (**V**), (**p**), (**c**) and (**q**) are known, then its Sales or Revenue Planned is:

$$S= S'[1+s][v+f+i]+[(Q+Vp/\{360[c-q]\})/l U]$$
$$/\{[1-t][1-d]\}$$

Rule-22259:

If both (*l*), (**U**), (**S**), (**Q**), (**s**), (**v**), (**f**), (**i**), (**t**), (**d**), (**V**), (**p**), (**c**) and (**q**) are known, then its Sales Past must be:

$$S'= \{S-[(Q+Vp/\{360[c-q]\})/l U]/\{[1-t][1-d]\}\}$$
$$/\{[1+s][v+f+i]\}\}$$

Rule-22260:

If both (*l*), (**U**), (**S**), (**S'**), (**Q**), (**v**), (**f**), (**i**), (**t**), (**d**), (**V**), (**p**), (**c**) and (**q**) are known, then its Sales or Revenue Planned is:

$$s= S/S'-1$$

or it can also be found as

$$s= \{S-[(Q+Vp/\{360[c-q]\})/l U]$$
$$/\{[1-t][1-d]\}\}/\{S'[v+f+i]\}\}-1$$

Steve Asikin ISBN 14: 978-1515158691, ISBN 10: 1515158691

Rule-22261:

If both (l), $(\mathbf{U})$, $(\mathbf{\$})$, $(\mathbf{\$'})$, $(\mathbf{s})$, $(\mathbf{Q})$, $(\mathbf{f})$, $(\mathbf{i})$, $(\mathbf{t})$, $(\mathbf{d})$, $(\mathbf{V})$, $(\mathbf{p})$, $(\mathbf{c})$ and $(\mathbf{q})$ are known, then its Variable Portion Planned is:

$$v = V/\$$$

or it can also be found as

$$v = \{\$ - [(Q + Vp/\{360[c-q]\})/\mathit{l}U]/\{[1-t][1-d]\}\} / \{\$'[1+s]\} - f - i$$

Rule-22262:

If both (l), $(\mathbf{U})$, $(\mathbf{\$})$, $(\mathbf{\$'})$, $(\mathbf{s})$, $(\mathbf{v})$, $(\mathbf{Q})$, $(\mathbf{i})$, $(\mathbf{t})$, $(\mathbf{d})$, $(\mathbf{V})$, $(\mathbf{p})$, $(\mathbf{c})$ and $(\mathbf{q})$ are known, then its Fixed Portion Planned is:

$$f = \{\$ - [(Q + Vp/\{360[c-q]\})/\mathit{l}U]/\{[1-t][1-d]\}\} / \{\$'[1+s]\} - v - i$$

Rule-22263:

If both (l), $(\mathbf{U})$, $(\mathbf{\$})$, $(\mathbf{\$'})$, $(\mathbf{s})$, $(\mathbf{v})$, $(\mathbf{f})$, $(\mathbf{Q})$, $(\mathbf{t})$, $(\mathbf{d})$, $(\mathbf{V})$, $(\mathbf{p})$, $(\mathbf{c})$ and $(\mathbf{q})$ are known, then its Interest Portion Planned is:

$$i = \{\$ - [(Q + Vp/\{360[c-q]\})/\mathit{l}U]/\{[1-t][1-d]\}\} / \{\$'[1+s]\} - v - f$$

Rule-22264:

If both (l), $(\mathbf{U})$, $(\mathbf{\$})$, $(\mathbf{\$'})$, $(\mathbf{s})$, $(\mathbf{v})$, $(\mathbf{f})$, $(\mathbf{i})$, $(\mathbf{Q})$, $(\mathbf{d})$, $(\mathbf{V})$, $(\mathbf{p})$, $(\mathbf{c})$ and $(\mathbf{q})$ are known, then its Tax Rate Planned is:

$$t = 1 - [(Q + Vp/\{360[c-q]\}) / (\{\$ - \$'[1+s][v+f+i]\}[1-d])$$

Steve Asikin ISBN 14: 978-1515158691, ISBN 10: 1515158691

Rule-22265:

If both (I), (**U**), (**$**), (**$'**), (**$**), (**v**), (**f**), (**i**), (**t**), (**Q**), (**V**), (**p**), (**c**) and (**q**) are known, then its Dividend Payout Planned is:

$$d= 1-[(Q+Vp/\{360[c-q]\})$$
$$/(\{$-$'[1+s][v+f+i]\}[1-t])$$

Rule-22266:

If both (I), (**U**), (**$**), (**$'**), (**$**), (**v**), (**f**), (**i**), (**t**), (**d**), (**Q**), (**p**), (**c**) and (**q**) are known, then its Variable Cost Planned is:

$$V= 360[c-q]$$
$$[IU+\{$-$'[1+s][v+f+i]\}[1-t][1-d])-Q]/p$$

Rule-22267:

If both (I), (**U**), (**$**), (**$'**), (**$**), (**v**), (**f**), (**i**), (**t**), (**d**), (**V**), (**Q**), (**c**) and (**q**) are known, then its Procured Inventory Days Planned:

$$p= 360[c-q]$$
$$[IU+\{$-$'[1+s][v+f+i]\}[1-t][1-d])-Q]/V$$

Rule-22268:

If both (I), (**U**), (**$**), (**$'**), (**$**), (**v**), (**f**), (**i**), (**t**), (**d**), (**V**), (**p**), (**Q**) and (**q**) are known, then its Current Ratio Planned is:

$$c= q+Vp/\{360[IU+\{$-$'[1+s][v+f+i]\}$$
$$[1-t][1-d])-Q]\}$$

Steve Asikin ISBN 14: 978-1515158691, ISBN 10: 1515158691

Rule-22269:

If both $(\textit{I})$, $(\textbf{U})$, $(\textbf{S})$, $(\textbf{S'})$, $(\textbf{s})$, $(\textbf{v})$, $(\textbf{f})$, $(\textbf{i})$, $(\textbf{t})$, $(\textbf{d})$, $(\textbf{V})$, $(\textbf{p})$, $(\textbf{c})$ and $(\textbf{Q})$ are known, then its Quick or Acid Test Ratio Planned is:

$$q= c\text{-}Vp/\{360[\textit{I}U+\{S\text{-}S'[1+s][v+f+i]\}$$
$$[1\text{-}t][1\text{-}d])\text{-}Q]\}$$

Rule-22270:

If both $(\textit{I})$, $(\textbf{U})$, $(\textbf{S})$, $(\textbf{S'})$, $(\textbf{s})$, $(\textbf{v})$, $(\textbf{f})$, $(\textbf{i})$, $(\textbf{t})$, $(\textbf{d})$, $(\textbf{V})$, $(\textbf{p})$, $(\textbf{c})$ and $(\textbf{q})$ are known, then its Quoted Longterm Debt Planned is:

$$Q= \textit{I}U+\{S\text{-}S'[1+s][v+f+i]\}[1\text{-}t][1\text{-}d])$$
$$-Svp/\{360[c\text{-}q]\}$$

Rule-22271:

If both $(\textbf{Q})$, $(\textbf{U})$, $(\textbf{S})$, $(\textbf{S'})$, $(\textbf{s})$, $(\textbf{v})$, $(\textbf{f})$, $(\textbf{i})$, $(\textbf{t})$, $(\textbf{d})$, $(\textbf{V})$, $(\textbf{p})$, $(\textbf{c})$ and $(\textbf{q})$ are known, then its Leverage or Gearing Ratio Planned is:

$$\textit{I}= (Q+Svp/\{360[c\text{-}q]\})$$
$$/(U+\{S\text{-}S'[1+s][v+f+i]\}[1\text{-}t][1\text{-}d])$$

Rule-22272:

If both $(\textit{I})$, $(\textbf{Q})$, $(\textbf{S})$, $(\textbf{S'})$, $(\textbf{s})$, $(\textbf{v})$, $(\textbf{f})$, $(\textbf{i})$, $(\textbf{t})$, $(\textbf{d})$, $(\textbf{V})$, $(\textbf{p})$, $(\textbf{c})$ and $(\textbf{q})$ are known, then its Utilized or Starting Capital must be:

$$U= (Q+Svp/\{360[c\text{-}q]\})/\textit{I}$$
$$-\{S\text{-}S'[1+s][v+f+i]\}[1\text{-}t][1\text{-}d]$$

Steve Asikin ISBN 14: 978-1515158691, ISBN 10: 1515158691

Rule-22273:

If both (**/**), (**U**), (**Q**), (**S'**), (**s**), (**v**), (**f**), (**i**), (**t**), (**d**), (**V**), (**p**), (**c**) and (**q**) are known, then its Sales or Revenue Planned is:

$$S = (/\{U-S'[1+s][v+f+i][1-t][1-d]\}-Q)$$
$$/(vp/\{360[c-q]\}-/[1-t][1-d])$$

Rule-22274:

If both (**/**), (**U**), (**S**), (**Q**), (**s**), (**v**), (**f**), (**i**), (**t**), (**d**), (**V**), (**p**), (**c**) and (**q**) are known, then its Sales Past must be:

$$S' = \{S-[(Q+Svp/\{360[c-q]\})//U]/\{[1-t][1-d]\}\}$$
$$/\{[1+s][v+f+i]\}$$

Rule-22275:

If both (**/**), (**U**), (**S**), (**S'**), (**Q**), (**v**), (**f**), (**i**), (**t**), (**d**), (**V**), (**p**), (**c**) and (**q**) are known, then its Sales Growth Planned is:

$$s = S/S'-1$$

or it can also be found as

$$s = \{S-[(Q+Svp/\{360[c-q]\})//U]/\{[1-t][1-d]\}\}$$
$$/\{S'[v+f+i]\}\}-1$$

Rule-22276:

If both (**/**), (**U**), (**S**), (**S'**), (**s**), (**Q**), (**f**), (**i**), (**t**), (**d**), (**V**), (**p**), (**c**) and (**q**) are known, then its Variable Portion Planned is:

$$v = [/U+\{S-S'[1+s][f+i]\}[1-t][1-d])-Q]$$
$$/(Sp/\{360[c-q]\}+S'/[1+s][1-t][1-d])$$

Steve Asikin ISBN 14: 978-1515158691, ISBN 10: 1515158691

Rule-22277:

If both (I), (U), (S), (S'), (s), (v), (Q), (i), (t), (d), (V), (p), (c) and (q) are known, then its Fixed Portion Planned is:

$$f = \{S-[(Q+Svp/\{360[c-q]\})/I \cdot U]/\{[1-t][1-d]\}\}$$
$$/\{S'[1+s]\}-v-i$$

Rule-22278:

If both (I), (U), (S), (S'), (s), (v), (f), (Q), (t), (d), (V), (p), (c) and (q) are known, then its Interest Portion Planned is:

$$i = \{S-[(Q+Svp/\{360[c-q]\})/I \cdot U]/\{[1-t][1-d]\}\}$$
$$/\{S'[1+s]\}-v-f$$

Rule-22279:

If both (I), (U), (S), (S'), (s), (v), (f), (i), (Q), (d), (V), (p), (c) and (q) are known, then its Tax Rate Planned is:

$$t = 1-[(Q+Svp/\{360[c-q]\})$$
$$/(\{S-S'[1+s][v+f+i]\}[1-d])$$

Rule-22280:

If both (I), (U), (S), (S'), (s), (v), (f), (i), (t), (Q), (V), (p), (c) and (q) are known, then its Dividend Payout Planned is:

$$d = 1-[(Q+Svp/\{360[c-q]\})$$
$$/(\{S-S'[1+s][v+f+i]\}[1-t])$$

Steve Asikin ISBN 14: 978-1515158691, ISBN 10: 1515158691

Rule-22281:

If both $(\textbf{\textit{I}})$, $(\textbf{U})$, $(\textbf{S})$, $(\textbf{S'})$, $(\textbf{s})$, $(\textbf{v})$, $(\textbf{f})$, $(\textbf{i})$, $(\textbf{t})$, $(\textbf{d})$, $(\textbf{V})$, $(\textbf{Q})$, $(\textbf{c})$ and $(\textbf{q})$ are known, then its Procured Inventory Days Planned:

$$\textbf{p}= 360[\textbf{c-q}]$$
$$[\textbf{\textit{I}U}+\{\textbf{S-S'}[1+\textbf{s}][\textbf{v+f+i}]\}[1\text{-}\textbf{t}][1\text{-}\textbf{d}])\text{-}\textbf{Q}]/[\textbf{Sv}]$$

Rule-22282:

If both $(\textbf{\textit{I}})$, $(\textbf{U})$, $(\textbf{S})$, $(\textbf{S'})$, $(\textbf{s})$, $(\textbf{v})$, $(\textbf{f})$, $(\textbf{i})$, $(\textbf{t})$, $(\textbf{d})$, $(\textbf{V})$, $(\textbf{p})$, $(\textbf{Q})$ and $(\textbf{q})$ are known, then its Current Ratio Planned is:

$$\textbf{c}= \textbf{q}+\textbf{Svp}/\{360[\textbf{\textit{I}U}+\{\textbf{S-S'}[1+\textbf{s}][\textbf{v+f+i}]\}$$
$$[1\text{-}\textbf{t}][1\text{-}\textbf{d}])\text{-}\textbf{Q}]\}$$

Rule-22283:

If both $(\textbf{\textit{I}})$, $(\textbf{U})$, $(\textbf{S})$, $(\textbf{S'})$, $(\textbf{s})$, $(\textbf{v})$, $(\textbf{f})$, $(\textbf{i})$, $(\textbf{t})$, $(\textbf{d})$, $(\textbf{V})$, $(\textbf{p})$, $(\textbf{c})$ and $(\textbf{Q})$ are known, then its Quick or Acid Test Ratio Planned is:

$$\textbf{q}= \textbf{c}\text{-}\textbf{Svp}/\{360[\textbf{\textit{I}U}+\{\textbf{S-S'}[1+\textbf{s}][\textbf{v+f+i}]\}$$
$$[1\text{-}\textbf{t}][1\text{-}\textbf{d}])\text{-}\textbf{Q}]\}$$

Rule-22284:

If both $(\textbf{\textit{I}})$, $(\textbf{U})$, $(\textbf{S})$, $(\textbf{S'})$, $(\textbf{s})$, $(\textbf{v})$, $(\textbf{f})$, $(\textbf{i})$, $(\textbf{t})$, $(\textbf{d})$, $(\textbf{V})$, $(\textbf{p})$, $(\textbf{c})$ and $(\textbf{q})$ are known, then its Quoted Longterm Debt Planned is:

$$\textbf{Q}= \textbf{\textit{I}U}+\{\textbf{S-S'}[1+\textbf{s}][\textbf{v+f+i}]\}[1\text{-}\textbf{t}][1\text{-}\textbf{d}])$$
$$\text{-}\textbf{S'vp}[1+\textbf{s}]/\{360[\textbf{c-q}]\}$$

Steve Asikin ISBN 14: 978-1515158691, ISBN 10: 1515158691

Rule-22285:

If both $(\mathbf{Q})$, $(\mathbf{U})$, $(\mathbf{\$})$, $(\mathbf{\$'})$, $(\mathbf{s})$, $(\mathbf{v})$, $(\mathbf{f})$, $(\mathbf{i})$, $(\mathbf{t})$, $(\mathbf{d})$, $(\mathbf{V})$, $(\mathbf{p})$, $(\mathbf{c})$ and $(\mathbf{q})$ are known, then its Leverage or Gearing Ratio Planned is:

$$\mathbf{\mathit{l}} = (\mathbf{Q} + \mathbf{\$'vp}[1+\mathbf{s}]/\{360[\mathbf{c}-\mathbf{q}]\})$$
$$/(\mathbf{U} + \{\mathbf{\$} - \mathbf{\$'}[1+\mathbf{s}][\mathbf{v}+\mathbf{f}+\mathbf{i}]\}[1-\mathbf{t}][1-\mathbf{d}])$$

Rule-22286:

If both $(\mathbf{\mathit{l}})$, $(\mathbf{Q})$, $(\mathbf{\$})$, $(\mathbf{\$'})$, $(\mathbf{s})$, $(\mathbf{v})$, $(\mathbf{f})$, $(\mathbf{i})$, $(\mathbf{t})$, $(\mathbf{d})$, $(\mathbf{V})$, $(\mathbf{p})$, $(\mathbf{c})$ and $(\mathbf{q})$ are known, then its Utilized or Starting Capital must be:

$$\mathbf{U} = (\mathbf{Q} + \mathbf{\$'vp}[1+\mathbf{s}]/\{360[\mathbf{c}-\mathbf{q}]\})/\mathbf{\mathit{l}}$$
$$-\{\mathbf{\$} - \mathbf{\$'}[1+\mathbf{s}][\mathbf{v}+\mathbf{f}+\mathbf{i}]\}[1-\mathbf{t}][1-\mathbf{d}]$$

Rule-22287:

If both $(\mathbf{\mathit{l}})$, $(\mathbf{U})$, $(\mathbf{Q})$, $(\mathbf{\$'})$, $(\mathbf{s})$, $(\mathbf{v})$, $(\mathbf{f})$, $(\mathbf{i})$, $(\mathbf{t})$, $(\mathbf{d})$, $(\mathbf{V})$, $(\mathbf{p})$, $(\mathbf{c})$ and $(\mathbf{q})$ are known, then its Sales or Revenue Planned is:

$$\mathbf{\$} = (\mathbf{Q} - \mathbf{\mathit{l}}\{\mathbf{U} - \mathbf{\$'}[1+\mathbf{s}][\mathbf{v}+\mathbf{f}+\mathbf{i}][1-\mathbf{t}][1-\mathbf{d}]\}$$
$$+\mathbf{\$'vp}[1+\mathbf{s}]/\{360[\mathbf{c}-\mathbf{q}]\})/\{\mathbf{\mathit{l}}[1-\mathbf{t}][1-\mathbf{d}]\}$$

Rule-22288:

If both $(\mathbf{\mathit{l}})$, $(\mathbf{U})$, $(\mathbf{\$})$, $(\mathbf{Q})$, $(\mathbf{s})$, $(\mathbf{v})$, $(\mathbf{f})$, $(\mathbf{i})$, $(\mathbf{t})$, $(\mathbf{d})$, $(\mathbf{V})$, $(\mathbf{p})$, $(\mathbf{c})$ and $(\mathbf{q})$ are known, then its Sales Past must be:

$$\mathbf{\$'} = [\mathbf{\mathit{l}}\{\mathbf{U} + \{\mathbf{\$} - \mathbf{\$'}[1+\mathbf{s}][\mathbf{f}+\mathbf{i}]\}[1-\mathbf{t}][1-\mathbf{d}]\} - \mathbf{Q}]$$
$$/[\mathbf{v}[1+\mathbf{s}](\mathbf{p}/\{360[\mathbf{c}-\mathbf{q}]\} + \mathbf{\mathit{l}}[1-\mathbf{t}][1-\mathbf{d}])]$$

Steve Asikin ISBN 14: 978-1515158691, ISBN 10: 1515158691

Rule-22289:

If both (**/**), (**U**), (**$**), (**$'**), (**Q**), (**v**), (**f**), (**i**), (**t**), (**d**), (**V**), (**p**), (**c**) and (**q**) are known, then its Sales Growth Planned is:

$s= \$/\$'-1$

or it can also be found as

$s= [/\textbf{U}+\{\$-\$'[1+s][f+i]\}[1-t][1-d])-\textbf{Q}]$
$/[\$'v(p/\{360[c-q]\}+/[1-t][1-d])]-1$

Rule-22290:

If both (**/**), (**U**), (**$**), (**$'**), (**s**), (**Q**), (**f**), (**i**), (**t**), (**d**), (**V**), (**p**), (**c**) and (**q**) are known, then its Variable Portion Planned is:

$v= [/\textbf{U}+\{\$-\$'[1+s][f+i]\}[1-t][1-d])-\textbf{Q}]$
$/[\$'[1+s](p/\{360[c-q]\}+/[1-t][1-d])]$

Rule-22291:

If both (**/**), (**U**), (**$**), (**$'**), (**s**), (**v**), (**Q**), (**i**), (**t**), (**d**), (**V**), (**p**), (**c**) and (**q**) are known, then its Fixed Portion Planned is:

$f= \{\$-[(\textbf{Q}+\$'vp[1+s]/\{360[c-q]\})/\textbf{/U}]$
$/\{[1-t][1-d]\}\}/\{\$'[1+s]\}-v-i$

Rule-22292:

If both (**/**), (**U**), (**$**), (**$'**), (**s**), (**v**), (**f**), (**Q**), (**t**), (**d**), (**V**), (**p**), (**c**) and (**q**) are known, then its Interest Portion Planned is:

$i=\{\$-[(\textbf{Q}+\$'vp[1+s]/\{360[c-q]\})/\textbf{/U}]$
$/\{[1-t][1-d]\}\}/\{\$'[1+s]\}-v-f$

Steve Asikin ISBN 14: 978-1515158691, ISBN 10: 1515158691

Rule-22293:

If both (f), (U), (S), (S'), (s), (v), (f), (i), (Q), (d), (V), (p), (c) and (q) are known, then its Tax Rate Planned is:

$$t = 1 - [(Q + S'vp[1+s]/\{360[c-q]\}) / (\{S-S'[1+s][v+f+i]\}[1-d])$$

Rule-22294:

If both (f), (U), (S), (S'), (s), (v), (f), (i), (t), (Q), (V), (p), (c) and (q) are known, then its Dividend Payout Planned is:

$$d = 1 - [(Q + S'vp[1+s]/\{360[c-q]\}) / (\{S-S'[1+s][v+f+i]\}[1-t])$$

Rule-22295:

If both (f), (U), (S), (S'), (s), (v), (f), (i), (t), (d), (V), (Q), (c) and (q) are known, then its Procured Inventory Days Planned:

$$p = 360[c-q][(U + \{S-S'[1+s][v+f+i]\}[1-t][1-d]) - Q] / \{S'v[1+s]\}$$

Rule-22296:

If both (f), (U), (S), (S'), (s), (v), (f), (i), (t), (d), (V), (p), (Q) and (q) are known, then its Current Ratio Planned is:

$$c = q + S'vp[1+s]/\{360[(U + \{S-S'[1+s][v+f+i]\}[1-t][1-d]) - Q]\}$$

Steve Asikin ISBN 14: 978-1515158691, ISBN 10: 1515158691

Rule-22297:

If both (**ʃ**), (**U**), (**\$**), (**\$'**), (**s**), (**v**), (**f**), (**i**), (**t**), (**d**), (**V**), (**p**), (**c**) and (**Q**) are known, then its Quick or Acid Test Ratio Planned is:

$$q = c\text{-}\$'vp[1+s]/\{360[\textbf{ʃ}U+\{\$\text{-}\$'[1+s][v+f+i]\}[1\text{-}t][1\text{-}d])\text{-}Q]\}$$

Steve Asikin ISBN 14: 978-1515158691, ISBN 10: 1515158691

CHAPTER-20:
G= GOODS or Trade Payable Optimization,

Rule-22298:
>   If both (**V**) and (**q**) are known, then its Goods or Trade
>   Account Payable Planned is:
> $G= Vg/360$

Rule-22299:
>   If both (**G**) and (**q**) are known, then its Variable Cost
>   Planned is:
> $V= 360G/g$

Rule-22300:
>   If both (**V**) and (**G**) are known, then its Quick or Acid
>   Test Ratio Planned is:
> $q= 360G/V$

Rule-22301:
>   If both (**$**), (**v**) and (**g**) are known, then its Goods or
>   Trade Account Payable Planned is:
> $G= \$vg/360$

Rule-22302:
>   If both (**G**), (**v**) and (**g**) are known, then its Sales or
>   Revenue Payable Planned is:
> $\$= 360G/[vg]$

Steve Asikin ISBN 14: 978-1515158691, ISBN 10: 1515158691

Rule-22303:

If both ($), (**G**) and (**g**) are known, then its Variable Portion Planned is:

$$v = 360G/[\$g]$$

Rule-22304:

If both ($), (**v**) and (**G**) are known, then its Goods or Trade Payable Days Planned:

$$g = 360G/[\$v]$$

Rule-22305:

If both ($'), (**v**), (**g**) and (**s**) are known, then its Goods or Trade Account Payable Planned is:

$$G = \$'vg[1+s]/360$$

Rule-22306:

If both (**G**), (**v**), (**g**) and (**s**) are known, then its Sales Past must be:

$$\$' = 360G/\{vg[1+s]\}$$

Rule-22307:

If both ($'), (**G**), (**g**) and (**s**) are known, then its Variable Portion Planned is:

$$v = 360G/\{\$'g[1+s]\}$$

Rule-22308:

If both ($'), (**v**), (**G**) and (**s**) are known, then its Goods or Trade Payable Days Planned:

$$g = 360G/\{\$'v[1+s]\}$$

Steve Asikin ISBN 14: 978-1515158691, ISBN 10: 1515158691

<u>Rule-22309</u>:

If both (**$'**), (**v**), (**g**) and (**G**) are known, then its Sales Growth Planned is:

$$s = 360G/[\$'vg]-1$$

Steve Asikin ISBN 14: 978-1515158691, ISBN 10: 1515158691

CHAPTER-21:

Z= ZERO TRADE or Current Debt Optimization,

Rule-22310:
> If both (**X**) and (**G**) are known, then its Zero Trade
> Current Debt Planned is:
> **Z= X-G**

Rule-22311:
> If both (**Z**) and (**G**) are known, then its Xpress or
> Current Debt Planned is:
> **X= Z+G**

Rule-22312:
> If both (**X**) and (**Z**) are known, then its Goods or Trade
> Account Payable Planned is:
> **G= X-Z**

Rule-22313:
> If both (**X**), (**V**) and (**g**) are known, then its Zero Trade
> Current Debt Planned is:
> **Z= X-Vg**/360

Rule-22314:
> If both (**Z**), (**V**) and (**g**) are known, then its Xpress or
> Current Debt Planned is:
> **X= Z+Vg**/360

Steve Asikin ISBN 14: 978-1515158691, ISBN 10: 1515158691

Rule-22315:

If both (**X**), (**Z**) and (**g**) are known, then its Variable Cost Planned is:
$$V= 360[X\text{-}Z]/g$$

Rule-22316:

If both (**X**), (**V**) and (**Z**) are known, then its Goods or Trade Payable Days Planned:
$$g= 360[X\text{-}Z]/V$$

Rule-22317:

If both (**X**), (**$**), (**v**) and (**g**) are known, then its Zero Trade Current Debt Planned is:
$$Z= X\text{-}\$vg/360$$

Rule-22318:

If both (**Z**), (**$**), (**v**) and (**g**) are known, then its Xpress or Current Debt Planned is:
$$X= Z+\$vg/360$$

Rule-22319:

If both (**X**), (**Z**), (**v**) and (**g**) are known, then its Sales or Revenue Planned is:
$$\$= 360[X\text{-}Z]/[vg]$$

Rule-22320:

If both (**X**), (**$**), (**Z**) and (**g**) are known, then its Variable Portion Planned is:
$$v= 360[X\text{-}Z]/[\$g]$$

Steve Asikin ISBN 14: 978-1515158691, ISBN 10: 1515158691

Rule-22321:

If both (**X**), (**$**), (**v**) and (**Z**) are known, then its Goods or Trade Payable Days Planned:

$\mathbf{g}$= 360[**X-Z**]/[**$v**]

Rule-22322:

If both (**X**), (**$'**), (**v**), (**g**) and (**s**) are known, then its Zero Trade Current Debt Planned is:

$\mathbf{Z}$= **X-$'vg**[1+**s**]/360

Rule-22323:

If both (**Z**), (**$'**), (**v**), (**g**) and (**s**) are known, then its Quoted Longterm Debt Planned is:

$\mathbf{X}$= **Z+$'vg**[1+**s**]/360

Rule-22324:

If both (**X**), (**Z**), (**v**), (**g**) and (**s**) are known, then its Sales Past must be:

$\mathbf{\$'}$= 360[**X-Z**]/{**vg**[1+**s**]}

Rule-22325:

If both (**X**), (**$'**), (**Z**), (**g**) and (**s**) are known, then its Variable Portion Planned is:

$\mathbf{v}$= 360[**X-Z**]/{**$'g**[1+**s**]}

Rule-22326:

If both (**X**), (**$'**), (**v**), (**Z**) and (**s**) are known, then its Goods or Trade Payable Days Planned:

$\mathbf{g}$= 360[**X-Z**]/{**$'v**[1+**s**]}

Steve Asikin ISBN 14: 978-1515158691, ISBN 10: 1515158691

Rule-22327:

    If both (X), (S'), (v), (g) and (Z) are known, then its Sales Growth Planned is:

$$s = 360[X\text{-}Z]/[S'vg]\text{-}1$$

Rule-22328:

    If both (P), (c), (q) and (G) are known, then its Zero Trade Current Debt Planned is:

$$Z = P/[c\text{-}q]\text{-}G$$

Rule-22329:

    If both (P), (c), (q) and (Z) are known, then its Goods or Trade Account Payable Planned is:

$$G = P/[c\text{-}q]\text{-}Z$$

Rule-22330:

    If both (Z), (c), (q) and (G) are known, then its Procured Inventory Planned is:

$$P = [c\text{-}q][G+Z]$$

Rule-22331:

    If both (P), (Z), (q) and (G) are known, then its Current Ratio Planned is:

$$c = P/[G+Z]+q$$

Rule-22332:

    If both (P), (c), (Z) and (G) are known, then its Quick or Acid Test Ratio Planned is:

$$q = c\text{-}P/[G+Z]$$

Steve Asikin ISBN 14: 978-1515158691, ISBN 10: 1515158691

Rule-22333:
> If both (**P**), (**c**), (**q**), (**V**) and (**g**) are known, then its Zero Trade Current Debt Planned is:
> $$Z= P/[c\text{-}q]\text{-}Vg/360$$

Rule-22334:
> If both (**P**), (**c**), (**q**), (**Z**) and (**g**) are known, then its Variable Cost Planned is:
> $$V= 360\{P/[c\text{-}q]\text{-}Z\}/g$$

Rule-22335:
> If both (**P**), (**c**), (**q**), (**V**) and (**Z**) are known, then its Goods or Trade Payable Days Planned:
> $$g= 360\{P/[c\text{-}q]\text{-}Z\}/V$$

Rule-22336:
> If both (**Z**), (**c**), (**q**), (**V**) and (**g**) are known, then its Procured Inventory Planned is:
> $$P= [c\text{-}q][Z+Vg/360]$$

Rule-22337:
> If both (**P**), (**Z**), (**q**), (**V**) and (**g**) are known, then its Current Ratio Planned is:
> $$c= P/[Z+Vg/360]+q$$

Rule-22338:
> If both (**P**), (**c**), (**Z**), (**V**) and (**g**) are known, then its Quick or Acid Test Ratio Planned is:
> $$q= c\text{-}P/[Z+Vg/360]$$

Steve Asikin ISBN 14: 978-1515158691, ISBN 10: 1515158691

Rule-22339:

If both (P), (c), (q), (S), (v) and (g) are known, then its Zero Trade Current Debt Planned is:

$Z = P/[c-q]-Svg/360$

Rule-22340:

If both (P), (c), (q), (Z), (v) and (g) are known, then its Sales or Revenue Planned is:

$S = 360\{P/[c-q]-Z\}/[vg]$

Rule-22341:

If both (P), (c), (q), (S), (Z) and (g) are known, then its Variable Portion Planned is:

$v = 360\{P/[c-q]-Z\}/[Sg]$

Rule-22342:

If both (P), (c), (q), (S), (v) and (Z) are known, then its Goods or Trade Payable Days Planned:

$g = 360\{P/[c-q]-Z\}/[Sv]$

Rule-22343:

If both (Z), (c), (q), (S), (v) and (g) are known, then its Procured Inventory Planned is:

$P = [c-q][Z+Svg/360]$

Rule-22344:

If both (P), (Z), (q), (S), (v) and (g) are known, then its Current Ratio Planned is:

$c = P/[Z+Svg/360]+q$

Steve Asikin ISBN 14: 978-1515158691, ISBN 10: 1515158691

Rule-22345:

If both (**P**), (**c**), (**Z**), (**$**), (**v**) and (**g**) are known, then its Quick or Acid Test Ratio Planned is:

$$q= c\text{-}P/[Z+\$vg/360]$$

Rule-22346:

If both (**P**), (**c**), (**q**), (**$'**), (**v**), (**g**) and (**s**) are known, then its Quoted Longterm Debt Planned is:

$$Z= P/[c\text{-}q]\text{-}\$'vg[1+s]/360$$

Rule-22347:

If both (**P**), (**c**), (**q**), (**Z**), (**v**), (**g**) and (**s**) are known, then its Sales Past must be:

$$\$'= 360\{P/[c\text{-}q]\text{-}Z\}/\{vg[1+s]\}$$

Rule-22348:

If both (**P**), (**c**), (**q**), (**$'**), (**Z**), (**g**) and (**s**) are known, then its Variable Portion Planned is:

$$v= 360\{P/[c\text{-}q]\text{-}Z\}/\{\$'g[1+s]\}$$

Rule-22349:

If both (**P**), (**c**), (**q**), (**$'**), (**v**), (**Z**) and (**s**) are known, then its Goods or Trade Payable Days Planned:

$$g= 360\{P/[c\text{-}q]\text{-}Z\}/\{\$'v[1+s]\}$$

Rule-22350:

If both (**P**), (**c**), (**q**), (**$'**), (**v**), (**g**) and (**Z**) are known, then its Sales Growth Planned is:

$$s= 360\{P/[c\text{-}q]\text{-}Z\}/[\$'vg]\text{-}1$$

Steve Asikin ISBN 14: 978-1515158691, ISBN 10: 1515158691

Rule-22351:
> If both (Z), (c), (q), (S'), (v), (g) and (s) are known,
> then its Procured Inventory Planned is:
> $$P= [c\text{-}q]\{Z+S'vg[1+s]/360\}$$

Rule-22352:
> If both (P), (Z), (q), (S'), (v), (g) and (s) are known,
> then its Current Ratio Planned is:
> $$c= P/\{Z+S'vg[1+s]/360\}+q$$

Rule-22353:
> If both (P), (c), (Z), (S'), (v), (g) and (s) are known,
> then its Quick or Acid Test Ratio Planned is:
> $$q= c\text{-}P/\{Z+S'vg[1+s]/360\}$$

Rule-22354:
> If both (V), (p), (c), (q) and (G) are known, then its
> Quoted Longterm Debt Planned is:
> $$Z= Vp/[c\text{-}q]\text{-}G$$

Rule-22355:
> If both (V), (p), (c), (q) and (Z) are known, then its
> Goods or Trade Account Payable Planned is:
> $$G= Vp/[c\text{-}q]\text{-}Z$$

Rule-22356:
> If both (Z), (p), (c), (q) and (G) are known, then its
> Variable Cost Planned is:
> $$V= [c\text{-}q][G+Z]/p$$

Steve Asikin ISBN 14: 978-1515158691, ISBN 10: 1515158691

Rule-22357:

If both (**V**), (**p**), (**c**), (**Z**) and (**G**) are known, then its Procured Inventory Days Planned:

$$p= [c-q][G+Z]/V$$

Rule-22358:

If both (**V**), (**p**), (**Z**), (**q**) and (**G**) are known, then its Current Ratio Planned is:

$$c= Vp/[G+Z]+q$$

Rule-22359:

If both (**V**), (**p**), (**c**), (**Z**) and (**G**) are known, then its Quick or Acid Test Ratio Planned is:

$$q= c-Vp/[G+Z]$$

Rule-22360:

If both (**V**), (**p**), (**c**), (**q**) and (**g**) are known, then its Zero Trade Current Debt Planned is:

$$Z= V\{p/[c-q]-g/360\}$$

Rule-22361:

If both (**Z**), (**p**), (**c**), (**q**) and (**g**) are known, then its Variable Cost Planned is:

$$V= Z/\{p/[c-q]-g/360\}$$

Rule-22362:

If both (**V**), (**p**), (**c**), (**q**) and (**Z**) are known, then its Goods or Trade Payable Days Planned:

$$g= 360\{Vp/[c-q]-Z\}/V$$

Steve Asikin ISBN 14: 978-1515158691, ISBN 10: 1515158691

Rule-22363:
>   If both (V), (Z), (c), (q) and (g) are known, then its
>   Procured Inventory Days Planned:
> $p = [c-q][Z+Vg/360]/V$

Rule-22364:
>   If both (V), (p), (Z), (q) and (g) are known, then its
>   Current Ratio Planned is:
> $c = Vp/[Z+Vg/360]+q$

Rule-22365:
>   If both (V), (p), (c), (Z) and (g) are known, then its
>   Quick or Acid Test Ratio Planned is:
> $q = c-Vp/[Z+Vg/360]$

Rule-22366:
>   If both (V), (p), (c), (q), $(\$)$, (v) and (g) are known,
>   then its Zero Trade Current Debt Planned is:
> $Z = Vp/[c-q]-\$vg/360$

Rule-22367:
>   If both (V), (p), (c), (q), (Z), (v) and (g) are known,
>   then its Sales or Revenue Planned is:
> $\$ = 360\{Vp/[c-q]-Z\}/[vg]$

Rule-22368:
>   If both (V), (p), (c), (q), $(\$)$, (Z) and (g) are known,
>   then its Variable Portion Planned is:
> $v = 360\{Vp/[c-q]-Z\}/[\$g]$

Steve Asikin ISBN 14: 978-1515158691, ISBN 10: 1515158691

<u>Rule-22369</u>:

If both **(V)**, **(p)**, **(c)**, **(q)**, **($)**, **(v)** and **(Z)** are known, then its Goods or Trade Payable Days Planned:

$$g= 360\{Vp/[c\text{-}q]\text{-}Z\}/[\$v]$$

<u>Rule-22370</u>:

If both **(Z)**, **(p)**, **(c)**, **(q)**, **($)**, **(v)** and **(g)** are known, then its Variable Cost Planned is:

$$V= [c\text{-}q][Z+\$vg/360]/p$$

<u>Rule-22371</u>:

If both **(V)**, **(Z)**, **(c)**, **(q)**, **($)**, **(v)** and **(g)** are known, then its Procured Inventory Days Planned:

$$p= [c\text{-}q][Z+\$vg/360]/V$$

<u>Rule-22372</u>:

If both **(V)**, **(p)**, **(Z)**, **(q)**, **($)**, **(v)** and **(g)** are known, then its Current Ratio Planned is:

$$c= Vp/[Z+\$vg/360]+q$$

<u>Rule-22373</u>:

If both **(V)**, **(p)**, **(c)**, **(Z)**, **($)**, **(v)** and **(g)** are known, then its Quick or Acid Test Ratio Planned is:

$$q= c\text{-}Vp/[Z+\$vg/360]$$

<u>Rule-22374</u>:

If both **(V)**, **(p)**, **(c)**, **(q)**, **($')**, **(v)**, **(g)** and **(s)** are known, then its Zero Trade Current Debt Planned is:

$$Z= Vp/[c\text{-}q]\text{-}\$'vg[1+s]/360$$

Steve Asikin ISBN 14: 978-1515158691, ISBN 10: 1515158691

Rule-22375:

If both (V), (p), (c), (q), (Z), (v), (g) and (s) are known, then its Sales Past must be:

$s' = 360\{Vp/[c-q]-Z\}/\{vg[1+s]\}$

Rule-22376:

If both (V), (p), (c), (q), (s'), (Z), (g) and (s) are known, then its Variable Portion Planned is:

$v = 360\{Vp/[c-q]-Z\}/\{s'g[1+s]\}$

Rule-22377:

If both (V), (p), (c), (q), (s'), (v), (Z) and (s) are known, then its Goods or Trade Payable Days Planned:

$g = 360\{Vp/[c-q]-Z\}/\{s'v[1+s]\}$

Rule-22378:

If both (V), (p), (c), (q), (s'), (v), (g) and (Z) are known, then its Sales Growth Planned is:

$s = 360\{Vp/[c-q]-Z\}/[s'vg]-1$

Rule-22379:

If both (Z), (p), (c), (q), (s'), (v), (g) and (s) are known, then its Variable Cost Planned is:

$V = [c-q]\{Z+s'vg[1+s]/360\}/p$

Rule-22380:

If both (V), (Z), (c), (q), (s'), (v), (g) and (s) are known, then its Procured Inventory Days Planned:

$p = [c-q]\{Z+s'vg[1+s]/360\}/V$

Steve Asikin ISBN 14: 978-1515158691, ISBN 10: 1515158691

Rule-22381:

If both (**V**), (**p**), (**Z**), (**q**), (**$'**), (**v**), (**g**) and (**s**) are known, then its Current Ratio Planned is:

$$c= Vp/\{Z+\$'vg[1+s]/360\}+q$$

Rule-22382:

If both (**V**), (**p**), (**c**), (**Z**), (**$'**), (**v**), (**g**) and (**s**) are known, then its Quick or Acid Test Ratio Planned is:

$$q= c-Vp/\{Z+\$'vg[1+s]/360\}$$

Rule-22383:

If both (**$**), (**v**), (**p**), (**c**), (**q**) and (**G**) are known, then its Zero Trade Current Debt Planned is:

$$Z= \$vp/[c-q]-G$$

Rule-22384:

If both (**$**), (**v**), (**p**), (**c**), (**q**) and (**Z**) are known, then its Goods or Trade Account Payable Planned is:

$$G= \$vp/[c-q]-Z$$

Rule-22385:

If both (**Z**), (**v**), (**p**), (**c**), (**q**) and (**G**) are known, then its Sales or Revenue Planned is:

$$\$= [c-q][G+Z]/[vp]$$

Rule-22386:

If both (**$**), (**Z**), (**p**), (**c**), (**q**) and (**G**) are known, then its Variable Portion Planned is:

$$v= [c-q][G+Z]/[\$p]$$

Steve Asikin ISBN 14: 978-1515158691, ISBN 10: 1515158691

Rule-22387:

If both **($)**, **(v)**, **(Z)**, **(c)**, **(q)** and **(G)** are known, then its Procured Inventory Days Planned:

$$p= [c\text{-}q][G\text{+}Z]/[\$v]$$

Rule-22388:

If both **($)**, **(v)**, **(p)**, **(Z)**, **(q)** and **(G)** are known, then its Current Ratio Planned is:

$$c= \$vp/[G\text{+}Z]\text{+}q$$

Rule-22389:

If both **($)**, **(v)**, **(p)**, **(c)**, **(Z)** and **(G)** are known, then its Quick or Acid Test Ratio Planned is:

$$q= c\text{-}\$vp/[G\text{+}Z]$$

Rule-22390:

If both **($)**, **(v)**, **(p)**, **(c)**, **(q)**, **(V)** and **(g)** are known, then its Zero Trade Current Debt Planned is:

$$Z= \$vp/[c\text{-}q]\text{-}Vg/360$$

Rule-22391:

If both **($)**, **(v)**, **(p)**, **(c)**, **(q)**, **(Z)** and **(g)** are known, then its Variable Cost Planned is:

$$V= 360\{\$vp/[c\text{-}q]\text{-}Z\}/g$$

Rule-22392:

If both **($)**, **(v)**, **(p)**, **(c)**, **(q)**, **(V)** and **(Z)** are known, then its Goods or Trade Payable Days Planned:

$$g= 360\{\$vp/[c\text{-}q]\text{-}Z\}/V$$

Steve Asikin ISBN 14: 978-1515158691, ISBN 10: 1515158691

Rule-22393:

If both (Z), (v), (p), (c), (q), (V) and (g) are known, then its Sales or Revenue Planned is:

$$S= [c-q][Z+Vg/360]/[vp]$$

Rule-22394:

If both (S), (Z), (p), (c), (q), (V) and (g) are known, then its Variable Portion Planned is:

$$v= [c-q][Z+Vg/360]/[Sp]$$

Rule-22395:

If both (S), (v), (Z), (c), (q), (V) and (g) are known, then its Procured Inventory Days Planned:

$$p= [c-q][Z+Vg/360]/[Sv]$$

Rule-22396:

If both (S), (v), (p), (Z), (q), (V) and (g) are known, then its Current Ratio Planned is:

$$c= Svp/[Z+Vg/360]+q$$

Rule-22397:

If both (S), (v), (p), (c), (Z), (V) and (g) are known, then its Quick or Acid Test Ratio Planned is:

$$q= c-Svp/[Z+Vg/360]$$

Rule-22398:

If both (S), (v), (p), (c), (q) and (g) are known, then its Zero Tade Current Debt Planned is:

$$Z= Sv\{p/[c-q]-g/360\}$$

Steve Asikin ISBN 14: 978-1515158691, ISBN 10: 1515158691

Rule-22399:

If both (Z), (v), (p), (c), (q) and (g) are known, then its Sales or Revenue Planned is:

$S= Z/(v\{p/[c-q]-g/360\})$

Rule-22400:

If both (S), (Z), (p), (c), (q) and (g) are known, then its Variable Portion Planned is:

$v= Z/(S\{p/[c-q]-g/360\})$

Rule-22401:

If both (S), (v), (p), (c), (q) and (Z) are known, then its Goods or Trade Payable Days Planned:

$g= 360\{Svp/[c-q]-Z\}/[Sv]$

Rule-22402:

If both (S), (v), (Z), (c), (q) and (g) are known, then its Procured Inventory Days Planned:

$p= [c-q][Z+Svg/360]/[Sv]$

Rule-22403:

If both (S), (v), (p), (Z), (q) and (g) are known, then its Current Ratio Planned is:

$c= Svp/[Z+Svg/360]+q$

Rule-22404:

If both (S), (v), (p), (c), (Z) and (g) are known, then its Quick or Acid Test Ratio Planned is:

$q= c-Svp/[Z+Svg/360]$

Steve Asikin ISBN 14: 978-1515158691, ISBN 10: 1515158691

Math Fin Law 6, Mathematical Financial Law, Public Listed Firm Rule No.19905-23237

Rule-22405:

If both (**$**), (**v**), (**p**), (**c**), (**q**), (**$'**), (**s**) and (**g**) are known, then its Zero Trade Current Debt Debt Planned is:

$$Z = \$vp/[c\text{-}q] - \$'vg[1+s]/360$$

Rule-22406:

If both (**$**), (**v**), (**p**), (**c**), (**q**), (**Z**), (**s**) and (**g**) are known, then its Sales Past must be:

$$\$' = 360\{\$vp/[c\text{-}q] - Z\}/\{vg[1+s]\}$$

Rule-22407:

If both (**$**), (**Z**), (**p**), (**c**), (**q**), (**$'**), (**s**) and (**g**) are known, then its Variable Portion Planned is:

$$v = Z/\{\$p/[c\text{-}q] - \$'g[1+s]/360\}$$

Rule-22408:

If both (**$**), (**v**), (**p**), (**c**), (**q**), (**$'**), (**s**) and (**Z**) are known, then its Goods or Trade Payable Days Planned:

$$g = 360\{\$vp/[c\text{-}q] - Z\}/\{\$'v[1+s]\}$$

Rule-22409:

If both (**$**), (**v**), (**p**), (**c**), (**q**), (**Z**), (**$'**) and (**g**) are known, then its Sales Growth Planned is:

$$s = 360\{\$vp/[c\text{-}q] - Z\}/[\$'vg] - 1$$

Rule-22410:

If both (**Z**), (**v**), (**p**), (**c**), (**q**), (**$'**) and (**g**) are known, then its Sales or Revenue Planned is:

$$\$ = [c\text{-}q]\{Z + \$'vg[1+s]/360\}/[vp]$$

564
Steve Asikin ISBN 14: 978-1515158691, ISBN 10: 1515158691

Rule-22411:
    If both **($)**, **(v)**, **(Z)**, **(c)**, **(q)**, **($')** and **(g)** are known,
    then its Procured Inventory Days Planned:
$$p = [c-q]\{Z+\$'vg[1+s]/360\}/[\$v]$$

Rule-22412:
    If both **($)**, **(v)**, **(p)**, **(Z)**, **(q)**, **($')** and **(g)** are known,
    then its Current Ratio Planned is:
$$c = \$vp/\{Z+\$'vg[1+s]/360\}+q$$

Rule-22413:
    If both **($)**, **(v)**, **(p)**, **(c)**, **(Z)**, **($')** and **(g)** are known,
    then its Quick or Acid Test Ratio Planned is:
$$q = c-\$vp/\{Z+\$'vg[1+s]/360\}$$

Rule-22414:
    If both **($')**, **(v)**, **(s)**, **(p)**, **(c)**, **(q)** and **(G)** are known,
    then its Zero Trade Current Debt Planned is:
$$Z = \$'vp[1+s]/[c-q]-G$$

Rule-22415:
    If both **($')**, **(v)**, **(s)**, **(p)**, **(c)**, **(q)** and **(Z)** are known,
    then its Goods or Trade Account Payable Planned is:
$$G = \$'vp[1+s]/[c-q]-Z$$

Rule-22416:
    If both **(Z)**, **(v)**, **(s)**, **(p)**, **(c)**, **(q)** and **(G)** are known,
    then its Sales Past must be:
$$\$' = [c-q][G+Z]/\{vp[1+s]\}$$

Steve Asikin ISBN 14: 978-1515158691, ISBN 10: 1515158691

Rule-22417:

If both (S'), (Z), (s), (p), (c), (q) and (G) are known, then its Variable Portion Planned is:

$$v= [c-q][G+Z]/\{S'p[1+s]\}$$

Rule-22418:

If both (S'), (v), (s), (Z), (c), (q) and (G) are known, then its Procured Inventory Days Planned:

$$p= [c-q][G+Z]/\{S'v[1+s]\}$$

Rule-22419:

If both (S'), (v), (Z), (p), (c), (q) and (G) are known, then its Sales Growth Planned is:

$$s= [c-q][G+Z]/[S'vp]-1$$

Rule-22420:

If both (S'), (v), (s), (p), (Z), (q) and (G) are known, then its Current Ratio Planned is:

$$c= S'vp[1+s]/[G+Z]+q$$

Rule-22421:

If both (S'), (v), (s), (p), (c), (Z) and (G) are known, then its Quick or Acid Test Ratio Planned is:21

$$q= c- S'vp[1+s]/[G+Z]$$

Rule-22422:

If both (S'), (v), (s), (p), (c), (q), (V) and (g) are known, then its Zero Trade Current Debt Planned is:

$$Z= S'vp[1+s]/[c-q]-Vg/360$$

Steve Asikin ISBN 14: 978-1515158691, ISBN 10: 1515158691

Rule-22423:

If both $(\textbf{S'})$, $(\textbf{v})$, $(\textbf{s})$, $(\textbf{p})$, $(\textbf{c})$, $(\textbf{q})$, $(\textbf{Z})$ and $(\textbf{g})$ are known, then its Variable Cost Planned is:

$V = 360\{\textbf{S'vp}[1+\textbf{s}]/[\textbf{c-q}]-\textbf{Z}\}/\textbf{g}$

Rule-22424:

If both $(\textbf{S'})$, $(\textbf{v})$, $(\textbf{s})$, $(\textbf{p})$, $(\textbf{c})$, $(\textbf{q})$, $(\textbf{V})$ and $(\textbf{Z})$ are known, then its Goods or Trade Payable Days Planned:

$g = 360\{\textbf{S'vp}[1+\textbf{s}]/[\textbf{c-q}]-\textbf{Z}\}/\textbf{V}$

Rule-22425:

If both $(\textbf{Z})$, $(\textbf{v})$, $(\textbf{s})$, $(\textbf{p})$, $(\textbf{c})$, $(\textbf{q})$, $(\textbf{V})$ and $(\textbf{g})$ are known, then its Sales Past must be:

$S' = [\textbf{c-q}][\textbf{Z+Vg}/360]/\{\textbf{vp}[1+\textbf{s}]\}$

Rule-22426:

If both $(\textbf{S'})$, $(\textbf{Z})$, $(\textbf{s})$, $(\textbf{p})$, $(\textbf{c})$, $(\textbf{q})$, $(\textbf{V})$ and $(\textbf{g})$ are known, then its Variable Portion Planned is:

$v = [\textbf{c-q}][\textbf{Z+Vg}/360]/\{\textbf{S'p}[1+\textbf{s}]\}$

Rule-22427:

If both $(\textbf{S'})$, $(\textbf{v})$, $(\textbf{s})$, $(\textbf{Z})$, $(\textbf{c})$, $(\textbf{q})$, $(\textbf{V})$ and $(\textbf{g})$ are known, then its Procured Inventory Days Planned:

$p = [\textbf{c-q}][\textbf{Z+Vg}/360]/\{\textbf{S'v}[1+\textbf{s}]\}$

Rule-22428:

If both $(\textbf{S'})$, $(\textbf{v})$, $(\textbf{Z})$, $(\textbf{p})$, $(\textbf{c})$, $(\textbf{q})$, $(\textbf{V})$ and $(\textbf{g})$ are known, then its Sales Growth Planned is:

$s = \textbf{S}/\textbf{S'}-1$

or it can also be found as

$s = [\textbf{c-q}][\textbf{Z+Vg}/360]/[\textbf{S'vp}]-1$

Steve Asikin ISBN 14: 978-1515158691, ISBN 10: 1515158691

Rule-22429:
If both $(\mathbf{S'})$, $(\mathbf{v})$, $(\mathbf{s})$, $(\mathbf{p})$, $(\mathbf{Z})$, $(\mathbf{q})$, $(\mathbf{V})$ and $(\mathbf{g})$ are known, then its Current Ratio Planned is:
$$c = S'vp[1+s]/[Z+Vg/360]+q$$

Rule-22430:
If both $(\mathbf{S'})$, $(\mathbf{v})$, $(\mathbf{s})$, $(\mathbf{p})$, $(\mathbf{c})$, $(\mathbf{Z})$, $(\mathbf{V})$ and $(\mathbf{g})$ are known, then its Quick or Acid Test Ratio Planned is:
$$q = c - S'vp[1+s]/[Z+Vg/360]$$

Rule-22431:
If both $(\mathbf{S'})$, $(\mathbf{v})$, $(\mathbf{p})$, $(\mathbf{s})$, $(\mathbf{c})$, $(\mathbf{q})$, $(\mathbf{S})$ and $(\mathbf{g})$ are known, then its Zero Trade Current Debt Planned is:
$$Z = S'vp[1+s]/[c-q]-Svg/360\}$$

Rule-22432:
If both $(\mathbf{S'})$, $(\mathbf{v})$, $(\mathbf{p})$, $(\mathbf{s})$, $(\mathbf{c})$, $(\mathbf{q})$, $(\mathbf{Z})$ and $(\mathbf{g})$ are known, then its Quoted Longterm Debt Planned is:
$$S = 360(S'vp[1+s]/[c-q]-Z)/[vg]$$

Rule-22433:
If both $(\mathbf{S'})$, $(\mathbf{Z})$, $(\mathbf{p})$, $(\mathbf{s})$, $(\mathbf{c})$, $(\mathbf{q})$, $(\mathbf{S})$ and $(\mathbf{g})$ are known, then its Variable Portion Planned is:
$$v = Z/\{S'p[1+s]/[c-q]-Sg/360\}$$

Rule-22434:
If both $(\mathbf{S'})$, $(\mathbf{v})$, $(\mathbf{p})$, $(\mathbf{s})$, $(\mathbf{c})$, $(\mathbf{q})$, $(\mathbf{S})$ and $(\mathbf{Z})$ are known, then its Goods or Trade Payable Planned is:
$$g = 360(S'vp[1+s]/[c-q]-Z)/[Sv]$$

Steve Asikin ISBN 14: 978-1515158691, ISBN 10: 1515158691

Rule-22435:

If both **(Z)**, **(v)**, **(p)**, **(s)**, **(c)**, **(q)**, **($)** and **(g)** are known, then its Sale Past must be:

$$S' = [c\text{-}q][Z + Svg/360]/\{vp[1+s]\}$$

Rule-22436:

If both **(S')**, **(v)**, **(Z)**, **(s)**, **(c)**, **(q)**, **($)** and **(g)** are known, then its Procured Inventory Days Planned:

$$p = [c\text{-}q][Z + Svg/360]/\{S'v[1+s]\}$$

Rule-22437:

If both **(S')**, **(v)**, **(p)**, **(Z)**, **(c)**, **(q)**, **($)** and **(g)** are known, then its Sales Growth Planned is:

$$s = S/S' - 1$$

or it can also be found as

$$s = [c\text{-}q][Z + Svg/360]/[S'vp] - 1$$

Rule-22438:

If both **(S')**, **(v)**, **(p)**, **(s)**, **(Z)**, **(q)**, **($)** and **(g)** are known, then its Current Ratio Planned is:

$$c = S'vp[1+s]/[Z + Svg/360] + q$$

Rule-22439:

If both **(S')**, **(v)**, **(p)**, **(s)**, **(c)**, **(Z)**, **($)** and **(g)** are known, then its Quick or Acid Test Ratio Planned is:

$$q = c - S'vp[1+s]/[Z + Svg/360]$$

Rule-22440:

If both **(S')**, **(v)**, **(s)**, **(p)**, **(c)**, **(q)** and **(g)** are known, then its Zero Trade Current Debt Planned is:

$$Z = S'v[1+s]\{p/[c\text{-}q] - g/360\}$$

Steve Asikin ISBN 14: 978-1515158691, ISBN 10: 1515158691

Rule-22441:

If both $(\mathbf{Z})$, $(\mathbf{v})$, $(\mathbf{s})$, $(\mathbf{p})$, $(\mathbf{c})$, $(\mathbf{q})$ and $(\mathbf{g})$ are known, then its Sales Past must be:

$\mathbf{S'} = \mathbf{Z}/(\mathbf{v}[1+\mathbf{s}]\{\mathbf{p}/[\mathbf{c}-\mathbf{q}]-\mathbf{g}/360\})$

Rule-22442:

If both $(\mathbf{S'})$, $(\mathbf{Z})$, $(\mathbf{s})$, $(\mathbf{p})$, $(\mathbf{c})$, $(\mathbf{q})$ and $(\mathbf{g})$ are known, then its Variable Portion Planned is:

$\mathbf{v} = \mathbf{Z}/(\mathbf{S'}[1+\mathbf{s}]\{\mathbf{p}/[\mathbf{c}-\mathbf{q}]-\mathbf{g}/360\})$

Rule-22443:

If both $(\mathbf{S'})$, $(\mathbf{v})$, $(\mathbf{s})$, $(\mathbf{p})$, $(\mathbf{c})$, $(\mathbf{q})$ and $(\mathbf{Z})$ are known, then its Goods or Trade Payable Days Planned:

$\mathbf{g} = 360\{\mathbf{S'vp}[1+\mathbf{s}]/[\mathbf{c}-\mathbf{q}]-\mathbf{Z}\}/\{\mathbf{S'v}[1+\mathbf{s}]\}$

Rule-22444:

If both $(\mathbf{S'})$, $(\mathbf{v})$, $(\mathbf{Z})$, $(\mathbf{p})$, $(\mathbf{c})$, $(\mathbf{q})$ and $(\mathbf{g})$ are known, then its Sales Growth Planned is:

$\mathbf{s} = \mathbf{Z}/(\mathbf{S'v}\{\mathbf{p}/[\mathbf{c}-\mathbf{q}]-\mathbf{g}/360\})-1$

Rule-22445:

If both $(\mathbf{S'})$, $(\mathbf{v})$, $(\mathbf{s})$, $(\mathbf{Z})$, $(\mathbf{c})$, $(\mathbf{q})$ and $(\mathbf{g})$ are known, then its Procured Inventory Days Planned:

$\mathbf{p} = [\mathbf{c}-\mathbf{q}]\{\mathbf{Z}+\mathbf{S'vg}[1+\mathbf{s}]/360\}/[\mathbf{Sv}]$

Rule-22446:

If both $(\mathbf{S'})$, $(\mathbf{v})$, $(\mathbf{s})$, $(\mathbf{p})$, $(\mathbf{Z})$, $(\mathbf{q})$ and $(\mathbf{g})$ are known, then its Current Ratio Planned is:

$\mathbf{c} = \mathbf{S'vp}[1+\mathbf{s}]/\{\mathbf{Z}+\mathbf{S'vg}[1+\mathbf{s}]/360\}+\mathbf{q}$

Steve Asikin ISBN 14: 978-1515158691, ISBN 10: 1515158691

Rule-22447:

If both (**$'**), (**v**), (**s**), (**p**), (**c**), (**Z**) and (**g**) are known, then its Quick or Acid Test Ratio Planned is:

$$q= c- \$'vp[1+s]/\{Z+\$'vg[1+s]/360\}$$

Steve Asikin ISBN 14: 978-1515158691, ISBN 10: 1515158691

CHAPTER-22:

J= JOB or Trade Account Receivable Optimization,

Rule-22448:
If both (**$**) and (**j**) are known, then its Job or Trade Account Receivable Planned is:
$$J= \$j/360$$

Rule-22449:
If both (**J**) and (**j**) are known, then its Sales or Revenue Planned is:
$$\$=360J/j$$

Rule-22450:
If both (**J**) and (**$**) are known, then its Job or Trade Account Receivable Days Planned:
$$j=360J/\$$$

Rule-22451:
If both (**$'**), (**j**) and (**s**) are known, then its Job or Trade Account Receivable Planned is:
$$J= \$'j[1+s]/360$$

Rule-22452:
If both (**J**), (**j**) and (**s**) are known, then its Sales Past must be:
$$\$'= 360J/\{j[1+s]\}$$

Steve Asikin ISBN 14: 978-1515158691, ISBN 10: 1515158691

Rule-22453:

    If both **($')**, **(J)** and **(s)** are known, then its Job or Trade Account Receivable Days Planned:

$$j = 360J/\{\$'[1+s]\}$$

Rule-22454:

    If both **($')**, **(j)** and **(J)** are known, then its Sales Growth Planned is:

$$s = 360J/[\$'j]-1$$

Steve Asikin ISBN 14: 978-1515158691, ISBN 10: 1515158691

CHAPTER-23:

C= CURRENT ASSETS, Optimization.

Rule-22455:
 If both (**X**) and (**c**) are known, then its Current Asset
 Planned is:
 $$C= Xc$$

Rule-22456:
 If both (**C**) and (**c**) are known, then its Xpress or
 Current Debt Planned is:
 $$X= C/c$$

Rule-22457:
 If both (**X**) and (**C**) are known, then its Current Ratio
 Planned is:
 $$c= X/C$$

Rule-22458:
 If both (**P**), (**c**) and (**q**) are known, then its Current
 Asset Planned is:
 $$C= Pc/[c\text{-}q]$$

Rule-22459:
 If both (**C**), (**c**) and (**q**) are known, then its Procured
 Inventory Planned is:
 $$P= C[c\text{-}q]/c$$

Steve Asikin ISBN 14: 978-1515158691, ISBN 10: 1515158691

Rule-22460:

If both (**C**), (**q**) and (**P**) are known, then its Current Ratio Planned is:

$$c= Cq/[C-P]$$

Rule-22461:

If both (**c**), (**C**) and (**P**) are known, then its Quick or Acid Test Ratio Planned is:

$$q= c-Pc/C$$

Rule-22462:

If both (**V**), (**p**), (**c**) and (**q**) are known, then its Current Asset Planned is:

$$C= Vpc/\{360[c-q]\}$$

Rule-22463:

If both (**C**), (**p**), (**c**) and (**q**) are known, then its Variable Cost Planned is:

$$V= 360C[c-q]/[pc]$$

Rule-22464:

If both (**V**), (**C**), (**c**) and (**q**) are known, then its Procured Inventory Days Planned:

$$p= 360C[c-q]/[Vc]$$

Rule-22465:

If both (**V**), (**p**), (**C**) and (**q**) are known, then its Sales Growth Planned is:

$$c= 360Cq/[360C-Vp]$$

Steve Asikin ISBN 14: 978-1515158691, ISBN 10: 1515158691

<u>Rule-22466</u>:

If both (**V**), (**p**), (**c**) and (**C**) are known, then its Quick or Acid Test Ratio Planned is:

$$q = c - Vpc/[360C]$$

<u>Rule-22467</u>:

If both (**$**), (**v**), (**p**), (**c**) and (**q**) are known, then its Current Asset Planned is:

$$C = \$vpc/\{360[c-q]\}$$

<u>Rule-22468</u>:

If both (**C**), (**v**), (**p**), (**c**) and (**q**) are known, then its Sales or Revenue Planned is:

$$\$ = 360[c-q]C/[vpc]$$

<u>Rule-22469</u>:

If both (**$**), (**C**), (**p**), (**c**) and (**q**) are known, then its Variable Portion Planned is:

$$v = 360[c-q]C/[\$pc]$$

<u>Rule-22470</u>:

If both (**$**), (**v**), (**C**), (**c**) and (**q**) are known, then its Procured Inventory Days Planned is:

$$p = 360[c-q]C/[\$vc]$$

<u>Rule-22471</u>:

If both (**$**), (**v**), (**p**), (**c**) and (**C**) are known, then its Quick or Acid Test Ratio Planned is:

$$q = c - \$vpc/[360C]$$

Steve Asikin ISBN 14: 978-1515158691, ISBN 10: 1515158691

Rule-22472:

If both ($), (v), (p), (C) and (q) are known, then its Current Ratio Planned is:
$$c= 360Cq/[360C- \$up]$$

Rule-22473:

If both ($'), (v), (p), (c), (s) and (q) are known, then its Current Asset Planned is:
$$C= \$upc[1+s]/\{360[c-q]\}$$

Rule-22474:

If both (C), (v), (p), (c), (s) and (q) are known, then its Sales Past must be:
$$\$'= 360[c-q]C/\{upc[1+s]\}$$

Rule-22475:

If both ($'), (C), (p), (c), (s) and (q) are known, then its Variable Portion Planned is:
$$v= 360[c-q]C/\{\$'pc[1+s]\}$$

Rule-22476:

If both ($'), (v), (p), (c), (C) and (q) are known, then its Sales Growth is:
$$s= 360[c-q]C/[\$'upc]-1$$

Rule-22477:

If both ($'), (v), (p), (C), (s) and (q) are known, then its Current Ratio Planned is:
$$c= 360Cq/\{360C-\$'up[1+s]\}$$

Steve Asikin ISBN 14: 978-1515158691, ISBN 10: 1515158691

Rule-22478:

    If both $(\textbf{S'})$, $(\textbf{v})$, $(\textbf{p})$, $(\textbf{c})$, $(\textbf{s})$ and $(\textbf{C})$ are known, then its Quick or Acid Test Ratio Planned is:

    $q = c\text{-}S'vpc[1+s]/[360\ C]$

Rule-22479:

    If both $(\textbf{S'})$, $(\textbf{v})$, $(\textbf{C})$, $(\textbf{c})$, $(\textbf{s})$ and $(\textbf{q})$ are known, then its Procured Inventory Days Planned:

    $p = 360C[c\text{-}q]/\{S'vc[1+s]\}$

Steve Asikin ISBN 14: 978-1515158691, ISBN 10: 1515158691

CHAPTER-24:

K=KIND of CASH, Optimization,

Rule-22480:
> If both (**C**), (**J**) and (**P**) are known, then its Kind of Cash Planned is:
> **K= C-J-P**

Rule-22481:
> If both (**K**), (**J**) and (**P**) are known, then its Current Asset Planned is:
> **C= K+J+P**

Rule-22482:
> If both (**C**), (**K**) and (**P**) are known, then its Job or Trade Account Receivable Planned is:
> **J= C-K-P**

Rule-22483:
> If both (**C**), (**J**) and (**K**) are known, then its Procured Inventory Planned is:
> **P= C-J-K**

Rule-22484:
> If both (**C**), (**J**), (**V**) and (**p**) are known, then its Kind of Cash Planned is:
> **K= C-J-Vp**/360

Steve Asikin ISBN 14: 978-1515158691, ISBN 10: 1515158691

Rule-22485:
>   If both (K), (J), (V) and (p) are known, then its
>   Current Asset Planned is:
> $C= K+J+Vp/360$

Rule-22486:
>   If both (C), (K), (V) and (p) are known, then its Job or
>   Trade Account Receivable Planned is:
> $J= C-K-Vp/360$

Rule-22487:
>   If both (C), (J), (K) and (p) are known, then its
>   Variable Cost Planned is:
> $V= 360[C-J-K]/p$

Rule-22488:
>   If both (C), (J), (V) and (K) are known, then its
>   Procured Inventory Days Planned:
> $p= 360[C-J-K]/V$

Rule-22489:
>   If both (C), (J), $(\$)$, (v) and (p) are known, then its
>   Kind of Cash Planned is:
> $K= C-J-\$vp/360$

Rule-22490:
>   If both (K), (J), $(\$)$, (v) and (p) are known, then its
>   Current Asset Planned is:
> $C= K+J+\$vp/360$

Steve Asikin ISBN 14: 978-1515158691, ISBN 10: 1515158691

Rule-22491:

If both (**C**), (**K**), (**$**), (**v**) and (**p**) are known, then its Job or Trade Account Raceivable Days Planned:
$$J= C-K-Svp/360$$

Rule-22492:

If both (**C**), (**J**), (**K**), (**v**) and (**p**) are known, then its Sales or Revenue Planned is:
$$S= 360[C-J-K]/[vp]$$

Rule-22493:

If both (**C**), (**J**), (**$**), (**K**) and (**p**) are known, then its Variable Portion Planned is:
$$v= 360[C-J-K]/[Sp]$$

Rule-22494:

If both (**C**), (**J**), (**$**), (**v**) and (**K**) are known, then its Procured Inventory Days Planned:
$$p= 360[C-J-K]/[Sv]$$

Rule-22495:

If both (**C**), (**J**), (**$'**), (**v**), (**p**) and (**s**) are known, then its Kind of Cash Planned is:
$$K= C-J-S'vp[1+s]/360$$

Rule-22496:

If both (**K**), (**J**), (**$'**), (**v**), (**p**) and (**s**) are known, then its Current Asset Planned Planned is:
$$C= K+J+S'vp[1+s]/360$$

581

Steve Asikin ISBN 14: 978-1515158691, ISBN 10: 1515158691

Rule-22497:

If both (**C**), (**K**), (**S'**), (**v**), (**p**) and (**s**) are known, then its Job or Trade Account Receivable Planned is:

$$J= C-K-S'vp[1+s]/360$$

Rule-22498:

If both (**C**), (**J**), (**K**), (**v**), (**p**) and (**s**) are known, then its Sales Past must be:

$$S'= 360[C-J-K]/\{vp[1+s]\}$$

Rule-22499:

If both (**C**), (**J**), (**S'**), (**K**), (**p**) and (**s**) are known, then its Variable Planned is:

$$v= 360[C-J-K]/\{S'p[1+s]\}$$

Rule-22500:

If both (**C**), (**J**), (**S'**), (**v**), (**K**) and (**s**) are known, then its Procured Inventory Days Planned:

$$p= 360[C-J-K]/\{S'v[1+s]\}$$

Rule-22501:

If both (**C**), (**J**), (**S'**), (**v**), (**p**) and (**K**) are known, then its Sales Growth Planned is:

$$s= 360[C-J-K]/[S'vp]-1$$

Rule-22502:

If both (**C**), (**S**), (**j**) and (**P**) are known, then its Kind of Cash Planned is:

$$K= C-Sj/360-P$$

Steve Asikin ISBN 14: 978-1515158691, ISBN 10: 1515158691

Rule-22503:

If both **(K)**, **($)**, **(j)** and **(P)** are known, then its Current Ratio Planned is:

$$C = K + \$j/360 + P$$

Rule-22504:

If both **(C)**, **(K)**, **(j)** and **(P)** are known, then its Sales or Revenue Planned is:

$$\$ = 360[C - K - P]/j$$

Rule-22505:

If both **(C)**, **($)**, **(K)** and **(P)** are known, then its Job or Trade Receivable Days Planned:

$$j = 360[C - K - P]/\$$$

Rule-22506:

If both **(C)**, **($)**, **(j)** and **(K)** are known, then its Procured Inventory Planned is:

$$P = C - \$j/360 - K$$

Rule-22507:

If both **(C)**, **($)**, **(j)**, **(V)** and **(p)** are known, then its Kind of Cash Planned is:

$$K = C - \$j/360 - Vp/360$$

Rule-22508:

If both **(K)**, **($)**, **(j)**, **(V)** and **(p)** are known, then its Current Asset Planned is:

$$C = K + \$j/360 + Vp/360$$

Steve Asikin ISBN 14: 978-1515158691, ISBN 10: 1515158691

Rule-22509:

If both (**C**), (**K**), (**j**), (**V**) and (**p**) are known, then its Sales or Revenue Planned is:

$$S= \{360[C\text{-}K]\text{-}Vp\}/j$$

Rule-22510:

If both (**C**), (**S**), (**K**), (**V**) and (**p**) are known, then its Job or Trade Receivable Days Planned:

$$j= \{360[C\text{-}K]\text{-}Vp\}/S$$

Rule-22511:

If both (**C**), (**S**), (**j**), (**K**) and (**p**) are known, then its Variable Cost Planned is:

$$V= 360[C\text{-}Sj/360\text{-}K]/p$$

Rule-22512:

If both (**C**), (**S**), (**j**), (**V**) and (**K**) are known, then its Procured Inventory Days Planned:

$$p= 360[C\text{-}Sj/360\text{-}K]/V$$

Rule-22513:

If both (**C**), (**S**), (**j**), (**v**) and (**p**) are known, then its Kind of Cash Planned is:

$$K= C\text{-}Sj/360\text{-}Svp/360$$

Rule-22514:

If both (**K**), (**S**), (**j**), (**v**) and (**p**) are known, then its Current Asset Planned is:

$$C= K+Sj/360+Svp/360$$

Steve Asikin ISBN 14: 978-1515158691, ISBN 10: 1515158691

Rule-22515:

   If both **(C)**, **(K)**, **(j)**, **(v)** and **(p)** are known, then its Sales or Revenue Planned is:

$$\$= 360[\mathbf{C+K}]/[\mathbf{j+vp}]$$

Rule-22516:

   If both **(C)**, **($)**, **(K)**, **(v)** and **(p)** are known, then its Job or Trade Receivable Days Planned:

$$\mathbf{j}= \{360[\mathbf{C-K}]-\mathbf{\$vp}\}/\$$$

Rule-22517:

   If both **(C)**, **($)**, **(j)**, **(K)** and **(p)** are known, then its Variable Portion Planned is:

$$\mathbf{v}= 360[\mathbf{C}-\$\mathbf{j}/360-\mathbf{K}]/[\$\mathbf{p}]$$

Rule-22518:

   If both **(C)**, **($)**, **(j)**, **(v)** and **(K)** are known, then its Procured Inventory Days Planned:

$$\mathbf{p}= 360[\mathbf{C}-\$\mathbf{j}/360-\mathbf{K}]/[\$\mathbf{v}]$$

Rule-22519:

   If both **(C)**, **($)**, **(j)**, **($')**, **(v)**, **(p)** and **(s)** are known, then its Kind of Cash Planned is:

$$\mathbf{K}= \mathbf{C}-\$\mathbf{j}/360-\$'\mathbf{vp}[1+\mathbf{s}]/360$$

Rule-22520:

   If both **(K)**, **($)**, **(j)**, **($')**, **(v)**, **(p)** and **(s)** are known, then its Current Asset Planned is:

$$\mathbf{C}= \mathbf{K}+\$\mathbf{j}/360+\$'\mathbf{vp}[1+\mathbf{s}]/360$$

Steve Asikin ISBN 14: 978-1515158691, ISBN 10: 1515158691

Rule-22521:
If both **(C)**, **(K)**, **(j)**, **(S')**, **(v)**, **(p)** and **(s)** are known, then its Sales or Revenue Planned is:
$$S= \{360[C\text{-}K]\text{-}S'vp[1+s]\}/j$$

Rule-22522:
If both **(C)**, **(S)**, **(K)**, **(S')**, **(v)**, **(p)** and **(s)** are known, then its Job or Trade Receivable Days Planned is:
$$j= \{360[C\text{-}K]\text{-}S'vp[1+s]\}/S$$

Rule-22523:
If both **(C)**, **(S)**, **(j)**, **(K)**, **(v)**, **(p)** and **(s)** are known, then its Sales Past must be:
$$S'= 360[C\text{-}Sj/360\text{-}K]/\{vp[1+s]\}$$

Rule-22524:
If both **(C)**, **(S)**, **(j)**, **(S')**, **(K)**, **(p)** and **(s)** are known, then its Variable Portion Planned is:
$$v= 360[C\text{-}Sj/360\text{-}K]/\{S'p[1+s]\}$$

Rule-22525:
If both **(C)**, **(S)**, **(j)**, **(S')**, **(v)**, **(K)** and **(s)** are known, then its Procured Inventory Days Planned:
$$p= 360[C\text{-}Sj/360\text{-}K]/\{S'v[1+s]\}$$

Rule-22526:
If both **(C)**, **(S)**, **(j)**, **(S')**, **(v)**, **(p)** and **(K)** are known, then its Sales Growth Planned is:
$$s= S/S'\text{-}1$$
or it can also be found as
$$s= 360[C\text{-}Sj/360\text{-}K]/[S'vp]\text{-}1$$

Steve Asikin ISBN 14: 978-1515158691, ISBN 10: 1515158691

Rule-22527:

If both (**C**), (**S'**), (**j**), (**s**) and (**P**) are known, then its Kind of Cash Planned is:
$$K= C-S'j[1+s]/360-P$$

Rule-22528:

If both (**K**), (**S'**), (**j**), (**s**) and (**P**) are known, then its Current Asset Planned is:
$$C= K+S'j[1+s]/360+P$$

Rule-22529:

If both (**C**), (**K**), (**j**), (**s**) and (**P**) are known, then its Sales Past must be:
$$S'= 360[C-K-P]/\{j[1+s]\}$$

Rule-22530:

If both (**C**), (**S'**), (**K**), (**s**) and (**P**) are known, then its Job or Tade Receivable Days Planned:
$$j= 360[C-K-P]/\{S'[1+s]\}$$

Rule-22531:

If both (**C**), (**S'**), (**j**), (**K**) and (**P**) are known, then its Sales Growth Planned is:
$$s= 360[C-K-P]/[S'j]-1$$

Rule-22532:

If both (**C**), (**S'**), (**j**), (**s**) and (**K**) are known, then its Procured Inventory Planned is:
$$P= C-S'j[1+s]/360-K$$

587

Steve Asikin ISBN 14: 978-1515158691, ISBN 10: 1515158691

Rule-22533:
> If both (**C**), (**$'**), (**j**), (**s**), (**V**) and (**p**) are known, then its Kind of Cash Planned is:
>
> $\mathbf{K} = \mathbf{C} - \mathbf{\$'j}[1+\mathbf{s}]/360 - \mathbf{Vp}/360$

Rule-22534:
> If both (**K**), (**$'**), (**j**), (**s**), (**V**) and (**p**) are known, then its Current Asset Planned is:
>
> $\mathbf{C} = \mathbf{K} + \mathbf{\$'j}[1+\mathbf{s}] + \mathbf{Vp}/360$

Rule-22535:
> If both (**C**), (**K**), (**j**), (**s**), (**V**) and (**p**) are known, then its Sales Past Planned:
>
> $\mathbf{\$'} = \{360[\mathbf{C}-\mathbf{K}] - \mathbf{Vp}\} / \{\mathbf{j}[1+\mathbf{s}]\}$

Rule-22536:
> If both (**C**), (**$'**), (**K**), (**s**), (**V**) and (**p**) are known, then its Job or Trade Receivable Days Planned:
>
> $\mathbf{j} = \{360[\mathbf{C}-\mathbf{K}] - \mathbf{Vp}\} / \{\mathbf{\$'}[1+\mathbf{s}]\}$

Rule-22537:
> If both (**C**), (**$'**), (**j**), (**K**), (**V**) and (**p**) are known, then its Sales Growth Planned is:
>
> $\mathbf{s} = \{360[\mathbf{C}-\mathbf{K}] - \mathbf{Vp}\} / [\mathbf{\$'j}] - 1$

Rule-22538:
> If both (**C**), (**$'**), (**j**), (**s**), (**Q**) and (**p**) are known, then its Variable Cost Planned is:
>
> $\mathbf{V} = 360[\mathbf{C} - \mathbf{\$'j}[1+\mathbf{s}]/360 - \mathbf{K}]/\mathbf{p}$

Steve Asikin ISBN 14: 978-1515158691, ISBN 10: 1515158691

Rule-22539:
   If both **(C)**, **(S')**, **(j)**, **(s)**, **(V)** and **(K)** are known, then
   its Procured Inventory Days Planned:
   $p = 360[C - S'j[1+s]/360 - K]/V$

Rule-22540:
   If both **(C)**, **(S')**, **(j)**, **(s)**, **(S)**, **(v)** and **(p)** are known,
   then its Kind of Cash Planned is:
   $K = C - S'j[1+s]/360 - Svp/360$

Rule-22541:
   If both **(K)**, **(S')**, **(j)**, **(s)**, **(S)**, **(v)** and **(p)** are known,
   then its Current Asset Planned is:
   $C = K + S'j[1+s]/360 + Svp/360$

Rule-22542:
   If both **(C)**, **(K)**, **(j)**, **(s)**, **(S)**, **(v)** and **(p)** are known,
   then its Sales Past must be:
   $S' = \{360[C-K] - Svp\}/\{j[1+s]\}$

Rule-22543:
   If both **(C)**, **(S')**, **(K)**, **(s)**, **(S)**, **(v)** and **(p)** are known,
   then its Job or Trade Receivable Days Planned:
   $j = \{360[C-K] - Svp\}/\{S'[1+s]\}$

Rule-22544:
   If both **(C)**, **(S')**, **(j)**, **(K)**, **(S)**, **(v)** and **(p)** are known,
   then its Sales Growth Planned is:
   $s = S/S' - 1$

   or it can also be found as
   $s = \{360[C-K] - Svp\}/[S'j] - 1$

Steve Asikin ISBN 14: 978-1515158691, ISBN 10: 1515158691

Rule-22545:

    If both (**C**), (**$'**), (**j**), (**s**), (**K**), (**v**) and (**p**) are known, then its Sales or Revenue Planned is:

$$\$= 360[\textbf{C-}\$\textbf{'j}[1+\textbf{s}]/360\textbf{-K}]/[\textbf{vp}]$$

Rule-22546:

    If both (**C**), (**$'**), (**j**), (**s**), (**$**), (**K**) and (**p**) are known, then its Variable Portion Planned is:

$$\textbf{v}= 360[\textbf{C-}\$\textbf{'j}[1+\textbf{s}]/360\textbf{-K}]/[\$\textbf{p}]$$

Rule-22547:

    If both (**C**), (**$'**), (**j**), (**s**), (**$**), (**v**) and (**K**) are known, then its Procured Inventory Days Planned:

$$\textbf{p}= 360[\textbf{C-}\$\textbf{'j}[1+\textbf{s}]/360\textbf{-K}]/[\$\textbf{v}]$$

Rule-22548:

    If both (**C**), (**$'**), (**j**), (**s**), (**v**) and (**p**) are known, then its Kind of Cash Planned is:

$$\textbf{K}= \textbf{C-}\$\textbf{'j}[1+\textbf{s}]/360\textbf{-}\$\textbf{'vp}[1+\textbf{s}]/360$$

Rule-22549:

    If both (**K**), (**$'**), (**j**), (**s**), (**v**) and (**p**) are known, then its Current Asset Planned is:

$$\textbf{C}= \textbf{K+}\$\textbf{'j}[1+\textbf{s}]/360\textbf{+}\$\textbf{'vp}[1+\textbf{s}]/360$$

Rule-22550:

    If both (**C**), (**Q**), (**j**), (**s**), (**v**) and (**p**) are known, then its Sales Past must be:

$$\$'= 360[\textbf{C-K}]/\{[\textbf{j+vp}][1+\textbf{s}]\}$$

Steve Asikin ISBN 14: 978-1515158691, ISBN 10: 1515158691

Rule-22551:

If both **(C)**, **($')**, **(K)**, **(s)**, **(v)** and **(p)** are known, then its Job or Trade Receivable Days Planned:

$j = \{360[\mathbf{C}\text{-}\mathbf{K}]\text{-}\mathbf{\$'vp}[1+s]\}/\{\mathbf{\$'}[1+s]\}$

Rule-22552:

If both **(C)**, **($')**, **(j)**, **(K)**, **(v)** and **(p)** are known, then its Sales Growth Planned:

$s = \$/\$'\text{-}1$

or it can also be found as

$s = 360[\mathbf{C}\text{-}\mathbf{K}]/\{\mathbf{\$'}[\mathbf{j}+\mathbf{vp}]\}\text{-}1$

Rule-22553:

If both **(C)**, **($')**, **(j)**, **(s)**, **(K)** and **(p)** are known, then its Variable Portion Planned is:

$v = 360[\mathbf{C}\text{-}\mathbf{\$'j}[1+s]/360\text{-}\mathbf{K}]/\{\mathbf{\$'p}[1+s]\}$

Rule-22554:

If both **(C)**, **($')**, **(j)**, **(s)**, **(v)** and **(K)** are known, then its Procured Inventory Days Planned:

$p = 360[\mathbf{C}\text{-}\mathbf{\$'j}[1+s]/360\text{-}\mathbf{K}]/\{\mathbf{\$'v}[1+s]\}$

Rule-22555:

If both **(X)**, **(c)**, **(J)** and **(P)** are known, then its Kind of Cash Planned:

$\mathbf{K} = \mathbf{Xc}\text{-}\mathbf{J}\text{-}\mathbf{P}$

Rule-22556:

If both **(K)**, **(c)**, **(J)** and **(P)** are known, then its Xpress or Current Debt Planned is:

$\mathbf{X} = \{\mathbf{K}+\mathbf{J}+\mathbf{P}\}/\mathbf{c}$

591

Steve Asikin ISBN 14: 978-1515158691, ISBN 10: 1515158691

Rule-22557:

> If both (**X**), (**K**), (**J**) and (**P**) are known, then its
> Current Ratio Planned is:
> $c= \{K+J+P\}/X$

Rule-22558:

> If both (**X**), (**c**), (**K**) and (**P**) are known, then its Job or
> Trade Account Receivable Planned is:
> $J= 360[Xc-K-P]$

Rule-22559:

> If both (**X**), (**c'**), (**J**) and (**K**) are known, then its
> Procured Inventory Planned is:
> $P= Xc-J-K$

Rule-22560:

> If both (**X**), (**c**), (**J**), (**V**) and (**p**) are known, then its
> Kind of Cash Planned is:
> $K= Xc-J-Vp/360$

Rule-22561:

> If both (**K**), (**c**), (**J**), (**V**) and (**p**) are known, then its
> Xpress or Current Debt Planned is:
> $X= \{K+J+Vp/360\}/c$

Rule-22562:

> If both (**X**), (**K**), (**J**), (**V**) and (**p**) are known, then its
> Current Ratio Planned is:
> $c= \{K+J+Vp/360\}/X$

Steve Asikin ISBN 14: 978-1515158691, ISBN 10: 1515158691

Rule-22563:

If both (X), (c), (K), (V) and (p) are known, then its Job or Trade Account Receivable Planned is:

$$J= \{360[Xc-K]-Vp\}$$

Rule-22564:

If both (X), (c), (J), (K) and (p) are known, then its Variable Cost Planned is:

$$V= 360[Xc-J-K]/p$$

Rule-22565:

If both (X), (c), (J), (V) and (p) are known, then its Procured Inventory Days Planned:

$$p= 360[Xc-J-K]/V$$

Rule-22566:

If both (X), (c), (J), (S), (v) and (p) are known, then its Kind of Cash Planned is:

$$K= Xc-J-Svp/360$$

Rule-22567:

If both (K), (c), (J), (S), (v) and (p) are known, then its Xpress or Current Debt Planned is:

$$X= \{K+J+Svp/360\}/c$$

Rule-225682:

If both (X), (K), (J), (S), (v) and (p) are known, then its Current Ratio Planned is:

$$c= \{K+J+Svp/360\}/X$$

Steve Asikin ISBN 14: 978-1515158691, ISBN 10: 1515158691

Rule-22569:

If both (**X**), (**c**), (**K**), (**$**), (**v**) and (**p**) are known, then its Job or Trade Account Receivable Planned is:

$J = \{360[Xc\text{-}K]\text{-}\$vp\}$

Rule-22570:

If both (**X**), (**c**), (**J**), (**K**), (**v**) and (**p**) are known, then its Sales or Revenue Planned is:

$\$ = 360[Xc\text{-}J\text{-}K]/[vp]$

Rule-22571:

If both (**X**), (**c**), (**J**), (**$**), (**K**) and (**p**) are known, then its Variable Portion Planned is:

$v = 360[Xc\text{-}J\text{-}K]/[\$p]$

Rule-22572:

If both (**X**), (**c**), (**J**), (**$**), (**v**) and (**K**) are known, then its Procured Inventory Days Planned:

$p = 360[Xc\text{-}J\text{-}K]/[\$v]$

Rule-22573:

If both (**X**), (**c**), (**J**), (**$'**), (**v**), (**p**) and (**s**) are known, then its Kind of Cash Planned is:

$K = Xc\text{-}J\text{-}\$'vp[1+s]/360$

Rule-22574:

If both (**K**), (**c**), (**J**), (**$'**), (**v**), (**p**) and (**s**) are known, then its Xpress or Current Debt Planned is:

$X = \{K+J+\$'vp[1+s]/360\}/c$

Steve Asikin ISBN 14: 978-1515158691, ISBN 10: 1515158691

Rule-22575:

    If both $(\mathbf{X})$, $(\mathbf{K})$, $(\mathbf{J})$, $(\mathbf{S'})$, $(\mathbf{v})$, $(\mathbf{p})$ and $(\mathbf{s})$ are known, then its Current Ratio Planned is:

$$c= \{\mathbf{K+J+S'vp}[1+s]/360\}/\mathbf{X}$$

Rule-22576:

    If both $(\mathbf{X})$, $(\mathbf{c})$, $(\mathbf{K})$, $(\mathbf{S'})$, $(\mathbf{v})$, $(\mathbf{p})$ and $(\mathbf{s})$ are known, then its Job or Trade Account Receivable Planned:

$$J= \{360[\mathbf{Xc\text{-}K}]\text{-}\mathbf{S'vp}[1+s]\}$$

Rule-22577:

    If both $(\mathbf{X})$, $(\mathbf{c})$, $(\mathbf{J})$, $(\mathbf{K})$, $(\mathbf{v})$, $(\mathbf{p})$ and $(\mathbf{s})$ are known, then its Sales Past must be:

$$S'= 360[\mathbf{Xc\text{-}J\text{-}K}]/\{\mathbf{vp}[1+s]\}$$

Rule-22578:

    If both $(\mathbf{X})$, $(\mathbf{c})$, $(\mathbf{J})$, $(\mathbf{S'})$, $(\mathbf{K})$, $(\mathbf{p})$ and $(\mathbf{s})$ are known, then its Variable Portion Planned is:

$$v= 360[\mathbf{Xc\text{-}J\text{-}K}]/\{\mathbf{S'p}[1+s]\}$$

Rule-22579:

    If both $(\mathbf{X})$, $(\mathbf{c})$, $(\mathbf{J})$, $(\mathbf{S'})$, $(\mathbf{v})$, $(\mathbf{K})$ and $(\mathbf{s})$ are known, then its Procured Inventory Days Planned:

$$p= 360[\mathbf{Xc\text{-}J\text{-}K}]/\{\mathbf{S'v}[1+s]\}$$

Rule-22580:

    If both $(\mathbf{X})$, $(\mathbf{c})$, $(\mathbf{J})$, $(\mathbf{S'})$, $(\mathbf{v})$, $(\mathbf{p})$ and $(\mathbf{K})$ are known, then its Sales Growth Planned is:

$$s= 360[\mathbf{Xc\text{-}J\text{-}K}]/[\mathbf{S'vp}]\text{-}1$$

Steve Asikin ISBN 14: 978-1515158691, ISBN 10: 1515158691

Rule-22581:
  If both (**X**), (**c**), (**$**), (**j**) and (**P**) are known, then its
  Kind of Cash Planned is:
  $K= Xc\text{-}\$j/360\text{-}P$

Rule-22582:
  If both (**K**), (**c**), (**$**), (**j**) and (**P**) are known, then its
  Xpress or Current Debt Planned is:
  $X= \{K+\$j/360+P\}/c$

Rule-22583:
  If both (**X**), (**K**), (**$**), (**j**) and (**P**) are known, then its
  Current Ratio Planned is:
  $c= \{K+\$j/360+P\}/X$

Rule-22584:
  If both (**X**), (**c**), (**K**), (**j**) and (**P**) are known, then its
  Sales or Revenue Planned is:
  $\$= 360[Xc\text{-}K\text{-}P]/j$

Rule-22585:
  If both (**X**), (**c**), (**$**), (**K**) and (**P**) are known, then its
  Job or Trade Receivable Days Planned:
  $j= 360[Xc\text{-}K\text{-}P]/\$$

Rule-22586:
  If both (**X**), (**c**), (**$**), (**j**) and (**K**) are known, then its
  Procured Inventory Planned is:
  $P= Xc\text{-}\$j/360\text{-}K$

Steve Asikin ISBN 14: 978-1515158691, ISBN 10: 1515158691

Rule-22587:

If both $(\mathbf{X})$, $(\mathbf{c})$, $(\mathbf{\$})$, $(\mathbf{j})$, $(\mathbf{V})$ and $(\mathbf{p})$ are known, then its Kind of Cash Planned is:

$\mathbf{K= Xc\text{-}\$j}/360\text{-}\mathbf{Vp}/360$

Rule-22588:

If both $(\mathbf{K})$, $(\mathbf{c})$, $(\mathbf{\$})$, $(\mathbf{j})$, $(\mathbf{V})$ and $(\mathbf{p})$ are known, then its Xpress or Current Debt Planned is:

$\mathbf{X= \{K+\$j}/360+\mathbf{Vp}/360\}/\mathbf{c}$

Rule-22589:

If both $(\mathbf{X})$, $(\mathbf{K})$, $(\mathbf{\$})$, $(\mathbf{j})$, $(\mathbf{V})$ and $(\mathbf{p})$ are known, then its Current Ratio Planned is:

$\mathbf{c= \{K+\$j}/360+\mathbf{Vp}/360\}/\mathbf{X}$

Rule-22590:

If both $(\mathbf{X})$, $(\mathbf{c})$, $(\mathbf{K})$, $(\mathbf{j})$, $(\mathbf{V})$ and $(\mathbf{p})$ are known, then its Sales or Revenue Planned is:

$\mathbf{\$= \{360[Xc\text{-}K]\text{-}Vp\}/j}$

Rule-22591:

If both $(\mathbf{X})$, $(\mathbf{c})$, $(\mathbf{\$})$, $(\mathbf{K})$, $(\mathbf{V})$ and $(\mathbf{p})$ are known, then its Job or Trade Receivable Days Planned:

$\mathbf{j= \{360[Xc\text{-}K]\text{-}Vp\}/\$}$

Rule-22592:

If both $(\mathbf{X})$, $(\mathbf{c})$, $(\mathbf{\$})$, $(\mathbf{j})$, $(\mathbf{Q})$ and $(\mathbf{p})$ are known, then its Variable Portion Planned is:

$\mathbf{V= 360[Xc\text{-}\$j}/360\text{-}\mathbf{K]/p}$

Steve Asikin ISBN 14: 978-1515158691, ISBN 10: 1515158691

Rule-22593:

If both (**X**), (**c**), (**$**), (**j**), (**V**) and (**K**) are known, then its Procured Inventory Days Planned:

$$\mathbf{p}= 360[\mathbf{Xc\text{-}\$j}/360\text{-}\mathbf{K}]/\mathbf{V}$$

Rule-22594:

If both (**X**), (**c**), (**$**), (**j**), (**v**) and (**p**) are known, then its Kind of Cash Planned is:

$$\mathbf{K}= \mathbf{Xc\text{-}\$j}/360\text{-}\mathbf{\$vp}/360$$

Rule-22595:

If both (**K**), (**c**), (**$**), (**j**), (**v**) and (**p**) are known, then its Xpress or Current Debt Planned is:

$$\mathbf{X}= \{\mathbf{K}+\mathbf{\$j}/360+\mathbf{\$vp}/360\}/\mathbf{c}$$

Rule-22596:

If both (**X**), (**K**), (**$**), (**j**), (**v**) and (**p**) are known, then its Current Ratio Planned is:

$$\mathbf{c}= \{\mathbf{K}+\mathbf{\$j}/360+\mathbf{\$vp}/360\}/\mathbf{X}$$

Rule-22597:

If both (**X**), (**c**), (**K**), (**j**), (**v**) and (**p**) are known, then its Sales or Revenue Planned is:

$$\mathbf{\$}= 360[\mathbf{Xc\text{-}K}]/[\mathbf{j}+\mathbf{vp}]$$

Rule-22598:

If both (**X**), (**c**), (**$**), (**K**), (**v**) and (**p**) are known, then its Job or Trade Receivable Days Planned:

$$\mathbf{j}= \{360[\mathbf{Xc\text{-}K}]\text{-}\mathbf{\$vp}\}/\mathbf{\$}$$

Steve Asikin ISBN 14: 978-1515158691, ISBN 10: 1515158691

Rule-22599:

If both (**X**), (**c**), (**$**), (**j**), (**K**) and (**p**) are known, then its Variable Portion Planned is:

$$v= 360[\textbf{Xc-\$j}/360\textbf{-K}]/[\textbf{\$p}]$$

Rule-22600:

If both (**X**), (**c**), (**$**), (**j**), (**v**) and (**K**) are known, then its Procured Inventory Days Planned:

$$p= 360[\textbf{Xc-\$j}/360\textbf{-K}]/[\textbf{\$v}]$$

Rule-22601:

If both (**X**), (**c**), (**$**), (**j**), (**$'**), (**v**), (**p**), and (**s**) are known, then its Kind of Cash Planned is:

$$K= \textbf{Xc-\$j}/360\textbf{-\$'vp}[1+s]/360$$

Rule-22602:

If both (**K**), (**c**), (**$**), (**j**), (**$'**), (**v**), (**p**) and (**s**) are known, then its Xpress or Current Debt Planned is:

$$X= \{\textbf{K+\$j}/360\textbf{+\$'vp}[1+s]/360\}/\textbf{c}$$

Rule-22603:

If both (**X**), (**K**), (**$**), (**j**), (**$'**), (**v**), (**p**) and (**s**) are known, then its Current Ratio Planned is:

$$c= \{\textbf{K+\$j}/360\textbf{+\$'vp}[1+s]/360\}/\textbf{X}$$

Rule-22604:

If both (**X**), (**c**), (**K**), (**j**), (**$'**), (**v**), (**p**) and (**s**) are known, then its Sales or Revenue Planned is:

$$\$= \{360[\textbf{Xc-K}]\textbf{-\$'vp}[1+s]\}/\textbf{j}$$

Steve Asikin ISBN 14: 978-1515158691, ISBN 10: 1515158691

Rule-22605:

If both (**X**), (**c**), (**$**), (**K**), (**$'**), (**v**), (**p**) and (**s**) are known, then its Job or Trade Receivable Days Planned:

$j = \{360[Xc\text{-}K]\text{-}\$'vp[1+s]\}/\$$

Rule-22606:

If both (**X**), (**c**), (**$**), (**j**), (**K**), (**v**), (**p**) and (**s**) are known, then its Sales Past must be:

$\$' = 360[Xc\text{-}\$j/360\text{-}K]/\{vp[1+s]\}$

Rule-22607:

If both (**X**), (**c**), (**$**), (**j**), (**$'**), (**K**), (**p**) and (**s**) are known, then its Variable Portion Planned is:

$v = 360[Xc\text{-}\$j/360\text{-}K]/\{\$'p[1+s]\}$

Rule-22608:

If both (**X**), (**c**), (**$**), (**j**), (**$'**), (**v**), (**K**) and (**s**) are known, then its Procured Inventory Days Planned:

$p = 360[Xc\text{-}\$j/360\text{-}K]/\{\$'v[1+s]\}$

Rule-22609:

If both (**X**), (**c**), (**$**), (**j**), (**$'**), (**v**), (**p**) and (**K**) are known, then its Sales Growth Planned is:

$s = \$/\$'\text{-}1$

or it can also be found as

$s = 360[Xc\text{-}\$j/360\text{-}K]/[\$'vp]\text{-}1$

Rule-22610:

If both (**X**), (**c**), (**$'**), (**j**), (**s**) and (**P**) are known, then its Procured Inventory Days Planned:

$K = Xc\text{-}\$'j[1+s]/360\text{-}P$

Steve Asikin ISBN 14: 978-1515158691, ISBN 10: 1515158691

Rule-22611:
> If both (**K**), (**c**), (**$'**), (**j**), (**s**) and (**P**) are known, then its Xpress or Current Debt Planned is:
> $$X= \{K+\$'j[1+s]/360+P\}/c$$

Rule-22612:
> If both (**X**), (**K**), (**$'**), (**j**), (**s**) and (**P**) are known, then its Current Ratio Planned is:
> $$c= \{K+\$'j[1+s]/360+P\}/X$$

Rule-22613:
> If both (**X**), (**c**), (**K**), (**j**), (**s**) and (**P**) are known, then its Sales Past must be:
> $$\$'= 360[Xc-K-P]/\{j[1+s]\}$$

Rule-22614:
> If both (**X**), (**c**), (**$'**), (**K**), (**s**) and (**P**) are known, then its Job or Trade Receivable Days Planned:
> $$j= 360[Xc-K-P]/\{\$'[1+s]\}$$

Rule-22615:
> If both (**X**), (**c**), (**$'**), (**j**), (**K**) and (**P**) are known, then its Sales Growth Planned is:
> $$s= 360[Xc-K-P]/[\$'j]-1$$

Rule-22616:
> If both (**X**), (**c**), (**$'**), (**j**), (**s**) and (**K**) are known, then its Procured Inventory Planned is:
> $$P= Xc-\$'j[1+s]/360-K$$

Steve Asikin ISBN 14: 978-1515158691, ISBN 10: 1515158691

Rule-22617:

If both $(\mathbf{X})$, $(\mathbf{c})$, $(\mathbf{S'})$, $(\mathbf{j})$, $(\mathbf{s})$, $(\mathbf{V})$ and $(\mathbf{p})$ are known, then its Kind of Cash Planned is:

$\mathbf{K} = \mathbf{X}\mathbf{c} - \mathbf{S'}\mathbf{j}[1+\mathbf{s}]/360 - \mathbf{V}\mathbf{p}/360$

Rule-22618:

If both $(\mathbf{K})$, $(\mathbf{c})$, $(\mathbf{S'})$, $(\mathbf{j})$, $(\mathbf{s})$, $(\mathbf{V})$ and $(\mathbf{p})$ are known, then its Xpress or Current Debt Planned is:

$\mathbf{X} = \{\mathbf{K} + \mathbf{S'}\mathbf{j}[1+\mathbf{s}] + \mathbf{V}\mathbf{p}/360\}/\mathbf{c}$

Rule-22619:

If both $(\mathbf{X})$, $(\mathbf{K})$, $(\mathbf{S'})$, $(\mathbf{j})$, $(\mathbf{s})$, $(\mathbf{V})$ and $(\mathbf{p})$ are known, then its Current Ratio Planned is:

$\mathbf{c} = \{\mathbf{K} + \mathbf{S'}\mathbf{j}[1+\mathbf{s}] + \mathbf{V}\mathbf{p}/360\}/\mathbf{X}$

Rule-22620:

If both $(\mathbf{X})$, $(\mathbf{c})$, $(\mathbf{K})$, $(\mathbf{j})$, $(\mathbf{s})$, $(\mathbf{V})$ and $(\mathbf{p})$ are known, then its Sales Past must be:

$\mathbf{S'} = \{360[\mathbf{X}\mathbf{c}-\mathbf{K}] - \mathbf{V}\mathbf{p}\}/\{\mathbf{j}[1+\mathbf{s}]\}$

Rule-22621:

If both $(\mathbf{X})$, $(\mathbf{c})$, $(\mathbf{S'})$, $(\mathbf{K})$, $(\mathbf{s})$, $(\mathbf{V})$ and $(\mathbf{p})$ are known, then its Job or Trade Receivable Days Planned:

$\mathbf{j} = \{360[\mathbf{X}\mathbf{c}-\mathbf{K}] - \mathbf{V}\mathbf{p}\}/\{\mathbf{S'}[1+\mathbf{s}]\}$

Rule-22622:

If both $(\mathbf{X})$, $(\mathbf{c})$, $(\mathbf{S'})$, $(\mathbf{j})$, $(\mathbf{K})$, $(\mathbf{V})$ and $(\mathbf{p})$ are known, then its Sales Growth Planned is:

$\mathbf{s} = \{360[\mathbf{X}\mathbf{c}-\mathbf{K}] - \mathbf{V}\mathbf{p}\}/[\mathbf{S'}\mathbf{j}] - 1$

Steve Asikin ISBN 14: 978-1515158691, ISBN 10: 1515158691

Rule-22623:
    If both (**X**), (**c**), (**S'**), (**j**), (**s**), (**K**) and (**p**) are known,
    then its Variable Portion Planned is:
      $V = 360[\mathbf{Xc}\text{-}\mathbf{S'j}[1+s]/360\text{-}\mathbf{K}]/\mathbf{p}$

Rule-22624:
    If both (**X**), (**c**), (**S'**), (**j**), (**s**), (**V**) and (**K**) are known,
    then its Procured Inventory Days Planned:
      $p = 360[\mathbf{Xc}\text{-}\mathbf{S'j}[1+s]/360\text{-}\mathbf{K}]/\mathbf{V}$

Rule-22625:
    If both (**X**), (**c**), (**S'**), (**j**), (**s**), (**S**), (**v**) and (**p**) are known,
    then its Kind of Cash Planned is:
      $K = \mathbf{Xc}\text{-}\mathbf{S'j}[1+s]/360\text{-}\mathbf{Svp}/360$

Rule-22626:
    If both (**K**), (**c**), (**S'**), (**j**), (**s**), (**S**), (**v**) and (**p**) are known,
    then its Xpress or Current Debt Planned is:
      $X = \{\mathbf{K}+\mathbf{S'j}[1+s]/360+\mathbf{Svp}/360\}/\mathbf{c}$

Rule-22627:
    If both (**X**), (**K**), (**S'**), (**j**), (**s**), (**S**), (**v**) and (**p**) are
    known, then its Current Ratio Planned is:
      $c = \{\mathbf{K}+\mathbf{S'j}[1+s]/360+\mathbf{Svp}/360\}/\mathbf{X}$

Rule-22628:
    If both (**X**), (**c**), (**Q**), (**j**), (**s**), (**S**), (**v**) and (**p**) are known,
    then its Sales Past must be:
      $S' = \{360[\mathbf{Xc}\text{-}\mathbf{K}]\text{-}\mathbf{Svp}\}/\{\mathbf{j}[1+s]\}$

Steve Asikin ISBN 14: 978-1515158691, ISBN 10: 1515158691

Rule-22629:

If both (X), (c), (S'), (K), (s), (S), (v) and (p) are known, then its Job or Trade Receivable Days Planned:

$$j = \{360[Xc-K]-Svp\}/\{S'[1+s]\}$$

Rule-22630:

If both (X), (c), (S'), (j), (K), (S), (v) and (p) are known, then its Sales Growth Planned is:

$$s = S/S'-1$$

or it can also be found as

$$s = \{360[Xc-K]-Svp\}/[S'j]-1$$

Rule-22631:

If both (X), (c), (S'), (j), (s), (K), (v) and (p) are known, then its Sales or Revenue Planned is:

$$S = 360[Xc-S'j[1+s]/360-K]/[vp]$$

Rule-22632:

If both (X), (c), (S'), (j), (s), (S), (K) and (p) are known, then its Variable Position Planned is:

$$v = 360[Xc-S'j[1+s]/360-K]/[Sp]$$

Rule-22633:

If both (X), (c), (S'), (j), (s), (S), (v) and (K) are known, then its Procured Inventory Days Planned:

$$p = 360[Xc-S'j[1+s]/360-K]/[Sv]$$

Steve Asikin ISBN 14: 978-1515158691, ISBN 10: 1515158691

Rule-22634:

If both (X), (c), (S'), (j), (s), (v) and (p) are known, then its Kind of Cash Planned is:

$K = Xc - S'j[1+s]/360 - S'vp[1+s]/360$

Rule-22635:

If both (K), (c), (S'), (j), (s), (v) and (p) are known, then its Xpress or Current Debt Planned is:

$X = \{K + S'j[1+s]/360 + S'vp[1+s]/360\}/c$

Rule-22636:

If both (X), (K), (S'), (j), (s), (v) and (p) are known, then its Current Ratio Planned is:

$c = \{K + S'j[1+s]/360 + S'vp[1+s]/360\}/X$

Rule-22637:

If both (X), (c), (K), (j), (s), (v) and (p) are known, then its Sales Past must be:

$S' = 360[Xc-K]/\{[1+s][j+vp]\}$

Rule-22638:

If both (X), (c), (S'), (K), (s), (v) and (p) are known, then its Job or Trade Receivable Days Planned:

$j = \{360[Xc-K] - S'vp[1+s]\}/\{S'[1+s]\}$

Rule-22639:

If both (X), (c), (S'), (j), (K), (v) and (p) are known, then its Sales Growth Planned is:

$s = 360[Xc-K]/\{S'[j+vp]\} - 1$

Steve Asikin ISBN 14: 978-1515158691, ISBN 10: 1515158691

Rule-22640:

If both $(\mathbf{X})$, $(\mathbf{c})$, $(\mathbf{S'})$, $(\mathbf{j})$, $(\mathbf{s})$, $(\mathbf{K})$ and $(\mathbf{p})$ are known, then its Variable Portion Planned is:

$$\mathbf{v} = 360[\mathbf{Xc\text{-}S'j}[1+\mathbf{s}]/360\text{-}\mathbf{K}]/\{\mathbf{S'p}[1+\mathbf{s}]\}$$

Rule-22641:

If both $(\mathbf{X})$, $(\mathbf{c})$, $(\mathbf{S'})$, $(\mathbf{j})$, $(\mathbf{s})$, $(\mathbf{v})$ and $(\mathbf{K})$ are known, then its Procured Inventory Days Planned:

$$\mathbf{p} = 360[\mathbf{Xc\text{-}S'j}[1+\mathbf{s}]/360\text{-}\mathbf{K}]/\{\mathbf{S'v}[1+\mathbf{s}]\}$$

Rule-22642:

If both $(\mathbf{P})$, $(\mathbf{c})$, $(\mathbf{q})$ and $(\mathbf{J})$ are known, then its Kind of Cash Planned is:

$$\mathbf{K} = \mathbf{Pc}/[\mathbf{c\text{-}q}]\text{-}\mathbf{J}\text{-}\mathbf{P}$$

Rule-22643:

If both $(\mathbf{K})$, $(\mathbf{c})$, $(\mathbf{q})$ and $(\mathbf{J})$ are known, then its Procured Inventory Planned is:

$$\mathbf{P} = [\mathbf{J}+\mathbf{K}]/\{\mathbf{c}/[\mathbf{c\text{-}q}]\text{-}1\}$$

Rule-22644:

If both $(\mathbf{P})$, $(\mathbf{K})$, $(\mathbf{q})$ and $(\mathbf{J})$ are known, then its Current Ratio Planned is:

$$\mathbf{c} = \mathbf{q}[\mathbf{K}+\mathbf{J}+\mathbf{P}]/[\mathbf{K}+\mathbf{J}]$$

Rule-22645:

If both $(\mathbf{P})$, $(\mathbf{c})$, $(\mathbf{K})$ and $(\mathbf{J})$ are known, then its Procured Inventory Days Planned:

$$\mathbf{q} = \mathbf{c}[\mathbf{K}+\mathbf{J}]/[\mathbf{K}+\mathbf{J}+\mathbf{P}]$$

Steve Asikin ISBN 14: 978-1515158691, ISBN 10: 1515158691

Rule-22646:
> If both (**P**), (**c**), (**q**) and (**K**) are known, then its Job or
> Trade Account Receivable Planned is:
> $J= Pc/[c-q]-K-P$

Rule-22647:
> If both (**P**), (**c**), (**q**), (**J**), (**V**) and (**p**) are known, then
> its Kind of Cash Planned:
> $K= Pc/[c-q]-J-Vp/360$

Rule-22648:
> If both (**K**), (**c**), (**q**), (**J**), (**V**) and (**p**) are known, then
> its Procured Inventory Planned is:
> $P= [c-q][J+K+Vp/360]/c$

Rule-22649:
> If both (**P**), (**K**), (**q**), (**J**), (**V**) and (**p**) are known, then
> its Pcurrent Ratio Planned is:
> $c= q[J+K+Vp/360]/\{[J+K+Vp/360]-P\}$

Rule-22650:
> If both (**P**), (**c**), (**K**), (**J**), (**V**) and (**p**) are known, then
> its Quick or Acid Test Ratio Planned is:
> $q= c\{[J+K+Vp/360]-P\}/[J+K+Vp/360]$

Rule-22651:
> If both (**P**), (**c**), (**q**), (**K**), (**V**) and (**p**) are known, then
> its Job or Trade Account Receivable Planned is:
> $J= Pc/[c-q]-K-Vp/360$

Steve Asikin ISBN 14: 978-1515158691, ISBN 10: 1515158691

Rule-22652:
  If both (**P**), (**c**), (**q**), (**J**), (**K**) and (**p**) are known, then
  its Variable Portion Planned is:
  $$V = 360\{Pc/[c-q]-K-J\}/p$$

Rule-22653:
  If both (**P**), (**c**), (**q**), (**J**), (**V**) and (**K**) are known, then
  its Procured Inventory Days Planned:
  $$p = 360\{Pc/[c-q]-K-J\}/V$$

Rule-22654:
  If both (**P**), (**c**), (**q**), (**J**), (**S**), (**v**) and (**p**) are known,
  then its Procured Inventory Days Planned is:
  $$K = Pc/[c-q]-J-Svp/360$$

Rule-22655:
  If both (**K**), (**c**), (**q**), (**J**), (**S**), (**v**) and (**p**) are known,
  then its Procured Inventory Planned is:
  $$P = [c-q][K+J+Svp/360]/c$$

Rule-22656:
  If both (**P**), (**K**), (**q**), (**J**), (**S**), (**v**) and (**p**) are known,
  then its Current Ratio Planned is:
  $$c = q[K+J+Svp/360]/\{[K+J+Svp/360]-P\}$$

Rule-22657:
  If both (**P**), (**c**), (**K**), (**J**), (**S**), (**v**) and (**p**) are known,
  then its Quick or Acid Test Ratio Planned is:
  $$q = c\{[K+J+Svp/360]-P\}/[K+J+Svp/360]$$

Steve Asikin ISBN 14: 978-1515158691, ISBN 10: 1515158691

Rule-22658:

If both (P), (c), (q), (K), (S), (v) and (p) are known, then its Job or Trade Account Receivable Planned is:

$J = Pc/[c-q]-K-Svp/360$

Rule-22659:

If both (P), (c), (q), (J), (K), (v) and (p) are known, then its Sales or Revenue Planned is:

$S = 360\{Pc/[c-q]-J-K\}/[vp]$

Rule-22660:

If both (P), (c), (q), (J), (S), (K) and (p) are known, then its Variable Portion Planned is:

$v = 360\{Pc/[c-q]-J-K\}/[Sp]$

Rule-22661:

If both (P), (c), (q), (J), (S), (v) and (K) are known, then its Procured Inventory Days Planned:

$p = 360\{Pc/[c-q]-J-K\}/[Sv]$

Rule-22662:

If both (P), (c), (q), (J), (S'), (v), (p) and (s) are known, then its Kind of Cash Planned is:

$K = Pc/[c-q]-J-S'vp[1+s]/360$

Rule-22663:

If both (K), (c), (q), (J), (S'), (v), (p) and (s) are known, then its Procured Inventory Planned is:

$P = [c-q]\{K+J+S'vp[1+s]/360\}/c$

Steve Asikin ISBN 14: 978-1515158691, ISBN 10: 1515158691

Rule-22664:
    If both $(\mathbf{P})$, $(\mathbf{K})$, $(\mathbf{q})$, $(\mathbf{J})$, $(\mathbf{S'})$, $(\mathbf{v})$, $(\mathbf{p})$ and $(\mathbf{s})$ are known, then its Current Ratio Planned is:
$$c = q\{K{+}J{+}S'vp[1{+}s]/360\}$$
$$/(\{K{+}J{+}S'vp[1{+}s]/360\}{-}P)$$

Rule-22665:
    If both $(\mathbf{P})$, $(\mathbf{c})$, $(\mathbf{K})$, $(\mathbf{J})$, $(\mathbf{S'})$, $(\mathbf{v})$, $(\mathbf{p})$ and $(\mathbf{s})$ are known, then its Quick or Acid Test Ratio Planned is:
$$q = c(\{K{+}J{+}S'vp[1{+}s]/360\}{-}P)$$
$$/\{K{+}J{+}S'vp[1{+}s]/360\}$$

Rule-22666:
    If both $(\mathbf{P})$, $(\mathbf{c})$, $(\mathbf{q})$, $(\mathbf{K})$, $(\mathbf{S'})$, $(\mathbf{v})$, $(\mathbf{p})$ and $(\mathbf{s})$ are known, then its Job or Trade Account Receivable Planned is:
$$J = Pc/[c{-}q]{-}K{-}S'vp[1{+}s]/360$$

Rule-22667:
    If both $(\mathbf{P})$, $(\mathbf{c})$, $(\mathbf{q})$, $(\mathbf{J})$, $(\mathbf{K})$, $(\mathbf{v})$, $(\mathbf{p})$ and $(\mathbf{s})$ are known, then its Sales Past must be:
$$S' = 360\{Pc/[c{-}q]{-}K{-}J\}/\{vp[1{+}s]\}$$

Rule-22668:
    If both $(\mathbf{P})$, $(\mathbf{c})$, $(\mathbf{q})$, $(\mathbf{J})$, $(\mathbf{S'})$, $(\mathbf{K})$, $(\mathbf{p})$ and $(\mathbf{s})$ are known, then its Variable Portion Planned is:
$$v = 360\{Pc/[c{-}q]{-}K{-}J\}/\{S'p[1{+}s]\}$$

Rule-22669:
    If both $(\mathbf{P})$, $(\mathbf{c})$, $(\mathbf{q})$, $(\mathbf{J})$, $(\mathbf{S'})$, $(\mathbf{v})$, $(\mathbf{K})$ and $(\mathbf{s})$ are known, then its Procured Inventory Days Planned:
$$p = 360\{Pc/[c{-}q]{-}K{-}J\}/\{S'v[1{+}s]\}$$

Steve Asikin ISBN 14: 978-1515158691, ISBN 10: 1515158691

Rule-22670:

> If both (P), (c), (q), (J), (S'), (v), (p) and (K) are known, then its Sales Growth Planned is:
>
> $s = 360\{Pc/[c-q]-K-J\}/[S'vp]-1$

Rule-22671:

> If both (P), (c), (q), (S) and (j) are known, then its Kind of Cash Planned:
>
> $K = Pc/[c-q]-Sj/360-P$

Rule-22672:

> If both (K), (c), (q), (S) and (j) are known, then its Procured Inventory Planned is:
>
> $P = [K+Sj/360]/\{c/[c-q]-1\}$

Rule-22673:

> If both (P), (K), (q), (S) and (j) are known, then its Current Ratio Planned is:
>
> $c = q[K+Sj/360+P]/[K+Sj/360]$

Rule-22674:

> If both (P), (c), (K), (S) and (j) are known, then its Quick or Acid Test Ratio Planned is:
>
> $q = c[K+Sj/360]/[K+Sj/360+P]$

Rule-22675:

> If both (P), (c), (q), (K) and (j) are known, then its Sales or Revenue Planned is:
>
> $S = 360\{Pc/[c-q]-K-P\}/j$

Steve Asikin ISBN 14: 978-1515158691, ISBN 10: 1515158691

Rule-22676:

    If both (**P**), (**c**), (**q**), (**$**) and (**j**) are known, then its Procured Inventory Days Planned:

$$j= 360\{Pc/[c\text{-}q]\text{-}K\text{-}P\}/\$$$

Rule-22677:

    If both (**P**), (**c**), (**q**), (**$**), (**j**), (**V**) and (**p**) are known, then its Kind of Cash Planned is:

$$K= Pc/[c\text{-}q]\text{-}\$j/360\text{-}Vp/360$$

Rule-22678:

    If both (**K**), (**c**), (**q**), (**$**), (**j**), (**V**) and (**p**) are known, then its Procured Inventory Planned is:

$$P= \{K+[\$j+Vp]/360\}/\{c/[c\text{-}q]\}$$

Rule-22679:

    If both (**P**), (**K**), (**q**), (**$**), (**j**), (**V**) and (**p**) are known, then its Current Ratio Planned is:

$$c= q\{K+[\$j+Vp]/360\}/(\{K+[\$j+Vp]/360\}\text{-}P)$$

Rule-22680:

    If both (**P**), (**c**), (**K**), (**$**), (**j**), (**V**) and (**p**) are known, then its Quick or Acid Test Ratio Planned is:

$$q= c(\{K+[\$j+Vp]/360\}\text{-}P)/\{K+[\$j+Vp]/360\}$$

Rule-22681:

    If both (**P**), (**c**), (**q**), (**K**), (**j**), (**V**) and (**p**) are known, then its Sales or Revenue Planned is:

$$\$= 360\{Pc/[c\text{-}q]\text{-}K\text{-}P\}/j$$

Steve Asikin ISBN 14: 978-1515158691, ISBN 10: 1515158691

Rule-22682:

If both (**P**), (**c**), (**q**), (**$**), (**K**), (**V**) and (**p**) are known, then its Job or Trade Receivable Days Planned:

$$j= 360\{\mathbf{Pc}/[\mathbf{c\text{-}q}]\text{-}\mathbf{K\text{-}P}\}/\mathbf{\$}$$

Rule-22683:

If both (**P**), (**c**), (**q**), (**$**), (**j**), (**K**) and (**p**) are known, then its Procured Inventory Days Planned:

$$V= 360(\mathbf{P}\{\mathbf{c}/[\mathbf{c\text{-}q}]\}\text{-}\mathbf{K\text{-}\$j})/\mathbf{p}$$

Rule-22684:

If both (**P**), (**c**), (**q**), (**$**), (**j**), (**V**) and (**K**) are known, then its Procured Inventory Days Planned:

$$p= 360(\mathbf{P}\{\mathbf{c}/[\mathbf{c\text{-}q}]\}\text{-}\mathbf{K\text{-}\$j})/\mathbf{V}$$

Rule-22685:

If both (**P**), (**c**), (**q**), (**$**), (**j**), (**v**) and (**p**) are known, then its Kind of Cash Planned is:

$$K= \mathbf{Pc}/[\mathbf{c\text{-}q}]\text{-}\mathbf{\$j}/360\text{-}\mathbf{\$vp}/360$$

Rule-22686:

If both (**K**), (**c**), (**q**), (**$**), (**j**), (**v**) and (**p**) are known, then its Procured Inventory Planned is:

$$P= \{\mathbf{K}+\mathbf{\$}[\mathbf{j}+\mathbf{vp}]/360\}/\{\mathbf{c}/[\mathbf{c\text{-}q}]\}$$

Rule-22687:

If both (**P**), (**K**), (**q**), (**$**), (**j**), (**v**) and (**p**) are known, then its Current Ratio Planned is:

$$c= \mathbf{q}\{\mathbf{K}+\mathbf{\$}[\mathbf{j}+\mathbf{vp}]/360\}/(\{\mathbf{K}+\mathbf{\$}[\mathbf{j}+\mathbf{vp}]/360\}\text{-}\mathbf{P})$$

Steve Asikin ISBN 14: 978-1515158691, ISBN 10: 1515158691

Rule-22688:

If both (**P**), (**c**), (**K**), (**$**), (**j**), (**v**) and (**p**) are known, then its Quick or Acid Test Ratio Planned is:

$$q= c(\{K+\$[j+vp]/360\}-P)/\{K+\$[j+vp]/360\}$$

Rule-22689:

If both (**P**), (**c**), (**q**), (**K**), (**j**), (**v**) and (**p**) are known, then its Sales or Revenue Planned is:

$$\$= 360\{Pc/[c-q]-K\}/[j+vp]$$

Rule-22690:

If both (**P**), (**c**), (**q**), (**$**), (**K**), (**v**) and (**p**) are known, then its Job or Trade Receivable Days Planned:

$$j= 360\{Pc/[c-q]-K\}/\$-vp$$

Rule-22691:

If both (**P**), (**c**), (**q**), (**$**), (**j**), (**K**) and (**p**) are known, then its Variable Portion Planned is:

$$v= (360\{Pc/[c-q]-K\}/\$-j)/p$$

Rule-22692:

If both (**P**), (**c**), (**q**), (**$**), (**j**), (**v**) and (**K**) are known, then its Procured Inventory Days Planned:

$$p= (360\{Pc/[c-q]-K\}/\$-j)/v$$

Rule-22693:

If both (**P**), (**c**), (**q**), (**$**), (**j**), (**$'**). (**v**), (**p**) and (**s**) are known, then its Kind of Cash Planned:

$$K= Pc/[c-q]-\$j/360-\$'vp[1+s]/360$$

Steve Asikin ISBN 14: 978-1515158691, ISBN 10: 1515158691

Rule-22694:

If both $(\mathbf{K})$, $(\mathbf{c})$, $(\mathbf{q})$, $(\mathbf{\$})$, $(\mathbf{j})$, $(\mathbf{\$'})$. $(\mathbf{v})$, $(\mathbf{p})$ and $(\mathbf{s})$ are known, then its Procured Inventory Planned is:

$\mathbf{P} = (\mathbf{K} + \{\mathbf{\$j} + \mathbf{\$'vp}[1+\mathbf{s}]\}/360)/\{\mathbf{c}/[\mathbf{c}-\mathbf{q}]\}$

Rule-22695:

If both $(\mathbf{P})$, $(\mathbf{K})$, $(\mathbf{q})$, $(\mathbf{\$})$, $(\mathbf{j})$, $(\mathbf{\$'})$. $(\mathbf{v})$, $(\mathbf{p})$ and $(\mathbf{s})$ are known, then its Current Ratio Planned is:

$\mathbf{c} = \mathbf{q}(\mathbf{K} + \{\mathbf{\$j} + \mathbf{\$'vp}[1+\mathbf{s}]\}/360)$
$\qquad /[(\mathbf{K} + \{\mathbf{\$j} + \mathbf{\$'vp}[1+\mathbf{s}]\}/360) - \mathbf{P}]$

Rule-22696:

If both $(\mathbf{P})$, $(\mathbf{c})$, $(\mathbf{K})$, $(\mathbf{\$})$, $(\mathbf{j})$, $(\mathbf{\$'})$. $(\mathbf{v})$, $(\mathbf{p})$ and $(\mathbf{s})$ are known, then its Quick or Acid Test Ratio Planned is:

$\mathbf{q} = \mathbf{c}[(\mathbf{K} + \{\mathbf{\$j} + \mathbf{\$'vp}[1+\mathbf{s}]\}/360) - \mathbf{P}]$
$\qquad / (\mathbf{K} + \{\mathbf{\$j} + \mathbf{\$'vp}[1+\mathbf{s}]\}/360)$

Rule-22697:

If both $(\mathbf{P})$, $(\mathbf{c})$, $(\mathbf{q})$, $(\mathbf{K})$, $(\mathbf{j})$, $(\mathbf{\$'})$. $(\mathbf{v})$, $(\mathbf{p})$ and $(\mathbf{s})$ are known, then its Sales or Revenue Planned:

$\mathbf{\$} = 360(\{\mathbf{Pc}/[\mathbf{c}-\mathbf{q}] - \mathbf{K}\} - \mathbf{\$'vp}[1+\mathbf{s}])/\mathbf{j}$

Rule-22698:

If both $(\mathbf{P})$, $(\mathbf{c})$, $(\mathbf{q})$, $(\mathbf{\$})$, $(\mathbf{K})$, $(\mathbf{\$'})$. $(\mathbf{v})$, $(\mathbf{p})$ and $(\mathbf{s})$ are known, then its Job or Trade Receivable Days Planned:

$\mathbf{j} = 360(\{\mathbf{Pc}/[\mathbf{c}-\mathbf{q}] - \mathbf{K}\} - \mathbf{\$'vp}[1+\mathbf{s}])/\mathbf{\$}$

Rule-22699:

If both $(\mathbf{P})$, $(\mathbf{c})$, $(\mathbf{q})$, $(\mathbf{\$})$, $(\mathbf{j})$, $(\mathbf{K})$. $(\mathbf{v})$, $(\mathbf{p})$ and $(\mathbf{s})$ are known, then its Procured Inventory Days Planned:

$\mathbf{\$'} = (360\{\mathbf{Pc}/[\mathbf{c}-\mathbf{q}] - \mathbf{K}\} - \mathbf{\$j})/\{\mathbf{vp}[1+\mathbf{s}]\}$

Steve Asikin ISBN 14: 978-1515158691, ISBN 10: 1515158691

Rule-22700:

If both (**P**), (**c**), (**q**), (**$**), (**j**), (**$'**). (**K**), (**p**) and (**s**) are known, then its Variable Portion Planned is:

$$v = (360\{Pc/[c-q]-K\}-\$j)/\{\$'p[1+s]\}$$

Rule-22701:

If both (**P**), (**c**), (**q**), (**$**), (**j**), (**$'**). (**v**), (**K**) and (**s**) are known, then its Procured Inventory Days Planned:

$$p = (360\{Pc/[c-q]-K\}-\$j)/\{\$'v[1+s]\}$$

Rule-22702:

If both (**P**), (**c**), (**q**), (**$**), (**j**), (**$'**). (**v**), (**p**) and (**K**) are known, then its Sales Growth Planned is:

$$s = \$/\$'-1$$

or it can also be found as

$$s = (360\{Pc/[c-q]-K\}-\$j)/[\$'vp]-1$$

Rule-22703:

If both (**P**), (**c**), (**q**), (**$'**), (**j**) and (**s**) are known, then its Kind of Cash Planned is:

$$K = Pc/[c-q]-\$'j[1+s]/360-P$$

Rule-22704:

If both (**K**), (**c**), (**q**), (**$'**), (**j**) and (**s**) are known, then its Procured Inventory Planned is:

$$P = \{K+\$'j[1+s]/360\}/\{c/[c-q]-1\}$$

Rule-22705:

If both (**P**), (**K**), (**q**), (**$'**), (**j**) and (**s**) are known, then its Current Ratio Planned is:

$$c = q(\{K+\$'j[1+s]/360\}/P-1)$$
$$/[(\{K+\$'j[1+s]/360\}/P-1)-1]$$

Steve Asikin ISBN 14: 978-1515158691, ISBN 10: 1515158691

Rule-22706:

If both $(\textbf{P})$, $(\textbf{c})$, $(\textbf{K})$, $(\textbf{S'})$, $(\textbf{j})$ and $(\textbf{s})$ are known, then its Quick or Acid Test Ratio Planned is:

$$q = c[(\{K+S'j[1+s]/360\}/P-1)-1]$$
$$/(\{K+S'j[1+s]/360\}/P-1)$$

Rule-22707:

If both $(\textbf{P})$, $(\textbf{c})$, $(\textbf{q})$, $(\textbf{K})$, $(\textbf{j})$ and $(\textbf{s})$ are known, then its Sales Past must be:

$$S' = 360(P\{c/[c-q]-1\}-K)/\{j[1+s]\}$$

Rule-22708:

If both $(\textbf{P})$, $(\textbf{c})$, $(\textbf{q})$, $(\textbf{S'})$, $(\textbf{K})$ and $(\textbf{s})$ are known, then its Job or Trade Receivable Days Planned:

$$j = 360(P\{c/[c-q]-1\}-K)/\{S'[1+s]\}$$

Rule-22709:

If both $(\textbf{P})$, $(\textbf{c})$, $(\textbf{q})$, $(\textbf{S'})$, $(\textbf{j})$ and $(\textbf{K})$ are known, then its Sales Growth Planned is:

$$s = 360(P\{c/[c-q]-1\}-K)/[S'j]-1$$

Rule-22710:

If both $(\textbf{P})$, $(\textbf{c})$, $(\textbf{q})$, $(\textbf{S'})$, $(\textbf{j})$, $(\textbf{s})$, $(\textbf{V})$ and $(\textbf{p})$ are known, then its Kind of Cash Planned is:

$$K = Pc/[c-q]-S'j[1+s]/360-Vp/360$$

Steve Asikin ISBN 14: 978-1515158691, ISBN 10: 1515158691

Rule-22711:

If both (K), (c), (q), (S'), (j), (s), (V) and (p) are known, then its Procured Inventory Planned is:
$$P= \{K+S'j[1+s]/360+Vp/360\}/\{c/[c-q]\}$$

Rule-22712:

If both (P), (K), (q), (S'), (j), (s), (V) and (p) are known, then its Current Ratio Planned is:
$$c= q(\{K+S'j[1+s]/360+Vp/360\}/P)$$
$$/[(\{K+S'j[1+s]/360+Vp/360\}/P)-1]$$

Rule-22713:

If both (P), (c), (K), (S'), (j), (s), (V) and (p) are known, then its Quick or Acid Test Ratio Planned is:
$$q= c[(\{K+S'j[1+s]/360+Vp/360\}/P)-1]$$
$$/(\{K+S'j[1+s]/360+Vp/360\}/P)$$

Rule-22714:

If both (P), (c), (q), (K), (j), (s), (V) and (p) are known, then its Sales Past must be:
$$S'= (360\{Pc/[c-q]-K\}-Vp)/\{j[1+s]\}$$

Rule-22715:

If both (P), (c), (q), (S'), (K), (s), (V) and (p) are known, then its Job or Trade Receivable Days Planned:
$$j= (360\{Pc/[c-q]-K\}-Vp)/\{S'[1+s]\}$$

Steve Asikin ISBN 14: 978-1515158691, ISBN 10: 1515158691

Rule-22716:
> If both (P), (c), (q), (S'), (j), (K), (V) and (p) are
> known, then its Sales Growth Planned is:
> $$s= (360\{Pc/[c\text{-}q]\text{-}K\}\text{-}Vp)/[S'j]\text{-}1$$

Rule-22717:
> If both (P), (c), (q), (S'), (j), (s), (K) and (p) are
> known, then its Variable Cost Planned is:
> $$V= (360\{Pc/[c\text{-}q]\text{-}K\}\text{-}S'j[1+s])/p$$

Rule-22718:
> If both (P), (c), (q), (S'), (j), (s), (V) and (K) are
> known, then its Procured Inventory Days Planned:
> $$p= (360\{Pc/[c\text{-}q]\text{-}K\}\text{-}S'j[1+s])/V$$

Rule-22719:
> If both (P), (c), (q), (S'), (j), (s), (S), (v) and (p) are
> known, then its Kind of Cash Planned is:
> $$K= Pc/[c\text{-}q]\text{-}S'j[1+s]/360\text{-}Svp/360$$

Rule-22720:
> If both (K), (c), (q), (S'), (j), (s), (S), (v) and (p) are
> known, then its Procured Inventory Planned is:
> $$P= \{K+S'j[1+s]/360+Svp/360\}/\{c/[c\text{-}q]\}$$

Rule-22721:
> If both (P), (K), (q), (S'), (j), (s), (S), (v) and (p) are
> known, then its Current Ratio Planned is:
> $$c= q(\{K+S'j[1+s]/360+Svp/360\}/P)$$
> $$/[(\{K+S'j[1+s]/360+Svp/360\}/P)\text{-}1]$$

Steve Asikin ISBN 14: 978-1515158691, ISBN 10: 1515158691

Rule-22722:

If both **(P)**, **(c)**, **(K)**, **(S')**, **(j)**, **(s)**, **(S)**, **(v)** and **(p)** are known, then its Quick or Acid Test Ratio Planned is:

$$q = c[(\{K + S'j[1+s]/360 + Svp/360\}/P) - 1]$$
$$/(\{K + S'j[1+s]/360 + Svp/360\}/P)$$

Rule-22723:

If both **(P)**, **(c)**, **(q)**, **(K)**, **(j)**, **(s)**, **(S)**, **(v)** and **(p)** are known, then its Sales Past must be:

$$S' = (360\{Pc/[c-q] - K\} - Svp)/\{j[1+s]\}$$

Rule-22724:

If both **(P)**, **(c)**, **(q)**, **(S')**, **(K)**, **(s)**, **(S)**, **(v)** and **(p)** are known, then its Job or Trade Receivable Days Planned:

$$j = (360\{Pc/[c-q] - K\} - Svp)/\{S'[1+s]\}$$

Rule-22725:

If both **(P)**, **(c)**, **(q)**, **(S')**, **(j)**, **(K)**, **(S)**, **(v)** and **(p)** are known, then its Sales Growth Planned is:

$$s = S/S' - 1$$

or it can also be found as

$$s = (360\{Pc/[c-q] - K\} - Svp)/[S'j] - 1$$

Rule-22726:

If both **(P)**, **(c)**, **(q)**, **(S')**, **(j)**, **(s)**, **(K)**, **(v)** and **(p)** are known, then its Sales or Revenue Planned is:

$$S = (360\{Pc/[c-q] - K\} - S'j[1+s])/[vp]$$

Steve Asikin ISBN 14: 978-1515158691, ISBN 10: 1515158691

Rule-22727:

If both (P), (c), (q), (S'), (j), (s), (S), (K) and (p) are known, then its Variable Portion Planned:
$$v= (360\{Pc/[c\text{-}q]\text{-}K\}\text{-}S'j[1+s])/[Sp]$$

Rule-22728:

If both (P), (c), (q), (S'), (j), (s), (S), (v) and (K) are known, then its Procured Inventory Days Planned:
$$p= (360\{Pc/[c\text{-}q]\text{-}K\}\text{-}S'j[1+s])/[Sv]$$

Rule-22729:

If both (P), (c), (q), (S'), (j), (s), (v) and (p) are known, then its Kind of Cash Planned is:
$$K= Pc/[c\text{-}q]\text{-}S'j[1+s]/360\text{-}S'vp[1+s]/360$$

Rule-22730:

If both (K), (c), (q), (S'), (j), (s), (v) and (p) are known, then its Procured Inventory Planned is:
$$P= \{K+S'j[1+s]/360+S'vp[1+s]/360\}/\{c/[c\text{-}q]\}$$

Rule-22731:

If both (P), (K), (q), (S'), (j), (s), (v) and (p) are known, then its Current Ratio Planned is:
$$c= q(\{K+S'j[1+s]/360+S'vp[1+s]/360\}/P)$$
$$/[(\{K+S'j[1+s]/360+S'vp[1+s]/360\}/P)\text{-}1]$$

Rule-22732:

If both (P), (c), (K), (S'), (j), (s), (v) and (p) are known, then its Quick or Acid Test Ratio Planned is:
$$q= c[(\{K+S'j[1+s]/360+S'vp[1+s]/360\}/P)\text{-}1]$$
$$/(\{K+S'j[1+s]/360+S'vp[1+s]/360\}/P)$$

Steve Asikin ISBN 14: 978-1515158691, ISBN 10: 1515158691

Rule-22733:
> If both (**P**), (**c**), (**q**), (**K**), (**j**), (**s**), (**v**) and (**p**) are known, then its Sales Past must be:
> $$S' = \{Pc/[c\text{-}q]\text{-}K\}/\{[1+s][j+vp]/360\}$$

Rule-22734:
> If both (**P**), (**c**), (**q**), (**S'**), (**K**), (**s**), (**v**) and (**p**) are known, then its Job or Trade Receivable Days Planned:
> $$j = (360\{Pc/[c\text{-}q]\text{-}K\}\text{-}S'vp[1+s])/\{S'[1+s]\}$$

Rule-22735:
> If both (**P**), (**c**), (**q**), (**S'**), (**j**), (**K**), (**v**) and (**p**) are known, then its Sales Growth Planned is:
> $$s = \{Pc/[c\text{-}q]\text{-}K\}/\{S'[j+vp]/360\}\text{-}1$$

Rule-22736:
> If both (**P**), (**c**), (**q**), (**S'**), (**j**), (**s**), (**K**) and (**p**) are known, then its Variable Portion Planned is:
> $$v = (360\{Pc/[c\text{-}q]\text{-}K\}\text{-}S'j[1+s])/\{S'p[1+s]\}$$

Rule-22737:
> If both (**P**), (**c**), (**q**), (**S'**), (**j**), (**s**), (**v**) and (**K**) are known, then its Procured Inventory Days Planned:
> $$p = (360\{Pc/[c\text{-}q]\text{-}K\}\text{-}S'j[1+s])/\{S'v[1+s]\}$$

Rule-22738:
> If both (**V**), (**p**), (**c**), (**q**), (**J**) and (**P**) are known, then its Kind of Cash Planned:
> $$K = Vpc/\{360[c\text{-}q]\}\text{-}J\text{-}P$$

Steve Asikin ISBN 14: 978-1515158691, ISBN 10: 1515158691

Rule-22739:

  If both (**K**), (**p**), (**c**), (**q**), (**J**) and (**P**) are known, then its Variable Cost Planned is:

  $$\mathbf{V} = 360[\mathbf{c\text{-}q}][\mathbf{K\text{+}J\text{+}P}]/[\mathbf{pc}]$$

Rule-22740:

  If both (**V**), (**K**), (**c**), (**q**), (**J**) and (**P**) are known, then its Procured Inventory Days Planned:

  $$\mathbf{p} = 360[\mathbf{c\text{-}q}][\mathbf{K\text{+}J\text{+}P}]/[\mathbf{Vc}]$$

Rule-22741:

  If both (**V**), (**p**), (**K**), (**q**), (**J**) and (**P**) are known, then its Current Ratio Planned is:

  $$\mathbf{c} = 360\mathbf{q}[\mathbf{K\text{-}J\text{-}P}]/\{360[\mathbf{K\text{-}J\text{-}P}]\text{-}\mathbf{Vp}\}$$

Rule-22742:

  If both (**V**), (**p**), (**c**), (**K**), (**J**) and (**P**) are known, then its Quick or Acid Test Ratio Planned is:

  $$\mathbf{q} = \mathbf{c}\{360[\mathbf{K\text{-}J\text{-}P}]\text{-}\mathbf{Vp}\}/\{360[\mathbf{K\text{-}J\text{-}P}]\}$$

Rule-22743:

  If both (**V**), (**p**), (**c**), (**q**), (**K**) and (**P**) are known, then its Job or Trade Receivable Days Planned is:

  $$\mathbf{J} = \mathbf{Vpc}/\{360[\mathbf{c\text{-}q}]\}\text{-}\mathbf{K\text{-}P}$$

Rule-22744:

  If both (**V**), (**p**), (**c**), (**q**), (**J**) and (**K**) are known, then its Procured Inventory Planned is:

  $$\mathbf{P} = \mathbf{Vpc}/\{360[\mathbf{c\text{-}q}]\}\text{-}\mathbf{J\text{-}K}$$

Steve Asikin ISBN 14: 978-1515158691, ISBN 10: 1515158691

<u>Rule-22745</u>:

  If both **(V)**, **(p)**, **(c)**, **(q)** and **(J)** are known, then its Kind of Cash Planned is:

  $$K= Vpc/\{360[c\text{-}q]\}\text{-}J\text{-}Vp/360$$

<u>Rule-22746</u>:

  If both **(K)**, **(p)**, **(c)**, **(q)** and **(J)** are known, then its Variable Portion Planned is:

  $$V= [K\text{+}J]/(p\{c/[c\text{-}q]\text{+}1\}/360)$$

<u>Rule-22747</u>:

  If both **(V)**, **(K)**, **(c)**, **(q)** and **(J)** are known, then its Procured Inventory Days Planned:

  $$p=[K\text{+}J]/(V\{c/[c\text{-}q]\text{+}1\}/360)$$

<u>Rule-22748</u>:

  If both **(V)**, **(p)**, **(K)**, **(q)** and **(J)** are known, then its Current Ratio Planned is:

  $$c= q\{360[K\text{+}J]\text{+}Vp\}/(\{360[K\text{+}J]\text{+}Vp\}\text{-}Vp)$$

<u>Rule-22749</u>:

  If both **(V)**, **(p)**, **(c)**, **(K)** and **(J)** are known, then its Quick or Acid Test Ratio Planned is:

  $$q= c(\{360[K\text{+}J]\text{+}Vp\}\text{-}Vp)/\{360[K\text{+}J]\text{+}Vp\}$$

<u>Rule-22750</u>:

  If both **(V)**, **(p)**, **(c)**, **(q)**, **(K)** and **(p)** are known, then its Job or Trade Account Receivable Planned is:

  $$J= Vpc/\{360[c\text{-}q]\}\text{-}K\text{-}Vp/360$$

Steve Asikin ISBN 14: 978-1515158691, ISBN 10: 1515158691

Rule-22751:

If both (V), (p), (c), (q), (J), (S) and (v) are known,
then its Kind of Cash Planned is:
$$K= Vpc/\{360[c-q]\}-J-Svp/360$$

Rule-22752:

If both (K), (p), (c), (q), (J), (S) and (v) are known,
then its Variable Cost Planned is:
$$V= [c-q]\{360[K+J]+Svp\}/[pc]$$

Rule-22753:

If both (V), (K), (c), (q), (J), (S) and (v) are known,
then its Procured Inventory Days Planned:
$$p= 360[J+ K]/\{Vc/[c-q]-Sv\}$$

Rule-22754:

If both (V), (p), (K), (q), (J), (S) and (v) are known,
then its Current Ratio Planned is:
$$c= q\{360[K+J]+Svp\}/(\{360[K+J]+Svp\}-Vp)$$

Rule-22755:

If both (V), (p), (c), (K), (J), (S) and (v) are known,
then its Quick or Acid Test Ratio Planned:
$$q= c(\{360[K+J]+Svp\}-Vp)/\{360[K+J]+Svp\}$$

Rule-22756:

If both (V), (p), (c), (q), (K), (S) and (v) are known,
then its Job or Trade Account Raceivable Planned is:
$$J= Vpc/\{360[c-q]\}-K-SVp/360$$

Steve Asikin ISBN 14: 978-1515158691, ISBN 10: 1515158691

Rule-22757:

If both (V), (p), (c), (q), (J), (K) and (v) are known, then its Sales or Revenue Planned is:

$S = \{Vpc/[c-q]-360[J+K]\}/[vp]$

Rule-22758:

If both (V), (p), (c), (q), (J), (S) and (K) are known, then its Procured Inventory Days Planned:

$v = V/S$

or it can also be found as

$v = \{Vpc/[c-q]-360[J+K]\}/[Sp]$

Rule-22759:

If both (V), (p), (c), (q), (J), (S'), (v) and (s) are known, then its Kind of Cash Planned is:

$K = Vpc/\{360[c-q]\}-J-S'vp[1+s]/360$

Rule-22760:

If both (K), (p), (c), (q), (J), (S'), (v) and (s) are known, then its Variable Cost Planned is:

$V = [c-q]\{360[K+J]+S'vp[1+s]\}/[pc]$

Rule-22761:

If both (V), (K), (c), (q), (J), (S'), (v) and (s) are known, then its Procured Inventory Days Planned:

$p = 360[K+J]/(Vc/[c-q]-S'v[1+s])$

Rule-22762:

If both (V), (p), (K), (q), (J), (S'), (v) and (s) are known, then its Current Ratio Planned is:

$c = q\{360[K+J]+S'vp[1+s]\}$
$/(\{360[K+J]+S'vp[1+s]\}-Vp)$

Steve Asikin ISBN 14: 978-1515158691, ISBN 10: 1515158691

Rule-22763:

> If both (V), (p), (c), (K), (J), (S'), (v) and (s) are
> known, then its Quick or Acid Test Ratio Planned is:
> $$q = c(\{360[K+J]+S'vp[1+s]\}-Vp)$$
> $$/\{360[K+J]+S'vp[1+s]\}$$

Rule-22764:

> If both (V), (p), (c), (q), (K), (S'), (v) and (s) are
> known, then its Job or Trade Account Receivable
> Planned is:
> $$J = Vpc/\{360[c-q]\}-K-S'vp[1+s]/360$$

Rule-22765:

> If both (V), (p), (c), (q), (J), (K), (v) and (s) are
> known, then its Sales Past must be:
> $$S' = \{Vpc/[c-q]-360[J+K]\}/\{vp[1+s]\}$$

Rule-22766:

> If both (V), (p), (c), (q), (J), (S'), (K) and (s) are
> known, then its Variable Portion Planned is:
> $$v = \{Vpc/[c-q]-360[J+K]\}/\{S'p[1+s]\}$$

Rule-22767:

> If both (V), (p), (c), (q), (J), (S'), (v) and (K) are
> known, then its Sales Growth Planned is:
> $$s = \{Vpc/[c-q]-360[J+K]\}/[S'vp]-1$$

Rule-22768:

> If both (V), (p), (c), (q), (S), (j) and (P) are known,
> then its Kind of Cash Planned is:
> $$K = Vpc/\{360[c-q]\}-Sj/360-P$$

Steve Asikin ISBN 14: 978-1515158691, ISBN 10: 1515158691

Rule-22769:

If both (**K**), (**p**), (**c**), (**q**), (**$**), (**j**) and (**P**) are known,
then its Variable Cost Planned is:
$$V = [c\text{-}q]\{360[K+P]+\$j\}/[pc]$$

Rule-22770:

If both (**V**), (**K**), (**c**), (**q**), (**$**), (**j**) and (**P**) are known,
then its Procured Inventory Days Planned:
$$p = [c\text{-}q]\{360[K+P]+\$j\}/[Vc]$$

Rule-22771:

If both (**V**), (**p**), (**K**), (**q**), (**$**), (**j**) and (**P**) are known,
then its Current Ratio Planned is:
$$c = q\{360[K+P]+\$j\}/(\{360[K+P]+\$j\}\text{-}Vp)$$

Rule-22772:

If both (**V**), (**p**), (**c**), (**K**), (**$**), (**j**) and (**P**) are known,
then its Quick or Acid Test Ratio Planned is:
$$q = c(\{360[K+P]+\$j\}\text{-}Vp)/\{360[K+P]+\$j\}$$

Rule-22773:

If both (**V**), (**p**), (**c**), (**q**), (**K**), (**j**) and (**P**) are known,
then its Sales or Revenue Planned is:
$$\$ = \{Vpc/[c\text{-}q]\text{-}360[K+P]\}/j$$

Rule-22774:

If both (**V**), (**p**), (**c**), (**q**), (**$**), (**K**) and (**P**) are known,
then its Job or Trade Receivable Days Planned:
$$j = \{Vpc/[c\text{-}q]\text{-}360[K+P]\}/\$$$

Steve Asikin ISBN 14: 978-1515158691, ISBN 10: 1515158691

Rule-22775:

    If both **(V)**, **(p)**, **(c)**, **(q)**, **($)**, **(j)** and **(K)** are known,
then its Procured Inventory Planned is:

$$P= Vpc/\{360[c\text{-}q]\}\text{-}\$j/360\text{-}K$$

Rule-22776:

    If both **(V)**, **(p)**, **(c)**, **(q)**, **($)** and **(j)** are known, then its
Kind of Cash Planned is:

$$K= Vpc/\{360[c\text{-}q]\}\text{-}\$j/360\text{-}Vp/360$$

Rule-22777:

    If both **(K)**, **(p)**, **(c)**, **(q)**, **($)** and **(j)** are known, then its
Variable Cost Planned is:

$$V= ([360K+\$j]/\{c/[c\text{-}q]\text{-}1\})/p$$

Rule-22778:

    If both **(V)**, **(K)**, **(c)**, **(q)**, **($)** and **(j)** are known, then its
Procured Inventory Days Planned:

$$p= ([360K+\$j]/\{c/[c\text{-}q]\text{-}1\})/V$$

Rule-22779:

    If both **(V)**, **(p)**, **(K)**, **(q)**, **($)** and **(j)** are known, then
its Current Ratio Planned is:

$$c= q[360K+Vp+\$j]/\{[360K+Vp+\$j]\text{-}Vpc\}$$

Rule-22780:

    If both **(V)**, **(p)**, **(c)**, **(K)**, **($)** and **(j)** are known, then its
Quick or Acid Test Ratio Planned is:

$$q= c\{[360K+Vp+\$j]\text{-}Vpc\}/[360K+Vp+\$j]$$

Steve Asikin ISBN 14: 978-1515158691, ISBN 10: 1515158691

Rule-22781:

    If both **(V)**, **(p)**, **(c)**, **(q)**, **(K)** and **(j)** are known, then its Sales or Revenue Planned is:

$$S= \{Vpc/[c-q]-Vp-360K\}/j$$

Rule-22782:

    If both **(V)**, **(p)**, **(c)**, **(q)**, **(S)** and **(K)** are known, then its Job or Trade Receivable Days Planned:

$$j= \{Vpc/[c-q]-Vp-360K\}/S$$

Rule-22783:

    If both **(V)**, **(p)**, **(c)**, **(q)**, **(S)**, **(j)** and **(v)** are known, then its Kind of Cash Planned is:

$$K= Vpc/\{360[c-q]\}-Sj/360-Svp/360$$

Rule-22784:

    If both **(K)**, **(p)**, **(c)**, **(q)**, **(S)**, **(j)** and **(v)** are known, then its Variable Cost Planned is:

$$V= [c-q][360K+Sj+Svp]/[pc]$$

Rule-22785:

    If both **(V)**, **(K)**, **(c)**, **(q)**, **(S)**, **(j)** and **(v)** are known, then its Procured Inventory Days Planned:

$$p= [360K+Sj]/\{Vc/[c-q]-Sv\}$$

Rule-22786:

    If both **(V)**, **(p)**, **(K)**, **(q)**, **(S)**, **(j)** and **(v)** are known, then its Current Ratio Planned is:

$$c= q[360K+Sj+Svp]/\{[360K+Sj+Svp]-Vp\}$$

Steve Asikin ISBN 14: 978-1515158691, ISBN 10: 1515158691

Rule-22787:

If both **(V)**, **(p)**, **(c)**, **(K)**, **($)**, **(j)** and **(v)** are known, then its Quick or Acid Test Ratio Planned is:

$$q= c\{[360K+Sj+Svp]-Vp\}/[360K+Sj+Svp]$$

Rule-22788:

If both **(V)**, **(p)**, **(c)**, **(q)**, **(K)**, **(j)** and **(v)** are known, then its Sales or Revenue Planned is:

$$S= \{Vpc/[c-q]-360K\}/[j-vp]$$

Rule-22789:

If both **(V)**, **(p)**, **(c)**, **(q)**, **($)**, **(K)** and **(v)** are known, then its Job or Trade Receivable Days Planned is:

$$j= \{Vpc/[c-q]-Svp-360K\}/S$$

Rule-22790:

If both **(V)**, **(p)**, **(c)**, **(q)**, **($)**, **(j)** and **(K)** are known, then its Variable Portion Planned is:

$$v= \{Vpc/[c-q]-Sj-360K\}/[Sp]$$

Rule-22791:

If both **(V)**, **(p)**, **(c)**, **(q)**, **($)**, **(j)**, **($')**, **(v)** and **(s)** are known, then its Kind of Cash Planned is:

$$K= Vpc/\{360[c-q]\}-Sj/360-S'vp[1+s]/360$$

Rule-22792:

If both **(K)**, **(p)**, **(c)**, **(q)**, **($)**, **(j)**, **($')**, **(v)** and **(s)** are known, then its Variable Cost Planned is:

$$V= \{360K+Sj+S'vp[1+s]\}/\{pc/[c-q]\}$$

Steve Asikin ISBN 14: 978-1515158691, ISBN 10: 1515158691

Rule-22793:

If both (V), (K), (c), (q), $(\$)$, (j), $(\$')$, (v) and (s) are known, then its Procured Inventory Days Planned:

$$p = [360K + \$j]/\{Vc/[c-q] - \$'v[1+s]\}$$

Rule-22794:

If both (V), (p), (K), (q), $(\$)$, (j), $(\$')$, (v) and (s) are known, then its Current Ratio Planned:

$$c = q\{360K + \$j - \$'vp[1+s]\}$$
$$/(\{360K + \$j - \$'vp[1+s]\} - Vp)$$

Rule-22795:

If both (V), (p), (c), (K), $(\$)$, (j), $(\$')$, (v) and (s) are known, then its Quick or Acid Test Ratio Planned is:

$$q = c(\{360K + \$j - \$'vp[1+s]\} - Vp)$$
$$/\{360K + \$j - \$'vp[1+s]\}$$

Rule-22796:

If both (V), (p), (c), (q), (K), (j), $(\$')$, (v) and (s) are known, then its Sales or Revenue Planned is:

$$\$ = \{Vpc/[c-q] - \$'vp[1+s] - 360K\}/j$$

Rule-22797:

If both (V), (p), (c), (q), $(\$)$, (K), $(\$')$, (v) and (s) are known, then its Job or Trade Receivable Days Planned:

$$j = \{Vpc/[c-q] - \$'vp[1+s] - 360K\}/\$$$

Rule-22798:

If both (V), (p), (c), (q), $(\$)$, (j), (K), (v) and (s) are known, then its Sales Past must be:

$$\$' = \{Vpc/[c-q] - \$j - 360K\}/\{vp[1+s]\}$$

Steve Asikin ISBN 14: 978-1515158691, ISBN 10: 1515158691

Rule-22799:

If both $(\mathbf{V})$, $(\mathbf{p})$, $(\mathbf{c})$, $(\mathbf{q})$, $(\mathbf{\$})$, $(\mathbf{j})$, $(\mathbf{\$'})$, $(\mathbf{K})$ and $(\mathbf{s})$ are known, then its Variable Portion Planned is:

$$\mathbf{v}= \{\mathbf{Vpc}/[\mathbf{c\text{-}q}]\text{-}\$\mathbf{j}\text{-}360\mathbf{K}\}/\{\$'\mathbf{p}[1+\mathbf{s}]\}$$

Rule-22800:

If both $(\mathbf{V})$, $(\mathbf{p})$, $(\mathbf{c})$, $(\mathbf{q})$, $(\mathbf{\$})$, $(\mathbf{j})$, $(\mathbf{\$'})$, $(\mathbf{v})$ and $(\mathbf{K})$ are known, then its Sales Growth Planned is:

$$\mathbf{s}= \{\mathbf{Vpc}/[\mathbf{c\text{-}q}]\text{-}\$\mathbf{j}\text{-}360\mathbf{K}\}/[\$'\mathbf{vp}]\text{-}1$$

Rule-22801:

If both $(\mathbf{V})$, $(\mathbf{p})$, $(\mathbf{c})$, $(\mathbf{q})$, $(\mathbf{\$'})$, $(\mathbf{j})$, $(\mathbf{s})$ and $(\mathbf{P})$ are known, then its Kind of Cash Planned is:

$$\mathbf{K}= \mathbf{Vpc}/\{360[\mathbf{c\text{-}q}]\}\text{-}\$'\mathbf{j}[1+\mathbf{s}]/360\text{-}\mathbf{P}$$

Rule-22802:

If both $(\mathbf{K})$, $(\mathbf{p})$, $(\mathbf{c})$, $(\mathbf{q})$, $(\mathbf{\$'})$, $(\mathbf{j})$, $(\mathbf{s})$ and $(\mathbf{P})$ are known, then its Variable Cost Planned is:

$$\mathbf{V}= [\mathbf{c\text{-}q}]\{360[\mathbf{K}+\mathbf{P}]+\$'\mathbf{j}[1+\mathbf{s}]\}/[\mathbf{pc}]$$

Rule-22803:

If both $(\mathbf{V})$, $(\mathbf{K})$, $(\mathbf{c})$, $(\mathbf{q})$, $(\mathbf{\$'})$, $(\mathbf{j})$, $(\mathbf{s})$ and $(\mathbf{P})$ are known, then its Procured Inventory Days Planned:

$$\mathbf{p}= [\mathbf{c\text{-}q}]\{360[\mathbf{K}+\mathbf{P}]+\$'\mathbf{j}[1+\mathbf{s}]\}/[\mathbf{Vc}]$$

Rule-22804:

If both $(\mathbf{V})$, $(\mathbf{p})$, $(\mathbf{K})$, $(\mathbf{q})$, $(\mathbf{\$'})$, $(\mathbf{j})$, $(\mathbf{s})$ and $(\mathbf{P})$ are known, then its Current Ratio Planned is:

$$\mathbf{c}= \mathbf{q}\{360[\mathbf{K}+\mathbf{P}]+\$'\mathbf{j}[1+\mathbf{s}]\}$$
$$/(\{360[\mathbf{K}+\mathbf{P}]+\$'\mathbf{j}[1+\mathbf{s}]\}\text{-}\mathbf{Vp})$$

Steve Asikin ISBN 14: 978-1515158691, ISBN 10: 1515158691

Rule-22805:

    If both $(\mathbf{V})$, $(\mathbf{p})$, $(\mathbf{c})$, $(\mathbf{K})$, $(\mathbf{S'})$, $(\mathbf{j})$, $(\mathbf{s})$ and $(\mathbf{P})$ are known, then its Quick or Acid Test Ratio Planned is:

$$\mathbf{q} = \mathbf{c}(\{360[\mathbf{K}+\mathbf{P}]+\mathbf{S'j}[1+\mathbf{s}]\}-\mathbf{Vp}) / \{360[\mathbf{K}+\mathbf{P}]+\mathbf{S'j}[1+\mathbf{s}]\}$$

Rule-22806:

    If both $(\mathbf{V})$, $(\mathbf{p})$, $(\mathbf{c})$, $(\mathbf{q})$, $(\mathbf{K})$, $(\mathbf{j})$, $(\mathbf{s})$ and $(\mathbf{P})$ are known, then its Sales Past must be:

$$\mathbf{S'} = \{\mathbf{Vpc}/[\mathbf{c}-\mathbf{q}]-360[\mathbf{K}+\mathbf{P}]\}/\{\mathbf{j}[1+\mathbf{s}]\}$$

Rule-22807:

    If both $(\mathbf{V})$, $(\mathbf{p})$, $(\mathbf{c})$, $(\mathbf{q})$, $(\mathbf{S'})$, $(\mathbf{K})$, $(\mathbf{s})$ and $(\mathbf{P})$ are known, then its Job or Trade Receivable Days Planned:

$$\mathbf{j} = \{\mathbf{Vpc}/[\mathbf{c}-\mathbf{q}]-360[\mathbf{K}+\mathbf{P}]\}/\{\mathbf{S'}[1+\mathbf{s}]\}$$

Rule-22808:

    If both $(\mathbf{V})$, $(\mathbf{p})$, $(\mathbf{c})$, $(\mathbf{q})$, $(\mathbf{S'})$, $(\mathbf{j})$, $(\mathbf{K})$ and $(\mathbf{P})$ are known, then its Sales Growth Planned:

$$\mathbf{s} = \{\mathbf{Vpc}/[\mathbf{c}-\mathbf{q}]-360[\mathbf{K}+\mathbf{P}]\}/[\mathbf{S'j}]-1$$

Rule-22809:

    If both $(\mathbf{V})$, $(\mathbf{p})$, $(\mathbf{c})$, $(\mathbf{q})$, $(\mathbf{S'})$, $(\mathbf{j})$, $(\mathbf{s})$ and $(\mathbf{K})$ are known, then its Procured Inventory Planned is:

$$\mathbf{P} = \mathbf{Vpc}/\{360[\mathbf{c}-\mathbf{q}]\}-\mathbf{S'j}[1+\mathbf{s}]/360-\mathbf{K}$$

Rule-22810:

    If both $(\mathbf{V})$, $(\mathbf{p})$, $(\mathbf{c})$, $(\mathbf{q})$, $(\mathbf{S'})$, $(\mathbf{j})$ and $(\mathbf{s})$ are known, then its Kind of Cash Planned:

$$\mathbf{K} = \mathbf{Vpc}/\{360[\mathbf{c}-\mathbf{q}]\}-\mathbf{S'j}[1+\mathbf{s}]/360-\mathbf{Vp}/360$$

Steve Asikin ISBN 14: 978-1515158691, ISBN 10: 1515158691

Rule-22811:
   If both (K), (p), (c), (q), (S'), (j) and (s) are known,
   then its Variable Cost Planned is:
   $V= \{360K+S'j[1+s]\}/(p\{c/[c-q]-1\})$

Rule-22812:
   If both (V), (K), (c), (q), (S'), (j) and (s) are known,
   then its Procured Inventory Days Planned:
   $p= \{360K+S'j[1+s]\}/(V\{c/[c-q]-1\})$

Rule-22813:
   If both (V), (p), (K), (q), (S'), (j) and (s) are known,
   then its Current Ratio Planned is:
   $c= q\{360K+S'j[1+s]+Vp\}/\{360K+S'j[1+s]\}$

Rule-22814:
   If both (V), (p), (c), (K), (S'), (j) and (s) are known,
   then its Quick or Acid Test Ratio Planned is:
   $q= c\{360K+S'j[1+s]\}/\{360K+S'j[1+s]+Vp\}$

Rule-22815:
   If both (V), (p), (c), (q), (K), (j) and (s) are known,
   then its Sales Past must be:
   $S'= \{Vpc/[c-q]-Vp-360K\}/\{j[1+s]\}$

Rule-22816:
   If both (V), (p), (c), (q), (S'), (K) and (s) are known,
   then its Job or Trade Receivable Days Planned:
   $j= \{Vpc/[c-q]-Vp-360K\}/\{S'[1+s]\}$

Steve Asikin ISBN 14: 978-1515158691, ISBN 10: 1515158691

Rule-22817:

If both (V), (p), (c), (q), (S'), (j) and (K) are known, then its Sales Growth Planned is:

$s= \{Vpc/[c-q]-Vp-360K\}/[S'j]-1$

Rule-22818:

If both (V), (p), (c), (q), (S'), (j), (s), (S) and (v) are known, then its Kind of Cash Planned:

$K= Vpc/\{360[c-q]\}-S'j[1+s]/360-Svp/360$

Rule-22819:

If both (K), (p), (c), (q), (S'), (j), (s), (S) and (v) are known, then its Variable Cost Planned is:

$V= \{360K+S'j[1+s]+Svp\}/\{pc/[c-q]\}$

Rule-22820:

If both (V), (K), (c), (q), (S'), (j), (s), (S) and (v) are known, then its Procured Inventory Days Planned:

$p= \{360K+S'j[1+s]\}/\{Vc/[c-q]-Sv\}$

Rule-22821:

If both (V), (p), (K), (q), (S'), (j), (s), (S) and (v) are known, then its Current Ratio Planned is:

$c= q\{360K+S'j[1+s]+Svp\}$
$/(\{360K+S'j[1+s]+Svp\}-Vp)$

Rule-22822:

If both (V), (p), (c), (K), (S'), (j), (s), (S) and (v) are known, then its Quick or Acid Test Ratio Planned is:

$q= c(\{360K+S'j[1+s]+Svp\}-Vp)$
$/\{360K+S'j[1+s]+Svp\}$

Steve Asikin ISBN 14: 978-1515158691, ISBN 10: 1515158691

Rule-22823:

If both (V), (p), (c), (q), (K), (j), (s), $(\$)$ and (v) are known, then its Sales Past must be:

$\$'=\{Vpc/[c-q]-\$vp-360K\}/\{j[1+s]\}$

Rule-22824:

If both (V), (p), (c), (q), $(\$')$, (K), (s), $(\$)$ and (v) are known, then its Job or Trade Receivable Days Planned:

$j=\{Vpc/[c-q]-\$vp-360K\}/\{\$'[1+s]\}$

Rule-22825:

If both (V), (p), (c), (q), $(\$')$, (j), (K), $(\$)$ and (v) are known, then its Sales Growth Planned:

$s=\$/\$'-1$

or it can also be found as

$s=\{Vpc/[c-q]-\$vp-360K\}/[\$'j]-1$

Rule-22826:

If both (V), (p), (c), (q), $(\$')$, (j), (s), (K) and (v) are known, then its Sales or Revenue Planned is:

$\$=\{Vpc/[c-q]-\$'j[1+s]-360K\}/\{vp\}$

Rule-22827:

If both (V), (p), (c), (q), $(\$')$, (j), (s), $(\$)$ and (K) are known, then its Variable Portion Planned is:

$v=\{Vpc/[c-q]-\$'j[1+s]-360K\}/\{\$p\}$

Rule-22828:

If both (V), (p), (c), (q), $(\$')$, (j), (s) and (v) are known, then its Kind of Cash Planned is:

$K=Vpc/\{360[c-q]\}-\$'j[1+s]/360-\$'vp[1+s]/360$

637

Steve Asikin ISBN 14: 978-1515158691, ISBN 10: 1515158691

Rule-22829:

If both (K), (p), (c), (q), (S'), (j), (s) and (v) are known, then its Variable Cost Planned is:

$V = [c\text{-}q]\{360K\text{-}S'j[1+s]\text{-}S'vp[1+s]\}/[pc]$

Rule-22830:

If both (V), (K), (c), (q), (S'), (j), (s) and (v) are known, then its Procured Inventory Days Planned:

$p = \{360K+S'j[1+s]\}/\{Vc/[c\text{-}q]\text{-}S'v[1+s]\}$

Rule-22831:

If both (V), (p), (K), (q), (S'), (j), (s) and (v) are known, then its Current Ratio Planned is:

$c = q\{360K+S'j[1+s]+S'vp[1+s]\}$
$/(\{360K+S'j[1+s]+S'vp[1+s]\}\text{-}Vp)$

Rule-22832:

If both (V), (p), (c), (K), (S'), (j), (s) and (v) are known, then its Quick or Acid Test Ratio Planned is:

$q = c(\{360K+S'j[1+s]+S'vp[1+s]\}\text{-}Vp)$
$/\{360K+S'j[1+s]+S'vp[1+s]\}$

Rule-22833:

If both (V), (p), (c), (q), (K), (j), (s) and (v) are known, then its Sales Past must be:

$S' = \{Vpc/[c\text{-}q]\text{-}360K\}/\{[1+s][j+vp]\}$

Rule-22834:

If both (V), (p), (c), (q), (S'), (K), (s) and (v) are known, then its Job or Trade Receivable Days Planned:

$j = \{Vpc/[c\text{-}q]\text{-}S'vp[1+s]\text{-}360K\}/\{S'[1+s]\}$

Steve Asikin ISBN 14: 978-1515158691, ISBN 10: 1515158691

Rule-22835:

   If both $(\mathbf{V})$, $(\mathbf{p})$, $(\mathbf{c})$, $(\mathbf{q})$, $(\mathbf{S'})$, $(\mathbf{j})$, $(\mathbf{K})$ and $(\mathbf{v})$ are known, then its Sales Growth Planned is:

$$s= \{\mathbf{Vpc}/[\mathbf{c\text{-}q}]\text{-}360\mathbf{K}\}/\{\mathbf{S'}[\mathbf{j}+\mathbf{vp}]\}\text{-}1$$

Rule-22836:

   If both $(\mathbf{V})$, $(\mathbf{p})$, $(\mathbf{c})$, $(\mathbf{q})$, $(\mathbf{S'})$, $(\mathbf{j})$, $(\mathbf{s})$ and $(\mathbf{K})$ are known, then its Variable Portion Planned is:

$$v= \{\mathbf{Vpc}/[\mathbf{c\text{-}q}]\text{-}\mathbf{S'j}[1+s]\text{-}360\mathbf{K}\}/\{\mathbf{S'p}[1+s]\}$$

Rule-22837:

   If both $(\mathbf{S})$, $(\mathbf{v})$, $(\mathbf{p})$, $(\mathbf{c})$, $(\mathbf{q})$, $(\mathbf{J})$ and $(\mathbf{P})$ are known, then its Kind of Cash Planned is:

$$\mathbf{K}= \mathbf{Svpc}/\{360[\mathbf{c\text{-}q}]\}\text{-}\mathbf{J}\text{-}\mathbf{P}$$

Rule-22838:

   If both $(\mathbf{K})$, $(\mathbf{v})$, $(\mathbf{p})$, $(\mathbf{c})$, $(\mathbf{q})$, $(\mathbf{J})$ and $(\mathbf{P})$ are known, then its Sales or Revenue Planned is:

$$S=\{360[\mathbf{c\text{-}q}][\mathbf{K}+\mathbf{J}+\mathbf{P}]\}/[\mathbf{vpc}]$$

Rule-22839:

   If both $(\mathbf{S})$, $(\mathbf{K})$, $(\mathbf{p})$, $(\mathbf{c})$, $(\mathbf{q})$, $(\mathbf{J})$ and $(\mathbf{P})$ are known, then its Variable Cost Planned is:

$$v=\{360[\mathbf{c\text{-}q}][\mathbf{K}+\mathbf{J}+\mathbf{P}]\}/[\mathbf{Spc}]$$

Rule-22840:

   If both $(\mathbf{S})$, $(\mathbf{v})$, $(\mathbf{K})$, $(\mathbf{c})$, $(\mathbf{q})$, $(\mathbf{J})$ and $(\mathbf{P})$ are known, then its Procured Inventory Days Planned:

$$p=\{360[\mathbf{c\text{-}q}][\mathbf{K}+\mathbf{J}+\mathbf{P}]\}/[\mathbf{Svc}]$$

Steve Asikin ISBN 14: 978-1515158691, ISBN 10: 1515158691

Rule-22841:
> If both **($)**, **(v)**, **(p)**, **(K)**, **(q)**, **(J)** and **(P)** are known,
> then its Current Ratio Planned is:
> $c = \{360[c\text{-}q][K\text{+}J\text{+}P]\}/[\$vp]$

Rule-22842:
> If both **($)**, **(v)**, **(p)**, **(c)**, **(q)**, **(K)** and **(P)** are known,
> then its Kind of Cash Planned is:
> $J = \$vpc/\{360[c\text{-}q]\}\text{-}K\text{-}P$

Rule-22843:
> If both **($)**, **(v)**, **(p)**, **(c)**, **(q)**, **(J)** and **(K)** are known,
> then its Procured Inventory Planned is:
> $P = \$vpc/\{360[c\text{-}q]\}\text{-}J\text{-}K$

Rule-22844:
> If both **($)**, **(v)**, **(p)**, **(c)**, **(q)**, **(J)** and **(V)** are known,
> then its Kind of Cash Planned is:
> $K = \$vpc/\{360[c\text{-}q]\}\text{-}J\text{-}Vp/360$

Rule-22845:
> If both **(K)**, **(v)**, **(p)**, **(c)**, **(q)**, **(J)** and **(V)** are known,
> then its Sales or Revenue Planned is:
> $\$ = \{360[c\text{-}q][K\text{+}J\text{+}Vp]\}/[vpc]$

Rule-22846:
> If both **($)**, **(K)**, **(p)**, **(c)**, **(q)**, **(J)** and **(V)** are known,
> then its Variable Portion Planned is:
> $v = \{360[c\text{-}q][K\text{+}J\text{+}Vp]\}/[\$pc]$

Steve Asikin ISBN 14: 978-1515158691, ISBN 10: 1515158691

Rule-22847:

If both **(S)**, **(v)**, **(K)**, **(c)**, **(q)**, **(J)** and **(V)** are known, then its Procured Inventory Days Planned:

$$p = 360[K+J]/\{Svc/[c-q]-V\}$$

Rule-22848:

If both **(S)**, **(v)**, **(p)**, **(K)**, **(q)**, **(J)** and **(V)** are known, then its Current Ratio Planned is:

$$c = \{360[c-q][K+J+Vp]\}/[Svp]$$

Rule-22849:

If both **(S)**, **(v)**, **(p)**, **(c)**, **(q)**, **(K)** and **(V)** are known, then its Job or Trade Account Receivable Planned is:

$$J = Svpc/\{360[c-q]\}-K-Vp$$

Rule-22850:

If both **(S)**, **(v)**, **(p)**, **(c)**, **(q)**, **(J)** and **(K)** are known, then its Variable Cost Planned is:

$$V = Svc/\{360[c-q]\}-[J+K]/p$$

Rule-22851:

If both **(S)**, **(v)**, **(p)**, **(c)**, **(q)** and **(J)** are known, then its Kind of Cash Planned is:

$$K = Svpc/\{360[c-q]\}-J-Svp/360$$

Rule-22852:

If both **(K)**, **(v)**, **(p)**, **(c)**, **(q)** and **(J)** are known, then its Sales or Revenue Planned is:

$$S = (360[K+J]/\{c/[c-q]-1\})/[vp]$$

Steve Asikin ISBN 14: 978-1515158691, ISBN 10: 1515158691

Rule-22853:

If both $(\$)$, $(\mathbf{K})$, $(\mathbf{p})$, $(\mathbf{c})$, $(\mathbf{q})$ and $(\mathbf{J})$ are known, then its Variable Portion Planned is:
$$\mathbf{v} = (360[\mathbf{K+J}]/\{\mathbf{c}/[\mathbf{c-q}]-1\})/[\$\mathbf{p}]$$

Rule-22854:

If both $(\$)$, $(\mathbf{v})$, $(\mathbf{K})$, $(\mathbf{c})$, $(\mathbf{q})$ and $(\mathbf{J})$ are known, then its Procured Inventory Days Planned:
$$\mathbf{p} = (360[\mathbf{K+J}]/\{\mathbf{c}/[\mathbf{c-q}]-1\})/[\$\mathbf{v}]$$

Rule-22855:

If both $(\$)$, $(\mathbf{v})$, $(\mathbf{p})$, $(\mathbf{K})$, $(\mathbf{q})$ and $(\mathbf{J})$ are known, then its Current Ratio Planned is:
$$\mathbf{c} = \mathbf{q}\{360[\mathbf{K+J}]+\$\mathbf{vp}\}/(\{360[\mathbf{K+J}]+\$\mathbf{vp}\}-\$\mathbf{vp})$$

Rule-22856:

If both $(\$)$, $(\mathbf{v})$, $(\mathbf{p})$, $(\mathbf{c})$, $(\mathbf{K})$ and $(\mathbf{J})$ are known, then its Quick or Acid Test Ratio Planned is:
$$\mathbf{q} = \mathbf{c}(\{360[\mathbf{K+J}]+\$\mathbf{vp}\}-\$\mathbf{vp})/\{360[\mathbf{K+J}]+\$\mathbf{vp}\}$$

Rule-22857:

If both $(\$)$, $(\mathbf{v})$, $(\mathbf{p})$, $(\mathbf{c})$, $(\mathbf{q})$ and $(\mathbf{K})$ are known, then its Job or Trade Account Receivable Planned is:
$$\mathbf{J} = \$\mathbf{vpc}/\{360[\mathbf{c-q}]\}-\mathbf{K}-\$\mathbf{vp}/360$$

Rule-22858:

If both $(\$)$, $(\mathbf{v})$, $(\mathbf{p})$, $(\mathbf{c})$, $(\mathbf{q})$, $(\mathbf{J})$, $(\$')$ and $(\mathbf{s})$ are known, then its Kind of Cash Planned is:
$$\mathbf{K} = \$\mathbf{vpc}/\{360[\mathbf{c-q}]\}-\mathbf{J}-\$'\mathbf{vp}[1+\mathbf{s}]/360$$

Steve Asikin ISBN 14: 978-1515158691, ISBN 10: 1515158691

Rule-22859:

If both (K), (v), (p), (c), (q), (J), (S') and (s) are known, then its Sales or Revenue Planned is:

$S= [c-q]\{360[K+J]+S'vp[1+s]\}/[vpc]$

Rule-22860:

If both (S), (K), (p), (c), (q), (J), (S') and (s) are known, then its Variable Portion Planned is:

$v= 360[K+J]= \{Sc/[c-q]-S'[1+s]\}/p$

Rule-22861:

If both (S), (v), (K), (c), (q), (J), (S') and (s) are known, then its Procured Inventory Days Planned:

$p= 360[K+J]= \{Sc/[c-q]-S'[1+s]\}/v$

Rule-22862:

If both (S), (v), (p), (K), (q), (J), (S') and (s) are known, then its Current Ratio Planned is:

$c= q\{360[K+J]+S'vp[1+s]\}$
$/(\{360[K+J]+S'vp[1+s]\}-Svp)$

Rule-22863:

If both (S), (v), (p), (c), (K), (J), (S') and (s) are known, then its Quick or Acid Test Ratio Planned is:

$q= c(\{360[K+J]+S'vp[1+s]\}-Svp)$
$/\{360[K+J]+S'vp[1+s]\}$

Steve Asikin ISBN 14: 978-1515158691, ISBN 10: 1515158691

Rule-22864:

If both $(\textbf{\$})$, $(\textbf{v})$, $(\textbf{p})$, $(\textbf{c})$, $(\textbf{q})$, $(\textbf{K})$, $(\textbf{\$'})$ and $(\textbf{s})$ are known, then its Job or Trade Account Receivable Planned is:

$$J= \textbf{\$vpc}/\{360[\textbf{c-q}]\}\textbf{-K-\$'vp}[1\textbf{+s}]/360$$

Rule-22865:

If both $(\textbf{\$})$, $(\textbf{v})$, $(\textbf{p})$, $(\textbf{c})$, $(\textbf{q})$, $(\textbf{J})$, $(\textbf{K})$ and $(\textbf{s})$ are known, then its Sales Past must be:

$$\textbf{\$'}= \{\textbf{\$vpc}/[\textbf{c-q}]-360[\textbf{J+K}]\}/\{\textbf{vp}[1\textbf{+s}]\}$$

Rule-22866:

If both $(\textbf{\$})$, $(\textbf{v})$, $(\textbf{p})$, $(\textbf{c})$, $(\textbf{q})$, $(\textbf{J})$, $(\textbf{\$'})$ and $(\textbf{K})$ are known, then its Sales Growth Planned is:

$$s= \textbf{\$/\$'}-1$$

or it can also be found as

$$s= \{\textbf{\$vpc}/[\textbf{c-q}]-360[\textbf{J+K}]\}/[\textbf{\$'vp}]-1$$

Rule-22867:

If both $(\textbf{\$})$, $(\textbf{v})$, $(\textbf{p})$, $(\textbf{c})$, $(\textbf{q})$, $(\textbf{j})$ and $(\textbf{P})$ are known, then its Kind of Cash Planned is:

$$K= \textbf{\$vpc}/\{360[\textbf{c-q}]\}\textbf{-\$j}/360\textbf{-P}$$

Rule-22868:

If both $(\textbf{K})$, $(\textbf{v})$, $(\textbf{p})$, $(\textbf{c})$, $(\textbf{q})$, $(\textbf{j})$ and $(\textbf{P})$ are known, then its Sales or Revenue Planned is:

$$\$= 360[\textbf{K+P}]= (\textbf{vpc}/[\textbf{c-q}]-\textbf{j})$$

Steve Asikin ISBN 14: 978-1515158691, ISBN 10: 1515158691

Rule-22869:

If both $(\$)$, (K), (p), (c), (q), (j) and (P) are known, then its Variable Portion Planned is:

$v= [c\text{-}q]\{360[K+P+Sj]\}/[Spc]$

Rule-22870:

If both $(\$)$, (v), (K), (c), (q), (j) and (P) are known, then its Procured Inventory Days Planned:

$p= [c\text{-}q]\{360[K+P+Sj]\}/[Svc]$

Rule-22871:

If both $(\$)$, (v), (p), (K), (q), (j) and (P) are known, then its Current Ratio Planned is:

$c= q\{360[K+P]+Sj\}/(\{360[K+P]+Sj\}\text{-}Svp)$

Rule-22872:

If both $(\$)$, (v), (p), (c), (K), (j) and (P) are known, then its Quick or Acid Test Ratio Planned is:

$q= c(\{360[K+P]+Sj\}\text{-}Svp)/\{360[K+P]+Sj\}$

Rule-22873:

If both $(\$)$, (v), (p), (c), (q), (K) and (P) are known, then its Job or Trade Receivable Days Planned:

$j= \{Svpc/[c\text{-}q]\text{-}360[K+P]\}/\$$

Rule-22874:

If both $(\$)$, (v), (p), (c), (q), (j) and (K) are known, then its Procured Inventory Planned is:

$P= Svpc/\{360[c\text{-}q]\}\text{-}Sj/360\text{-}K$

Steve Asikin ISBN 14: 978-1515158691, ISBN 10: 1515158691

Rule-22875:

If both **($)**, **(v)**, **(p)**, **(c)**, **(q)**, **(j)** and **(V)** are known, then its Kind of Cash Planned is:

$K= \$vpc/\{360[c\text{-}q]\}\text{-}\$j/360\text{-}Vp/360$

Rule-22876:

If both **(K)**, **(v)**, **(p)**, **(c)**, **(q)**, **(j)** and **(V)** are known, then its Sales or Revenue Planned is:

$\$= [c\text{-}q][360K+\$j+Vp]/[vpc]$

Rule-22877:

If both **($)**, **(K)**, **(p)**, **(c)**, **(q)**, **(j)** and **(V)** are known, then its Variable Portion Planned is:

$v= [c\text{-}q][360K+\$j+Vp]/[\$pc]$

Rule-22878:

If both **($)**, **(v)**, **(K)**, **(c)**, **(q)**, **(j)** and **(V)** are known, then its Procured Inventory Days Planned:

$p= [360K+\$j]/\{\$vc/[c\text{-}q]\text{ -}V\}$

Rule-22879:

If both **($)**, **(v)**, **(p)**, **(K)**, **(q)**, **(j)** and **(V)** are known, then its Current Ratio Planned is:

$c= q[360K+\$j+Vp]/\{[360K+\$j+Vp]\text{-}\$vp\}$

Rule-22880:

If both **($)**, **(v)**, **(p)**, **(c)**, **(K)**, **(j)** and **(V)** are known, then its Quick or Acid Test Ratio Planned is:

$q= c\{[360K+\$j+Vp]\text{-}\$vp\}/[360K+\$j+Vp]$

Steve Asikin ISBN 14: 978-1515158691, ISBN 10: 1515158691

Rule-22881:

If both (S), (v), (p), (c), (q), (K) and (V) are known, then its Job or Trade Receivable Days Planned:

$j = \{Svpc/[c-q]-Vp-360K\}/S$

Rule-22882:

If both (S), (v), (p), (c), (q), (j) and (K) are known, then its Vartiable Cost Planned is:

$V = \{Svpc/[c-q]-Sj-360K\}/p$

Rule-22883:

If both (S), (v), (p), (c), (q) and (j) are known, then its Kind of Cash Planned is:

$K = Svpc/\{360[c-q]\}-Sj/360-Svp/360$

Rule-22884:

If both (Q), (v), (p), (c), (q) and (j) are known, then its Sales or Revenue Planned is:

$S = 360K/\{v(pc/[c-q]-p-j\}$

Rule-22885:

If both (S), (K), (p), (c), (q) and (j) are known, then its Variable Portion Planned is:

$v = [360K+Sj]/(Sp\{c/[c-q]-1\})$

Rule-22886:

If both (S), (v), (K), (c), (q) and (j) are known, then its Procured Inventory Days Planned:

$p = [360K+Sj]/(Sv\{c/[c-q]-1\})$

Steve Asikin ISBN 14: 978-1515158691, ISBN 10: 1515158691

Rule-22887:

If both $(\$)$, $(\mathbf{v})$, $(\mathbf{p})$, $(\mathbf{K})$, $(\mathbf{q})$ and $(\mathbf{j})$ are known, then its Current Ratio Planned is:

$$c= q[360K+\$j+\$vp]/\{[360K+\$j+\$vp]-\$vpc\}$$

Rule-22888:

If both $(\$)$, $(\mathbf{v})$, $(\mathbf{p})$, $(\mathbf{c})$, $(\mathbf{K})$ and $(\mathbf{j})$ are known, then its Quick or Acid Test Ratio Planned is:

$$q= c\{[360K+\$j+\$vp]-\$vpc\}/[360K+\$j+\$vp]$$

Rule-22889:

If both $(\$)$, $(\mathbf{v})$, $(\mathbf{p})$, $(\mathbf{c})$, $(\mathbf{q})$ and $(\mathbf{K})$ are known, then its Job or Trade Receivable Days Planned:

$$j= [\$vpc/[c-q]-\$vp-360K]/\$$$

Rule-22890:

If both $(\$)$, $(\mathbf{v})$, $(\mathbf{p})$, $(\mathbf{c})$, $(\mathbf{q})$, $(\mathbf{j})$, $(\$')$ and $(\mathbf{s})$ are known, then its Kind of Cash Planned is:

$$K= \$vpc/\{360[c-q]\}-\$j/360-\$'vp[1+s]/360$$

Rule-22891:

If both $(\mathbf{K})$, $(\mathbf{v})$, $(\mathbf{p})$, $(\mathbf{c})$, $(\mathbf{q})$, $(\mathbf{j})$, $(\$')$ and $(\mathbf{s})$ are known, then its Sales or Revenue Planned is:

$$\$= \{360K-\$'vp[1+s]\}/\{vpc/[c-q]-j\}$$

Rule-22892:

If both $(\$)$, $(\mathbf{K})$, $(\mathbf{p})$, $(\mathbf{c})$, $(\mathbf{q})$, $(\mathbf{j})$, $(\$')$ and $(\mathbf{s})$ are known, then its Variable Portion Planned is:

$$v= [360K-\$j]/(p\{\$c/[c-q]-\$'[1+s]\})$$

Steve Asikin ISBN 14: 978-1515158691, ISBN 10: 1515158691

Rule-22893:

If both $(\$)$, $(\mathbf{v})$, $(\mathbf{K})$, $(\mathbf{c})$, $(\mathbf{q})$, $(\mathbf{j})$, $(\$')$ and $(\mathbf{s})$ are known, then its Procured Inventory Days Planned:

$$p= [360K\text{-}\$j]/(v\{\$c/[c\text{-}q]\text{-}\$'[1+s]\})$$

Rule-22894:

If both $(\$)$, $(\mathbf{v})$, $(\mathbf{p})$, $(\mathbf{K})$, $(\mathbf{q})$, $(\mathbf{j})$, $(\$')$ and $(\mathbf{s})$ are known, then its Current Ratio Planned is:

$$c= q\{360K\text{-}\$j\text{-}\$'vp[1+s]\}$$
$$/(\{360K\text{-}\$j\text{-}\$'vp[1+s]\}\text{-}\$vpc)$$

Rule-22895:

If both $(\$)$, $(\mathbf{v})$, $(\mathbf{p})$, $(\mathbf{c})$, $(\mathbf{K})$, $(\mathbf{j})$, $(\$')$ and $(\mathbf{s})$ are known, then its Quick or Acid Test Ratio Planned is:

$$q= c(\{360K\text{-}\$j\text{-}\$'vp[1+s]\}\text{-}\$vpc)$$
$$/\{360K\text{-}\$j\text{-}\$'vp[1+s]\}$$

Rule-22896:

If both $(\$)$, $(\mathbf{v})$, $(\mathbf{p})$, $(\mathbf{c})$, $(\mathbf{q})$, $(\mathbf{K})$, $(\$')$ and $(\mathbf{s})$ are known, then its Job or Trade Receivable Days Planned:

$$j= \{\$vpc/[c\text{-}q]\text{-}\$'vp[1+s]\text{-}360K\}/\$$$

Rule-22897:

If both $(\$)$, $(\mathbf{v})$, $(\mathbf{p})$, $(\mathbf{c})$, $(\mathbf{q})$, $(\mathbf{j})$, $(\mathbf{K})$ and $(\mathbf{s})$ are known, then its Sales Past must be:

$$\$'= \{\$vpc/[c\text{-}q]\text{-}\$j\text{-}360K\}/\{vp[1+s]\}$$

Steve Asikin ISBN 14: 978-1515158691, ISBN 10: 1515158691

Rule-22898:

If both $(\$)$, (v), (p), (c), (q), (j), $(\$')$ and (K) are known, then its Sales Growth is:

$s = \$/\$' - 1$

or it can also be found as

$s = \{\$vpc/[c-q]-\$j-360K\}/[\$'vp] - 1$

Rule-22899:

If both $(\$)$, (v), (p), (c), (q), (j), $(\$')$, (s) and (P) are known, then its Kind of Cash Planned is:

$K = \$vpc/\{360[c-q]\} - \$'j[1+s]/360 - P$

Rule-22900:

If both (K), (v), (p), (c), (q), (j), $(\$')$, (s) and (P) are known, then its Sales or Revenue Planned is:

$\$ = [c-q]\{360[K+P]+\$'j[1+s]\}/[vpc]$

Rule-22901:

If both $(\$)$, (K), (p), (c), (q), (j), $(\$')$, (s) and (P) are known, then its Variable Portion Planned is:

$v = [c-q]\{360[K+P]+\$'j[1+s]\}/[\$pc]$

Rule-22902:

If both $(\$)$, (v), (K), (c), (q), (j), $(\$')$, (s) and (P) are known, then its Procured Inventory Days Planned:

$p = [c-q]\{360[K+P]+\$'j[1+s]\}/[\$vc]$

Steve Asikin ISBN 14: 978-1515158691, ISBN 10: 1515158691

Rule-22903:
>   If both $(\$)$, (υ), (p), (K), (q), (j), $(\$')$, (s) and (P) are known, then its Current Ratio Planned is:
> $$c= q\{360[K+P]+\$'j[1+s]\}$$
> $$/(\{360[K+P]+\$'j[1+s]\}-\$upc)$$

Rule-22904:
>   If both $(\$)$, (υ), (p), (c), (K), (j), $(\$')$, (s) and (P) are known, then its Quick or Acid Test Ratio Planned is:
> $$q= c(\{360[K+P]+\$'j[1+s]\}-\$upc)$$
> $$/\{360[K+P]+\$'j[1+s]\}$$

Rule-22905:
>   If both $(\$)$, (υ), (p), (c), (q), (j), (K), (s) and (P) are known, then its Sales Past must be:
> $$\$'= \{\$upc/[c-q]-360[K+P]\}/\{j[1+s]\}$$

Rule-22906:
>   If both $(\$)$, (υ), (p), (c), (q), (K), $(\$')$, (s) and (P) are known, then its Job or Trade Receivable Days Planned is:
> $$j= \{\$upc/[c-q]-360[K+P]\}/\{\$'[1+s]\}$$

Rule-22907:
>   If both $(\$)$, (υ), (p), (c), (q), (j), $(\$')$, (K) and (P) are known, then its Sales Growth Planned is:
> $$s= \$/\$'-1$$
>   or it can also be found as
> $$s= \{\$upc/[c-q]-360[K+P]\}/[\$'j]-1$$

Steve Asikin ISBN 14: 978-1515158691, ISBN 10: 1515158691

Rule-22908:

If both $(\$)$, $(\mathbf{v})$, $(\mathbf{p})$, $(\mathbf{c})$, $(\mathbf{q})$, $(\mathbf{j})$, $(\$')$, $(\mathbf{s})$ and $(\mathbf{K})$ are known, then its Procured Inventory Planned is:

$\mathbf{P} = \$vpc/\{360[\mathbf{c}\text{-}\mathbf{q}]\}\text{-}\$'j[1+s]/360\text{-}\mathbf{K}$

Rule-22909:

If both $(\$)$, $(\mathbf{v})$, $(\mathbf{p})$, $(\mathbf{c})$, $(\mathbf{q})$, $(\mathbf{j})$, $(\$')$, $(\mathbf{s})$ and $(\mathbf{V})$ are known, then its Kind of Cash Planned is:

$\mathbf{K} = \$vpc/\{360[\mathbf{c}\text{-}\mathbf{q}]\}\text{-}\$'j[1+s]/360\text{-}\mathbf{Vp}/360$

Rule-22910:

If both $(\mathbf{K})$, $(\mathbf{v})$, $(\mathbf{p})$, $(\mathbf{c})$, $(\mathbf{q})$, $(\mathbf{j})$, $(\$')$, $(\mathbf{s})$ and $(\mathbf{V})$ are known, then its Sales or Revenue Planned is:

$\$ = [\mathbf{c}\text{-}\mathbf{q}]\{360\mathbf{K}\text{-}\$'j[1+s]\text{-}\mathbf{Vp}\}/[vpc]$

Rule-22911:

If both $(\$)$, $(\mathbf{K})$, $(\mathbf{p})$, $(\mathbf{c})$, $(\mathbf{q})$, $(\mathbf{j})$, $(\$')$, $(\mathbf{s})$ and $(\mathbf{V})$ are known, then its Kind of Cash Planned is:

$\mathbf{v} = \mathbf{V}/\$$

or it can also be found as

$\mathbf{v} = [\mathbf{c}\text{-}\mathbf{q}]\{360\mathbf{K}\text{-}\$'j[1+s]\text{-}\mathbf{Vp}\}/[\$pc]$

Rule-22912:

If both $(\$)$, $(\mathbf{v})$, $(\mathbf{K})$, $(\mathbf{c})$, $(\mathbf{q})$, $(\mathbf{j})$, $(\$')$, $(\mathbf{s})$ and $(\mathbf{V})$ are known, then its Procured Inventory Days Planned:

$\mathbf{p} = \{360\mathbf{K}+\$'j[1+s]\}/\{\$vc/[\mathbf{c}\text{-}\mathbf{q}]\text{-}\mathbf{V}\}$

Steve Asikin ISBN 14: 978-1515158691, ISBN 10: 1515158691

Rule-22913:

If both $(\$)$, $(\mathbf{v})$, $(\mathbf{p})$, $(\mathbf{K})$, $(\mathbf{q})$, $(\mathbf{j})$, $(\$')$, $(\mathbf{s})$ and $(\mathbf{V})$ are known, then its Current Ratio Planned is:

$$c = \mathbf{q}\{360\mathbf{K}\text{-}\$'\mathbf{j}[1+\mathbf{s}]\text{-}\mathbf{Vp}\}$$
$$/(\{360\mathbf{K}\text{-}\$'\mathbf{j}[1+\mathbf{s}]\text{-}\mathbf{Vp}\}\text{-}\$\mathbf{vpc})$$

Rule-22914:

If both $(\$)$, $(\mathbf{v})$, $(\mathbf{p})$, $(\mathbf{c})$, $(\mathbf{K})$, $(\mathbf{j})$, $(\$')$, $(\mathbf{s})$ and $(\mathbf{V})$ are known, then its Quick or Acid Test Ratio Planned is:

$$q = \mathbf{c}(\{360\mathbf{K}\text{-}\$'\mathbf{j}[1+\mathbf{s}]\text{-}\mathbf{Vp}\}\text{-}\$\mathbf{vpc})$$
$$/\{360\mathbf{K}\text{-}\$'\mathbf{j}[1+\mathbf{s}]\text{-}\mathbf{Vp}\}$$

Rule-22915:

If both $(\$)$, $(\mathbf{v})$, $(\mathbf{p})$, $(\mathbf{c})$, $(\mathbf{q})$, $(\mathbf{j})$, $(\mathbf{Q})$, $(\mathbf{s})$ and $(\mathbf{V})$ are known, then its Sales Past must be:

$$\$' = \{\$\mathbf{vpc}/[\mathbf{c}\text{-}\mathbf{q}]\text{-}\mathbf{Vp}\text{-}360\,\mathbf{K}\}/\{\mathbf{j}[1+\mathbf{s}]\}$$

Rule-22916:

If both $(\$)$, $(\mathbf{v})$, $(\mathbf{p})$, $(\mathbf{c})$, $(\mathbf{q})$, $(\mathbf{K})$, $(\$')$, $(\mathbf{s})$ and $(\mathbf{V})$ are known, then its Job or Trade Receivable Days Planned is:

$$j = \{\$\mathbf{vpc}/[\mathbf{c}\text{-}\mathbf{q}]\text{-}\mathbf{Vp}\text{-}360\,\mathbf{K}\}/\{\$'[1+\mathbf{s}]\}$$

Rule-22917:

If both $(\$)$, $(\mathbf{v})$, $(\mathbf{p})$, $(\mathbf{c})$, $(\mathbf{q})$, $(\mathbf{j})$, $(\$')$, $(\mathbf{s})$ and $(\mathbf{V})$ are known, then its Kind of Cash Planned is:

$$s = \$/\$'\text{-}1$$

or it can also be found as

$$s = \{\$\mathbf{vpc}/[\mathbf{c}\text{-}\mathbf{q}]\text{-}\mathbf{Vp}\text{-}360\,\mathbf{K}\}/[\$'\mathbf{j}]\text{-}1$$

Steve Asikin ISBN 14: 978-1515158691, ISBN 10: 1515158691

Rule-22918:

    If both $(\$)$, (v), (p), (c), (q), (j), $(\$')$, (s) and (K) are known, then its Variable Cost Planned is:

$$V = \{\$vpc/[c-q]-\$'j[1+s] -360\ K\}/p$$

Rule-22919:

    If both $(\$)$, (v), (p), (c), (q), $(\$')$, (j) and (s) are known, then its Kind of Cash Planned is:

$$K= \$vpc/\{360[c-q]\}-\$'j[1+s]/360-\$vp/360$$

Rule-22920:

    If both (K), (v), (p), (c), (q), $(\$')$, (j) and (s) are known, then its Sales or Revenue Planned is:

$$\$= \{360K+\$'j[1+s]\}/(vp\{c/[c-q]-1\})$$

Rule-22921:

    If both $(\$)$, (K), (p), (c), (q), $(\$')$, (j) and (s) are known, then its Variable Portion Planned is:

$$v= \{360K+\$'j[1+s]\}/(\$p\{c/[c-q]-1\})$$

Rule-22922:

    If both $(\$)$, (v), (K), (c), (q), $(\$')$, (j) and (s) are known, then its Procured Inventory Days Planned is:

$$p= \{360K+\$'j[1+s]\}/(\$v\{c/[c-q]-1\})$$

Rule-22923:

    If both $(\$)$, (v), (p), (K), (q), $(\$')$, (j) and (s) are known, then its Current Ratio Planned is:

$$c= q\{360K+\$'j[1+s]+\$vp\}$$
$$/(\{360K+\$'j[1+s]+\$vp\}-\$vpc)$$

Steve Asikin ISBN 14: 978-1515158691, ISBN 10: 1515158691

Rule-22924:
  If both $(\$)$, $(\mathbf{v})$, $(\mathbf{p})$, $(\mathbf{c})$, $(\mathbf{K})$, $(\$')$, $(\mathbf{j})$ and $(\mathbf{s})$ are
  known, then its Quick or Acid Test Ratio Planned is:
  $$q= c(\{360K+\$'j[1+s]+\$vp\}-\$vpc)$$
  $$/\{360K+\$'j[1+s]+\$vp\}$$

Rule-22925:
  If both $(\$)$, $(\mathbf{v})$, $(\mathbf{p})$, $(\mathbf{c})$, $(\mathbf{q})$, $(\mathbf{Q})$, $(\mathbf{j})$ and $(\mathbf{s})$ are
  known, then its Sales Past must be:
  $$\$'= \{\$vpc/[c\text{-}q] -\$vp\text{-}360K\}/\{j[1+s]\}$$

Rule-22926:
  If both $(\$)$, $(\mathbf{v})$, $(\mathbf{p})$, $(\mathbf{c})$, $(\mathbf{q})$, $(\$')$, $(\mathbf{K})$ and $(\mathbf{s})$ are
  known, then its Job or Trade Receivable Days
  Planned is:
  $$j= \{\$vpc/[c\text{-}q] -\$vp\text{-}360K\}/\{\$'[1+s]\}$$

Rule-22927:
  If both $(\$)$, $(\mathbf{v})$, $(\mathbf{p})$, $(\mathbf{c})$, $(\mathbf{q})$, $(\$')$, $(\mathbf{j})$ and $(\mathbf{Q})$ are
  known, then its Sales Growth Planned is:
  $$s= \$/\$'\text{-}1$$
  or it can also be found as
  $$s= \{\$vpc/[c\text{-}q] -\$vp\text{-}360K\}/[\$'j]\text{-}1$$

Rule-22928:
  If both $(\$)$, $(\mathbf{v})$, $(\mathbf{p})$, $(\mathbf{c})$, $(\mathbf{q})$, $(\$')$, $(\mathbf{j})$ and $(\mathbf{s})$ are
  known, then its Kind of Cash Planned is:
  $$K= \$vpc/\{360[c\text{-}q]\}\text{-}\$'j[1+s]/360\text{-}\$'vp[1+s]/360$$

Steve Asikin ISBN 14: 978-1515158691, ISBN 10: 1515158691

Rule-22929:

If both (K), (v), (p), (c), (q), (S'), (j) and (s) are known, then its Sales or Revenue Planned is:

$S= [c-q]\{360K+S'[1+s][j+vp]\}/[vpc]$

Rule-22930:

If both (S), (K), (p), (c), (q), (S'), (j) and (s) are known, then its Variable Portion Planned is:

$v=\{360K+S'j[1+s]\}/(p\{Sc/[c-q]-S'[1+s]\})$

Rule-22931:

If both (S), (v), (K), (c), (q), (S'), (j) and (s) are known, then its Procured Inventory Days Planned:

$p=\{360K+S'j[1+s]\}/(v\{Sc/[c-q]-S'[1+s]\})$

Rule-22932:

If both (S), (v), (p), (K), (q), (S'), (j) and (s) are known, then its Current Ratio Planned is:

$c= q\{360K+S'j[1+s]+S'vp[1+s]\}$
$/(\{360K+S'j[1+s]+S'vp[1+s]\}-Svpc)$

Rule-22933:

If both (S), (v), (p), (c), (K), (S'), (j) and (s) are known, then its Quick or Acid Test Ratio Planned is:

$q= c(\{360K+S'j[1+s]+S'vp[1+s]\}-Svpc)$
$/\{360K+S'j[1+s]+S'vp[1+s]\}$

Rule-22934:

If both (S), (v), (p), (c), (q), (K), (j) and (s) are known, then its Sales Past must be:

$S'= \{Svpc/[c-q]- 360K\}/\{j[1+s]+vp[1+s]\}$

Steve Asikin ISBN 14: 978-1515158691, ISBN 10: 1515158691

Rule-22935:

    If both $(\$)$, $(\mathbf{v})$, $(\mathbf{p})$, $(\mathbf{c})$, $(\mathbf{q})$, $(\$')$, $(\mathbf{K})$ and $(\mathbf{s})$ are known, then its Job Receivable Days Planned:

    $j = \{\$vpc/[c-q]-\$'vp[1+s]-360K\}/\{\$'[1+s]\}$

Rule-22936:

    If both $(\$)$, $(\mathbf{v})$, $(\mathbf{p})$, $(\mathbf{c})$, $(\mathbf{q})$, $(\$')$, $(\mathbf{j})$ and $(\mathbf{s})$ are known, then its Kind of Cash Planned is:

    $s = \$/\$'-1$

        or it can also be found as

    $s = \{\$vpc/[c-q] -360K\}/\{\$'[j+vp]\}-1$

Rule-22937:

    If both $(\$')$, $(\mathbf{v})$, $(\mathbf{p})$, $(\mathbf{c})$, $(\mathbf{s})$, $(\mathbf{q})$, $(\mathbf{J})$ and $(\mathbf{P})$ are known, then its Kind of Cash Planned is:

    $K = \$'vpc[1+s]/\{360[c-q]\}-J-P$

Rule-22938:

    If both $(\mathbf{K})$, $(\mathbf{v})$, $(\mathbf{p})$, $(\mathbf{c})$, $(\mathbf{s})$, $(\mathbf{q})$, $(\mathbf{J})$ and $(\mathbf{P})$ are known, then its Sales Past must be:

    $\$' = 360[c-q][K+J+P]/\{vpc[1+s]\}$

Rule-22939:

    If both $(\$')$, $(\mathbf{K})$, $(\mathbf{p})$, $(\mathbf{c})$, $(\mathbf{s})$, $(\mathbf{q})$, $(\mathbf{J})$ and $(\mathbf{P})$ are known, then its Variable Portion Planned is:

    $v = 360[c-q][K+J+P]/\{\$'pc[1+s]\}$

Rule-22940:

    If both $(\$')$, $(\mathbf{v})$, $(\mathbf{K})$, $(\mathbf{c})$, $(\mathbf{s})$, $(\mathbf{q})$, $(\mathbf{J})$ and $(\mathbf{P})$ are known, then its Procured Inventory Days Planned:

    $p = 360[c-q][K+J+P]/\{\$'vc[1+s]\}$

Steve Asikin ISBN 14: 978-1515158691, ISBN 10: 1515158691

Rule-22941:
If both (**S'**), (**v**), (**p**), (**K**), (**s**), (**q**), (**J**) and (**P**) are known, then its Current Ratio Planned is:
$$c= 360q[K+J+P]/\{360[K+J+P]-S'vpc[1+s]\}$$

Rule-22942:
If both (**S'**), (**v**), (**p**), (**c**), (**K**), (**q**), (**J**) and (**P**) are known, then its Sales Growth Planned is:
$$s= 360[c-q][K+J+P]/[S'vpc]-1$$

Rule-22943:
If both (**S'**), (**v**), (**p**), (**c**), (**s**), (**K**), (**J**) and (**P**) are known, then its Quick or Acid Test Ratio Planned is:
$$q= c\{360[K+J+P]-S'vpc[1+s]\}/\{360[K+J+P]\}$$

Rule-22944:
If both (**S'**), (**v**), (**p**), (**c**), (**s**), (**q**), (**K**) and (**P**) are known, then its Jobs or Trade Account Receivable Planned is:
$$J= S'vpc[1+s]/\{360[c-q]\}-K-P$$

Rule-22945:
If both (**S'**), (**v**), (**p**), (**c**), (**s**), (**q**), (**J**) and (**K**) are known, then its Procured Inventory Planned is:
$$P= S'vpc[1+s]/\{360[c-q]\}-J-K$$

Rule-22946:
If both (**S'**), (**v**), (**p**), (**c**), (**s**), (**q**), (**J**) and (**V**) are known, then its Kind of Cash Planned is:
$$K= S'vpc[1+s]/\{360[c-q]\}-J-Vp/360$$

Steve Asikin ISBN 14: 978-1515158691, ISBN 10: 1515158691

Rule-22947:

If both (K), (v), (p), (c), (s), (q), (J) and (V) are known, then its Sales Past must be:
$$S'= [c\text{-}q]\{360[K+J]+Vp\}/\{vpc[1+s]\}$$

Rule-22948:

If both (S'), (K), (p), (c), (s), (q), (J) and (V) are known, then its Variable Portion Planned is:
$$v= [c\text{-}q]\{360[K+J]+Vp\}/\{S'pc[1+s]\}$$

Rule-22949:

If both (S'), (v), (K), (c), (s), (q), (J) and (V) are known, then its Procured Inventory Days Planned:
$$p= \{360[K+J]\}/\{S'vc[1+s]/[c\text{-}q]\text{-}V\}$$

Rule-22950:

If both (S'), (v), (p), (K), (s), (q), (J) and (V) are known, then its Current Ratio Planned is:
$$c= q\{360[K+J]+Vp\}/(\{360[K+J]+Vp\}\text{-}S'vp[1+s])$$

Rule-22951:

If both (S'), (v), (p), (c), (K), (q), (J) and (V) are known, then its Sales Growth Planned is:
$$s= [c\text{-}q]\{360[K+J]+Vp\}/[S'vpc]\text{-}1$$

Rule-22952:

If both (S'), (v), (p), (c), (s), (q), (J) and (V) are known, then its Kind of Cash Planned is:
$$q= c(\{360[K+J]+Vp\}\text{-}S'vp[1+s])/\{360[K+J]+Vp\}$$

Steve Asikin ISBN 14: 978-1515158691, ISBN 10: 1515158691

Rule-22953:

If both ($), (v), (p), (c), (s), (q), (K) and (V) are known, then its Job or Trade Account Receivable Planned is:

$$J= S'vpc[1+s]/\{360[c-q]\}-K-Vp/360$$

Rule-22954:

If both ($), (v), (p), (c), (s), (q), (J) and (K) are known, then its Variable Cost Planned is:

$$V= \{S'vpc[1+s]/[c-q]-360[J+K]\}/p$$

Rule-22955:

If both ($), (v), (p), (c), (s), (q), (J) and ($) are known, then its Kind of Cash Planned is:

$$K= S'vpc[1+s]/\{360[c-q]\}-J-Svp/360$$

Rule-22956:

If both (K), (v), (p), (c), (s), (q), (J) and ($) are known, then its Sales Past must be:

$$S'= [c-q]\{360[K+J]+Svp\}/\{vpc[1+s]\}$$

Rule-22957:

If both ($), (K), (p), (c), (s), (q), (J) and ($) are known, then its Variable Portion Planned is:

$$v= 360[K+J]/(p\{S'c[1+s]/[c-q]-S\})$$

Rule-22958:

If both ($), (v), (K), (c), (s), (q), (J) and ($) are known, then its Procured Inventory Days Planned:

$$p= 360[K+J]/(v\{S'c[1+s]/[c-q]-S\})$$

Steve Asikin ISBN 14: 978-1515158691, ISBN 10: 1515158691

Rule-22959:

If both (S'), (v), (p), (K), (s), (q), (J) and (S) are known, then its Current Ratio Planned is:

$$c = q\{360[K+J]+Svp\}$$
$$/(\{360[K+J]+Svp\}-S'vp\ c[1+s])$$

Rule-22960:

If both (S'), (v), (p), (c), (K), (q), (J) and (S) are known, then its Sales Growth Planned is:

$$s = S/S'-1$$

or it can also be found as

$$s = [c-q]\{360[K+J]+Svp\}/[S'vpc]-1$$

Rule-22961:

If both (S'), (v), (p), (c), (s), (K), (J) and (S) are known, then its Quick or Acid Test Ratio Planned is:

$$q = c(\{360[K+J]+Svp\}-S'vp\ c[1+s])$$
$$/\{360[K+J]+Svp\}$$

Rule-22962:

If both (S'), (v), (p), (c), (s), (q), (K) and (S) are known, then its Job or Trade Account Receivable Planned is:

$$J = S'vpc[1+s]/\{360[c-q]\}-K-Svp/360$$

Rule-22963:

If both (S'), (v), (p), (c), (s), (q), (J) and (K) are known, then its Sales or Revenue Planned is:

$$S = \{S'vpc[1+s]/[c-q]-360[K+J]\}/[vp]$$

Steve Asikin ISBN 14: 978-1515158691, ISBN 10: 1515158691

Rule-22964:

    If both (S'), (v), (p), (c), (s), (q) and (J) are known, then its Kind of Cash Planned is:

$$K = S'vpc[1+s]/\{360[c-q]\}-J-S'vp[1+s]/360$$

Rule-22965:

    If both (K), (v), (p), (c), (s), (q) and (J) are known, then its Sales Past must be:

$$S' = \{360[K+J]/(vp[1+s]\{c/[c-q]-1\})$$

Rule-22966:

    If both (S'), (K), (p), (c), (s), (q) and (J) are known, then its Variable Portion Planned is:

$$v = \{360[K+J]/(S'p[1+s]\{c/[c-q]-1\})$$

Rule-22967:

    If both (S'), (v), (K), (c), (s), (q) and (J) are known, then its Procured Inventory Days Planned:

$$p = \{360[K+J]/(S'v[1+s]\{c/[c-q]-1\})$$

Rule-22968:

    If both (S'), (v), (p), (c), (K), (q) and (J) are known, then its Sales Growth Planned is:

$$s = \{360[K+J]/(S'vp\{c/[c-q]-1\})-1$$

Rule-22969:

    If both (S'), (v), (p), (K), (s), (q) and (J) are known, then its Current Ratio Planned is:

$$c = q\{360[K+J]+S'vp[1+s]\}$$
$$/(\{360[K+J]+S'vp[1+s]\}-S'vpc[1+s])$$

Steve Asikin ISBN 14: 978-1515158691, ISBN 10: 1515158691

Rule-22970:

If both $(\textbf{\$'})$, $(\textbf{v})$, $(\textbf{p})$, $(\textbf{c})$, $(\textbf{s})$, $(\textbf{K})$ and $(\textbf{J})$ are known, then its Quick or Acid Test Ratio Planned is:

$$q= c(\{360[K+J]+\$'vp[1+s]\}-\$'vpc[1+s])$$
$$/\{360[K+J]+\$'vp[1+s]\}$$

Rule-22971:

If both $(\textbf{\$'})$, $(\textbf{v})$, $(\textbf{p})$, $(\textbf{c})$, $(\textbf{s})$, $(\textbf{q})$ and $(\textbf{K})$ are known, then its Job or Trade Account Receivable Planned is:

$$J= \$'vpc[1+s]/\{360[c-q]\}-K-\$'vp[1+s]/360$$

Rule-22972:

If both $(\textbf{\$'})$, $(\textbf{v})$, $(\textbf{p})$, $(\textbf{c})$, $(\textbf{s})$, $(\textbf{q})$, $(\textbf{\$})$, $(\textbf{j})$ and $(\textbf{P})$ are known, then its Kind of Cash Planned is:

$$K= \$'vpc[1+s]/\{360[c-q]\}-\$j/360-P$$

Rule-22973:

If both $(\textbf{K})$, $(\textbf{v})$, $(\textbf{p})$, $(\textbf{c})$, $(\textbf{s})$, $(\textbf{q})$, $(\textbf{\$})$, $(\textbf{j})$ and $(\textbf{P})$ are known, then its Sales Past must be:

$$\$'= [c-q]\{360[K+P]+\$j\}/\{vpc[1+s]\}$$

Rule-22974:

If both $(\textbf{\$'})$, $(\textbf{K})$, $(\textbf{p})$, $(\textbf{c})$, $(\textbf{s})$, $(\textbf{q})$, $(\textbf{\$})$, $(\textbf{j})$ and $(\textbf{P})$ are known, then its Variable Portion Planned is:

$$v= [c-q]\{360[K+P]+\$j\}/\{\$'pc[1+s]\}$$

Rule-22975:

If both $(\textbf{\$'})$, $(\textbf{v})$, $(\textbf{K})$, $(\textbf{c})$, $(\textbf{s})$, $(\textbf{q})$, $(\textbf{\$})$, $(\textbf{j})$ and $(\textbf{P})$ are known, then its Procured Inventory Days Planned:

$$p= [c-q]\{360[K+P]+\$j\}/\{\$'vc[1+s]\}$$

Steve Asikin ISBN 14: 978-1515158691, ISBN 10: 1515158691

Rule-22976:

If both (**$'**), (**υ**), (**p**), (**K**), (**s**), (**q**), (**$**), (**j**) and (**P**) are known, then its Current Ratio Planned is:

c= **q**{360[**K**+**P**]-**$j**}/({360[**K**+**P**]-**$j**}-**$'υpc**[1+**s**])

Rule-22977:

If both (**$'**), (**υ**), (**p**), (**c**), (**K**), (**q**), (**$**), (**j**) and (**P**) are known, then its Kind of Cash Planned is:

s= **$/$'**-1

or it can also be found as

s= [**c**-**q**]{360[**K**+**P**]+**$j**}/[**$'υpc**]-1

Rule-22978:

If both (**$'**), (**υ**), (**p**), (**c**), (**s**), (**K**), (**$**), (**j**) and (**P**) are known, then its Quick or Acid Test Ratio Planned is:

q= **c**({360[**K**+**P**]+**$j**}-**$'υp**[1+**s**])/{360[**K**+**P**]+**$j**}

Rule-22979:

If both (**$'**), (**υ**), (**p**), (**c**), (**s**), (**q**), (**K**), (**j**) and (**P**) are known, then its Sales or Revenue Planned is:

$**= {**$'υpc[1+**s**]/[**c**-**q**]-360[**K**+**P**]}/**j**

Rule-22980:

If both (**$'**), (**υ**), (**p**), (**c**), (**s**), (**q**), (**$**), (**K**) and (**P**) are known, then its Jobs or Trade Receivable Days Planned is:

j= {**$'υpc**[1+**s**]/[**c**-**q**]-360[**K**+**P**]}/**$**

Steve Asikin ISBN 14: 978-1515158691, ISBN 10: 1515158691

Rule-22981:

If both (S'), (v), (p), (c), (s), (q), (S), (j) and (K) are known, then its Procured Inventory Planned is:

$P = S'vpc[1+s]/\{360[c-q]\}-Sj/360-K$

Rule-22982:

If both (S'), (v), (p), (c), (s), (q), (S), (j) and (V) are known, then its Kind of Cash Planned is:

$K = S'vpc[1+s]/\{360[c-q]\}-Sj/360-Vp/360$

Rule-22983:

If both (K), (v), (p), (c), (s), (q), (S), (j) and (V) are known, then its Sales Past must be:

$S' = [c-q]\{360K+Sj+Vp\}/\{vpc[1+s]\}$

Rule-22984:

If both (S'), (K), (p), (c), (s), (q), (S), (j) and (V) are known, then its Variable Portion Planned is:

$v = [c-q]\{360K+Sj+Vp\}/\{S'pc[1+s]\}$

Rule-22985:

If both (S'), (v), (K), (c), (s), (q), (S), (j) and (V) are known, then its Procured Inventory Days Planned:

$p = [360K+Sj]/\{S'vc[1+s]/[c-q]-V\}$

Rule-22986:

If both (S'), (v), (p), (c), (K), (q), (S), (j) and (V) are known, then its Sales Griwth Planned is:

$s = S/S'-1$

or it can also be found as

$s = [c-q]\{360K+Sj+Vp\}/[S'vpc]-1$

Steve Asikin ISBN 14: 978-1515158691, ISBN 10: 1515158691

Rule-22987:

If both $(\textbf{S'})$, $(\textbf{v})$, $(\textbf{p})$, $(\textbf{K})$, $(\textbf{s})$, $(\textbf{q})$, $(\textbf{S})$, $(\textbf{j})$ and $(\textbf{V})$ are known, then its Current Ratio Planned is:

$c= q[360K+Sj+Vp]/([360K+Sj+Vp]-S'vp[1+s])$

Rule-22988:

If both $(\textbf{S'})$, $(\textbf{v})$, $(\textbf{p})$, $(\textbf{c})$, $(\textbf{s})$, $(\textbf{K})$, $(\textbf{S})$, $(\textbf{j})$ and $(\textbf{V})$ are known, then its Quick or Acid Test Ratio Planned is:

$q= c([360K+Sj+Vp]-S'vp[1+s])/[360K+Sj+Vp]$

Rule-22989:

If both $(\textbf{S'})$, $(\textbf{v})$, $(\textbf{p})$, $(\textbf{c})$, $(\textbf{s})$, $(\textbf{q})$, $(\textbf{K})$, $(\textbf{j})$ and $(\textbf{V})$ are known, then its Sales or Revenue Planned is:

$S= \{S'vpc[1+s]/[c-q] -Vp-360K\}/j$

Rule-22990:

If both $(\textbf{S'})$, $(\textbf{v})$, $(\textbf{p})$, $(\textbf{c})$, $(\textbf{s})$, $(\textbf{q})$, $(\textbf{S})$, $(\textbf{K})$ and $(\textbf{V})$ are known, then its Job or Trade Receivable Days Planned:

$j= \{S'vpc[1+s]/[c-q] -Vp-360K\}/S$

Rule-22991:

If both $(\textbf{S'})$, $(\textbf{v})$, $(\textbf{p})$, $(\textbf{c})$, $(\textbf{s})$, $(\textbf{q})$, $(\textbf{S})$, $(\textbf{j})$ and $(\textbf{K})$ are known, then its Variable Cost Planned is:

$V= S'vpc[1+s]/[c-q]-Sj-360K\}/p$

Rule-22992:

If both $(\textbf{S'})$, $(\textbf{v})$, $(\textbf{p})$, $(\textbf{c})$, $(\textbf{s})$, $(\textbf{q})$, $(\textbf{S})$ and $(\textbf{j})$ are known, then its Kind of Cash Planned is:

$K= S'vpc[1+s]/\{360[c-q]\}-Sj/360-Svp/360$

Steve Asikin ISBN 14: 978-1515158691, ISBN 10: 1515158691

Rule-22993:
> If both (K), (v), (p), (c), (s), (q), $(\$)$ and (j) are
> known, then its Sales Past must be:
> $\$' = [c-q]\{360K+\$[j+vp]\}/\{vpc[1+s]\}$

Rule-22994:
> If both $(\$')$, (K), (p), (c), (s), (q), $(\$)$ and (j) are
> known, then its Variable Portion Planned is:
> $v = [360K+\$j]/(p\{\$'c[1+s]/[c-q]-\$\})$

Rule-22995:
> If both $(\$')$, (v), (K), (c), (s), (q), $(\$)$ and (j) are
> known, then its Procured Inventory Days Planned:
> $p = [360K+\$j]/(v\{\$'c[1+s]/[c-q]-\$\})$

Rule-22996:
> If both $(\$')$, (v), (p), (K), (s), (q), $(\$)$ and (j) are
> known, then its Current Ratio Planned is:
> $c = q\{360K+\$[j+vp]\}$
> $/(\{360K+\$[j+vp]\}-\$'vpc[1+s])$

Rule-22997:
> If both $(\$')$, (v), (p), (c), (s), (q), $(\$)$ and (j) are
> known, then its Kind of Cash Planned is:
> $s = \$/\$'-1$
>
> > or it can also be found as
> $s = [c-q]\{360K+\$[j+vp]\}/[\$'vpc]-1$

Steve Asikin ISBN 14: 978-1515158691, ISBN 10: 1515158691

Rule-22998:
> If both (**$'**), (**v**), (**p**), (**c**), (**s**), (**K**), (**$**) and (**j**) are known, then its Quick or Acid Test Ratio Planned is:
>
> $$q= c(\{360K+\$[j+vp]\}-\$'vpc[1+s])$$
> $$/\{360K+\$[j+vp]\}$$

Rule-22999:
> If both (**$'**), (**v**), (**p**), (**c**), (**s**), (**q**), (**K**) and (**j**) are known, then its Sales or Revenue Planned is:
>
> $$\$= \{\$'vpc[1+s]/[c-q]- 360K\}/[j+vp]$$

Rule-23000:
> If both (**$'**), (**v**), (**p**), (**c**), (**s**), (**q**), (**$**) and (**K**) are known, then its Job or Trade Receivable Days Planned is:
>
> $$j = \{\$'vpc[1+s]/[c-q]-360K-\$vp\}/\$$$

Rule-23001:
> If both (**$'**), (**v**), (**p**), (**c**), (**s**), (**q**), (**$**) and (**j**) are known, then its Kind of Cash Planned is:
>
> $$K= \$'vpc[1+s]/\{360[c-q]\}-\$j/360-\$'vp[1+s]/360$$

Rule-23002:
> If both (**K**), (**v**), (**p**), (**c**), (**s**), (**q**), (**$**) and (**j**) are known, then its Sales Past must be:
>
> $$\$'= \{360K+\$j\}/(vp[1+s]\{c/[c-q]-1\})$$

Steve Asikin ISBN 14: 978-1515158691, ISBN 10: 1515158691

Rule-23003:

If both $(\$')$, (K), (p), (c), (s), (q), $(\$)$ and (j) are known, then its Variable Portion Planned is:

$v= \{360K+\$j\}/(\$'p[1+s]\{c/[c-q]-1\})$

Rule-23004:

If both $(\$')$, (v), (K), (c), (s), (q), $(\$)$ and (j) are known, then its Procured Inventory Days Planned:

$p= \{360K+\$j\}/(\$'v[1+s]\{c/[c-q]-1\})$

Rule-23005:

If both $(\$')$, (v), (p), (c), (K), (q), $(\$)$ and (j) are known, then its Sales Growth Planned is:

$s= \$/\$'-1$

or it can also be found as

$s= \{360K+\$j\}/(\$'vp\{c/[c-q]-1\})-1$

Rule-23006:

If both $(\$')$, (v), (p), (K), (s), (q), $(\$)$ and (j) are known, then its Current Ratio Planned is:

$c= q\{360K+\$j+\$'vp[1+s]\}/\{360K+\$j\}$

Rule-23007:

If both $(\$')$, (v), (p), (c), (s), (K), $(\$)$ and (j) are known, then its Quick or Acid Test Ratio Planned is:

$q= c\{360K+\$j\}/\{360K+\$j+\$'vp[1+s]\}$

Rule-23008:

If both $(\$')$, (v), (p), (c), (s), (q), (K) and (j) are known, then its Sales or Revenue Planned is:

$\$= (\$'vp\{c/[c-q]-1\}-360K)/j$

Steve Asikin ISBN 14: 978-1515158691, ISBN 10: 1515158691

Rule-23009:

If both $(\$')$, (v), (p), (c), (s), (q), $(\$)$ and (K) are known, then its Job or Trade Receivable Days Planned:

$j = (\$'vp\{c/[c-q]-1\}-360K)/\$$

Rule-23010:

If both $(\$')$, (v), (p), (c), (s), (q), (j) and (P) are known, then its Kind of Cash Planned is:

$K = \$'vpc[1+s]/\{360[c-q]\}-\$'j[1+s]/360-P$

Rule-23011:

If both (K), (v), (p), (c), (s), (q), (j) and (P) are known, then its Sales Past must be:

$\$' = 360[K+P]/([1+s]\{vpc/[c-q]-j\})$

Rule-23012:

If both $(\$')$, (K), (p), (c), (s), (q), (j) and (P) are known, then its Variable Portion Planned is:

$v = \{360[K+P]+\$'j[1+s]\}/\{\$'pc[1+s]/[c-q]\}$

Rule-23013:

If both $(\$')$, (v), (K), (c), (s), (q), (j) and (P) are known, then its Procured Inventory Days Planned:

$p = \{360[K+P]+\$'j[1+s]\}/\{\$'vc[1+s]/[c-q]\}$

Rule-23014:

If both $(\$')$, (v), (p), (K), (s), (q), (j) and (P) are known, then its Current Ratio Planned is:

$c = q\{360[K+P]+\$'j[1+s]\}$
$/(\{360[K+P]+\$'[1+s][j-vp])$

`

Steve Asikin ISBN 14: 978-1515158691, ISBN 10: 1515158691

Rule-23015:

If both (S'), (v), (p), (c), (K), (q), (j) and (P) are known, then its Sales Growth Planned is:

$s = 360[K+P]/(S'\{vpc/[c-q]-j\})-1$

Rule-23016:

If both (S'), (v), (p), (c), (s), (K), (j) and (P) are known, then its Quick or Acid Test Ratio Planned is:

$q = c(\{360[K+P]+S'[1+s][j-vp]\})$
$/\{360[K+P]+S'j[1+s]\}$

Rule-23017:

If both (S'), (v), (p), (c), (s), (q), (j) and (P) are known, then its Kind of Cash Planned is:

$j = \{S'vpc[1+s]/[c-q]- 360[K+P]\}/\{S'[1+s]\}$

Rule-23018:

If both (S'), (v), (p), (c), (s), (q), (j) and (K) are known, then its Procured Inventory Planned is:

$P = S'vpc[1+s]/\{360[c-q]\}-S'j[1+s]/360-K$

Rule-23019:

If both (S'), (v), (p), (c), (s), (q), (j) and (V) are known, then its Kind of Cash Planned is:

$K = S'vpc[1+s]/\{360[c-q]\}-S'j[1+s]/360-Vp/360$

Rule-23020:

If both (K), (v), (p), (c), (s), (q), (j) and (V) are known, then its Sales Past must be:

$S' = [360K+Vp]/([1+s]\{vpc/[c-q]-j\})$

Steve Asikin ISBN 14: 978-1515158691, ISBN 10: 1515158691

Rule-23021:
   If both $(\textbf{S'})$, $(\textbf{K})$, $(\textbf{p})$, $(\textbf{c})$, $(\textbf{s})$, $(\textbf{q})$, $(\textbf{j})$ and $(\textbf{V})$ are
   known, then its Variable Portion Planned is:
   $$\textbf{v} = [\textbf{c-q}]\{360\textbf{K}+\textbf{S'j}[1+\textbf{s}]+\textbf{Vp}\}/\{\textbf{S'pc}[1+\textbf{s}]\}$$

Rule-23022:
   If both $(\textbf{S'})$, $(\textbf{v})$, $(\textbf{K})$, $(\textbf{c})$, $(\textbf{s})$, $(\textbf{q})$, $(\textbf{j})$ and $(\textbf{V})$ are
   known, then its Procured Inventory Days Planned:
   $$\textbf{p} = \{360\textbf{K}+\textbf{S'j}[1+\textbf{s}]\}/\{\textbf{S'vc}[1+\textbf{s}]/[\textbf{c-q}]-\textbf{V}\}$$

Rule-23023:
   If both $(\textbf{S'})$, $(\textbf{v})$, $(\textbf{p})$, $(\textbf{K})$, $(\textbf{s})$, $(\textbf{q})$, $(\textbf{j})$ and $(\textbf{V})$ are
   known, then its Current Ratio Planned is:
   $$\textbf{c} = \textbf{q}\{360\textbf{K}+\textbf{S'j}[1+\textbf{s}]+\textbf{Vp}\}$$
   $$/(\{360\textbf{K}+\textbf{S'j}[1+\textbf{s}]+\textbf{Vp}\}-\textbf{S'vpc}[1+\textbf{s}])$$

Rule-23024:
   If both $(\textbf{S'})$, $(\textbf{v})$, $(\textbf{p})$, $(\textbf{c})$, $(\textbf{K})$, $(\textbf{q})$, $(\textbf{j})$ and $(\textbf{V})$ are
   known, then its Sales Growth Planned is:
   $$\textbf{s} = [360\textbf{K}+\textbf{Vp}]/(\textbf{S'}\{\textbf{vpc}/[\textbf{c-q}]-\textbf{j}\})-1$$

Rule-23025:
   If both $(\textbf{S'})$, $(\textbf{v})$, $(\textbf{p})$, $(\textbf{c})$, $(\textbf{s})$, $(\textbf{K})$, $(\textbf{j})$ and $(\textbf{V})$ are
   known, then its Quick or Acid Test Ratio Planned is:
   $$\textbf{q} = \textbf{c}(\{360\textbf{K}+\textbf{S'j}[1+\textbf{s}]+\textbf{Vp}\}-\textbf{S'vpc}[1+\textbf{s}])$$
   $$/\{360\textbf{K}+\textbf{S'j}[1+\textbf{s}]+\textbf{Vp}\}$$

Rule-23026:
   If both $(\textbf{S'})$, $(\textbf{v})$, $(\textbf{p})$, $(\textbf{c})$, $(\textbf{K})$, $(\textbf{q})$, $(\textbf{j})$ and $(\textbf{V})$ are
   known, then its Sales Growth Planned is:
   $$\textbf{s} = [360\textbf{K}+\textbf{Vp}]/(\textbf{S'}\{\textbf{vpc}/[\textbf{c-q}]-\textbf{j}\})-1$$

Steve Asikin ISBN 14: 978-1515158691, ISBN 10: 1515158691

Rule-23027:

If both (**$'**), (**ᴠ**), (**p**), (**c**), (**s**), (**q**), (**K**) and (**V**) are known, then its Job or Trade Receivable Days Planned:

$$j= \{\$'\mathbf{vpc}[1+s]/[c-q]-Vp-360K\}/\{\$'[1+s]\}$$

Rule-23028:

If both (**$'**), (**ᴠ**), (**p**), (**c**), (**s**), (**q**), (**j**) and (**K**) are known, then its Variable Cost Planned is:

$$V= (\$'\mathbf{vpc}[1+s]/[c-q]-\$'j[1+s]-360K)/p$$

Rule-23029:

If both (**$'**), (**ᴠ**), (**p**), (**c**), (**s**), (**q**), (**j**) and (**$**) are known, then its Kind of Cash Planned is:

$$K= \$'\mathbf{vpc}[1+s]/\{360[c-q]\}-\$'j[1+s]/360-\$\mathbf{vp}/360$$

Rule-23029:

If both (**K**), (**ᴠ**), (**p**), (**c**), (**s**), (**q**), (**j**) and (**$**) are known, then its Sales Past must be:

$$\$'= [360K+\$\mathbf{vp}]/([1+s]\{\mathbf{vpc}/[c-q]-j\})$$

Rule-23030:

If both (**$'**), (**K**), (**p**), (**c**), (**s**), (**q**), (**j**) and (**$**) are known, then its Variable Portion Planned is:

$$\mathbf{v}= \{360K-\$'j[1+s]\}/(p\{\$'c[1+s]/[c-q]-\$\})$$

Rule-23031:

If both (**$'**), (**ᴠ**), (**K**), (**c**), (**s**), (**q**), (**j**) and (**$**) are known, then its Procured Inventory Days Planned:

$$p= \{360K-\$'j[1+s]\}/(\mathbf{v}\{\$'c[1+s]/[c-q]-\$\})$$

Steve Asikin ISBN 14: 978-1515158691, ISBN 10: 1515158691

Rule-23032:

If both **($'$)**, **(v)**, **(p)**, **(K)**, **(s)**, **(q)**, **(j)** and **($\$$)** are known, then its Current Ratio Planned is:

$$c = q\{360K + \$'j[1+s] + \$vp\}$$
$$/(\{360K + \$'j[1+s] + \$vp\} - \$'vp[1+s])$$

Rule-23033:

If both **($'$)**, **(v)**, **(p)**, **(c)**, **(K)**, **(q)**, **(j)** and **($\$$)** are known, then its Kind of Cash Planned is:

$$s = \$/\$' - 1$$

or it can also be found as

$$s = [360K + \$vp]/(\$'vpc/[c-q] - j\}) - 1$$

Rule-23034:

If both **($'$)**, **(v)**, **(p)**, **(c)**, **(s)**, **(K)**, **(j)** and **($\$$)** are known, then its Quick or Acid Test Ratio Planned is:

$$q = c(\{360K + \$'j[1+s] + \$vp\} - \$'vp[1+s])$$
$$/\{360K + \$'j[1+s] + \$vp\}$$

Rule-23035:

If both **($'$)**, **(v)**, **(p)**, **(c)**, **(s)**, **(q)**, **(j)** and **($\$$)** are known, then its Kind of Cash Planned is:

$$j = \{\$'vpc[1+s]/[c-q] - \$vp - 360K\}/\{\$'[1+s]\}$$

Rule-23036:

If both **($'$)**, **(v)**, **(p)**, **(c)**, **(s)**, **(q)**, **(j)** and **(K)** are known, then its Sales or Revenue Planned is:

$$\$ = \{\$'vpc[1+s]/[c-q] - \$'j[1+s] - 360K\}/\{vp\}$$

Steve Asikin ISBN 14: 978-1515158691, ISBN 10: 1515158691

Rule-23037:

If both $(\mathbf{S'})$, $(\mathbf{v})$, $(\mathbf{p})$, $(\mathbf{c})$, $(\mathbf{s})$, $(\mathbf{q})$ and $(\mathbf{j})$ are known,
then its Kind of Cash Planned is:

$$K= S'vpc[1+s]/\{360[c-q]\}-S'j[1+s]/360$$
$$-S'vp[1+s]/360$$

Rule-23038:

If both $(\mathbf{K})$, $(\mathbf{v})$, $(\mathbf{p})$, $(\mathbf{c})$, $(\mathbf{s})$, $(\mathbf{q})$ and $(\mathbf{j})$ are known,
then its Sales Past must be:

$$S'= 360K/([1+s]\{vpc/[c-q]-j-vp\})$$

Rule-23039:

If both $(\mathbf{S'})$, $(\mathbf{K})$, $(\mathbf{p})$, $(\mathbf{c})$, $(\mathbf{s})$, $(\mathbf{q})$ and $(\mathbf{j})$ are known,
then its Variable Portion Planned is:

$$v= 360K-S'j[1+s]/(S'p[1+s]\{c/[c-q]-1\})$$

Rule-23040:

If both $(\mathbf{S'})$, $(\mathbf{v})$, $(\mathbf{K})$, $(\mathbf{c})$, $(\mathbf{s})$, $(\mathbf{q})$ and $(\mathbf{j})$ are known,
then its Procured Inventory Days Planned:

$$p= 360K-S'j[1+s]/(S'v[1+s]\{c/[c-q]-1\})$$

Rule-23041:

If both $(\mathbf{S'})$, $(\mathbf{v})$, $(\mathbf{p})$, $(\mathbf{K})$, $(\mathbf{s})$, $(\mathbf{q})$ and $(\mathbf{j})$ are known,
then its Current Ratio Planned is:

$$c= q\{360K+S'j[1+s]+S'vp[1+s]\}/\{360K+S'j[1+s]\}$$

Rule-23042:

If both $(\mathbf{S'})$, $(\mathbf{v})$, $(\mathbf{p})$, $(\mathbf{c})$, $(\mathbf{K})$, $(\mathbf{q})$ and $(\mathbf{j})$ are known,
then its Sales Growth Planned is:

$$s= 360K/(S'\{vpc/[c-q]-j-vp\})-1$$

Steve Asikin ISBN 14: 978-1515158691, ISBN 10: 1515158691

Rule-23043:

If both **($')**, **(v)**, **(p)**, **(c)**, **(s)**, **(K)** and **(j)** are known,
then its Quick or Acid Test Ratio Planned is:
$$q= c\{360K+\$'j[1+s]\}/\{360K+\$'j[1+s]+\$'vp[1+s]\}$$

Rule-23044:

If both **($')**, **(v)**, **(p)**, **(c)**, **(s)**, **(q)** and **(K)** are known,
then its Job or Trade Receivable Days Planned is:
$$j= \{\$'vpc[1+s]/[c-q]-\$'vp[1+s]-360K\}/\{\$'[1+s]\}$$

Steve Asikin ISBN 14: 978-1515158691, ISBN 10: 1515158691

CHAPTER-25:

N=NON CURRENT ASSETS, Optimization,

Rule-23045:

   If both (**W**) and (**C**) are known, then its Non Current
Asset Planned is:

   N= W-C

Rule-23046:

   If both (**N**) and (**C**) are known, then its Wealth or
Total Asset Planned is:

   W= N+C

Rule-23047:

   If both (**N**) and (**W**) are known, then its Current Asset
Planned is:

   C= W-N

Rule-23048:

   If both (**W**), (**K**), (**J**) and (**P**) are known, then its Non
Current Asset Planned is:

   N= W-K-J-P

Rule-23049:

   If both (**N**), (**K**), (**J**) and (**P**) are known, then its
Wealth or Total Asset Planned is:

   W= N+K+J+P

Steve Asikin ISBN 14: 978-1515158691, ISBN 10: 1515158691

Rule-23050:

If both (W), (N), (J) and (P) are known, then its Kind of Cash Planned is:

$$K = W-N-J-P$$

Rule-23051:

If both (W), (K), (N) and (P) are known, then its Job or Trade Account Receivable Planned is:

$$J = W-K-N-P$$

Rule-23052:

If both (W), (K), (J) and (N) are known, then its Non Procured Inventory Planned is:

$$P = W-K-J-N$$

Rule-23053:

If both (W), (K), (J), (V) and (p) are known, then its Non Current Asset Planned is:

$$N = W-K-J-Vp/360$$

Rule-23054:

If both (N), (K), (J), (V) and (p) are known, then its Wealth or Total Asset Planned is:

$$W = N+K+J+Vp/360$$

Rule-23055:

If both (W), (N), (J), (V) and (p) are known, then its Kind of Cash Planned is:

$$K = W-N-J-Vp/360$$

Steve Asikin ISBN 14: 978-1515158691, ISBN 10: 1515158691

Rule-23056:
> If both **(W)**, **(K)**, **(N)**, **(V)** and **(p)** are known, then its Job or Trade Account Receivable Planned is:
>
> $J= W\text{-}K\text{-}N\text{-}Vp/360$

Rule-23057:
> If both **(W)**, **(K)**, **(J)**, **(N)** and **(p)** are known, then its Variable Portion Planned is:
>
> $V= 360[W\text{-}K\text{-}J\text{-}N]/p$

Rule-23058:
> If both **(W)**, **(K)**, **(J)**, **(V)** and **(K)** are known, then its Procured Inventory Days Planned:
>
> $p= 360[W\text{-}K\text{-}J\text{-}N]/p$

Rule-23059:
> If both **(W)**, **(K)**, **(J)**, **(S)**, **(v)** and **(p)** are known, then its Non Current Asset Planned is:
>
> $N= W\text{-}K\text{-}J\text{-}Svp/360$

Rule-23060:
> If both **(N)**, **(K)**, **(J)**, **(S)**, **(v)** and **(p)** are known, then its Wealth or Total Asset Planned is:
>
> $W= N\text{+}K\text{+}J\text{+}Svp/360$

Rule-23061:
> If both **(W)**, **(N)**, **(J)**, **(S)**, **(v)** and **(p)** are known, then its Kind of Cash Planned is:
>
> $K= W\text{-}N\text{-}J\text{-}Svp/360$

Steve Asikin ISBN 14: 978-1515158691, ISBN 10: 1515158691

Rule-23062:
> If both (W), (K), (N), (S), (v) and (p) are known, then its Job or Trade Account Receivables Planned is:
> $J = W\text{-}K\text{-}N\text{-}Svp/360$

Rule-23063:
> If both (W), (K), (J), (N), (v) and (p) are known, then its Sales or Revenue Planned is:
> $S = 360[W\text{-}K\text{-}J\text{-}N]/[vp]$

Rule-23064:
> If both (W), (K), (J), (S), (N) and (p) are known, then its Variable Portion Planned is:
> $v = 360[W\text{-}K\text{-}J\text{-}N]/[Sp]$

Rule-23065:
> If both (W), (K), (J), (S), (v) and (N) are known, then its Procured Inventory Days Planned:
> $p = 360[W\text{-}K\text{-}J\text{-}N]/[Sv]$

Rule-23066:
> If both (W), (K), (J), (S'), (v), (p) and (s) are known, then its Non Current Asset Planned is:
> $N = W\text{-}K\text{-}J\text{-}S'vp[1+s]/360$

Rule-23067:
> If both (N), (K), (J), (S'), (v), (p) and (s) are known, then its Wealth or Total Asset Planned is:
> $W = N+K+J+S'vp[1+s]/360$

Steve Asikin ISBN 14: 978-1515158691, ISBN 10: 1515158691

Rule-23068:

    If both **(W)**, **(N)**, **(J)**, **(S')**, **(v)**, **(p)** and **(s)** are known, then its Kind of Cash Planned is:

    $K= W-N-J-S'vp[1+s]/360$

Rule-23069:

    If both **(W)**, **(K)**, **(N)**, **(S')**, **(v)**, **(p)** and **(s)** are known, then its Job or Trade Account Receivable Planned is:

    $J= W-K-N-S'vp[1+s]/360$

Rule-23070:

    If both **(W)**, **(K)**, **(J)**, **(Q)**, **(v)**, **(p)** and **(s)** are known, then its Sales Past must be:

    $S'= 360[W-K-J-N]/\{vp[1+s]\}$

Rule-23071:

    If both **(W)**, **(K)**, **(J)**, **(S')**, **(N)**, **(p)** and **(s)** are known, then its Variable Portion Planned is:

    $v= 360[W-K-J-N]/\{S'p[1+s]\}$

Rule-23072:

    If both **(W)**, **(K)**, **(J)**, **(S')**, **(v)**, **(N)** and **(s)** are known, then its Procured Inventory Days Planned:

    $p= 360[W-K-J-N]/\{S'v[1+s]\}$

Rule-23073:

    If both **(W)**, **(K)**, **(J)**, **(S')**, **(v)**, **(p)** and **(N)** are known, then its Sales Growth Planned is:

    $s= 360[W-K-J-N]/[S'vp]-1$

Steve Asikin ISBN 14: 978-1515158691, ISBN 10: 1515158691

Rule-23074:
> If both (**W**), (**K**), (**J**), (**$**), (**j**) and (**P**) are known, then
> its Non Current Asset Planned is:
> $N= W-K-\$j/360-P$

Rule-23075:
> If both (**N**), (**K**), (**J**), (**$**), (**j**) and (**P**) are known, then
> its Wealth or Total Asset Planned is:
> $W= N+K+\$j/360+P$

Rule-23076:
> If both (**W**), (**N**), (**J**), (**$**), (**j**) and (**P**) are known, then
> its Kind of Cash Planned is:
> $K= W-N-\$j/360-P$

Rule-23077:
> If both (**W**), (**K**), (**J**), (**N**), (**j**) and (**P**) are known, then
> its Sales or Revenue Planned is:
> $\$= 360[W-K-N-P]/j$

Rule-23078:
> If both (**W**), (**K**), (**J**), (**$**), (**N**) and (**P**) are known, then
> its Job or Trade Receivable Days Planned:
> $j= 360[W-K-N-P]/\$$

Rule-23079:
> If both (**W**), (**K**), (**J**), (**$**), (**j**) and (**N**) are known, then
> its Procured Inventory Planned is:
> $P= W-K-\$j/360-N$

Steve Asikin ISBN 14: 978-1515158691, ISBN 10: 1515158691

Rule-23080:

If both (W), (K), (S), (j), (V) and (p) are known, then its Non Current Asset Planned is:

$N = W - K - Sj/360 - Vp/360$

Rule-23081:

If both (N), (K), (S), (j), (V) and (p) are known, then its Wealth or Total Asset Planned is:

$W = N + K + Sj/360 + Vp/360$

Rule-23082:

If both (W), (N), (S), (j), (V) and (p) are known, then its Kind of Cash Planned is:

$K = W - N - Sj/360 - Vp/360$

Rule-23083:

If both (W), (K), (N), (j), (V) and (p) are known, then its Sales or Revenue Planned is:

$S = \{360[W-K-N]-Vp\}/j$

Rule-23084:

If both (W), (K), (S), (N), (V) and (p) are known, then its Job or Trade Receivable Days Planned:

$j = \{360[W-K-N]-Vp\}/S$

Rule-23085:

If both (W), (K), (S), (j), (N) and (p) are known, then its Variable Cost Planned is:

$V = \{360[W-K-N]-Sj\}/p$

Steve Asikin ISBN 14: 978-1515158691, ISBN 10: 1515158691

Rule-23086:

If both (**W**), (**K**), (**$**), (**j**), (**V**) and (**N**) are known, then its Procured Inventory Days Planned:

$$p= \{360[\textbf{W-K-N}]\text{-}\textbf{Sj}\}/\textbf{V}$$

Rule-23087:

If both (**W**), (**K**), (**$**), (**j**), (**v**) and (**p**) are known, then its Non Current Asset Planned is:

$$N= \textbf{W-K-Sj}/360\text{-}\textbf{Svp}/360$$

Rule-23088:

If both (**N**), (**K**), (**$**), (**j**), (**v**) and (**p**) are known, then its Wealth or Total Asset Planned is:

$$W= \textbf{N+K+Sj}/360+\textbf{Svp}/360$$

Rule-23089:

If both (**W**), (**N**), (**$**), (**j**), (**v**) and (**p**) are known, then its Kind of Cash Planned is:

$$K= \textbf{W-N-Sj}/360\text{-}\textbf{Svp}/360$$

Rule-23090:

If both (**W**), (**K**), (**N**), (**j**), (**v**) and (**p**) are known, then its Sales or Revenue Planned is:

$$S= 360[\textbf{W-K-N}]/[\textbf{j+vp}]$$

Rule-23091:

If both (**W**), (**K**), (**$**), (**N**), (**v**) and (**p**) are known, then its Job or Trade Receivable Days Planned:

$$j= \{360[\textbf{W-K-N}]\text{-}\textbf{Svp}\}/\textbf{S}$$

Steve Asikin ISBN 14: 978-1515158691, ISBN 10: 1515158691

Rule-23092:

If both (W), (K), (S), (j), (N) and (p) are known, then its Variable Portion Planned is:

$$v = \{360[W-K-N]-Sj\}/[Sp]$$

Rule-23093:

If both (W), (K), (S), (j), (v) and (N) are known, then its Procured Inventory Days Planned:

$$p = \{360[W-K-N]-Sj\}/[Sv]$$

Rule-23094:

If both (W), (K), (S), (j), (S'), (v), (p) and (s) are known, then its Non Current Asset Planned is:

$$N = W-K-Sj/360-S'vp[1+s]/360$$

Rule-23095:

If both (N), (K), (S), (j), (S'), (v), (p) and (s) are known, then its Wealth or Total Asset Planned is:

$$W = N+K+Sj/360+S'vp[1+s]/360$$

Rule-23096:

If both (W), (N), (S), (j), (S'), (v), (p) and (s) are known, then its Kind of Cash Planned is:

$$K = W-N-Sj/360-S'vp[1+s]/360$$

Rule-23097:

If both (W), (K), (N), (j), (S'), (v), (p) and (s) are known, then its Sales or Revenue Planned is:

$$S = \{360[W-K-N]-S'vp[1+s]\}/j$$

Steve Asikin ISBN 14: 978-1515158691, ISBN 10: 1515158691

Rule-23098:

If both **(W)**, **(K)**, **($)**, **(N)**, **($')**, **(v)**, **(p)** and **(s)** are known, then its Job or Trade Receivable Days Planned:

$$j= \{360[W\text{-}K\text{-}N]\text{-}\$'vp[1+s]\}/\$$$

Rule-23099:

If both **(W)**, **(K)**, **($)**, **(j)**, **(N)**, **(v)**, **(p)** and **(s)** are known, then its Sales Past must be:

$$\$'= \{360[W\text{-}K\text{-}N]\text{-}\$j\}/\{vp[1+s]\}$$

Rule-23100:

If both **(W)**, **(K)**, **($)**, **(j)**, **($')**, **(N)**, **(p)** and **(s)** are known, then its Variable Portion Planned is:

$$v= \{360[W\text{-}K\text{-}N]\text{-}\$j\}/\{\$'p[1+s]\}$$

Rule-23101:

If both **(W)**, **(K)**, **($)**, **(j)**, **($')**, **(v)**, **(N)** and **(s)** are known, then its Procured Inventory Days Planned:

$$p= \{360[W\text{-}K\text{-}N]\text{-}\$j\}/\{\$'v[1+s]\}$$

Rule-23102:

If both **(W)**, **(K)**, **($)**, **(j)**, **($')**, **(v)**, **(p)** and **(N)** are known, then its Sales Growth Planned is:

$$s= \$/\$'\text{-}1$$

or it can also be found as

$$s= \{360[W\text{-}K\text{-}N]\text{-}\$j\}/[\$'vp]\text{-}1$$

Rule-23103:

If both $(\mathbf{W})$, $(\mathbf{K})$, $(\mathbf{S'})$, $(\mathbf{j})$, $(\mathbf{s})$ and $(\mathbf{P})$ are known, then its Non Current Asset Planned is:

$\mathbf{N} = \mathbf{W} - \mathbf{K} - \mathbf{S'j}[1+\mathbf{s}]/360 - \mathbf{P}$

Rule-23104:

If both $(\mathbf{N})$, $(\mathbf{K})$, $(\mathbf{S'})$, $(\mathbf{j})$, $(\mathbf{s})$ and $(\mathbf{P})$ are known, then its Wealth or Total Asset Planned is:

$\mathbf{W} = \mathbf{N} + \mathbf{K} + \mathbf{S'j}[1+\mathbf{s}]/360 + \mathbf{P}$

Rule-23105:

If both $(\mathbf{W})$, $(\mathbf{N})$, $(\mathbf{S'})$, $(\mathbf{j})$, $(\mathbf{s})$ and $(\mathbf{P})$ are known, then its Kind of Cash Planned is:

$\mathbf{K} = \mathbf{W} - \mathbf{N} - \mathbf{S'j}[1+\mathbf{s}]/360 - \mathbf{P}$

Rule-23106:

If both $(\mathbf{W})$, $(\mathbf{K})$, $(\mathbf{N})$, $(\mathbf{j})$, $(\mathbf{s})$ and $(\mathbf{P})$ are known, then its Sales Past must be:

$\mathbf{S'} = 360[\mathbf{W} - \mathbf{K} - \mathbf{N} - \mathbf{P}]/\{\mathbf{j}[1+\mathbf{s}]\}$

Rule-23107:

If both $(\mathbf{W})$, $(\mathbf{K})$, $(\mathbf{S'})$, $(\mathbf{N})$, $(\mathbf{s})$ and $(\mathbf{P})$ are known, then its Job or Trade Receivable Days Planned:

$\mathbf{j} = 360[\mathbf{W} - \mathbf{K} - \mathbf{N} - \mathbf{P}]/\{\mathbf{S'}[1+\mathbf{s}]\}$

Rule-23108:

If both $(\mathbf{W})$, $(\mathbf{K})$, $(\mathbf{S'})$, $(\mathbf{j})$, $(\mathbf{N})$ and $(\mathbf{P})$ are known, then its Sales Growth Planned is:

$\mathbf{s} = 360[\mathbf{W} - \mathbf{K} - \mathbf{N} - \mathbf{P}]/[\mathbf{S'j}] - 1$

Steve Asikin ISBN 14: 978-1515158691, ISBN 10: 1515158691

Rule-23109:

If both **(W)**, **(K)**, **(S')**, **(j)**, **(s)** and **(N)** are known, then its Procured Inventory Planned is:

P$= 360[\mathbf{W}\text{-}\mathbf{K}\text{-}\mathbf{N}]\text{-}\mathbf{S'j}[1+\mathbf{s}]$

Rule-23110:

If both **(W)**, **(K)**, **(S')**, **(j)**, **(s)**, **(V)** and **(p)** are known, then its Non Current Asset Planned is:

N$= \mathbf{W}\text{-}\mathbf{K}\text{-}\mathbf{S'j}[1+\mathbf{s}]/360\text{-}\mathbf{Vp}/360$

Rule-23111:

If both **(N)**, **(K)**, **(S')**, **(j)**, **(s)**, **(V)** and **(p)** are known, then its Wealth or Total Asset Planned is:

W$= \mathbf{N}+\mathbf{K}+\mathbf{S'j}[1+\mathbf{s}]/360+\mathbf{Vp}/360$

Rule-23112:

If both **(W)**, **(N)**, **(S')**, **(j)**, **(s)**, **(V)** and **(p)** are known, then its Kind of Cash Planned is:

K$= \mathbf{W}\text{-}\mathbf{N}\text{-}\mathbf{S'j}[1+\mathbf{s}]/360\text{-}\mathbf{Vp}/360$

Rule-23113:

If both **(W)**, **(K)**, **(N)**, **(j)**, **(s)**, **(V)** and **(p)** are known, then its Sales Past must be:

S'$= \{360[\mathbf{W}\text{-}\mathbf{K}\text{-}\mathbf{N}]\text{-}\mathbf{Vp}\}/\{\mathbf{j}[1+\mathbf{s}]\}$

Rule-23114:

If both **(W)**, **(K)**, **(S')**, **(j)**, **(s)**, **(V)** and **(p)** are known, then its Non Current Asset Planned is:

j$= \{360[\mathbf{W}\text{-}\mathbf{K}\text{-}\mathbf{N}]\text{-}\mathbf{Vp}\}/\{\mathbf{S'}[1+\mathbf{s}]\}$

Steve Asikin ISBN 14: 978-1515158691, ISBN 10: 1515158691

Rule-23115:

If both (W), (K), (S'), (j), (N), (V) and (p) are known, then its Sales Growth Planned is:

$s= \{360[W-K-N]-Vp\}/[S'j]-1$

Rule-23116:

If both (W), (K), (S'), (j), (s), (N) and (p) are known, then its Variable Cost Planned is:

$V= \{360[W-K-N]-S'j[1+s]\}/p$

Rule-23117:

If both (W), (K), (S'), (j), (s), (V) and (N) are known, then its Procured Inventory Days Planned:

$p= \{360[W-K-N]-S'j[1+s]\}/S$

Rule-23118:

If both (W), (K), (S'), (j), (s), (S), (v) and (p) are known, then its Non Current Asset Planned is:

$N= W-K-S'j[1+s]/360-Svp/360$

Rule-23119:

If both (N), (K), (S'), (j), (s), (S), (v) and (p) are known, then its Wealth or Total Asset Planned is:

$W= N+K+S'j[1+s]/360+Svp/360$

Rule-23120:

If both (W), (N), (S'), (j), (s), (S), (v) and (p) are known, then its Kind of Cash Planned is:

$K= W-N-S'j[1+s]/360-Svp/360$

Steve Asikin ISBN 14: 978-1515158691, ISBN 10: 1515158691

Rule-23121:

If both $(\mathbf{W})$, $(\mathbf{K})$, $(\mathbf{N})$, $(\mathbf{j})$, $(\mathbf{s})$, $(\mathbf{\$})$, $(\mathbf{v})$ and $(\mathbf{p})$ are known, then its Sales Past must be:
$$\mathbf{\$'}= \{360[\mathbf{W\text{-}K\text{-}N}]\text{-}\mathbf{Svp}\}/\{\mathbf{j}[1+\mathbf{s}]\}$$

Rule-23122:

If both $(\mathbf{W})$, $(\mathbf{K})$, $(\mathbf{\$'})$, $(\mathbf{N})$, $(\mathbf{s})$, $(\mathbf{\$})$, $(\mathbf{v})$ and $(\mathbf{p})$ are known, then its Job or Trade Receivable Days Planned:
$$\mathbf{j}= \{360[\mathbf{W\text{-}K\text{-}N}]\text{-}\mathbf{Svp}\}/\{\mathbf{\$'}[1+\mathbf{s}]\}$$

Rule-23123:

If both $(\mathbf{W})$, $(\mathbf{K})$, $(\mathbf{\$'})$, $(\mathbf{j})$, $(\mathbf{N})$, $(\mathbf{\$})$, $(\mathbf{v})$ and $(\mathbf{p})$ are known, then its Sales Growth Planned is:
$$\mathbf{s}= \{360[\mathbf{W\text{-}K\text{-}N}]\text{-}\mathbf{Svp}\}/[\mathbf{\$'j}]\text{-}1$$

Rule-23124:

If both $(\mathbf{W})$, $(\mathbf{K})$, $(\mathbf{\$'})$, $(\mathbf{j})$, $(\mathbf{s})$, $(\mathbf{N})$, $(\mathbf{v})$ and $(\mathbf{p})$ are known, then its Sales or Revenue Planned is:
$$\mathbf{\$}= \{360[\mathbf{W\text{-}K\text{-}N}]\text{-}\mathbf{\$'j}[1+\mathbf{s}]\}/\{\mathbf{vp}\}$$

Rule-23125:

If both $(\mathbf{W})$, $(\mathbf{K})$, $(\mathbf{\$'})$, $(\mathbf{j})$, $(\mathbf{s})$, $(\mathbf{\$})$, $(\mathbf{N})$ and $(\mathbf{p})$ are known, then its Variable Portion Planned is:
$$\mathbf{v}= \{360[\mathbf{W\text{-}K\text{-}N}]\text{-}\mathbf{\$'j}[1+\mathbf{s}]\}/\{\mathbf{\$p}\}$$

Rule-23126:

If both $(\mathbf{W})$, $(\mathbf{K})$, $(\mathbf{\$'})$, $(\mathbf{j})$, $(\mathbf{s})$, $(\mathbf{\$})$, $(\mathbf{v})$ and $(\mathbf{N})$ are known, then its Procured Inventory Days Planned is:
$$\mathbf{p}= \{360[\mathbf{W\text{-}K\text{-}N}]\text{-}\mathbf{\$'j}[1+\mathbf{s}]\}/\{\mathbf{\$v}\}$$

Steve Asikin ISBN 14: 978-1515158691, ISBN 10: 1515158691

Rule-23127:

If both $(\mathbf{W})$, $(\mathbf{K})$, $(\mathbf{S'})$, $(\mathbf{j})$, $(\mathbf{s})$, $(\mathbf{v})$ and $(\mathbf{p})$ are known, then its Non Current Asset Planned is:

$\mathbf{N} = \mathbf{W}-\mathbf{K}-\mathbf{S'j}[1+\mathbf{s}]/360-\mathbf{S'vp}[1+\mathbf{s}]/360$

Rule-23128:

If both $(\mathbf{N})$, $(\mathbf{K})$, $(\mathbf{S'})$, $(\mathbf{j})$, $(\mathbf{s})$, $(\mathbf{v})$ and $(\mathbf{p})$ are known, then its Wealth or Total Asset Planned is:

$\mathbf{W} = \mathbf{N}+\mathbf{K}+\mathbf{S'j}[1+\mathbf{s}]/360+\mathbf{S'vp}[1+\mathbf{s}]/360$

Rule-23129:

If both $(\mathbf{W})$, $(\mathbf{N})$, $(\mathbf{S'})$, $(\mathbf{j})$, $(\mathbf{s})$, $(\mathbf{v})$ and $(\mathbf{p})$ are known, then its Kind of Cash Planned is:

$\mathbf{K} = \mathbf{W}-\mathbf{N}-\mathbf{S'j}[1+\mathbf{s}]/360-\mathbf{S'vp}[1+\mathbf{s}]/360$

Rule-23130:

If both $(\mathbf{W})$, $(\mathbf{K})$, $(\mathbf{N})$, $(\mathbf{j})$, $(\mathbf{s})$, $(\mathbf{v})$ and $(\mathbf{p})$ are known, then its Sales Past must be:

$\mathbf{S'} = 360[\mathbf{W}-\mathbf{K}-\mathbf{N}]/\{[1+\mathbf{s}][\mathbf{j}+\mathbf{vp}]\}$

Rule-23131:

If both $(\mathbf{W})$, $(\mathbf{K})$, $(\mathbf{S'})$, $(\mathbf{N})$, $(\mathbf{s})$, $(\mathbf{v})$ and $(\mathbf{p})$ are known, then its Job or Trade Receivable Days Planned:

$\mathbf{j} = (360[\mathbf{W}-\mathbf{K}-\mathbf{N}]/\{\mathbf{S'}[1+\mathbf{s}]\}-\mathbf{vp}$

Rule-23132:

If both $(\mathbf{W})$, $(\mathbf{K})$, $(\mathbf{S'})$, $(\mathbf{j})$, $(\mathbf{N})$, $(\mathbf{v})$ and $(\mathbf{p})$ are known, then its Sales Growth Planned is:

$\mathbf{s} = 360[\mathbf{W}-\mathbf{K}-\mathbf{N}]/\{\mathbf{S'}[\mathbf{j}+\mathbf{vp}]\}-1$

Steve Asikin ISBN 14: 978-1515158691, ISBN 10: 1515158691

Rule-23133:
  If both (**W**), (**K**), (**S'**), (**j**), (**s**), (**N**) and (**p**) are known,
  then its Variable Portion Planned is:
  $v = (360[\mathbf{W}\text{-}\mathbf{K}\text{-}\mathbf{N}]/\{\mathbf{S'}[1+\mathbf{s}]\}\text{-}\mathbf{j})/\mathbf{p}$

Rule-23134:
  If both (**W**), (**K**), (**S'**), (**j**), (**s**), (**v**) and (**N**) are known,
  then its Procured Inventory Days Planned:
  $p = (360[\mathbf{W}\text{-}\mathbf{K}\text{-}\mathbf{N}]/\{\mathbf{S'}[1+\mathbf{s}]\}\text{-}\mathbf{j})/\mathbf{v}$

Rule-23135:
  If both (**E**), (**I**) and (**C**) are known, then its Non
  Current Asset Planned is:
  $N = \mathbf{E} I \text{-} \mathbf{C}$

Rule-23136:
  If both (**N**), (**I**) and (**C**) are known, then its Equity or
  Capital Planned is:
  $E = [\mathbf{N}+\mathbf{C}]/I$

Rule-23137:
  If both (**E**), (**N**) and (**C**) are known, then its Leverage
  or Gearing Ratio Planned is:
  $I = [\mathbf{N}+\mathbf{C}]/\mathbf{E}$

Rule-23138:
  If both (**E**), (**I**) and (**N**) are known, then its Current
  Asset Planned is:
  $C = \mathbf{E} I \text{-} \mathbf{N}$

Steve Asikin ISBN 14: 978-1515158691, ISBN 10: 1515158691

Rule-23139:

If both (**E**), (**/**), (**K**), (**J**) and (**P**) are known, then its Non Current Asset Planned is:

$$N= E/-K-J-P$$

Rule-23140:

If both (**N**), (**/**), (**K**), (**J**) and (**P**) are known, then its Equity or Capital Planned is:

$$E= [N+K+J+P]//$$

Rule-23141:

If both (**E**), (**N**), (**K**), (**J**) and (**P**) are known, then its Leverage or Gearing Ratio Planned is:

$$/= [N+K+J+P]/E$$

Rule-23142:

If both (**E**), (**/**), (**N**), (**J**) and (**P**) are known, then its Kind of Cash Planned is:

$$K= E/-N-J-P$$

Rule-23143:

If both (**E**), (**/**), (**K**), (**N**) and (**P**) are known, then its Job or Trade Account Receivable Planned is:

$$J= E/-K-N-P$$

Rule-23144:

If both (**E**), (**/**), (**K**), (**J**) and (**N**) are known, then its Procured Inventory Planned is:

$$P= E/-K-J-N$$

Steve Asikin ISBN 14: 978-1515158691, ISBN 10: 1515158691

Rule-23145:
> If both $(\mathbf{E})$, $(\mathbf{I})$, $(\mathbf{K})$, $(\mathbf{J})$, $(\mathbf{V})$ and $(\mathbf{p})$ are known, then its Non Current Asset Planned is:
> $\mathbf{N} = \mathbf{E}/\mathbf{K}-\mathbf{J}-\mathbf{Vp}/360$

Rule-23146:
> If both $(\mathbf{N})$, $(\mathbf{I})$, $(\mathbf{K})$, $(\mathbf{J})$, $(\mathbf{V})$ and $(\mathbf{p})$ are known, then its Equity or Capital Planned is:
> $\mathbf{E} = [\mathbf{N}+\mathbf{K}+\mathbf{J}+\mathbf{Vp}/360]/\mathbf{I}$

Rule-23147:
> If both $(\mathbf{E})$, $(\mathbf{N})$, $(\mathbf{K})$, $(\mathbf{J})$, $(\mathbf{V})$ and $(\mathbf{p})$ are known, then its Leverage or Gearing Ratio Planned is:
> $\mathbf{I} = [\mathbf{N}+\mathbf{K}+\mathbf{J}+\mathbf{Vp}/360]/\mathbf{E}$

Rule-23148:
> If both $(\mathbf{E})$, $(\mathbf{I})$, $(\mathbf{N})$, $(\mathbf{J})$, $(\mathbf{V})$ and $(\mathbf{p})$ are known, then its Kind of Cash Planned is:
> $\mathbf{K} = \mathbf{E}/\mathbf{N}-\mathbf{J}-\mathbf{Vp}/360$

Rule-23149:
> If both $(\mathbf{E})$, $(\mathbf{I})$, $(\mathbf{K})$, $(\mathbf{N})$, $(\mathbf{V})$ and $(\mathbf{p})$ are known, then its Job or Trade Account Receivable Planned is:
> $\mathbf{J} = \mathbf{E}/\mathbf{K}-\mathbf{N}-\mathbf{Vp}/360$

Rule-23150:
> If both $(\mathbf{E})$, $(\mathbf{I})$, $(\mathbf{K})$, $(\mathbf{J})$, $(\mathbf{N})$ and $(\mathbf{p})$ are known, then its Variable Portion Planned is:
> $\mathbf{V} = 360[\mathbf{E}/\mathbf{K}-\mathbf{J}-\mathbf{N}]/\mathbf{p}$

Steve Asikin ISBN 14: 978-1515158691, ISBN 10: 1515158691

Rule-23151:
 If both **(E)**, **(ꞁ)**, **(K)**, **(J)**, **(V)** and **(N)** are known, then
 its Procured Inventory Days Planned:
 $\mathbf{p} = 360[\mathbf{E/K\text{-}J\text{-}N}]/\mathbf{p}$

Rule-23152:
 If both **(E)**, **(ꞁ)**, **(K)**, **(J)**, **($)**, **(v)** and **(p)** are known,
 then its Non Current Asset Planned is:
 $\mathbf{N} = \mathbf{E/K\text{-}J\text{-}\$vp}/360$

Rule-23153:
 If both **(N)**, **(ꞁ)**, **(K)**, **(J)**, **($)**, **(v)** and **(p)** are known,
 then its Equity or Capital Planned is:
 $\mathbf{E} = [\mathbf{N+K+J+\$vp}/360]/\mathbf{ꞁ}$

Rule-23154:
 If both **(E)**, **(N)**, **(K)**, **(J)**, **($)**, **(v)** and **(p)** are known,
 then its Leverage or Gearing Ratio Planned is:
 $\mathbf{ꞁ} = [\mathbf{N+K+J+\$vp}/360]/\mathbf{E}$

Rule-23155:
 If both **(E)**, **(ꞁ)**, **(N)**, **(J)**, **($)**, **(v)** and **(p)** are known,
 then its Kind of Cash Planned is:
 $\mathbf{K} = \mathbf{E/N\text{-}J\text{-}\$vp}/360$

Rule-23156:
 If both **(E)**, **(ꞁ)**, **(K)**, **(N)**, **($)**, **(v)** and **(p)** are known,
 then its Job or Trade Account Receivable Planned is:
 $\mathbf{J} = \mathbf{E/K\text{-}N\text{-}\$vp}/360$

Steve Asikin ISBN 14: 978-1515158691, ISBN 10: 1515158691

Rule-23157:

> If both (**E**), (**/**), (**K**), (**J**), (**N**), (**v**) and (**p**) are known, then its Sales or Revenue Planned is:
> $$S= 360[E\text{/}K\text{-}J\text{-}N]/[vp]$$

Rule-23158:

> If both (**E**), (**/**), (**K**), (**J**), (**S**), (**N**) and (**p**) are known, then its Variable Portion Planned is:
> $$v= 360[E\text{/}K\text{-}J\text{-}N]/[Sp]$$

Rule-23159:

> If both (**E**), (**/**), (**K**), (**J**), (**S**), (**v**) and (**N**) are known, then its Procured Inventory Days Planned:
> $$p= 360[E\text{/}K\text{-}J\text{-}N]/[Sv]$$

Rule-23160:

> If both (**E**), (**/**), (**K**), (**J**), (**S'**), (**v**), (**p**) and (**s**) are known, then its Non Current Asset Planned is:
> $$N= E\text{/}K\text{-}J\text{-}S\text{'}vp[1+s]/360$$

Rule-23161:

> If both (**N**), (**/**), (**K**), (**J**), (**S'**), (**v**), (**p**) and (**s**) are known, then its Equity or Capital Planned is:
> $$E= [N+K+J+S\text{'}vp[1+s]/360]/\text{/}$$

Rule-23162:

> If both (**E**), (**N**), (**K**), (**J**), (**S'**), (**v**), (**p**) and (**s**) are known, then its Leverage or Gearing Ratio Planned is:
> $$\text{/}= [N+K+J+S\text{'}vp[1+s]/360]/E$$

Steve Asikin ISBN 14: 978-1515158691, ISBN 10: 1515158691

Rule-23163:

    If both (E), (I), (N), (J), (S'), (v), (p) and (s) are known, then its Kind of Cash Planned is:

$$K= EIN\text{-}J\text{-}S'vp[1+s]/360$$

Rule-23164:

    If both (E), (I), (K), (N), (S'), (v), (p) and (s) are known, then its Job or Trade Account Receivable Days Planned:

$$J= EIK\text{-}N\text{-}S'vp[1+s]/360$$

Rule-23165:

    If both (E), (I), (K), (J), (N), (v), (p) and (s) are known, then its Sales Past must be:

$$S'= 360[EIK\text{-}J\text{-}N]/\{vp[1+s]\}$$

Rule-23166:

    If both (E), (I), (K), (J), (S'), (N), (p) and (s) are known, then its Variable Portion Planned is:

$$v= 360[EIK\text{-}J\text{-}N]/\{S'p[1+s]\}$$

Rule-23167:

    If both (E), (I), (K), (J), (S'), (v), (N) and (s) are known, then its Procured Inventory Days Planned:

$$p= 360[EIK\text{-}J\text{-}N]/\{S'v[1+s]\}$$

Rule-23168:

    If both (E), (I), (K), (J), (S'), (v), (p) and (N) are known, then its Sales Growth Planned is:

$$s= 360[EIK\text{-}J\text{-}N]/[S'vp]\text{-}1$$

Steve Asikin ISBN 14: 978-1515158691, ISBN 10: 1515158691

Rule-23169:

If both (**E**), (**/**), (**K**), (**$**), (**j**) and (**P**) are known, then its Non Current Asset Planned is:

$$N= E/K\text{-}\$j/360\text{-}P$$

Rule-23170:

If both (**N**), (**/**), (**K**), (**$**), (**j**) and (**P**) are known, then its Equity or Capital Planned is:

$$E= [N+K+\$j/360+P]//$$

Rule-23171:

If both (**E**), (**N**), (**K**), (**$**), (**j**) and (**P**) are known, then its Leverage or Gearing Ratio Planned is:

$$/= [N+K+\$j/360+P]/E$$

Rule-23172:

If both (**E**), (**/**), (**N**), (**$**), (**j**) and (**P**) are known, then its Kind of Cash Planned is:

$$K= E/N\text{-}\$j/360\text{-}P$$

Rule-23173:

If both (**E**), (**/**), (**K**), (**N**), (**j**) and (**P**) are known, then its Sales or Revenue Planned is:

$$\$= 360[E/K\text{-}N\text{-}P]/j$$

Rule-23174:

If both (**E**), (**/**), (**K**), (**$**), (**N**) and (**P**) are known, then its Job or Trade Receivable Days Planned is:

$$j= 360[E/K\text{-}N\text{-}P]/\$$$

Steve Asikin ISBN 14: 978-1515158691, ISBN 10: 1515158691

Rule-23175:
> If both (**E**), (**/**), (**K**), (**$**), (**j**) and (**N**) are known, then
> its Procured Inventory Planned is:
> $P= E/K-Sj/360-N$

Rule-23176:
> If both (**E**), (**/**), (**K**), (**$**), (**j**), (**V**) and (**p**) are known,
> then its Non Current Asset Planned is:
> $N= E/K-Sj/360-Vp/360$

Rule-23177:
> If both (**N**), (**/**), (**K**), (**$**), (**j**), (**V**) and (**p**) are known,
> then its Equity or Capital Planned is:
> $E= [N+K+Sj/360+Vp/360]/I$

Rule-23178:
> If both (**E**), (**N**), (**K**), (**$**), (**j**), (**V**) and (**p**) are known,
> then its Leverage or Gearing Ratio Planned is:
> $I= [N+K+Sj/360+Vp/360]/E$

Rule-23179:
> If both (**E**), (**/**), (**N**), (**$**), (**j**), (**V**) and (**p**) are known,
> then its Kind of Cash Planned is:
> $K= E/N-Sj/360-Vp/360$

Rule-23180:
> If both (**E**), (**/**), (**K**), (**N**), (**j**), (**V**) and (**p**) are known,
> then its Sales or Revenue Planned is:
> $S= \{360[E/K-N]-Vp\}/j$

Steve Asikin ISBN 14: 978-1515158691, ISBN 10: 1515158691

Rule-23181:

If both **(E)**, **(I)**, **(K)**, **($)**, **(N)**, **(V)** and **(p)** are known,
then its Job or Trade Receivable Planned is:
$j= \{360[E/K-N]-Vp\}/\$$

Rule-23182:

If both **(E)**, **(I)**, **(K)**, **($)**, **(j)**, **(N)** and **(p)** are known,
then its Variable Portion Planned is:
$V= \{360[E/K-N]-\$j\}/p$

Rule-23183:

If both **(E)**, **(I)**, **(K)**, **($)**, **(j)**, **(V)** and **(N)** are known,
then its Procured Inventory Days Planned is:
$p= \{360[E/K-N]-\$j\}/V$

Rule-23184:

If both **(E)**, **(I)**, **(K)**, **($)**, **(j)**, **(v)** and **(p)** are known,
then its Non Current Asset Planned is:
$N= E/K-\$j/360-\$vp/360$

Rule-23185:

If both **(N)**, **(I)**, **(K)**, **($)**, **(j)**, **(v)** and **(p)** are known,
then its Equity or Capital Planned is:
$E= [N+K+\$j/360+\$vp/360]/I$

Rule-23186:

If both **(E)**, **(N)**, **(K)**, **($)**, **(j)**, **(v)** and **(p)** are known,
then its Leverage or Gearing Ratio Planned is:
$I= [N+K+\$j/360+\$vp/360]/E$

Steve Asikin ISBN 14: 978-1515158691, ISBN 10: 1515158691

Rule-23187:

If both (E), (I), (N), (S), (j), (v) and (p) are known, then its Kind of Cash Planned is:

$K= EI\text{-}N\text{-}Sj/360\text{-}Svp/360$

Rule-23188:

If both (E), (I), (K), (N), (j), (v) and (p) are known, then its Sal;es or Revenue Planned is:

$S= 360[EI\text{-}K\text{-}N]/[j+vp]$

Rule-23189:

If both (E), (I), (K), (S), (N), (v) and (p) are known, then its Job Receivable Days Planned is:

$j= \{360[EI\text{-}K\text{-}N]\text{-}Svp\}/S$

Rule-23190:

If both (E), (I), (K), (S), (j), (N) and (p) are known, then its Variable Cost Planned is:

$v= \{360[EI\text{-}K\text{-}N]\text{-}Sj\}/[Sp]$

Rule-23191:

If both (E), (I), (K), (S), (j), (v) and (N) are known, then its Procured Inventory Days Planned:

$p= \{360[EI\text{-}K\text{-}N]\text{-}Sj\}/[Sv]$

Rule-23192:

If both (E), (I), (K), (S), (j), (S'), (v), (p) and (s) are known, then its Non Current Asset Planned is:

$N= EI\text{-}K\text{-}Sj/360\text{-}S'vp[1+s]/360$

Steve Asikin ISBN 14: 978-1515158691, ISBN 10: 1515158691

Rule-23193:

If both $(\mathbf{N})$, $(\mathbf{I})$, $(\mathbf{K})$, $(\mathbf{S})$, $(\mathbf{j})$, $(\mathbf{S'})$, $(\mathbf{v})$, $(\mathbf{p})$ and $(\mathbf{s})$ are known, then its Equity or Capital Planned is:

$$\mathbf{E} = [\mathbf{N}+\mathbf{K}+\mathbf{S}\mathbf{j}/360+\mathbf{S'}\mathbf{v}\mathbf{p}[1+\mathbf{s}]/360]/\mathbf{I}$$

Rule-23194:

If both $(\mathbf{E})$, $(\mathbf{N})$, $(\mathbf{K})$, $(\mathbf{S})$, $(\mathbf{j})$, $(\mathbf{S'})$, $(\mathbf{v})$, $(\mathbf{p})$ and $(\mathbf{s})$ are known, then its Leverage or Gearing Ratio Planned is:

$$\mathbf{I} = [\mathbf{N}+\mathbf{K}+\mathbf{S}\mathbf{j}/360+\mathbf{S'}\mathbf{v}\mathbf{p}[1+\mathbf{s}]/360]/\mathbf{E}$$

Rule-23195:

If both $(\mathbf{E})$, $(\mathbf{I})$, $(\mathbf{N})$, $(\mathbf{S})$, $(\mathbf{j})$, $(\mathbf{S'})$, $(\mathbf{v})$, $(\mathbf{p})$ and $(\mathbf{s})$ are known, then its Kind of Cash Planned is:

$$\mathbf{K} = \mathbf{E}\mathbf{I}\text{-}\mathbf{N}\text{-}\mathbf{S}\mathbf{j}/360\text{-}\mathbf{S'}\mathbf{v}\mathbf{p}[1+\mathbf{s}]/360$$

Rule-23196:

If both $(\mathbf{E})$, $(\mathbf{I})$, $(\mathbf{K})$, $(\mathbf{N})$, $(\mathbf{j})$, $(\mathbf{S'})$, $(\mathbf{v})$, $(\mathbf{p})$ and $(\mathbf{s})$ are known, then its Sales or Revenue Planned is:

$$\mathbf{S} = \{360[\mathbf{E}\mathbf{I}\text{-}\mathbf{K}\text{-}\mathbf{N}]\text{-}\mathbf{S'}\mathbf{v}\mathbf{p}[1+\mathbf{s}]\}/\mathbf{j}$$

Rule-23197:

If both $(\mathbf{E})$, $(\mathbf{I})$, $(\mathbf{K})$, $(\mathbf{S})$, $(\mathbf{N})$, $(\mathbf{S'})$, $(\mathbf{v})$, $(\mathbf{p})$ and $(\mathbf{s})$ are known, then its Job or Trade Receivable Days Planned:

$$\mathbf{j} = \{360[\mathbf{E}\mathbf{I}\text{-}\mathbf{K}\text{-}\mathbf{N}]\text{-}\mathbf{S'}\mathbf{v}\mathbf{p}[1+\mathbf{s}]\}/\mathbf{S}$$

Rule-23198:

If both $(\mathbf{E})$, $(\mathbf{I})$, $(\mathbf{K})$, $(\mathbf{S})$, $(\mathbf{j})$, $(\mathbf{N})$, $(\mathbf{v})$, $(\mathbf{p})$ and $(\mathbf{s})$ are known, then its Sales Past must be:

$$\mathbf{S'} = \{360[\mathbf{E}\mathbf{I}\text{-}\mathbf{K}\text{-}\mathbf{N}]\text{-}\mathbf{S}\mathbf{j}\}/\{\mathbf{v}\mathbf{p}[1+\mathbf{s}]\}$$

Steve Asikin ISBN 14: 978-1515158691, ISBN 10: 1515158691

Rule-23199:

   If both $(\mathbf{E})$, $(\mathbf{/})$, $(\mathbf{K})$, $(\mathbf{S})$, $(\mathbf{j})$, $(\mathbf{S'})$, $(\mathbf{N})$, $(\mathbf{p})$ and $(\mathbf{s})$ are known, then its Variable Portion Planned is:

$$\mathbf{v} = \{360[\mathbf{E/K\text{-}N}]\text{-}\mathbf{Sj}\}/\{\mathbf{S'p}[1+\mathbf{s}]\}$$

Rule-23200:

   If both $(\mathbf{E})$, $(\mathbf{/})$, $(\mathbf{K})$, $(\mathbf{S})$, $(\mathbf{j})$, $(\mathbf{S'})$, $(\mathbf{v})$, $(\mathbf{N})$ and $(\mathbf{s})$ are known, then its Procured Inventory Planned is:

$$\mathbf{p} = \{360[\mathbf{E/K\text{-}N}]\text{-}\mathbf{Sj}\}/\{\mathbf{S'v}[1+\mathbf{s}]\}$$

Rule-23201:

   If both $(\mathbf{E})$, $(\mathbf{/})$, $(\mathbf{K})$, $(\mathbf{S})$, $(\mathbf{j})$, $(\mathbf{S'})$, $(\mathbf{v})$, $(\mathbf{p})$ and $(\mathbf{N})$ are known, then its Sales Growth Planned is:

$$\mathbf{s} = \{360[\mathbf{E/K\text{-}N}]\text{-}\mathbf{Sj}\}/[\mathbf{S'vp}]\text{-}1$$

Rule-23202:

   If both $(\mathbf{E})$, $(\mathbf{/})$, $(\mathbf{K})$, $(\mathbf{S'})$, $(\mathbf{j})$, $(\mathbf{s})$ and $(\mathbf{P})$ are known, then its Non Current Asset Planned is:

$$\mathbf{N} = \mathbf{E/K\text{-}S'j}[1+\mathbf{s}]/360\text{-}\mathbf{P}$$

Rule-23203:

   If both $(\mathbf{N})$, $(\mathbf{/})$, $(\mathbf{K})$, $(\mathbf{S'})$, $(\mathbf{j})$, $(\mathbf{s})$ and $(\mathbf{P})$ are known, then its Equity or Capital Planned is:

$$\mathbf{E} = [\mathbf{N+K+S'j}[1+\mathbf{s}]/360+\mathbf{P}]/\mathbf{/}$$

Rule-23204:

   If both $(\mathbf{E})$, $(\mathbf{N})$, $(\mathbf{K})$, $(\mathbf{S'})$, $(\mathbf{j})$, $(\mathbf{s})$ and $(\mathbf{P})$ are known, then its Leverage or Gearing Ratio Planned is:

$$\mathbf{/} = [\mathbf{N+K+S'j}[1+\mathbf{s}]/360+\mathbf{P}]/\mathbf{E}$$

Steve Asikin ISBN 14: 978-1515158691, ISBN 10: 1515158691

Rule-23205:

    If both **(E)**, **(/)**, **(N)**, **(S')**, **(j)**, **(s)** and **(P)** are known,
then its Kind of Cash Planned is:

    $K = E/N-S'j[1+s]/360-P$

Rule-23206:

    If both **(E)**, **(/)**, **(K)**, **(N)**, **(j)**, **(s)** and **(P)** are known,
then its Sales Past must be:

    $S' = 360[E/K-N-P]/\{j[1+s]\}$

Rule-23207:

    If both **(E)**, **(/)**, **(K)**, **(S')**, **(N)**, **(s)** and **(P)** are known,
then its Job or Trade Receivable Days Planned is:

    $j = 360[E/K-N-P]/\{S'[1+s]\}$

Rule-23208:

    If both **(E)**, **(/)**, **(K)**, **(S')**, **(j)**, **(N)** and **(P)** are known,
then its Sales Growth Planned is:

    $s = 360[E/K-N-P]/[S'j]-1$

Rule-23209:

    If both **(E)**, **(/)**, **(K)**, **(S')**, **(j)**, **(s)** and **(N)** are known,
then its Procured Inventory Planned is:

    $P = 360[E/K-N]-S'j[1+s]$

Rule-23210:

    If both **(E)**, **(/)**, **(K)**, **(S')**, **(j)**, **(s)**, **(V)** and **(p)** are
known, then its Non Current Asset Planned is:

    $N = E/K-S'j[1+s]/360-Vp/360$

Steve Asikin ISBN 14: 978-1515158691, ISBN 10: 1515158691

Rule-23211:
   If both **(N)**, **(/)**, **(K)**, **(S')**, **(j)**, **(s)**, **(V)** and **(p)** are known, then its Equity or Capital Planned is:
   $$E= [N+K+S'j[1+s]/360+Vp/360]/I$$

Rule-23212:
   If both **(E)**, **(N)**, **(K)**, **(S')**, **(j)**, **(s)**, **(V)** and **(p)** are known, then its Leverage or Gearing Ratio Planned is:
   $$I= [N+K+S'j[1+s]/360+Vp/360]/E$$

Rule-23213:
   If both **(E)**, **(/)**, **(N)**, **(S')**, **(j)**, **(s)**, **(V)** and **(p)** are known, then its Kind of Cash Planned is:
   $$K= EI-N-S'j[1+s]/360-Vp/360$$

Rule-23214:
   If both **(E)**, **(/)**, **(K)**, **(N)**, **(j)**, **(s)**, **(V)** and **(p)** are known, then its Sales Past must be:
   $$S'= \{360[EI-K-N]-Vp\}/\{j[1+s]\}$$

Rule-23215:
   If both **(E)**, **(/)**, **(K)**, **(S')**, **(N)**, **(s)**, **(V)** and **(p)** are known, then its Job or Trade Receivable Days Planned is:
   $$j= \{360[EI-K-N]-Vp\}/\{S'[1+s]\}$$

Rule-23216:
   If both **(E)**, **(/)**, **(K)**, **(S')**, **(j)**, **(N)**, **(V)** and **(p)** are known, then its Sales Growth Planned is:
   $$s= \{360[EI-K-N]-Vp\}/[S'j]-1$$

Steve Asikin ISBN 14: 978-1515158691, ISBN 10: 1515158691

Rule-23217:
> If both $(\mathbf{E})$, $(\mathbf{\textit{I}})$, $(\mathbf{K})$, $(\mathbf{\$'})$, $(\mathbf{j})$, $(\mathbf{s})$, $(\mathbf{V})$ and $(\mathbf{p})$ are known, then its Non Current Asset Planned is:
> $\mathbf{V} = \{360[\mathbf{E}\textit{I}\mathbf{\text{-}K\text{-}N}]\text{-}\mathbf{\$'j}[1+\mathbf{s}]\}/\mathbf{p}$

Rule-23218:
> If both $(\mathbf{E})$, $(\mathbf{\textit{I}})$, $(\mathbf{K})$, $(\mathbf{\$'})$, $(\mathbf{j})$, $(\mathbf{s})$, $(\mathbf{V})$ and $(\mathbf{N})$ are known, then its Procured Inventory Days Planned:
> $\mathbf{p} = \{360[\mathbf{E}\textit{I}\mathbf{\text{-}K\text{-}N}]\text{-}\mathbf{\$'j}[1+\mathbf{s}]\}/\mathbf{\$}$

Rule-23219:
> If both $(\mathbf{E})$, $(\mathbf{\textit{I}})$, $(\mathbf{K})$, $(\mathbf{\$'})$, $(\mathbf{j})$, $(\mathbf{s})$, $(\mathbf{\$})$, $(\mathbf{v})$ and $(\mathbf{p})$ are known, then its Non Current Asset Planned is:
> $\mathbf{N} = \mathbf{E}\textit{I}\mathbf{\text{-}K}\text{-}\mathbf{\$'j}[1+\mathbf{s}]/360\text{-}\mathbf{\$vp}/360$

Rule-23220:
> If both $(\mathbf{N})$, $(\mathbf{\textit{I}})$, $(\mathbf{K})$, $(\mathbf{\$'})$, $(\mathbf{j})$, $(\mathbf{s})$, $(\mathbf{\$})$, $(\mathbf{v})$ and $(\mathbf{p})$ are known, then its Equity or Capital Planned is:
> $\mathbf{E} = [\mathbf{N}+\mathbf{K}+\mathbf{\$'j}[1+\mathbf{s}]/360+\mathbf{\$vp}/360]/\textit{I}$

Rule-23221:
> If both $(\mathbf{E})$, $(\mathbf{N})$, $(\mathbf{K})$, $(\mathbf{\$'})$, $(\mathbf{j})$, $(\mathbf{s})$, $(\mathbf{\$})$, $(\mathbf{v})$ and $(\mathbf{p})$ are known, then its Leverage or Gearing Ratio Planned is:
> $\textit{I} = [\mathbf{N}+\mathbf{K}+\mathbf{\$'j}[1+\mathbf{s}]/360+\mathbf{\$vp}/360]/\mathbf{E}$

Rule-23222:
> If both $(\mathbf{E})$, $(\mathbf{\textit{I}})$, $(\mathbf{N})$, $(\mathbf{\$'})$, $(\mathbf{j})$, $(\mathbf{s})$, $(\mathbf{\$})$, $(\mathbf{v})$ and $(\mathbf{p})$ are known, then its Kind of Cash Planned is:
> $\mathbf{K} = \mathbf{E}\textit{I}\mathbf{\text{-}N}\text{-}\mathbf{\$'j}[1+\mathbf{s}]/360\text{-}\mathbf{\$vp}/360$

Steve Asikin ISBN 14: 978-1515158691, ISBN 10: 1515158691

Rule-23223:
>   If both $(\mathbf{E})$, $(\mathbf{I})$, $(\mathbf{K})$, $(\mathbf{N})$, $(\mathbf{j})$, $(\mathbf{s})$, $(\mathbf{\$})$, $(\mathbf{v})$ and $(\mathbf{p})$ are known, then its Sales Past must be:
> $$\mathbf{\$'} = \{360[\mathbf{E/K\text{-}N}]\text{-}\mathbf{\$vp}\}/\{\mathbf{j}[1+\mathbf{s}]\}$$

Rule-23224:
>   If both $(\mathbf{E})$, $(\mathbf{I})$, $(\mathbf{K})$, $(\mathbf{\$'})$, $(\mathbf{N})$, $(\mathbf{s})$, $(\mathbf{\$})$, $(\mathbf{v})$ and $(\mathbf{p})$ are known, then its Job Receivable Days Planned is:
> $$\mathbf{j} = \{360[\mathbf{E/K\text{-}N}]\text{-}\mathbf{\$vp}\}/\{\mathbf{\$'}[1+\mathbf{s}]\}$$

Rule-23225:
>   If both $(\mathbf{E})$, $(\mathbf{I})$, $(\mathbf{K})$, $(\mathbf{\$'})$, $(\mathbf{j})$, $(\mathbf{N})$, $(\mathbf{\$})$, $(\mathbf{v})$ and $(\mathbf{p})$ are known, then its Sales Growth Planned is:
> $$\mathbf{s} = \{360[\mathbf{E/K\text{-}N}]\text{-}\mathbf{\$vp}\}/[\mathbf{\$'j}]\text{-}1$$

Rule-23226:
>   If both $(\mathbf{E})$, $(\mathbf{I})$, $(\mathbf{K})$, $(\mathbf{\$'})$, $(\mathbf{j})$, $(\mathbf{s})$, $(\mathbf{N})$, $(\mathbf{v})$ and $(\mathbf{p})$ are known, then its Sales or Revenue Planned is:
> $$\mathbf{\$} = \{360[\mathbf{E/K\text{-}N}]\text{-}\mathbf{\$'j}[1+\mathbf{s}]\}/\{\mathbf{vp}\}$$

Rule-23227:
>   If both $(\mathbf{E})$, $(\mathbf{I})$, $(\mathbf{K})$, $(\mathbf{\$'})$, $(\mathbf{j})$, $(\mathbf{s})$, $(\mathbf{\$})$, $(\mathbf{N})$ and $(\mathbf{p})$ are known, then its Variable Portion Planned is:
> $$\mathbf{v} = \{360[\mathbf{E/K\text{-}N}]\text{-}\mathbf{\$'j}[1+\mathbf{s}]\}/\{\mathbf{\$p}\}$$

Rule-23228:
>   If both $(\mathbf{E})$, $(\mathbf{I})$, $(\mathbf{K})$, $(\mathbf{\$'})$, $(\mathbf{j})$, $(\mathbf{s})$, $(\mathbf{\$})$, $(\mathbf{v})$ and $(\mathbf{N})$ are known, then its Procured Inventory Days Planned:
> $$\mathbf{p} = \{360[\mathbf{E/K\text{-}N}]\text{-}\mathbf{\$'j}[1+\mathbf{s}]\}/\{\mathbf{\$v}\}$$

Steve Asikin ISBN 14: 978-1515158691, ISBN 10: 1515158691

Rule-23229:

    If both (**E**), (**I**), (**K**), (**S'**), (**j**), (**s**), (**v**) and (**p**) are known, then its Non Current Asset Planned is:

$$N= EI\text{-}K\text{-}S\text{'}j[1+s]/360\text{-}S\text{'}vp[1+s]/360$$

Rule-23230:

    If both (**N**), (**I**), (**K**), (**S'**), (**j**), (**s**), (**v**) and (**p**) are known, then its Equity or Capital Planned is:

$$E= [N+K+S\text{'}j[1+s]/360+S\text{'}vp[1+s]/360]/I$$

Rule-23231:

    If both (**E**), (**N**), (**K**), (**S'**), (**j**), (**s**), (**v**) and (**p**) are known, then its Leverage or Gearing Ratio Planned is:

$$I= [N+K+S\text{'}j[1+s]/360+S\text{'}vp[1+s]/360]/E$$

Rule-23232:

    If both (**E**), (**I**), (**N**), (**S'**), (**j**), (**s**), (**v**) and (**p**) are known, then its Kind of Cash Planned is:

$$K= EI\text{-}N\text{-}S\text{'}j[1+s]/360\text{-}S\text{'}vp[1+s]/360$$

Rule-23233:

    If both (**E**), (**I**), (**K**), (**N**), (**j**), (**s**), (**v**) and (**p**) are known, then its Sales Past must be:

$$S\text{'}= 360[EI\text{-}K\text{-}N]/\{[1+s][j+vp]\}$$

Rule-23234:

    If both (**E**), (**I**), (**K**), (**S'**), (**N**), (**s**), (**v**) and (**p**) are known, then its Job or Trade Receivable Days Planned:

$$j= (360[EI\text{-}K\text{-}N]/\{S\text{'}[1+s]\}\text{-}vp$$

Steve Asikin ISBN 14: 978-1515158691, ISBN 10: 1515158691

Rule-23235:

If both $(\mathbf{E})$, $(\mathbf{/})$, $(\mathbf{K})$, $(\mathbf{S'})$, $(\mathbf{j})$, $(\mathbf{N})$, $(\mathbf{v})$ and $(\mathbf{p})$ are known, then its Sales GrowthPlanned is:

$$s= 360[\mathbf{E/K\text{-}N}]/\{\mathbf{S'[j+vp]}\}-1$$

Rule-23236:

If both $(\mathbf{E})$, $(\mathbf{/})$, $(\mathbf{K})$, $(\mathbf{S'})$, $(\mathbf{j})$, $(\mathbf{s})$, $(\mathbf{N})$ and $(\mathbf{p})$ are known, then its Variable Portion Planned is:

$$v= (360[\mathbf{E/K\text{-}N}]/\{\mathbf{S'}[1+\mathbf{s}]\}-\mathbf{j})/\mathbf{p}$$

Rule-23237:

If both $(\mathbf{E})$, $(\mathbf{/})$, $(\mathbf{K})$, $(\mathbf{S'})$, $(\mathbf{j})$, $(\mathbf{s})$, $(\mathbf{v})$ and $(\mathbf{N})$ are known, then its Procured Inventory Days Planned:

$$p= (360[\mathbf{E/K\text{-}N}]/\{\mathbf{S'}[1+\mathbf{s}]\}-\mathbf{j})/\mathbf{v}$$

Steve Asikin ISBN 14: 978-1515158691, ISBN 10: 1515158691

The Rest will be Continued in
The Series of Book

MATH FIN LAW 7
Mathematical Financial Law

At
http: //www.Amazon.Com

Steve Asikin ISBN 14: 978-1515158691, ISBN 10: 1515158691

REFERENCES:

A. Books:

1)Amin, A.R. & Asikin, S.(2011a- 110 pages), *CEO Time Matrix: Productivity Management Skill that Increase CEO Effectiveness by 12400%*, Seattle (US): Amazon Createspace. http://www.amazon.com/CEO-Time-Matrix-Productivity-Effectiveness/dp/1461066069/ref=sr_1_1?s=books&ie=UTF8&qid=1387091880&sr=1-1&keywords=ceo+time+matrix

2)Amin, A.R. & Asikin, S. (2011b- 292 pages), *The Celestial Management: Spiritual Wisdom for All Human Beings*, Seattle (US): Amazon Createspace. http://www.amazon.com/Celestial-Management-Islamic-Wisdom-Beings/dp/1456450700/ref=sr_1_1?s=books&ie=UTF8&qid=1387092036&sr=1-1&keywords=celestial+management

3)Amin, A.R. & Asikin, S. (2014- 76 pages), *No Urgency CEO: A High-Tech Book in Modern Management*, Seattle (US): Amazon Createspace. http://www.amazon.com/No-Urgency-CEO-Hi-Tech-Management/dp/1489509356/ref=sr_1_1?s=books&ie=UTF8&qid=1396440688&sr=1-1&keywords=no+urgency

4)Asikin, S. (2009- 548 pages), *Arigato, Obrigado... (Thanks): Portuguese-Japan 1550-1647 Historic Novel*, Seattle (US): Amazon Createspace. http://www.amazon.com/Arigato-Obrigado-Thanks-Portuguese-Japan-1550-1647/dp/1453786368/ref=sr_1_1?s=books&ie=UTF8&qid=1396442162&sr=1-1&keywords=arigato

5)Asikin, S. (2010a- 290 pages), *Investment ALGEBRA: Strategic Profitable Comprehensive Business Finance Engineering Technology*, Seattle (US): Amazon Createspace. http://www.amazon.co.uk/Investment-ALGEBRA-Profitable-Comprehensive-Engineering/dp/1453856811/ref=tmm_pap_title_0?ie=UTF8&qid=1385371309&sr=1-1,

6)Asikin, S. (2010b- 224 pages), *FINANCE Architecture: VISUAL art, for Corporate Finance's Beauty, Safety and Simplicity*, **Seattle** (US): Amazon Createspace. http://www.amazon.co.uk/FINANCE-Architecture-Corporate-Finances-Simplicity/dp/1453829547/ref=sr_1_14?s=books&ie=UTF8&qid=1385371842&sr=1-14&keywords=finance+architecture

7)Asikin, S. (2010c- 150 pages), *Geometric FINANCE: Useful concepts that work better than Economic Noble Laureates'* Seattle (US): Amazon Createspace. http://www.amazon.co.uk/Magic-Geometric-FINANCE-concepts-Laureates/dp/1453790284/ref=tmm_pap_title_0,

Steve Asikin ISBN 14: 978-1515158691, ISBN 10: 1515158691

8)**Asikin, S.** (2010d- 294 pages), *TREASURY Planology: Strategic Profitable Comprehensive Business Finance Engineering Technology*, Seattle (US): Amazon Createspace. http:/www.amazon.co.uk/TREASURY-PLANOLOGY-Steve-Asikin-ebook/dp/B00G88MTFW/ref=sr_1_1?s=books&ie=UTF8&qid=1385403690&sr=1-1&keywords=Treasury+Planology.

9)**Asikin, S.** (2011a- 352 pages), *State Economics: Comprehensive Macro-Micro Economics' Simple Fiscal-Monetary Export-Import Accouting, Integrated Supply-Demand Managerial ... Mathematical Engineering Visual* . Seattle (US): Amazon Createspace. http:/www.amazon.co.uk/STATE-ECONOMICS-Steve-Asikin-ebook/dp/B00G89Q5SS, http:/www.amazon.co.uk State-Economics-Steve-Asikin-ebook/dp/B00B8PLQDI,

10)**Asikin, S.** (2011b- 540 pages), *Terima Kasih: Novel Internasional Sejarah Jepang-Portugis 1550-1647*, Seattle (US): Amazon Createspace. http:/www.amazon.com/Terima-Kasih-Internasional-Jepang-Portugis-1550-1647/dp/146351106X/ref=sr_1_1?s=books&ie=UTF8&qid=1396442361&sr=1-1&keywords=terima,

11)**Asikin, S.** (2013a- 824 pages), *Econometric Rex: 888 Marketing Promo 6666 Six Sigma Parametric Theories (of 10 Data)*, Seattle (US): Amazon Createspace. http:/www.amazon.co.uk/Econometric-Rex-Marketing-Parametric-Theories/dp/1482745259,

12)**Asikin, S.** (2013b- 260 pages), *Technoeconomistat: Economic &Statistical Technology for Financial Planning*, Seattle (US): Amazon Createspace. http:/www.amazon.co.uk/Technoeconomistat-Economic-Statistical-Technology-Financial/dp/1482304244,

13)**Asikin, S.** (2014a- 190 pages), *Calculus Accountancy: Leibnitz Newton Pacioli's Polynomial Quantitative Finance Risk Modelling Optimization*, Seattle (US): Amazon Createspace. http:/www.amazon.com/Calculus-Accountancy-Polynomial-Quantitative-Optimization/dp/1495463028/ref=sr_1_1?s=books&ie=UTF8&qid=1396441338&sr=1-1&keywords=calculus+accountancy,

14)**Asikin, S.**(2014b- 318 pages), *No Parametric: No Parametric: Hyperbolic Octahedron Central 3-Dimensional Orthogonal Risk Distribution Topology*, Seattle (US): Amazon Createspace. http:/www.amazon.com/Parametric-Hyperbolic-Octahedron-3-Dimensional-Distribution/dp/1496089731/ref=sr_1_1?s=books&ie=UTF8&qid=1396441543&sr=1-1&keywords=No+Parametric,

15)**Asikin, S.** (2014c- 138 pages), *Quick Help: Religion vs Real-Legion Philosophy*, Seattle (US): Amazon Createspace. http:/www.amazon.com/Quick-Help-Religion-Real-Legion-Philosophy/dp/1497466520/ref=sr_1_2?s=books&ie=UTF8&qid=1396441740&sr=1-2&keywords=quick+help,

Steve Asikin ISBN 14: 978-1515158691, ISBN 10: 1515158691

16) Asikin, S. (2014d- 824 pages), *Anne of Denmark:*
King James I the Bible, Historic Plays Drama, Seattle (US): Amazon
Createspace. http:/www.amazon.com/Anne-Denmark-James-Bible-
Drama/dp/149299734X/ref=sr_1_1?s=books&ie=UTF8&qid=1396441857&sr=1-
1&keywords=anne+asikin,

17) Asikin, S. (2014e- 184 pages), *Constructive Spells: Positive Spiritual
Mentality Personal Development*: Seattle (US): Amazon Createspace.
http:/www.amazon.com/Constructive-Spells-Spiritual-Mentality-
Development/dp/1497498899/ref=sr_1_1?s=books&ie=UTF8&qid=1397664512&sr=1-
1&keywords=constructive+spells;

18) Asikin, S. (2014f- 466 pages), *Elizabeth of Russia:*
Another Virgin Queen Gloariana: A Romanov Plays Drama, Seattle (US):
Amazon Createspace. http:/www.amazon.com/Elizabeth-RUSSIA-Another-Gloriana-
Romanov/dp/1505542480/ref=sr_1_4?s=books&ie=UTF8&qid=1423888482&sr=1-
4&keywords=elizabeth+russia;

19) Asikin, S. (2014f- 708 pages), *Shinmen Munisai:*
Miyamoto Musashi's Father, A Sword Samurai Plays Drama, Seattle (US):
Amazon Createspace. http:/www.amazon.com/Shinmen-Munisai-Miyamoto-Musashis-
Samurai/dp/1507857799/ref=sr_1_1?s=books&ie=UTF8&qid=1428094280&sr=1-
1&keywords=munisai;

20) Asikin, S. (2015a-767 pages), *Math Fin Law-1:*
Mathematical Financial Law for Public Listed Firms Rule 1-4441, Seattle
(US): Amazon Createspace. http://www.amazon.com/Math-Fin-Law-Mathematical-
Financial-ebook/dp/B00WLNX3Y4/ref=asap_bc?ie=UTF8;

21) Asikin, S. (2015b-813 pages), *Math Fin Law-2:*
Mathematical Financial Law for Public Listed Firms Rule 4442-
8712, Seattle (US): Amazon Createspace. http://www.amazon.com/Math-
Fin-Law-Mathematical-No-4442-
8712/dp/1512285080/ref=sr_1_1?s=books&ie=UTF8&qid=1432732199&sr=1-
1&keywords=math+fin+law+2;

22) Asikin, S. (2015c-813 pages), *Math Fin Law-3:*
Mathematical Financial Law for Public Listed Firms Rule 8713-
12575, Seattle (US): Amazon Createspace. http://www.amazon.com/Math-
Fin-Law-Mathematical-8712-
12575/dp/1514276763/ref=sr_1_1?s=books&ie=UTF8&qid=1433982149&sr=1-
1&keywords=math+fin+law+3

23) Asikin, S. (2015d-813 pages), *Math Fin Law-4:*
Mathematical Financial Law for Public Listed Firms Rule 12576-
16333, Seattle (US): Amazon Createspace. http://www.amazon.com/Math-
Fin-Law-Mathematical-12576-
16333/dp/1514685132/ref=sr_1_1?s=books&ie=UTF8&qid=1435393628&sr=1-
1&keywords=math+fin+law+4

Steve Asikin ISBN 14: 978-1515158691, ISBN 10: 1515158691

24) **Asikin, S.** (2015d-816 pages), *Math Fin Law-5:*
Mathematical Financial Law for Public Listed Firms Rule 16334-
19904, Seattle (US): Amazon Createspace. http://www.amazon.com/Math-
Fin-Law-Mathematical-No-16334-
19904/dp/1515073009/ref=sr_1_1?s=books&ie=UTF8&qid=1437224693&sr=1-
1&keywords=math+fin+law+5

25)**Asikin, T.& Asikin, S.** (2013- 828 pages), *448 Poems:*
KARAOKE Song Book: 24 Hours Story of 30 Languages, Seattle (US):
Amazon Createspace. http://www.amazon.com/448-Poems-KARAOKE-Song-
Book/dp/1480034363/ref=sr_1_1?s=books&ie=UTF8&qid=1396442034&sr=1-
1&keywords=448+poems,

26)**Brigham, EF &Gapenski, LC,** . (1988- Textbook), *Financial Management:*
Theory and Practice (Study Guide), NY (US): Dryden Press.
http:/www.amazon.com/Financial-Management-Theory-Practice-
Study/dp/0030125391/ref=sr_1_1?s=books&ie=UTF8&qid=1423808035&sr=1-1 ;

27)**Nikisa, Evets.** (2014g- 828 pages), *Wonderful Live Spells:*
144-Steps Recreative Nice Children Story, Succesful Development:
Seattle (US): Amazon Createspace. http:/www.amazon.com/WONDERFUL-Live-
Spells-Recreative-
Development/dp/1499248423/ref=sr_1_1?s=books&ie=UTF8&qid=1418698379&sr=1-
1&keywords=wonderful+asikin

28)**Niswonger CR, Fess PE & Warren CE.** (1993-Textbook), *Accounting*
Principles: Seattle (US): South Western Publishing Co.
http://www.amazon.com/Accounting-Principles-Chapters-C-Rollin-
Niswonger/dp/0538818603

29)**Paccioli, L. (1989-Textbook),** *Summa de Arithmetica Geometria Proportioni*
et Proportionalita, Venice 1494. http:/www.amazon.com/Arithmetica-Geometria-
Proportioni-Proportionalita-
Venice/dp/B00T4HXTIW/ref=sr_1_fkmr0_2?s=books&ie=UTF8&qid=1429597966&sr=1-2-
fkmr0&keywords=summa+mhematica+arithmetica

30)**Weston JF, Copeland TE & Shastri, K.** (2004-Textbook), *Financial*
Theory and Corporate Policy: NY (US): Addison Wesley.
http://www.amazon.com/Financial-Corporate-Copeland-Kuldeep-
Paperback/dp/B008YSUE4M/ref=sr_1_1?s=books&ie=UTF8&qid=1429381700&sr=1-
1&keywords=weston+copeland

Steve Asikin ISBN 14: 978-1515158691, ISBN 10: 1515158691

B. Intellectual Properties:

1)**Indonesian Banker Certification 1993**, *Sertifikasi Bankir Indonesia 1993*;
2)**MATHFINPLAN** (Mathematical Financial Planning);
   Perencanaan Keuangan Matematis
3)**BANK MATHPLAN** (Bank's Mathematical Planning);
   Perencanaan Matematis Bank
4)**TECHNO-ECONOMISTAT** (Technology for Economical Statistics);
   Teknologi Statistik untuk Ekonomi
5)**FINANCIAL GEOMETRY** (IT Program); *Ilmu Geometri Keuangan*
6)**BAK DOBEL** (Car Gasoline Tank: Left and Right Side Fuel Fill-in).
   Tangki Mobil yang bisa diisi dari Kiri atau Kanan
7)**Mister GO-CHENG** (Cheap Food Franchising). *Waralaba Pangan Murah*

Steve Asikin ISBN 14: 978-1515158691, ISBN 10: 1515158691

C. SCIENTIFIC PAPERS:

1)**Eichhornia Crassipes**, as Anti Polutant and Fertile Sediment, *Pemanfaatan Eceng Gondok untuk Menetralkan Polusi dan Pengurukan Subur pada Genangan Air*, LKIR, Lomba Karya Ilmiah Remaja, LIPI (1979)

2)**Archimedes Catesian Buoyancy Law**, *Penyimpangan Hukum Archimedes pada Model Pelampung Cartesian*, Lomba Karya Ilmiah Remaja, Depdikbud-RI (1980)

3)**SURAMADU**, Surabaya-Madura Inter Island Bridge, Design Supervisor, *Sumarsono Sumali Engineering Contractor*, Tambak-V, Tandes, Surabaya (1982),

4)**GAZOLINE Taxi-Pool Depot**, Jemur Handayani, Surabaya, *Sumarsono Sumali Engineering Contractor*, Tambak-V, Tandes, Surabaya (1982),

5)**SOEKARNO-HATTA**, Jakarta International Airport, Garuda Maintenance Facilities, *Junior Architect*, As Bulit Drawings, Coinstruction Management, Aerospatiale-Encona Engineering (1983-1985),

6)**ADISUCIPTO**, Yogya International Airport, *Junior Architect*, Preliminary Technical Drawings, Budiono Surasno-Encona Engineering (1986),

7)**BLOK-M PLAZA**, "New Garden Hall", Kebayoran Baru Supermall, *Preliminary Design Architect*, Danisworo-James Ferry- Jasa Ferry-Encona (1986),

8)**NORTH JAKARTA PUBLIC LIBRARY**, Semper, Tanjung Priok, *Chief Architect*, Agung Darma Sarana-Asep Hermawan (1987),

9)**The Banker's Trainer**: A Manual for The Training Head in Banking. *Profesi Bankir Pelatih, Tuntunan bagi Kepala Diklat Perbankan*, (1200 Pages, 1993),

10)**Indonesian Bankers Certification**, a Strategic Move for the National Banking Improvement. *Sertifikasi Bankir Indonesia, Langkah Strategis Pembenahan Perbankan Nasional*, (KOMPAS, 20 Juli 1993).

11)**Indonesian Rupiah's Exchange Rate**, on a Quiet but Steady Movement. *Nilai Tukar Rupiah, Diam-diam Meningkat Pesat*, (MANAJEMEN, Jan-Feb 1994).

12)**The Bank's Credit**, for the Best Practices. *Kredit Praktis* (200 pages, 1994),

13)**Down to Earth Credit Meeting Credo**, *Kiat Rapat Pemberian Kredit yang Realistik*. (MANAJEMEN, Sep-Oct 1994).

14)**SCHIPHOL**, Amsterdam International Airport, 1st Anthony Fokker Weg, Sloten, *Airport Engineer*, Samsung- Fokker-Beurlin, BV (1996),

15)**Ultrapeneurship: The Secret of Starting Business Without Capital**. Ultrapreneur: *Kiat Usaha Tanpa Modal*, (MANAJEMEN, Nov-Dec 1996),

16)**Recipe from the Corporate Raider's Kitchen**, *Menguak Dapur Corporate Raiders*, (MANAJEMEN, Jan-Feb 1997).

17)**The 10-Tips and Tricks of Engineered Conglomeration**. *Sepuluh Rekayasa Keuangan Konglomerat*, (MANAJEMEN, Mar-Apr 1997),

18)**The Management of Succession in Five Parts of the World**. *Manajemen Suksesi di Lima bagian Dunia*, (MANAJEMEN, Mei-Jun 1997),

19)**Auditing**: The Global Language for the 2003's Human Resources. *Audit : Bahasa Global untuk SDM 2003*, (MANAJEMEN, Jul-Aug 1997),

Steve Asikin ISBN 14: 978-1515158691, ISBN 10: 1515158691

20)The Elected Parliament 1997-2002, and the Hope of Professionals. *DPR 1997–2002 dan Harapan Profesional*, (MANAJEMEN, Sep-Okt 1997),

21)Disappearance of World's Soft Currencies, and Proposition for Global Economic Thinking. *Lenyapnya Mata Uang Lunak Dunia dan Usulan Pemikiran Ekonomi Global*, (MANAJEMEN, Nov-Dec 1997),

22)The Human Resources and Its Accountability, in Critical Situation. *Akuntabilitas Sumberdaya Manusia di Masa Krisis*, (MANAJEMEN, Jan-Feb 1998), '

23)The Investment Economic Model, and the Failure of Growth Economic Theories. *Model Ekonomi Investasi dan Kegagalan Teori Ekonomi Pertumbuhan*, (MANAJEMEN, Nov 1998),

24)The Exact Equilibrium of Capital Budgeting, *Persamaan-persamaan akurat dalam Penganggaran Modal Perusahaan*. (MANAGEMENT EXPOSE, November 1998),

25)The Corruptive Economic Model, and Its Solution for Long Term Debt's Problem. *Model Ekonomi Korupsi dan Penyelesaian Utang Jangka Panjang*, (MANAJEMEN, Dec 1998),

26)The Self Equity Budgeting, and Operations Research in Investments. *Penganggaran Modal Sendiri dan Riset Operasi dalam Investasi*, (MANAJEMEN, Jan 1999),

27)Actual Impact of Inflation, and the Need of Geometric Approach in Statistic. *Pengaruh Riil Inflasi dan Perlunya Statistika Geometri*, (MANAJEMEN, Feb 1999),

28)Multiplication of Money, to Overcome the Crisis. *Melipat Gandakan Uang untuk Mengusir Krisis*, (MANAJEMEN, Apr 1999),

29)The Interactive Mathematical Model, for Multi Constraints Financial Budgeting, *Model Matematika Ekonomi Interaktif bagi Penganggaran Keuangan Multi Kendala*. (SCRIPTA ECONOMICA, Vol. 2, April 1999)

30)Application of Ethics, for the Management Accounting Practices. *Aplikasi Etika dalam Praktek Akuntansi Manajemen*, (MANAJEMEN, May 1999),

31)The Curriculum of Magistrate of Management in Finance: Its Needs and Practices. *Kurikulum Magister Manajemen Keuangan 1988-1998: Antara Kebutuhan dan Praktek*, (MANAJEMEN, Jun 1999),

32)The Attractiveness, Personal Power and Management. *Pesona Pribadi: "Personal Power" dan Manajemen*, (MANAJEMEN, Jul 1999),

33)The Market Research, for Magistrate Degree in Management. *Soal Jawab Riset Pasar untuk Pascasarjana Manajemen* (126 Pages, 1999),

34)The Thesis Manual, Question and Answer for Magistrate Degree in Management. *Soal Jawab Tugas Akhir untuk Pascasarjana Manajemen* (126 Pages, 1999),

35)The Risk Management, Question and Answers for Magistrate Degree in Management. *Soal Jawab Manajemen Resiko untuk Pascasarjana Manajemen* (126 Pages, 1999),

Steve Asikin ISBN 14: 978-1515158691, ISBN 10: 1515158691

36)**The Credit Manual**, Question and Answers for Magistrate Degree in Management. *Soal Jawab Manajemen Kredit untuk Pascasarjana Manajemen* (126 Pages, 1999),

37)**Enhancing the Morality Structure**. *Solidkan Struktur Moralitas,* (MANAJEMEN, Oct 1999)

38)**Performance of PT. Astra Otoparts**, Plc, in its Press Releases and Public Financial Reports. *Kinerja PT. Astra Otoparts dalam Siaran Pers dan Laporan Keuangan Publik,* (MANAJEMEN, Oct 1999)

39)**The Management of Conflicts**, for the New Indonesian's "Rainbow Colored" Cabinet. *Manajemen Konflik dan Kabinet Pelangi,* (MANAJEMEN, Dec 1999),

40)**Technoeconometric**: An Engineering Elaboration to Matrix Multi Variate Econometrics, *Tekno Ekonometrik sebagai sumbangan teknologi untuk Ekonometrika Matrik Multi Variat.* (MANAGEMENT EXPOSE, November 1999),

41)**Technoeconometric** : A Mathematical Shortcut for Non Linear Bivariate Econometrics, *Tekno Ekonometrik sebagai Usulan Jalan Pintas Matematik untuk Ekonometrika non Linear Bivariat.* (SCRIPTA ECONOMICA, Vol. 2, No. 3, Desember 1999),

42)**Developing "New Image"**, for the New Indonesia. *Membangun Citra Indonesia Baru,* (MANAJEMEN, Jan 2000),

43)**Segregations of Duties**, the Needs of Both Indonesian Governments and Private Sectors. *Pemerintah dan Swasta Perlu Berbagi Tugas,* MANAJEMEN, Mar 2000),

44)**The 812.500 Financial Models**, for the Autonomous Country Governments. *Tentang 812.500 Model Keuangan Otonomi Daerah,* (MANAJEMEN, Mar 2000),

45)**The Forecast of 10 Most Attractive Share Values**, in the Jakarta Stock Exchange, April 2000. *Prakiraan Harga Sepuluh Saham Unggulan di Bursa Efek Jakarta, April 2000,* (MANAJEMEN, Mar 2000),

46)**Let Us Merchandise the Intelectual Properties!**. *Jual "Intelectual Property" Saja!* (MANAJEMEN, Apr 2000),

47)**The Forecast of ten Most Attractive Share Values**, in the Jakarta Stock Exchange May 2000. *Prakiraan Harga Sepuluh Saham Unggulan di Bursa Efek Jakarta Mei 2000,* (MANAJEMEN, Apr 2000),

48)**HEATHROW**, London International Airport, Second Runway, *Q-Basic Algorithmic Engineer*, Beasley-Krishnamoorthy, Imperial College, Cambridge, Post Doctoral Project (2000),

49)**Why the Banking Sector Should be Healthy**, Before Others?. *Mengapa Perbankan Harus Sehat?* (MANAJEMEN, Mei 2000),

50)**Recycling of Wastes as a Negative Costing**. *Memanfaatkan Sampah atau Negative Costing,* (MANAJEMEN, Agt 2000),

51)**The Guidance for Indonesia Corporate Restructuring**. *Kiat Restrukturisasi Perusahaan Indonesia,* (MANAJEMEN, Agt 2000),

52)**Redefinition of the Ethical Comprehension**. *Pemaknaan Kembali atas Pemahaman Etika,* (MANAJEMEN, Sep 2000),

Steve Asikin ISBN 14: 978-1515158691, ISBN 10: 1515158691

53)**The Critical View the 1999's Indonesians Central Bank Financial Statement**. *Analisa Laporan Keuangan Bank Indonesia 1999*, (MANAJEMEN, Oct 2000),

54)**Becoming an "Achiever"**, not Just Common "Worker". *Bukan Sekadar "Kerja", Tapi Menghasilkan Karya!* (MANAJEMEN, Feb 2001),

55)**Ten Steps to Real Wealth!** *Sepuluh Langkah Menjadi Kaya*, (MANAJEMEN, Jun 2001),

56)**Creative Financial Breakthrough**, *Seminar Sehari tentang Terobosan Keuangan yang Kreatif*, (Gramedia-PPM dan MANAJEMEN, Jun 2001),

57)**The Marketing Practices of Financial Softwares**, for Publics Listed Corporation. *Praktek Pemasaran Program Perangkat Lunak untuk Keuangan Perusahaan Publik* (Univ Satyagama, Jakarta Juni 24, 2001).

58)**The Mathematical Financial Planning**, *Perancangan Keuangan Secara Eksak dengan Matematik.* (MANAJEMEN, Jul 2001),

59)**The Real Research and Development, not Rework and Destruction**. *Litbang yang Bukan "Sulit Berkembang"*, (MANAJEMEN, Jul 2001),

60)**The Treatise on the Advantages and Disadvantages**, of Revolutionary Learning Methods. *Baik Buruknya: Revolusi Cara Belajar*, (MANAJEMEN, Jul 2001),

61)**Better Make an Indonesian "Corruption Prevention" Body**, not Just a "Corruption Watch". *Buat Indonesian Corruption Prevention (ICP), Bukan Indonesian Corruption Watch, ICW!* (MANAJEMEN, Okt 2001),

62)**Preparation of the Worlds Business Leaders**, for our Century and the Century After. *Mempersiapkan Manusia-manusia yang akan menjadi Khafilah Bisnis Dunia Abad ini dan Abad Mendatang.* (Univ Satyagama – Gregorio Araneta Univ, Jakarta, Oktober 27, 2001).

63)**The Good Corporate Governance**, is not just a Morale Condition. *Kepemerintahan Perusahaan yang Baik, Bukan Sekedar Moralitas*, (MANAJEMEN, Nov 2001),

64)**Corporate Healthiness and Cleanliness as a Challenge**. *Bersih, Sehat Perusahaan, sebuah Tantangan* (MANAJEMEN, Jan 2002),

65)**Developing the Cultural Base for Future Competitions**. *Budaya untuk Bersaing di Masa Depan* (MANAJEMEN, Jan 2002),

66)**"Catapult", "Bullet" or "Ballistic Missiles" Approach**, for Corporate Healthiness and Cleanliness. *Pola "Ketapel", "Peluru Biasa" atau "Rudal" untuk Keuangan Perusahaan yang Bersih Sehat.* (MANAJEMEN, Feb 2002),

67)**The Mathematical Invesment Solution Model**, for Indonesian Foreign Debt and Its Corruption, *Solusi Matematika Investasi bagi Penyelesaian Masalah Utang Luar Negeri dan Korupsi di Indonesia.* (MANAGEMENT EXPOSE, Mar 2002),

68)**The Attitude, Skills and Knowledge (ASK)**, in Educational System Approach. *Sikap, Keterampilan dan Pengetahuan Menurut Pendekatan Sistemik Pendidikan*, (MANAJEMEN, Apr 2002),

Steve Asikin ISBN 14: 978-1515158691, ISBN 10: 1515158691

69)**Strategic Treatise on Regional Autonomy**, in Futurization, Digitalization, Securitization and Globalization Era in Accordance to the Awakening of the Mercantile Nations. *Kajian Strategik Otonomi Daerah : Era Futurisasi, Digitalisasi, Strukturisasi, dan Globalisasi dalam Kebangkitan Negara-Negara Dagang.* (Training for Provincial Government Regional Management, Indonesian Association of Provincial Governments, *Pelatihan Pengembangan Manajemen Pemerintah Daerah bagi Perangkat Pemerintah Propinsi, Angkatan II*, (APPSI), June 27, 2002.

70)**The Suggestion for the Doctoral Course**, for Islamic Financial Management System. *Usulan-usulan Perbaikan dalam Kurikulum Doktor Manajemen Keuangan Islam.* (December, 17-19, 2002, Canadian International Development Agency - CIDA, ADSGM, Universiti Utara Malaysia, Kedah, Da'arul Sintok).

71)**The Defensive Strategy for the Market Leader**, a Nokia Cellular Telephone's Case. *Strategi Bertahan Sang Pemimpin Pasar dengan Menggunakan Multi Merk, Kasus Telepon Seluler Nokia.* (March, 12, 2004, Univ Tarumanegara, Jakarta).

Steve Asikin ISBN 14: 978-1515158691, ISBN 10: 1515158691

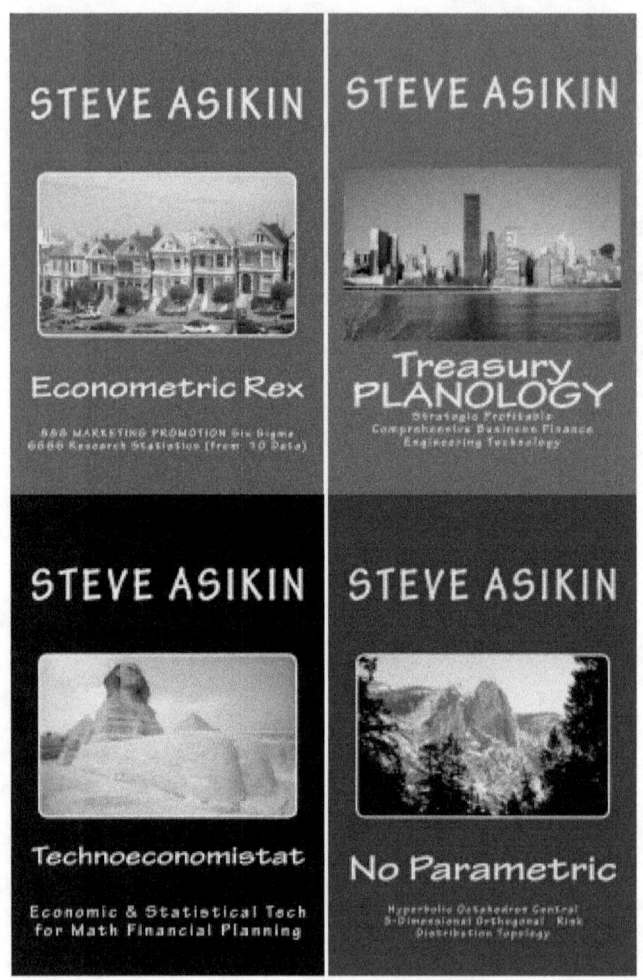

Steve Asikin ISBN 14: 978-1515158691, ISBN 10: 1515158691

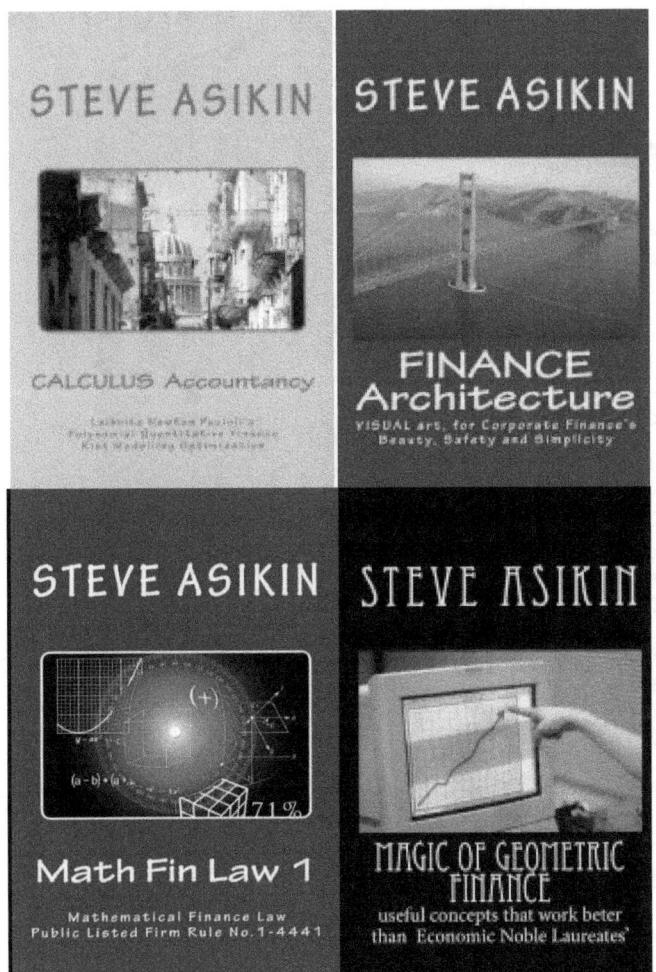

Steve Asikin ISBN 14: 978-1515158691, ISBN 10: 1515158691

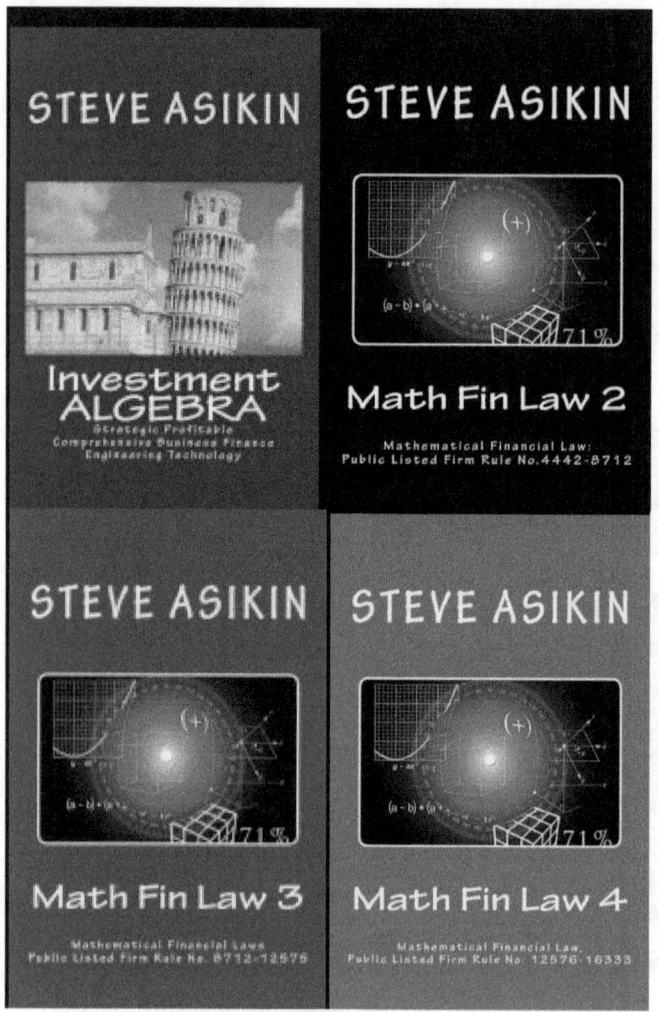

Steve Asikin ISBN 14: 978-1515158691, ISBN 10: 1515158691

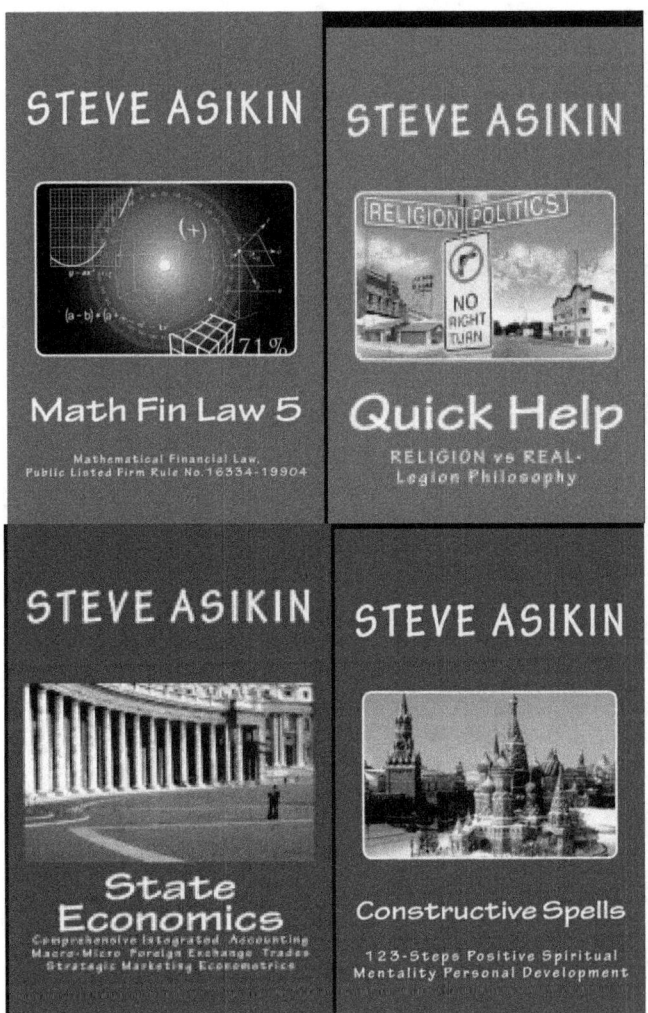

Steve Asikin ISBN 14: 978-1515158691, ISBN 10: 1515158691

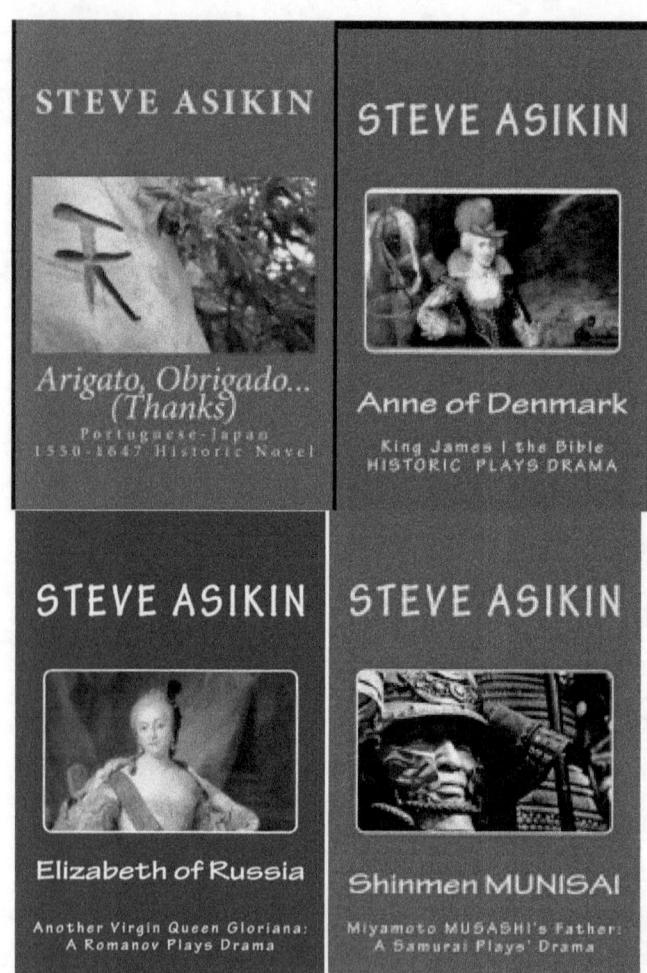

Steve Asikin ISBN 14: 978-1515158691, ISBN 10: 1515158691

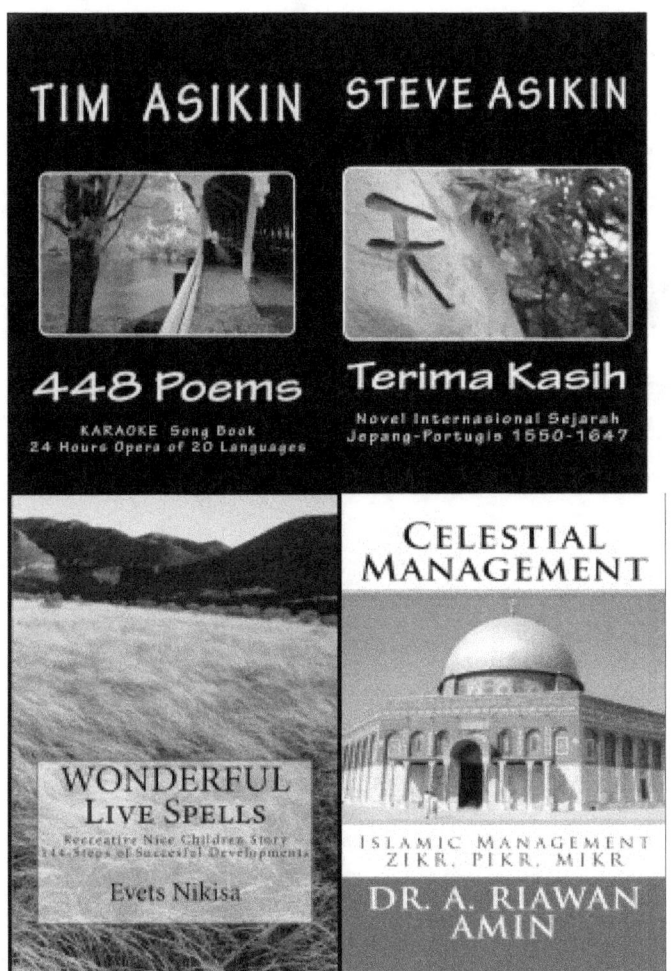

Steve Asikin ISBN 14: 978-1515158691, ISBN 10: 1515158691

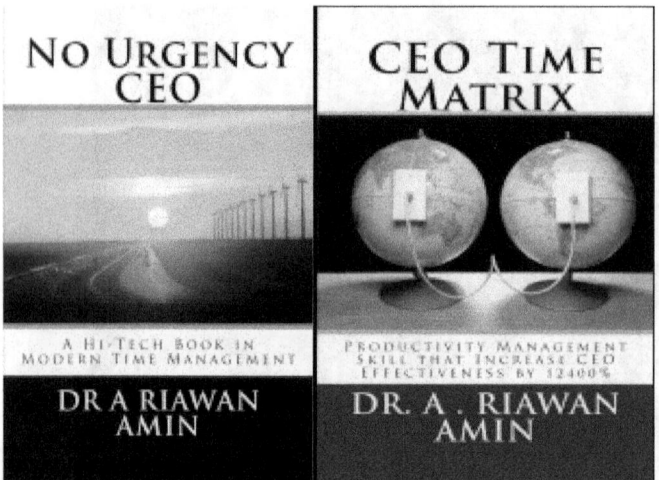

Steve Asikin ISBN 14: 978-1515158691, ISBN 10: 1515158691

www.ingramcontent.com/pod-product-compliance
Lightning Source LLC
Chambersburg PA
CBHW051847170526

45168CB00001B/6